T0319993

NAVIGATING A CHANGING WORLD

Canada's International Policies in an Age of Uncertainties

The negotiation of the Canada–United States Free Trade Agreement in 1985–8 initiated a period of substantially increased North American, and later global, economic integration. However, events since the election of Donald Trump in 2016 have created the potential for major policy shifts arising from NAFTA's renegotiation and from continuing political uncertainties in the United States and with Canada's other major trading partners.

Navigating a Changing World draws together scholars from both countries to examine Canada–U.S. policy relations, the evolution of various processes for regulating market and human movements across national borders, and the specific application of these dynamics to a cross-section of policy fields with significant implications for Canadian public policy. This book explores the impact of territorial institutions and extraterritorial forces – institutional, economic, and technological, among others – on interactions across national borders, both within North America and, where relevant, in broader economic relationships affecting the movement of goods, services, people, and capital. Above all, *Navigating a Changing World* represents the first major study to address Canada's international policy relations within and beyond North America since the elections of Justin Trudeau in 2015 and Donald Trump in 2016 and the renegotiation of NAFTA.

GEOFFREY HALE is a professor in the Department of Political Science at the University of Lethbridge.

GREG ANDERSON is a professor in the Department of Political Science at the University of Alberta.

Navigating a Changing World

Canada's International Policies in an Age of Uncertainties

EDITED BY GEOFFREY HALE
AND GREG ANDERSON

UNIVERSITY OF TORONTO PRESS
Toronto Buffalo London

© University of Toronto Press 2021
Toronto Buffalo London
utorontopress.com
Printed in the U.S.A.

ISBN 978-1-4875-0818-0 (cloth)   ISBN 978-1-4875-3771-5 (EPUB)
ISBN 978-1-4875-2571-2 (paper)   ISBN 978-1-4875-3770-8 (PDF)

Library and Archives Canada Cataloguing in Publication

Title: Navigating a changing world : Canada's international policies in an age
   of uncertainties / edited by Geoffrey Hale and Greg Anderson.
Description: Includes bibliographical references and index.
Identifiers: Canadiana (print) 20210090367 | Canadiana (ebook)
   20210090464 | ISBN 9781487508180 (cloth) | ISBN 9781487525712 (paper) |
   ISBN 9781487537715 (EPUB) | ISBN 9781487537708 (PDF)
Subjects: LCSH: Canada – Foreign economic relations – United States. |
   LCSH: United States – Foreign economic relations – Canada. | LCSH:
   Canada – Foreign economic relations.
Classification: LCC FC250 .N38 2021 | DDC 327.71073 – dc23

This book has been published with the help of a grant from the Federation
for the Humanities and Social Sciences, through the Awards to Scholarly
Publications Program, using funds provided by the Social Sciences and
Humanities Research Council of Canada.

University of Toronto Press acknowledges the financial assistance to its
publishing program of the Canada Council for the Arts and the Ontario Arts
Council, an agency of the Government of Ontario.

Canada Council   Conseil des Arts
for the Arts   du Canada

ONTARIO ARTS COUNCIL
CONSEIL DES ARTS DE L'ONTARIO
an Ontario government agency
un organisme du gouvernement de l'Ontario

Funded by the   Financé par le
Government   gouvernement
of Canada   du Canada
Canada

Contents

Figures and Tables

Figures

Tables

Preface

Canada's international policy relations have evolved substantially since work began on this volume in 2013 under the auspices of the *Borders in Globalization* (BiG) project – an international, multi-disciplinary research collaboration of academics and policy practitioners centred in North America but extending around the world. This volume extends the editors' previous work on cross-border, North American and international governance during the past fifteen years.[1] We appreciate opportunities for the sharing of ideas contained therein at periodic conferences organized by BiG's tireless coordinators, Emmanuel Brunet-Jailly and Victor Konrad, and project manager Nicole Bates-Eamer. Tannis Schilk has provided invaluable administrative, editorial and research assistance to the project including, but certainly not limited to coordination of an author's workshop at the Banff Centre in April 2018. Anonymous reviewers have provided constructive suggestions as part of the editorial press. Dan Quinlan, Leah Connor, and the editorial staff of the University of Toronto Press have been exceptionally helpful in bringing this project to publication. The Social Sciences and Humanities Research Council of Canada has provided invaluable financial assistance in facilitating research for this volume.

NOTE

1 Monica Gattinger and Geoffrey Hale, eds., *Borders and Bridges: Canada's Policy Relations in North America* (Toronto: Oxford University Press, 2010); Greg Anderson and Christopher Sands, eds., *Forgotten Partnership Redux: Canada-U.S. Relations in the 21st Century* (New York: Cambria Press, 2011); Geoffrey Hale, *So Near and Yet So Far: The Public and Hidden Faces of Canada-U.S. Relations* (Vancouver: UBC Press, 2012); Brian Bow and Greg Anderson, eds., *Regional Governance in Post-NAFTA North America: Building without Architecture* (New York: Routledge, 2015).

NAVIGATING A CHANGING WORLD

Canada's International Policies
in an Age of Uncertainties

1 Canada at the Crossroads: Canada's International Policy Relations in an Era of Political and Economic Uncertainty

GEOFFREY HALE AND GREG ANDERSON

Canada is at a crossroads. Thirty years after ratifying the Canada-U.S. Free Trade Agreement, national governments are revisiting many of the basic assumptions that have shaped the interaction of domestic and international policy networks and shaped Canada's international policy relations, particularly with the United States, during most of this period.

Political leaders and foreign ministries continue to shape and manage relationships between and among national governments. However, economic globalization – the progressive reduction of barriers in movements of goods, services, capital, and people across national borders, resulting in growing interdependence among nations – has had the paradoxical effect of decentralizing many of these relationships so that governments no longer function as "unitary actors" in their relations with other countries. Rather, dozens of federal (and sometimes provincial) departments and agencies maintain ongoing transgovernmental relationships with counterparts in other countries.[1] These links reflect both the realities of cross-border and broader international interdependence and the phenomenon of "intermesticity": the blurring of traditional distinctions between domestic and international policies.[2]

Canada is a trading nation. Canada's international trade accounted for the fourth-largest share of GDP among G-20 nations in 2018 (66 per cent), exceeded only by Germany (87 per cent), Korea (83 per cent), and Mexico (80 per cent).[3] The growth of bilateral and regional preferential trade agreements (PTAs) since the late 1990s, the expansion of global and North American supply and value chains, along with recent trends towards protectionism and the use of trade policies as instruments of geopolitical competition, especially in the United States and China, have significantly increased the importance of borders and border-related policies for Canadian firms, governments, and individuals.[4]

These realities have been reinforced by many countries' responses to the 2020 COVID pandemic and its trade-related shocks.[5]

This volume explores the evolving nature of Canada's international policy relations during a time of political and economic uncertainty not seen since at least the break-up of the post–Second World War economic order in the 1970s. For most of the post-war era, Canadian policymaking was reliably predicated on several pillars, nearly all of which were underwritten by American leadership: a prominent U.S. role in international political institutions like NATO, the UN, or the G-7, a rules-based multilateral trading system, and the incremental spread of liberal democratic institutions around the world.

However, in recent years, a series of broader trends have called into question many of the core assumptions underpinning Canada's international relations since the 1990s. Some of these developments reflect wider international developments from the financial crisis of 2007–9, the subsequent retrenchment of U.S. power under the Obama administration, and the emergence of an increasingly multipolar global system. Others reflect the growing populist reaction against economic globalization – symbolized by the rise of the Trump administration. Table 1.1 summarizes key elements in the erosion of the conventional wisdom guiding Canada's international economic policies since the mid-1980s. Its two columns summarize the core expectations and assumptions that guided Canada's embrace of North American integration and other international economic policies during the 1990s (and, to a lesser extent, after 2000), and the combination of economic, societal, and political factors – in Canada and elsewhere, especially the United States – that have led observers to challenge these assumptions in recent years.

First, previous trends towards North American and global economic integration can no longer be taken for granted. Although international trade and investment remain critically important to Canada's prosperity, its economic integration in regional markets has declined relative to GDP since 2000 and stagnated as a share of global markets since 2005. Similarly, net inward flows of foreign direct investment (FDI) to Canada have declined steadily in recent years, as noted in figure 1.1, while Canada has been a net exporter of capital since the late 1990s. These trends partly reflect declining terms of trade (and selected barriers to exports) in major sectors (e.g., automotive, energy), but also the needs of institutional investors, such as major pension funds, to diversify their investments beyond Canada's relatively narrow markets – even before the economic disruptions occasioned by the global pandemic of 2020.

Moreover, shifting political patterns in the U.S. and other major industrial countries have contributed to the erosion of the multilateral

Table 1.1. Shifting Assumptions Governing Canada's International Economic Policies

That Was Then (1990s)	This Is Now (late 2010s, 2020s)
Trends towards growing regional economic integration will continue, leading to increased global economic integration.	Canada's economic integration in regional markets has declined relative to GDP since 2000, and stagnated as share of global markets since 2005.
The United States and European Union are committed to the broadening and deepening of international economic institutions, which will encourage continued growth of market-based economic activity.	Erosion of multilateral trade regimes; strategic trade competition among major powers (US, EU, China); populist reactions to uneven distribution of economic costs, benefits threatening disruption of international system.
Canada will continue to depend on high levels of foreign direct investment (FDI) for its economic growth and competitiveness in international markets.	Net inward flows of FDI have generally declined in recent years, while Canada has been a net exporter of capital since the late 1990s.
Canada's major industries, particularly automotive, industrial materials, and diverse energy sectors, will depend on increased specialization and integration within North American supply chains to increase their productivity and competitiveness.	U.S. trade policies appear to be aimed at repatriation of industrial production, negotiation of managed-trade agreements in multiple sectors, limited political will in the U.S. for cooperative policy responses despite USMCA agreement.
Importance of trade facilitation points towards bilateral cooperation on border management, balancing law enforcement with facilitation of lawful trade, travel.	Expanded security requirements require extensive cross-border cooperation, but competing trade, immigration priorities, and basic differences in legal systems limit likelihood of growing convergence.
Differences in national immigration systems and priorities limit potential for convergence, but may be managed through trade provisions allowing greater flexibility for business travel, and recruitment of skilled personnel, with distinct national policies to address labour market needs.	Persistent national policy divergence on immigration issues; growing pressures of informal migration due to social disruptions, and conflicts in many developing countries; societal reactions to labour market and income trends limit; use of trade policies to facilitate mobility.
Canada has capacity for policy choice, depending on mix of bilateral, international cooperation and domestic policy coherence, but requiring substantial restructuring of domestic policies to emerging, U.S.-led international economic order.	Canada faces difficult trade-offs in adjusting to growing U.S. protectionism, unilateralism, and growing divergence of domestic policy regimes among major trading partners. Trade diversification critical, but independent Canadian leverage limited, as in 1970s.
Domestic policy adjustment, containment of inter-regional conflicts facilitated by expanding cross-border trade and investment relations.	Constraints on North American economic integration increase risks of zero-sum politics among diverse provincial and regional economies.

Figure 1.1. Shifting Trends in Foreign Direct Investment

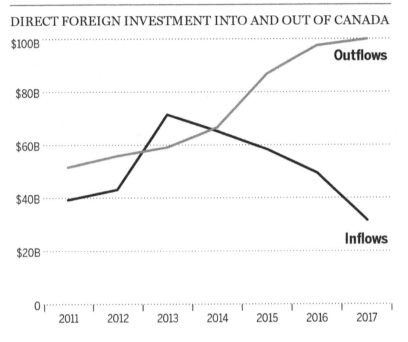

TAKE THE MONEY AND RUN

DIRECT FOREIGN INVESTMENT INTO AND OUT OF CANADA

Note: Annual figures generated by author from quarterly data. All figures are net, i.e., foreign direct investment in Canada is equal to new investment minus past investment moved out of Canada (by foreign owners). Outflows are equal to new investments abroad minus investments brought back to Canada (by Canadian owners).
Source: Statistics Canada; *National Post.*

trade regime, reinforcing pre-existing trends towards strategic competition among major economic powers. The spread of populist reactions against the uneven distribution of globalization's economic costs and benefits across industrial countries, symbolized by the twin 2016 shockwaves of Britain's Brexit vote and Donald Trump's election, have triggered major policy changes that appear to be aimed at the repatriation of industrial production and the negotiation of managed trade agreements. Sustained challenges to the North American Free Trade Agreement (NAFTA) and its intensely contested transformation into the United States–Mexico–Canada Agreement (USMCA) have driven home the disruptive potential for several major Canadian industries, including the automotive, aerospace, and industrial materials sectors, which continue to depend on integration within international supply chains for their productivity and competitiveness.[6] Similar challenges emerged from disruptions in international supply chains for pharmaceuticals, medical, and related protective equipment during the global pandemic. These trends are likely to continue, with changes affecting specific industries, following the 2020 election of President Joe Biden.

At the same time, the realities of economic interdependence impose certain disciplines on governments, particularly in the upper half of North America. These realities informed very different border measures affecting international travel (very restrictive) and trade (more accommodative) in both Canada and the United States during the 2020 global pandemic, although public health disruptions in multiple countries (including Canada) did have spillover effects on global supply chains. The Trudeau government secured negotiated restrictions on non-essential travel with the United States, while maintaining cross-border trade flows necessary to maintain food, medical, and other supply chains critical to containing the substantial economic effects of the wider public health emergency.[7]

International economic integration involves people – not just trade and investment. Previous trade agreements have sought to provide greater flexibility for international business travel and the recruitment of skilled personnel, while providing for distinct national policies to recruit immigrants to address domestic labour market needs. However, the growing divergence of Canadian and U.S. immigration policies have largely ruled out such cooperative measures, while increasing pressures leading to rising levels of irregular migration to Canada.

Renewed, cross-partisan trends towards U.S. protectionism and the growing divergence of domestic policy regimes among Canada's major trading partners create serious challenges for policymakers – recalling similar disruptions during the 1970s. Although successive governments have sought to diversify Canada's trade relations through an expanded network of trade agreements, Canada remains heavily dependent on the U.S. market. In 2018, 73.8 per cent of Canada's goods exports (70.8 per cent of total exports) went to the United States, which was the source of 64.4 per cent of Canada's goods imports (62.3 per cent of total imports).[8] Since the 1980s, expanded cross-border trade and investment with the United States have been critical in facilitating domestic policy adjustments and in managing inter-regional conflicts such as global energy shocks and related domestic conflicts during the 1970s and 1980s. However, the potential disruption of North American trade regimes inherent in NAFTA's renegotiation came with significant risks of spillover into domestic politics that will remain for some time to come. Growing conflict among Canada's diverse regional economies has been reinforced by diverging responses to climate change within North America. All these factors increase the challenges of international cooperation – but also the necessity of developing creative domestic responses to a changing political and economic environment.

These processes demonstrate that globalization is not a "single entity." Its dynamics reveal cross-cutting patterns of economic, political, and cultural relationships.[9] Much of Canada's experience of globalization reflects patterns of North American (or subcontinental) regionalism, including the emergence of cross-border regions with different levels of integration and interdependence, depending on economic and/or social (including environmental) relationships involved.[10] In many cases, these relationships substantially predate the closer economic integration of the 1990s and post-9/11 security preoccupations, and have helped to shape governmental and societal responses to them.[11]

A critical factor in managing the trade-offs between domestic and international policy dynamics has been the need to navigate what American political scientist James Rosenau has labelled the politics of "fragmegration" – the simultaneous interaction of integrative and fragmenting dynamics at multiple levels of analysis.[12] From a policy perspective, these cross-cutting dynamics include major asymmetries in the allocation of powers (and relative decentralization of) Canadian, U.S. (and Mexican) federal systems, similarities and differences in national, subnational, and sectoral regulatory systems and market structures, and the functional decentralization of bilateral (and broader international) policy relations within each government.[13]

The political and economic disruptions arising across major industrial countries, if less so in Canada, as a result of the financial crisis and recession of 2007–9, have changed the environment for international policy cooperation substantially. The initial retrenchment of American power after 2008 under President Barack Obama, described at one point as "leading from behind,"[14] may have been intended to accommodate a more collaborative style of international leadership. However, it also papered over growing public discontent in the United States (along with Europe and numerous emerging economies) over the disruptions associated with globalization, limited economic growth, and growing social and economic disparities between their beneficiaries and sizeable elements of society "left behind." The resulting political explosions, the rise of populist and nationalist movements in numerous countries (including "Brexit" and the Trump phenomenon) have challenged

Figure 1.2. The Continuum of Policy Relations and Policy Instruments for Regulatory Cooperation (IRCs)

Conflict   Independence   Harmonization

←---→

Parallelism Cooperation Coordination

Managed-   Dialogue / informal exchanges of
trade   information
agreements

"Voluntary"   Transgovernmental networks (may be
export   semi-formalized through MOUs
restraints
/ import
quotas

              Intergovernmental   Umbrella
              organizations (more   regulatory
              formalized soft law)   partnerships

Expansion of national
security provisions

Arbitrary product
specifications

Tightening of domestic
procurement restrictions

              Soft law provisions   via specific agreements or via supra-
              (e.g., codes of conduct,   conventions   national
              guidelines, roadmaps,   or joint
              etc.)   institutions

                       Recognition of international standards
                       (e.g., regulatory recognition of ISOs)

                       Regional agreements
                       with regulatory
                       provisions

                       Formal requirements
                       to consider "IRC" when
                       developing regulations

                       Mutual recognition agreements

Sources: Gattinger and Hale (2010, 13); OECD (2013, 24–5); Hale (2017).

political and economic assumptions underpinning Canada's international policy relations – not the least of which have flowed from the harried, multi-stage NAFTA revisions approved by the U.S. Congress in late 2019 and the Canadian Parliament in early 2020.

These developments challenge Canada's international policy relations, the continued capacity of Canadian governments to engage the United States and other countries in mutually beneficial policy cooperation, and the conditions necessary to enable such outcomes, at home and abroad. As during the 1970s, they also raise questions of the conditions necessary for Canadian governments to pursue international policy relationships in ways that provide widespread benefits to all regions, rather than setting the competing interests of provinces, regions, and other major socio-economic groups against one another in destructive zero-sum or negative-sum games. The centrality of federalism to Canadian and North American governances discussed by Anderson and Jones in chapter 2 is critical to managing these trade-offs among regional, societal, and related governmental interests.

Patterns of transgovernmental relations – "the interaction of subunits of governments on those occasions when they act relatively autonomously from higher authority in international politics"[15] – vary widely across departments and types of regulatory activities applying to various industry and policy sectors and subsectors. Figure 1.2 outlines a continuum of international policy cooperation, along policy instruments often associated with its different stages. It reflects varying degrees of independent, parallel or cooperative policy-making, while frequently accommodating agencies' efforts to preserve their capacity for independent action within their respective jurisdictions. The interaction of regulatory relations in each country with domestic interest group competition may result in protectionist or retaliatory measures intended to enhance or preserve the status of particular economic or social interests, or in periodic pressures from central agencies or major stakeholders to reconcile differences and facilitate closer coordination of regulatory activities.

These findings suggest that the significant regional and sectoral variations in economic interdependence and differences in structuring domestic and cross-border policy and regulatory regimes reflect differentiated integration: "the process whereby ... states, or substate units, opt to move at different speeds and/or towards different objectives"[16] in developing common or complementary policies in particular fields or sectors. Historically, these policies have reflected different degrees of regulatory parallelism, cooperation, or coordination informed by four broad realities confronting Canadian policymaking and frame the core questions this volume aims to highlight:[17]

- the extent to which market movements in particular sectors and subsectors are subject to international or domestic governance and related processes for border management, including the relative symmetry or asymmetries of interdependence in particular economic sectors and subsectors;[18]
- the relative institutional symmetry (or asymmetries) and concentration (or diffusion) of legislative and regulatory responsibilities within national (or subnational) governments, regulatory and administrative agencies;[19]
- degrees of similarity or differentiation between national and subnational government objectives and priorities, including relative unity or diffusion of policy focus: the number and relative importance of particular policy priorities in the mandates of particular agencies;[20]
- degrees of similarity or differentiation between policy instruments and settings used to implement – and often coordinate – these objectives, and the strengths and weaknesses of mechanisms for intergovernmental coordination.[21]

These framing realities, or constraints, help to explore the extent to which expressions of national or (in Canada) subnational sovereignty are framed primarily in terms of autonomous governmental authority or the influence of interdependence (including, but not limited to external policy pressures) on the exercise of such authority.

This book emerges from *Borders in Globalization*, an interdisciplinary research project that explores the extent to which policies shaping cross-border interactions and the management of borders across a wide range of policy fields are primarily territorially based or predominantly aterritorial – concepts that are explained below. Both sets of emphases may affect the definition and enforcement of borders – whether understood as physical or geographical constructs (if not necessarily those drawn on maps) that establish territorial boundaries between different political communities, as legal concepts and processes that facilitate but may also constrain the application of government authority, or that help to structure the rights and responsibilities of individuals and organizations in relation to states and one another. Other forms of borders involving cultural, environmental, and other factors[22] that transcend national boundaries are beyond the immediate context for this book. Consequently, some authors in this volume describe processes for defining, negotiating, or structuring such authority in contrast to (or in cooperation with) other jurisdictions as "bordering."

These distinctions matter for one simple reason: *governance.* That is a single word for a complex, evolving set of relationships. This volume addresses a cross-section of cross-border and broader international policy issues that address three core policy questions or themes: To what extent are shifting patterns of territorial and/or aterritorial governance generating significant challenges for Canadians? In what ways are the assumptions upon which policies are predicated being challenged? Where are the challenges to these assumptions coming from?

Territorially based policies and processes reflect the priorities of national (and sometimes subnational) governments and related domestic political forces. They involve the use of legal, regulatory, and administrative instruments by national and subnational governments to establish distinctive policies (or "the capacity for choice") in response to domestic imperatives, whether political, economic, or societal (including environmental). They may also involve the projection of authority in strategic sectors – for example, national security, control of technologically sensitive sectors, specific natural resources, the firms that exploit them (often involving provincial governments), or the assertion of control over land use or specific environmental risks by subnational governments (including First Nations and U.S. tribal governments) in both countries. Such policies often attempt to strike a particular balance among competing domestic, intermestic, regional, or subnational interests within Canada. However, they also may involve elements of bureaucratic territoriality: the attempted or actual projection of authority over a particular policy domain or, more colloquially, "turf."

Conversely, *aterritorial policies and processes* reflect the interaction (and varying mixtures) of external and domestic political, regulatory, but also frequently economic and/or technological developments and processes. They may be products of a sovereign state's ability to project or enforce its policy choices, including but not limited to the capacity to project national jurisdiction or enforce its legal authority beyond its territorial boundaries, as with the extraterritorial application of criminal, national security, or even human rights laws.[23] Alternatively, they may be reactive or defensive – responding to or paralleling U.S. or other powers' policy initiatives, often within international regimes that enable them to secure some degree of mutual recognition by other governments. Aterritorial approaches to policymaking may also arise from the workings of cross-border networks of governmental and nongovernment bodies seeking to develop common frames of reference and objectives to guide policymaking – as with assorted "soft law" agreements or through international nongovernmental bodies.[24] However, they may also reflect broader trends in capital flows, industrial organization (as discussed by Brendan Sweeney in chapter 16), technology (ranging from fracking to e-commerce), imbalances of skilled labour (as discussed by Meredith Lilly in chapter 7), public health and food safety issues as addressed by William Kerr and Jill Hobbs in chapter 18, and environmental networks, as discussed by Debora VanNijnatten and Carolyn Johns in chapter 11, along with domestic echoes of and reactions to such trends.

Aterritorial policy regimes may enhance national policymaking capacity by enabling reciprocity or complementarity of national policies that enhance the efficiency and/or predictability of national policy or administrative processes in Canada and other major industrial nations, often facilitating investment, trade, and travel. Examples include WTO requirements for national treatment of investors in domestic regulatory processes, cooperative processes for the development of technical standards through the International Organization for Standardization and other international agencies, subject to final domestic approval, and international agreements governing recognition of travel documents and principles for defining and managing flows of refugees and asylum-seekers. Given the tendency of bureaucratic and entrenched interest group "insiders" to dominate such processes, it often takes a special effort by governments to demonstrate to Canadians the benefits that arise from such processes – or to facilitate domestic adjustments to their outcomes.

However, as discussed by trade theorists, aterritorial regimes may also substantially constrain national discretion and that of citizens and domestic economic actors by increasing the complexity and unpredictability of trade flows – particularly when dealing with the "spaghetti bowl" of international trade agreements. Such measures may result in the external application of particular countries' criminal law and regulatory policies, particularly by major economic powers, as with global spillovers of U.S. anti-corruption, financial regulatory, and technology transfer policies discussed in chapters 4 and 19, among others. For example, the 2018 U.S. extradition request for a Huawei executive accused of violating U.S. financial disclosure laws in relation to alleged sanctions violations under its extradition treaty with Canada[25] embroiled this country in spreading geopolitical tensions between the United States and China. Moreover, these tensions have overflowed into debates on the prospective exclusion of Huawei equipment from

Canada's next generation ("5G") wireless telecommunications systems on national security grounds, and whether or not to follow the United States, Australia, and Great Britain in imposing such restrictions.[26] Formally, Canadian telecommunications policies, whether technical or security-related, are territorial. However, the practical implications of any likely decision could well have repercussions on security and trade relations with both the United States and China.

In practice, territorially and a-territorially based policies frequently overlap and/or function along a continuum of international policy co-operation, resulting in a "both-and" rather than "either-or" set of policy relationships. The greater the decentralization of major policy functions, whether within a single government or across different jurisdictions within a particular country, the more complex the *overall* mixture of territorial and a-territorial governance functions and practices is likely to become. However, as Slaughter[27] and others have noted, decentralization may also reduce coordination problems by allowing policymakers and societal actors (including but not limited to businesses) to develop specialized responses to particular challenges. Managing these activities requires simultaneous bordering processes to manage differences and the overlap of legal authority across and within national jurisdictions, while balancing the interaction of policy autonomy, discretion, and interdependence in economic and social spheres. In such settings, governmental policies do not exist in isolation. They frequently involve the assertion of shared, distinct, or overlapping identities and interests by individuals, businesses, and other organizations, with potential for inclusion in or exclusion from concepts of public or institutional interests defined by governments. They may also interact with private or nongovernmental standards developed by businesses and nongovernmental organizations, sometimes supported by international "branding" processes.

The dynamism and complexity of the policy landscape confronting Canadians is reflected in the diversity of the contributions to this volume. Each of the chapters we describe in brief below represents a particular analytical outlook, in a specific policy domain, and at a different level of analysis. What unites them is their common effort to grapple with the conflicting, contradictory, and increasingly exogenous variables we describe above. The mix of territorial and a-territorial forces bearing down on different policy domains may be idiosyncratic, necessitating, in some cases, equally idiosyncratic responses from state and non-state actors unaccustomed to such responses in the midst of growing degrees of inter-vulnerability, fragmegration, and intermesticity.

Outline of Volume

This volume is organized in four parts. Part 1 addresses overarching is-sues of domestic and cross-border or wider international governance and market flows, including trade and investment policies. Part 2 explores five intermestic policy fields that shape the territorial, cross-border, and wider international infrastructure and related regimes that supporting aspect of market and human movements (including both travel and migration) and environmental regimes. Part 3 explores the dynamics of Canada's principal cross-border regions – the "Pacific Northwest," Prairies–Great Plains, Eastern Canada and the Northeastern United States, the Great Lakes region, and the Arctic at different theoretical and practical levels of analysis. Part 4 explores the challenges of transnational economic and policy relationships in four specific sectors central to Canada's interna-tional economic relations: the automotive, energy, agri-food (particularly food safety), and aerospace industries. Each sector is deeply integrated within North American and global economic networks, but characterized by varied regulatory systems and exposure to political, economic, and broader societal challenges to their viability and continued expansion.

Horizontal Policy Systems

In chapter 2, Greg Anderson and David Jones emphasize the persistent importance of federalism in providing a territorial context for North American governance. While noting differences between federalism's explicitly territorial processes and other forms of governance, they note that many such initiatives are frequently constrained by legislated terms of delegation and subject to revision by participating govern-ments. It contrasts the very different patterns of and contexts for North American and European federalism, and the role of federalism as in-cubator of policy innovation and safety valves for differences in policy preferences across regions.

In chapter 3, Christopher Kukucha examines the persistent two-pronged approach of Canada and other countries towards trade policies in recent decades. Policymakers simultaneously pursue "offensive" or a-territorially driven negotiating strategies and policy instruments in sectors in which Canada enjoys potential comparative advantage. At the same time, they employ "defensive," reciprocal, or more territorially based strategies, including targeted industrial policies, in a number of strategic or politically significant sectors. Kukucha reviews the effects of economic and market forces, particularly the emergence of global value chains (GVCs), in creating path-dependent export and import relationships, which adapt incrementally but slowly to changing overall patterns of investment and trade, whatever the continuing efforts of firms or sectors to develop new markets. He also assesses the influence of competing societal, intellectual, and normative influences on the evolution of trade policies, while noting persistent gaps between rhetoric and reality.

The interaction of trade policies and domestic regulatory systems is a critical factor in facilitating, constraining, and defining the terms of economic and, in some cases, social interactions and integration across national borders. Geoffrey Hale explores factors contributing to and constraining bilateral (Canada-U.S.) and wider international regulatory cooperation in recent years in chapter 4. He reviews a cross-section of approaches: horizontal, crisis-driven, and incremental, agency-specific, that characterize such cooperation.

Greg Anderson outlines Canadian foreign investment policies in the broader context of comparable U.S. and European policies since the 1970s in chapter 5. He notes the relative importance of foreign direct investment (FDI) flows across major industrial countries and North America, summarizes sources of controversy over FDI and its implications for policy territoriality, and the parallel evolution of foreign investment regimes grounded in national security for both Canada and the United States.

Border Management, Transportation, and Other Meso-Policy Systems

In chapter 6, Geoffrey Hale provides an overview of the implications of evolving border policies and practices for cross-cutting patterns of cross-border trade and travel between Canada and related uncertainties arising from NAFTA's renegotiation. It summarizes trends in travel and freight movements during this period and adaptation to post-2001 border security measures. It outlines the realities of major trade corridors embodied in the Canadian federal government's pursuit of regionally varied gateway and "beyond-the-border" strategies, and summarizes current debates around rapidly expanding forms of cross-border trade including e-commerce and digital trade.

Meredith Lilly examines changing visa policies and rules surrounding temporary skilled labour mobility in Canada and the United States, their function within the broader context of national immigration policies, and their responses to evolving policy challenges in chapter 7. Lilly discusses the trade-offs associated with temporary worker programs, suggesting that Canadian governments have been relatively successful in balancing the opportunities and risks associated with such policies, providing a potential model for U.S. policymakers.

In chapter 8, Geoffrey Hale explores the context for Canada's transportation and related infrastructure policies in four key subsectors central to the structuring of cross-border transportation and related infrastructure regimes: trucking, railways, airlines, and the border infrastructure – including ports and airports — on which they depend to access international markets. He notes major variations in the extent and relative formality of regulatory cooperation, which is heavily dependent on relative centralization or decentralization of domestic regimes and the extent of North American or broader international interdependence in the operations of each subsector.

Barry Prentice and John Coleman explore the persistence of regulatory limits on cabotage – the unrestricted provision of domestic services by foreign-based transportation providers – across four major subsectors as an important non-tariff barrier, and its implications for transportation costs and economic development in Canada-U.S. border regions in chapter 9. They also consider the broader implications of cabotage for shippers and haulers in each subsector.

In chapter 10, Geoffrey Hale explores the very different worlds of "homeland" and critical infrastructure security as reflections of the different mixes of policy regimes governing these fields, with case studies of the electricity distribution and pipeline sectors. Since the 9/11 terrorist attacks, bilateral cooperation on security issues has become an underlying condition for relatively unimpeded access to U.S. markets for most Canadian industries. However, regimes governing the protection of critical infrastructure in both countries remain highly decentralized and domestically focused, notwithstanding their vulnerabilities, as networked systems, to international risks.

Debora VanNijnatten and Carolyn Johns address the "bordering" of environmental policy within North America, in the context of broader continental and international developments in chapter 11. They examine the architectures that have been erected to manage environmental problems that do not adhere to conventional borders and look at the degree to which policymakers have sought to fashion this architecture to ecosystems and to the economic dynamics underlying environmental problems.

Cross-Border Regions

Patricia Dewey Lambert explores cross-border regionalism – the interaction of provinces and states in cross-border cooperation in chapter 12. She focuses on the Pacific Northwest Economic Region – a public-private sector partnership bringing together legislative, governmental, diverse private sector, academic, and other societal actors. She develops a conceptual framework for studying the institutional structure and design of such cross-border entities.

Stephen Tomblin explains the progressive decline of cross-border coordination among Eastern provincial premiers and Northeastern governors in recent years in chapter 13. He suggests that this trend reflects the persistent institutional power and territorial priorities of the region's premiers and governors, ongoing interjurisdictional conflicts *within* each country, the fragmentation of regional institutions, the absence of effective leadership, and limited incentives to pursue cross-border regionalism in each country.

Kathryn Friedman addresses cross-border policymaking in the Great Lakes region through a case study of the Buffalo-Niagara-Hamilton cross-border sub-region in chapter 14. She uses an interdisciplinary approach bridging international law and international relations, urban planning, and public policy. The chapter focuses on recent efforts to cultivate cross-border cooperation among sub-regional stakeholders, while noting persistent institutional barriers to broadening and deepening such cooperation.

In chapter 15, Carolyn James explores Canada's bilateral relations with the United States in the Arctic, exploring ways of reconciling Canadian concerns over sovereignty with ongoing U.S. security concerns. Following an extensive analysis of the history of bilateral cooperation in the Arctic and other boundary waters, she argues that an institutionalized, binational mechanism is essential to handle such issues in the Arctic.

Sectoral and Subsectoral Issues

Brendan Sweeney explores the ongoing restructuring of Canada's automotive sector, reflecting the interaction of emerging international systems for organizing production with domestic (and cross-border) policy forces in chapter 16. He examines how these factors have influenced the size and organization of Canada's automotive industry in the twenty-first century, together with Canada's position within broader international automotive industry production networks.

In chapter 17, Monica Gattinger depicts a Canadian energy sector increasingly bound by a complex mix of physical, administrative, social, and political "borders" that have introduced uncertainty about one of Canada's major economic sectors. She cleverly deploys the acronym MESS (*m*arkets, *e*nvironment, *s*ecurity, and *s*ocial acceptance) to unpack the evolving, but generally thickening barriers to

Canadian policymaking. Some of those barriers are market-driven, a-territorial factors (technology in particular) that have upended the energy sector and forced Canada's hand. Other, territorial components focused on the environment, security, or social acceptance are, Gattinger argues, increasingly complex, severely limiting Canada's capacities to adapt to evolving market pressures and public expectations.

William Kerr and Jill Hobbs examine the question of whether borders are thickening for trade in agri-food products, particularly with respect to food safety, in chapter 18. Following its introduction, the chapter explains the nature of NAFTA's a-territorial architecture in the context of agriculture and food, and then examines how border issues pertaining to food safety and agri-food markets have evolved since the 1990s.

Canada's aerospace industry has consistently been at the forefront of both technological development and security considerations arising from the deep integration of U.S. and Canadian defence industries. In chapter 19, Mathilde Bourgeon and Élisabeth Vallet explore the challenges of navigating U.S. International Traffic in Arms Regulations (ITAR) governing the acquisition and use of defence- and security-related technology by Canadian aerospace firms, and the implications of these issues for sectoral hiring and other human resource policies and practices.

The concluding chapter explores factors influencing Canada's continuing "capacity for choice" in the contemporary environment of international policy relations, and key factors shaping the development of intermestic policy relations in the highly fragmented world of international trade, investment, and human mobility. It discusses challenges and uncertainties in the ongoing efforts of governments, industries, and firms to diversify Canada's trade relations, while managing relations on multiple fronts with its large, often distracted, and deeply divided neighbour to the south.

NOTES

1 For example, see Dieudonné Mouafo, Jeff Heynen, and Nadia Ponce Morales, *Building Cross-Border Links: A Compendium of Canada-US Government Collaboration* (Ottawa: Canada School of Public Service, 2004).

2 Bayless Manning, "The Congress, the Executive and Intermestic Affairs: Three Proposals," *Foreign Affairs* 55, no. 2 (1977): 306–24; Geoffrey Hale, *So Near Yet So Far: The Public and Hidden Worlds of Canada-U.S. Relations* (Vancouver: UBC Press, 2012), 4, 13.

3   World Bank, "Trade (% of GDP)" (Washington, DC: 2019). https://
    data.worldbank.org/indicator/ NE.TRD.GNFS.ZS.

4   Stephen Tapp, Robert Wolfe, and Ari VanAssche, eds., *Redesigning Trade
    Policies for New Global Realities* (Montreal: Institute for Research in Public
    Policy and McGill-Queen's University Press, 2017); Keith Johnson and
    Elias Groll, "China Raises Threat of Rare-Earths Cutoff to U.S.," *Foreign
    Policy*, 21 May 2019.

5   Douglas Bollyky and Chad P. Bown, "The Tragedy of Vaccine National-
    ism," *Foreign Affairs*, 27 July 2020; Carmen Reinhart and Vincent Reinhart,
    "The Pandemic Depression," *Foreign Affairs*, September–October 2020.

6   Douglas Porter, ed., *The Day after NAFTA: Economic Impact Analysis*
    (Toronto: BMO Capital Markets Economics, 20 November 2017). https://
    economics.bmocapitalmarkets.com/economics/reports/20171127
    / BMO%20Economics%20Special%20Report%20-%20The%20Day
    %20After%20NAFTA.pdf.

7   Christopher Sands, "Canada and the U.S. Closed the Border the Proper
    Way – Cooperatively and with Logical Exceptions," *Ottawa Citizen*, 19
    March 2020; U.S. Department of Homeland Security. "19 CFR Chapter 1:
    Notification of Temporary Travel Restrictions Applicable to Land Ports
    of Entry and Ferries "Service between the United States and Canada"
    (Washington, DC: 20 March 2020).

8   Statistics Canada, "Balance of International Payments, by Current and
    Capital Account, Annual," table 36-10-0007-01 (formerly CANSIM table
    376-0036) (Ottawa: 1 March 2019).

9   James N. Rosenau, *Distant Proximities: Dynamics beyond Globalization*
    (Princeton, NJ: Princeton University Press, 2003).

10 Emmanuel Brunet-Jailly, ed., *Borderlands: Comparing Border Security in
    North America and Europe* (Ottawa: University of Ottawa Press, 2007);
    Canada, Policy Research Institute, *The Emergence of Cross-Border Regions
    between Canada and the United States: Reaping the Promise and Public Value of
    Cross-Border Regional Relationships – Final Report* (Ottawa: Industry Canada,
    2008), http://publications.gc.ca/collections/collection_2009/policyresearch
    /PH4-31-2-2008E.pdf; Victor Konrad and Heather N. Nicol, *Beyond Walls:
    Reinventing the Canada–United States Borderlands* (Aldershot, UK: Ashgate,
    2008).

11 Reginald C. Stuart, *Dispersed Relations: Americans and Canadians in Upper
    North America* (Washington, DC, and Baltimore, MD: Wilson Center Press
    and Johns Hopkins University Press, 2007); Kyle Conway and Timothy
    Pasch, eds., *Beyond the Border: Tensions across the Forty-Ninth Parallel in the
    Great Plains and Prairies* (Montreal and Kingston: McGill-Queen's Univer-
    sity Press, 2013).

12 Rosenau, *Distant Proximities*, 9.

13 Greg Anderson and Christopher Sands, "Fragmegration, Federalism, and Canada–United States Relations," in *Borders and Bridges: Canada's Policy Relations in North America*, ed. Monica Gattinger and Geoffrey Hale, 41–58 (New York: Oxford University Press, 2010); Geoffrey Hale and Monica Gattinger, "Variable Geometry and Traffic Circles: Navigating Canada's Policy Relations in North America," in Gattinger and Hale, *Borders and Bridges*, 362–82; Brian Bow and Greg Anderson, "Building without Architecture: Regional Governance in Post-NAFTA North America," in *Regional Governance in Post-NAFTA North America: Building without Architecture*, 1–18 (New York: Routledge, 2015).

14 Ryan Lizza, "Leading from Behind," *New Yorker*, 26 April 2011.

15 Robert O. Keohane and Joseph S. Nye, "Transgovernmental Relations and International Organizations," *World Politics* 27, no. 1 (October 1974): 41; see also Geoffrey Hale, "Transnationalism, Transgovernmentalism, and Canada-U.S. Relations in the 21st Century," *American Review of Canadian Studies* 43, no. 4 (December 2013): 494–511.

16 Kenneth Dyson and Angelos Sepos, "Differentiation as Design Principle and as Tool in the Political Management of European Integration," in *Which Europe? The Politics of Differentiated Integration*, ed. Dyson and Sepos (Basingstoke, UK: Palgrave Macmillan, 2010), 4.

17 Hale and Gattinger, "Variable Geometry and Traffic Circles," 371–5; Bernard Hoekman, "International Regulatory Cooperation in a Supply Chain World," in *Redesigning Canadian Trade Policies for a Supply Chain World*, ed. Stephen Tapp, Ari VanAssche, and Robert Wolfe, 365–94 (Montreal: Institute for Research in Public Policy and McGill-Queen's University Press, 2017).

18 Porter, *Day after NAFTA*; Dan Ciuriak, Lucy Ciuriak, Ali Dadkhah, and Jingliang Xiao, "The NAFTA Renegotiation: What If the U.S. Walks Away," Working Paper (Toronto: C.D. Howe Institute, 28 November 2017).

19 Hoekman, "International Regulatory Cooperation," 368.

20 Dyson and Sepos, "Differentiation as Design Principle," 3–4; Paul Sando, "Water and Political Relations between the Upper Plains States and the Prairie Provinces: What Works, What Doesn't, and What's All Wet," in *Beyond the Border: Tensions across the Forty-Ninth Parallel in the Great Plains and Prairies*, ed. Kyle Conway and Timothy Pasch, 133–50 (Montreal and Kingston: McGill-Queen's University Press, 2013).

21 Dyson and Sepos, "Differentiation as Design Principle," 4; Hoekman, "International Regulatory Cooperation"; Geoffrey Hale, "Regulatory Cooperation in North America: Diplomacy Navigating Asymmetries," *American Review of Canadian Studies* 49, no. 1 (March 2019): 123–49.

22 Victor Konrad, "Towards a Theory of Borders in Motion," *Journal of Borderland Studies* 30, no. 1 (2015): 1–17.

23 Examples include the U.S. Foreign Corrupt Practices Act or Canada's Corruption of Foreign Public Officials Act, U.S. International Traffic in Arms Regulations (ITAR), discussed in chapter 19, targeted financial or travel sanctions against actual or alleged human rights abusers ("Magnitzky" laws), the use of domestic financial sector or securities regulations to enforce international economic sanctions, requirements for prescreening of foreign travellers seeking entry to the United States or Canada, and the criminal laws placing sanctions on "sex tourism" or the sexual exploitation of children beyond a country's territorial boundaries.

24 Anne-Marie Slaughter, *A New World Order* (Princeton, NJ: Princeton University Press, 2004); John J. Kirton and Jenilee Guebert, "Soft Law, Regulatory Coordination and Convergence in North America," in Gattinger and Hale, *Borders and Bridges*, 59–76; Tim Büthe and Walter Mattli, *The New Global Rulers: The Privatization of Regulation in the World Economy* (Princeton, NJ: Princeton University Press, 2011); Lawrence Herman, "The New Multilateralism: The Shift to Private Global Regulation," Commentary #360 (Toronto: C.D. Howe Institute, August 2012).

25 Robert Fife and Steven Chase, "Canada Arrests Huawei's Global Financial Officer in Vancouver," *Globe and Mail*, 6 December 2018, A1.

26 Richard Fadden, "For the Security of Canadians, Huawei Should Be Banned from Our 5G Networks," *Globe and Mail*, 21 January 2019, A13; Konrad Yakabuski, "Australia – Caught between the U.S. and China – Has Banned Huawei. Why Can't Canada?" *Globe and Mail*, 17 May 2019; Paul Waldie, "Britain to Purge Huawei from 5G Network by 2027 in Abrupt Policy U-turn," *Globe and Mail*, 15 July 2020.

27 Slaughter, *New World Order*.

PART ONE

Overarching Issues

2 The Great Unravelling? The Construction and Deconstruction of North America's Governance Architecture

GREG ANDERSON AND DAVID JONES

Between 2002 and 2011, the University of Toronto's Stephen Clarkson penned three masterful books on North America, the most poignantly titled of which was *Does North America Exist?* (2008).[1] In almost exactly the same period, American University's Robert Pastor put forward several visions for a more unified continent culminating what he called *The North American Idea* (2011).[2] As we begin implementing the "revision," "modernization," or "renegotiation" (pick your term) of the North American Free Trade Agreement (NAFTA), now inelegantly termed the United States–Mexico–Canada Agreement (USMCA), their work has never been more important. Yet their passing – Pastor in 2014 and Clarkson in 2016 – has left a large void in those thinking about the future governance of North America.

Pastor and Clarkson frequently disagreed with each other but shared a goal of understanding the governance that is modern North America – a patchwork shaped by the ideas of scholars, the policy mechanisms of governments, and a growing plethora of stakeholders that seems to lack a cohesive vision or direction. Hence both the question posed by Clarkson and the objective outlined by Pastor. Yet it's worth noting that the vision of a cohesive North America was not invented by NAFTA, or some fantasy concocted by scholars. Indeed, the original Indigenous inhabitants of North America had a more cohesive outlook on what the continent was, arguably flowing from its original name: Turtle Island.

Unfortunately, for most of the last three decades, NAFTA has cast a long shadow over the way we think about governance in North America. Indeed, many of us have had NAFTA on the brain,[3] imprisoned perhaps by the historical context in which NAFTA came into existence. The end of the Cold War seemingly ushered in a dramatic political and economic liberalization in a number of regions that has coloured our thinking about contemporary North America. In the Western Hemisphere, the

authoritarianism of the 1980s in some countries was giving way to democracy, economic modernization anchored upon Washington Consensus reforms, and an alphabet soup of regional configurations – NAFTA,
MERCOSUR, CARICOM, and the stillborn FTAA. Yet it was the dec-
ades-long deepening of the European project and the neoclassical stages of
integration that had the strongest impact on thinking about North American governance. While many acknowledged Europe and North America were not the same, Europe's successively deeper stages of economic
integration struck many as the direction North America would inevitably
go.[4] Indeed, the ink on NAFTA had not even dried when proponents of
deeper stages of integration began thinking about "next steps."

None of it has worked out that way. Instead, NAFTA became a *piñata*
symbolizing nearly every conceivable fissure in the architecture of
continental governance. Why aren't more jobs being created? Wasn't
NAFTA supposed to cure that? Environment not being cleaned up?
It must be NAFTA's fault. Wage rates not rising? Could be NAFTA.
Denied entry by U.S. Customs and Immigration? NAFTA. The sun is
blocked out by clouds? My coffee tastes bitter? My lawn isn't green
enough? Surely NAFTA's responsible. The more serious efforts to take
stock of NAFTA have pointed to the incongruence between expectations and realities surrounding NAFTA.[5]

Unfortunately, when we think about the governance of North America, NAFTA has been the dominant piece of governance architecture
structuring the way we think about the continent. Forms of cross-border,
regional, and transnational "governance" predated NAFTA and will
continue their evolution in wake of the ratification of the USMCA by
all three countries.

The basic argument of this chapter is that we shouldn't forget about
a much older, still potent framework of governance shaping the North
American space: federalism.

The point of departure for this chapter is the bewildering body of
literature in comparative politics and international political economy
attempting to identify, systematize, and understand the patchwork
of interactions we simplistically term "governance." Importantly, this
chapter attempts to reassert federalism as an essential conceptual starting point when considering governance in North America. The meaning of "governance" is certainly contestable. But for this chapter we
borrow some basic criteria for assessing what governance is: authority,
decision-making, and accountability.[6] Hence, when we claim to be ob-
serving "governance," we want to ask three questions: (1) who has, or
where is, the authority? (2) who, or what, makes the decisions? (3) who,
or what, is accountable?

A slightly different formulation tries to assess these questions in terms of whether the answers are found in entities that are territorial or aterritorial. In other words, where do we situate the authority centres of North American governance? Are those authority centres one and the same wherever we look: local, regional, national, trans-border, private, public, non-profit? In what areas is that authority constrained or bounded by political lines on a map? On which to they transcend or supersede those same lines?

As we observe patterns of governance in North America arising from the integration and security initiatives since the 1990s, our answer to most of these questions is that the state remains the most important starting point for analysis; in many cases it is both the beginning and end point. Indeed, the evolution of those patterns of governance is heavily territorial in that they are more firmly anchored in the theory and experience with federalism than with the centralizing impulses of economic integration.

The case for federalism as pivotal to understanding the architecture of North American governance will unfold in four parts. This chapter will first review some of the bewildering typologies of governance most closely connected to the European project, offering a critique of their applicability to understanding patterns of governance generally, and the North American setting specifically. Second, we will take a look back at the ups and downs of state-driven integration projects in North America, starting with Reciprocity in 1854, suggesting there have always been important territorially based limitations on pooled continental governance. Third, this chapter will discuss the hitherto underappreciated differences in conceptions of federalism in North America and Europe as models of governance. And finally, we'll look at federalism as both an "incubator of democracy" and driver of policy in North America.

Old Governance Wine in New Bottles?

Scholars looking at the ever-expanding complexity of governance mechanisms in the post-war European project have understandably grasped for new methods and concepts to understand and systematize what is going on: a stew of supranational, central state, subnational, regional, and localized governance structures – to say nothing of the countless non-state entities – interacting with one another. Yet the crux of it all is the apparent growth in governance actors challenging the primacy of the state as the sovereign locus of decision-making power. In other words, a shift from governance bounded by territorially defined institutions towards those unbound by the constraints of traditional sovereign authority: aterritoriality.

This juxtaposition of territoriality vs. aterritoriality in governance is part of a long-standing set of debates among scholars of international political economy (IPE), arguably dating to Adam Smith and his critique of the monopolistic qualities of state power in economic management. From there, a mountain of IPE literature has grappled with the rise of a seemingly endless challengers to state power, particularly those flowing from the post-war expansion of the global economy.[7]

To begin with, scholars are continuously debating the precise impact of greater economic openness on the state. A common place, certainly in the popular literature, is to assume that as states become more open and interconnected, they are beholden to a proliferating range of external actors – an "electronic herd" or "golden straitjacket" limiting the policy latitude of the state.[8] Indeed, one simple measure of inter-vulnerability to globalization and the loss of policy sovereignty is connected to a state's relative openness to, or dependence upon, open markets as its economic lifeline. As a state's overall trade (exports + imports) grows as a percentage of gross domestic product (GDP), the state is said to be more open. Table 2.1 depicts the relationship of trade to GDP for each North American state since 1975. In 2015, just over a quarter of the U.S. economy was tied to trade, whereas Canada (at 65.4 per cent of GDP) and Mexico (at 72.9 per cent of GDP) were far more open to (dependent upon) trade as a percentage of their economies.

With greater openness to trade comes a necessary decline in areas of policy sovereignty connected to trade: first and foremost are disciplines on the use of tariffs and non-tariff barriers. And with ever-deeper stages of integration comes greater pooling of sovereignty away from territorially defined borders.[9] One paradox in this straightforward characterization is that greater openness to (dependence on) the global economy is strongly correlated with increased demand for social spending.[10] In other words, advanced welfare states have been expanding alongside exposure to the global economy, and decision-making connected to that exposure made far away from domestic governance authorities. Hence a growing body of scholarship has probed the nuances of these relationships asking how, where, under what regime types, and to whom globalization's benefits accrue or fall most heavily.[11]

The post-war global economy expanded in tandem with a host of new institutional arrangements to govern different kinds of activity that had been the exclusive purview of the state. In trade, the General Agreement on Tariffs and Trade (GATT) and its successive "rounds" regularly advanced new "disciplines" on the state's policy latitude, a

Table 2.1. NAFTA Asymmetries ($U.S. Billions)

	1975	1987	1994	2000	2012	2015
Canada GDP	$170	$420	$560	$770	$1,780	$1,550
% of North American GDP	9.0%	7.9%	7.0%	6.5%	9.3%	7.5%
Exports + imports as % GDP	47.0%	53.0%	67.0%	85.0%	63.0%	65.4%
Mexico GDP	$88	$140	$420	$581	$1,178	$1,144
% of North American GDP	4.7%	2.6%	5.9%	5.2%	6.1%	5.5%
Exports + imports as % GDP	17.0%	33.0%	38.0%	64.0%	66.0%	72.9%
United States GDP	$1,600	$4,700	$7,017	$9,764	$16,240	$17,946
% of North American GDP	86.0%	89.3%	87.7%	88.2%	84.5%	86.9%
Exports + imports as % GDP	16.0%	19.0%	22.0%	26.0%	29.0%	28.1%

Source: World Bank.

question that begged to be asked was who was making trade policy, and where was it being made? The GATT was a consensus-driven body, but what did it mean that consensus on trade policy was being reached by member states in Geneva rather than member state capitals?

It all generated a kind of ambiguity around governance that prompted scholars to coin a number of terms to describe some of the cross-cutting, frequently contradictory impulses in contemporary governance, such as "intermestic," "fragmegration," or "glocalization."[12]

In 2003 Liesbet Hooghe and Gary Marks attempted to collect some of these proliferating threads and weave them into a simplified typology of governance. Indeed, they too marvel at the volume of governance concepts beyond the now-pedestrian "supranational," "multilevel," or "nongovernmental." Among these are "polycentric," "multi-perspectival," "condominio," "nested," "multi-centred," "bundled," "multi-jurisdictional," "matrix decision-making," "spheres of authority," "non-hierarchical," "non-territorial," or "governance by networks."[13]

Hooghe and Marks then develop a two-category typology of governance. In Type I multilevel governance are the mechanisms of traditional statist governance: a limited number of levels, well-defined jurisdictional responsibilities, general in purpose, organized around different functions and policy areas.[14] As Hooghe and Marks write, "The intellectual foundation for Type I governance is federalism, which is concerned with power sharing among a limited number of governments operating at just a few levels. Federalism is concerned chiefly with the relationship between central government and a tier of nonintersecting subnational governments. The unit of analysis is the individual government rather than the individual policy."[15]

Type II multilevel governance, they argue, is a much deeper box into which Hooghe and Marks toss nearly everything else. Indeed, this form of "multi-level governance is potentially vast rather than limited, in which jurisdictions are not aligned on just a few levels but operate at numerous territorial scales, in which jurisdictions are task-specific rather than general-purpose, and where jurisdictions are intended to be flexible rather than durable."[16]

Hooghe and Marks then go on to describe a bottomless pit of governance arrangements that fit this category, the overwhelming majority of which are local: everything from school boards, to toll-road operators, to port authorities.

Yet lumping the varieties of governance into two categories – essentially old vs. new – raises some scepticism about the conceptual utility of doing so.

An important line of research about governance in the global economy has sought understanding of the variety of ways in which non-state or transnational variables are challenging the dominance of the state at virtually every level, such as the decisions of bond-rating agencies, or hedge-fund or pension-fund managers, which may affect state credit rating and borrowing costs.

Curiously, the newish forms of Type II governance identified by Hooghe and Marks are not spontaneous challengers to the predominance of the state or the state's traditional modes of governance. In every instance, the Type II governance structures they identify are the direct offspring of the state. Several have quasi-independence, or "arm's length" qualities but are still creatures of, and responsible to, the state. Scholars, including contributors to this volume, have attempted to apply a number of the concepts identified by Hooghe and Marks to the North American setting. Yet the evolving nomenclature around non-traditional governance too often confuses the mere outsourcing of authority, decision-making, and responsibility to creations of the state as something entirely new. In other words, literature on multilevel governance may be mis-specifying the extension of traditional, territorial forms governance as something new that challenges the state.

Take, for example, the Port Authority of New York and New Jersey, to which Hooghe and Marks point as an example of an interstate "special district," implying an organically derived, authority and independence from New York and New Jersey – a sort of challenge to the authority of those two states. Nothing could be further from the truth. The Port Authority of New York and New Jersey was initiated by both states but ultimately authorized through congressional legislation in 1934: a delegation of article I, section 8 commerce powers. The authority may have a quasi-Independent single-purpose mandate, separate from the day-to-day management by branches of state government, but it nevertheless remains entirely accountable to both states.

Labelling as Type II any governance structures not formally a part of a traditional state-centric bureaucracy isn't particularly helpful methodologically. Generations of social scientists have spilled a lot of ink exploring the structures, mechanisms, and consequences of Type I governance. But how are we supposed to begin examining a basket full of Type II governance in which "there is no up or under, no lower or higher, no dominant class or actor, but a wide range of public and private actors who collaborate and compete in shifting coalitions."[17] By dumping all non-traditional governance mechanisms into a box labelled Type II, Hooghe and Marks don't add much to our conceptual clarity, nor do they point to an identifiable research program that might begin systematizing any of it.

Finally, Hooghe and Marks suggest that the Type II governance structures they've stumbled upon are "radical departures from the centralized state."[18] The problem is that while some of their identified examples are entities of a disaggregated or increasingly far-flung state – some critics might even call it the neoliberal state – little of what they've identified is far removed from the state. They point to extensive cooperation along the U.S.-Mexican border by state governments as typifying Type II governance: a non-standard form of governance across international borders.[19] Yet this kind of cross-border cooperation among states and provinces has always existed and is as robust as ever.[20]

At the heart of this cooperation are federal systems in which governance authority is shared, divided, and diffused among two levels of territorial government: central and state/provincial. In short, those seeing a significant shift towards aterritorial forms of governance in North America would do well not to overlook federalism as a major starting point.

Models of Federalism

Scholars of North American integration have fallen into a similar conceptual and terminological trap – sometimes seeing something old and labelling it new, or labelling things that are not really there. As we've watched the European project evolve, observers have sometimes over-emphasized the inexorability of ever-deeper neoclassical stages of integration in North America: a kind of deepening of the central state via the pooling of sovereignty. Indeed, whether focused on Europe or North America, the assumption is that the traditional authority, decision-making, and accountability functions of the sovereign state have been upended by populist revolts seeking to reassert those prerogatives: Brexit and the nationalism represented by President Trump being two prominent cases.

Daniel Elazar points to a major fallacy of this over-emphasis in his descriptions of how Europe and the United States have conceptualized and experienced federalism. In post-war Europe, the idea of federalism in European integration has been overtaken by more technical concepts swirling around the neoclassical stages of integration: supranational, community, and subsidiarity, for example. Indeed, the functional descent into the very process of integration in Europe has in many ways become a valued end in itself.[21] By contrast, "in the United States, the federal form of government was conceived to embody the fundamental republican and democratic principles which inhered in American civil society."[22] Furthermore, whereas the centralized state reigned supreme in Europe after 1648, the origins of the United States in the mid-eighteenth century turned the notion of centralized authority upside down by explicitly dividing power and authority territorially.[23] "American federalism ... vested sovereignty in the people to prevent the development of a centralized, reified the state by making all governments no more than governments of delegated powers whose scope the people could define and change as they pleased."[24]

Elazar depicts a European Union struggling with uneven governance wherein sovereign member states are constantly wrestling with which bits of authority to cede to supranational authority on the basis of different functional rationales. "In the United States, we've had the establishment of a modern federation as an effective way to liberalize the political and social order and promote democratic republicanism. Europe, on the other hand, is leading to a post-modern confederation designed to promote integration of previously politically sovereign states to better handle the problems of globalization, end devastating inter-state conflicts and empower ethnic groupings previously submerged in the statism of the nation-state."[25]

In other words, the centralized statism of Europe has had a more difficult time adjusting to the broad trajectory of post-war economic openness and the growing demands being made upon the state. The functional approach of Europe to these challenges has fostered ever-more complex patterns of governance to which scholars have readily attached new labels.

Fans of federalism point to the flexibility flowing from the inherent competitiveness enshrined in the explicit division of powers among different levels and orders of government. One of the most famous proponents of this idea was U.S. Supreme Court Justice Louis Brandeis, whose dissent in *New State Ice Co. v Liebmann* (1932) described the experimentation emerging from the states as vital to democracy: "To stay experimentation in things social and economic is a grave responsibility. Denial of the right to experiment may be fraught with serious consequences to the nation. It is one of the happy incidents of the federal system that a single courageous State may, if its citizens choose, serve as a laboratory, and try novel social and economic experiments without risk to the rest of the country."[26]

Is federalism the magic bullet for dealing with the pressures of globalization seemingly hastening the "retreat of the state"[27] or the "disaggregated state"?[28] Possibly not. Indeed, others point to more centralized, unitary states as being equally adept at preserving regional, cultural, and linguistic diversity.[29] Yet that critique is tempered with the observation that "federalism is no more than a constitutional legal fiction which can be given whatever content seems appropriate at the moment.... A constitutional form which can so quickly be changed from one kind of function to another has, of course, almost no real content of its own and is no more than a reflection of the more profound features of the political culture."[30]

The point here is that federalism's weaknesses as a stable, coherent system of governance are also its strengths. Moreover, it is mostly through the lens of federalism in North America that we can really appreciate the governance patterns that have evolved alongside North America's own experiments with neoclassical integration.

As Michael Burgess notes, experiments with federalism have been responding to the shifting sands of governance since the thirteenth century.[31] Moreover, Burgess also reminds us that "federalism is a multidimensional phenomenon and seeks in federation the constitutional and institutional practices that protect, promote and preserve the assortment of interests, identities and beliefs that naturally inhere in all societies."[32] In general terms, the pressures on the Westphalian state arising from the push and pull of globalization are indeed reordering governance "upwards beyond the state and downwards within it."[33] The question is how much and what kind of up and down reordering is really new in the context of the fluidity baked into federal systems such that they behave as "laboratories of democracy."

Hence some might argue the governance dynamism of federal states is a sign of the presence of aterritoriality, or the "unraveling of the central state" (borrowing from Hooghe and Marks). Yet how much aterritoriality, multilevel governance, or Type II is genuinely different from the functional fluidity of governance that's part of federalism's DNA? Perhaps not as much as analysts often want to see.

Trade That Comes and Goes

As noted earlier in the chapter, the initial threat of NAFTA's dissolution in the wake of Donald Trump's 2016 election was full of echoes of 1866 and the decision by Congress not to extend the 1854 Reciprocity Treaty – an agreement generally seen by Canadians as beneficial, but one many Americans had reservations about from the start.[34] One important claim we are making about patterns of governance then as now is that while the end of NAFTA, or the somewhat less liberal NAFTA embodied by parts of the USMCA, would usher in a period of increased non-tariff and potential tariff barriers, durable patterns of cross-border governance anchored in federalism will live on.

Like Reciprocity, the NAFTA was a shallow preferences arrangement – shallow in terms of the lowest stage of integration with the fewest strictures on state sovereignty (virtually none, in fact). As some of the examples we outline below suggest, a number of those barriers to deeper stages of integration are the result of federalism. Yet those same limitations on the centralizing pressures of economic integration also generate incentives for a broad range of policy initiatives by sub-federal governments. Moreover, because NAFTA entailed no pooled governance structures, its demise would not have required many to be torn down.

As we summarized above, Hooghe and Marks identify Type I governance structures as flowing from the relatively familiar structures of federal states. We appreciate their efforts to toss everything that doesn't look exactly like federalism into a box labelled Type II. Indeed, we share their bewilderment at the proliferation of terms and categories purporting to be fundamentally new in governance.

However, to our minds, categorizing all of this activity as Type II governance represents little that is actually new when cast against the elemental questions we posed at the outset about "governance": who has the authority, who decides, and who is accountable? We are hard-pressed to find many North American examples of the kind of aterritorial governance Hooghe and Marks label Type II. Indeed, wherever we look in North America, considerable authority, decision-making,

and accountability remain anchored in territorial governance structures, many of which flow from the dynamics of federalism.

The Port Authority of New York and New Jersey may appear to have
an independence from day-to-day management by the state, but it is
a delegated authority wherein the ultimate authority resides with the
state. Any restructuring of the Port Authority's mandate or adjustment
due to mismanagement would all require the sanction of the state.
New York and New Jersey – with congressional blessing – pooled some
sovereign authority but are both ultimately the location of authority,
decision-making, and accountability.

The institutional experience of twenty-five years of economic integration anchored in NAFTA is even more limited and failed to generate
the kind of delegated, aterritorial, cross-border governance that could
reasonably be classified as Type II. Figure 2.1 is a snapshot of the array
of cross-border initiatives, associations, pacts, and forums to which different levels of government (federal, state, provincial, territorial, etc.)
are parties. As far as we are aware, none of these entities would exist
without the participation of some level of government in both countries. Indeed, every single one is driven by state actors and for reasons
we would normally associate with the dynamics of federalism.

Organizations like the Council of State Governments, chapters of the
National Governors' Association, Canada's Council of the Federation,
the cross-border interaction of them, or policy initiatives such as the
Regional Greenhouse Gas Initiative or Western Climate Initiative are
all by-products of the traditional policy entrepreneurialism that inheres
to federalism.

Back to the Future of Federalism

Elazar's observation about the stark differences in the very conception
of federalism in Europe and the United States – Europe as a process
that is a valued end in itself, the United States originally conceived
as a republican and federal form of government from the outset – is
worth revisiting.[35] In both Canada and the United States, the diffusion
of power to different levels of government has been an important limitation on the proliferation of forms of governance.

One of the best examples of the way in which federalism shapes
or limits the evolution of cross-border governance is in the stark division of powers within the Canadian Constitution – specifically,
Section 92A of the Constitution Act 1867. There, Canada's provinces
are assigned exclusive jurisdiction over laws affecting management
of non-renewable resources – a sizable portion of Canada's overall

Figure 2.1. Cross-Border, Subnational Interactions

- Ontario and Quebec became associate members of the Council of Great Lakes Governors (Wisconsin, Ohio, Indiana, Michigan, New York, Minnesota, Illinois, Pennsylvania).
- Many provinces are associate or affiliate members of regional Councils of State Governments (CSG): CSG West (AB, BC), CSG East (NB, NF, ON, QC, PEI), CSG Midwest (AB, SK, MB, ON), CSG National (ON, QC).
- British Columbia, Alberta, Saskatchewan, Northwest Territories, and Yukon are full members of the Pacific Northwest Economic Region (that includes AK, WA, OR, ID, MT).
- Annual meeting of New England Governors–Eastern Canadian Premiers, established 1973 (ME, RI, CT, VT, MA, NH; QC, NB, NF, NS, PEI)
- Annual meeting of Western Governors–Western Premiers (AK, AZ, CO, HI, KA, NE, NM, OR, TX, WA, CA, ID, MT, NV, ND, OK, SD, UT, WY; MB, SK, AB, BC, YK, NWT)
- Southeastern U.S. States–Canadian Provinces Alliance (AL, GA, MS, NC, SC, TN; NF, NS, PEI, NB, QC, ON, MB).
- British Columbia and Washington organize joint full Cabinet meetings.
- Alberta has a formal government-to-government consultative mechanism with the State of Montana (Montana-Alberta Advisory Council).
- Quebec maintains several offices in the United States, including New York and Washington, DC.
- Alberta and Ontario have co-located an office within the Canadian Embassy in Washington, DC.
- Ontario also maintains a presence in two other U.S. locations.
- Manitoba has regularly maintained local representation to monitor events in Washington.

trade – and also, therefore, the environment. Hence, Canada frequently includes "carveout" or exceptions to text in trade agreements that could entail Ottawa having to impose international legal requirements in areas of provincial jurisdiction. A prime example of this kind of reservation is within NAFTA's Side Agreements on Labour and the Environment. Specifically, Clause (7) of Annex 41 to the Environmental Side Agreement states, "Canada shall use its best efforts to make this Agreement applicable to as many of its provinces as possible." The side-agreements' successor provisions in USMCA Chapters 23 and 24 are silent on these issues.

A similar story can be told in government procurement. In early 2009, the Obama administration pursued a large fiscal stimulus package in response to the worsening financial crisis that began in late 2007.[36] As part of that package, significant preferences were given to American firms when bidding on infrastructure projects let by the government; the "buy American" preferences in government procurement. Canadian firms would have been shut out of that procurement were it not for a 2010 bilateral agreement on government procurement. The fact was that Ottawa had only ever managed to get the limited participation of the provinces in the WTO's Government Procurement Agreement (GPA). Indeed, as the annexes to the GPA from 1994 state, many provinces were unwilling to fully open their procurement markets to foreign firms, while others were unwilling to open them at all.[37] In short, the constitutional prerogatives of the provinces – specifically Section 92 of the Constitution Act 1867 – make government procurement a story anchored more in the territorially based distribution of power under federalism than anything connected to governance towards or away from the central state. Only with the 2010 U.S.-Canada Agreement on Government Procurement was reciprocal access to each other's government procurement markets granted: thirty-seven U.S. states, all thirteen Canadian provinces and territories.[38]

The history of jurisprudence over commerce in the United States – and the Commerce Clause of Article I, Section 8, in particular – has permitted much greater centralization of federal control over international trade and investment relative to the states. However, that same federal power, and the asymmetries of national power between the United States and most of its trading partners, have limited Washington's willingness to pool sovereignty in the construction of a whole range of aterritorial governance mechanisms. America's reluctance to pool sovereignty institutionally can be seen in every post-war economic agreement it has concluded, including both NAFTA and its successor the USMCA.[39] However, while analysts have periodically identified a range of variables shaping the trajectory of (some would say limits of) North American integration, few have focused on federalism as one of them.

Federalism is constantly pushing the bounds of what governance looks like, often taking initiative on issues of local importance in which the central state may be out of touch or out of its depth. Hence, governance frequently spreads from one jurisdiction to another, seeming to become aterritorial by transcending international borders on issues ranging from health care, fuel efficiency standards, provision of social services, or climate change mitigation.

Networks, Nested or Otherwise?

One of our concerns about the literature on governance is that scholars have themselves become a little too entrepreneurial in their efforts to see new patterns of governance where there is nothing really new. We read about the proliferation of cross-border policy networks, epistemic communities, or the activities of private actors that are somehow embedded in policymaking. Once again, we struggle to see where new forms of "governance" exist in these descriptive categorizations.

To be genuinely different as a form of governance, we would need to see epistemic communities – academics, civil society organizations, business associations – begin to supersede or compete with the state in basic functions of "governance." We do not deny the capacity of some of these groupings to exert influence on policymaking, or the need of policymakers for stakeholder input to inform policy decisions. However, for any of these collaborative enterprises to be considered Type II, aterritorial, we'd need to see the locus of authority, decision-making, and accountability over policy shift from the state to these groupings.

The Mutual Recognition Antidote?

One area in which forms of governance have evolved in directions that seem to challenge, supersede, or transcend the state's governance authority is the growth of private professional associations: doctors, lawyers, accountants, and a host of trades who set their own professional standards through training, examination, or apprenticeship requirements that effectively control entry within territorially defined boundaries. Indeed, the refusal of many such associations to recognize credentials across state or provincial lines has been noted as a significant barrier to internal labour mobility, to say nothing of the impact on the flow of labour across international borders.

Importantly, the strength of these private professional associations is sometimes seen as a form of governance implicitly challenging the state's regulatory authority: state or provincial medical associations lobbying governments over a range of issues from standards- or regulatory-setting power to compensation.

Advocates of deeper North American integration have pointed optimistically to mutual recognition agreements (MRAs) as a mechanism wherein governance networks distinct from the state could expand, perhaps deepening integration where the state was unable or unwilling to lead. It is conceivable that territorially bound private associations could spontaneously agree to recognize equivalent credentials given in other jurisdictions. However, progress on MRAs has been extremely slow and limited, some say as the result of the state's unwillingness to put pressure on private professional associations. Consider Engineers Canada, the national self-regulating body representing nearly 300,000 engineers. The only MRA ever concluded between Canada and the United States covering engineers does so with only a single state; Texas.[40] Similarly, in 2014, all eleven of regulatory members of the Canadian Architectural Licensing Authorities signed an MRA with forty (only forty) U.S. member boards of the National Council of Architectural Registration Boards.[41] In other words, architects seeking to work across borders continue to face professional barriers in several jurisdictions. Finally, in November 2017, North America's public accounting bodies managed to ink an MRA allowing credentials earned through their respective associations to be honoured in all three countries – a genuinely trilateral MRA.[42]

Each of these professions governs the terms of entry to the profession, the educational or testing standards required for entry, and has developed extensive codes of professional conduct. Yet even if the patchwork qualities of MRAs were to become more complete and comprehensive across more North American jurisdictions, we'd still be hard pressed to categorize this as a form of aterritorial self-regulation of professional standards as the kind of governance that fundamentally challenges the state; indeed, in this case both the problem and the solution are defined by the actions of territorially defined actors.

Moreover, the independent governance that doctors, accountants, or electricians appear to have is sometimes less than meets the eye. Such professional associations may lobby the state for certain changes to audit or construction standards, but it's still the sovereign state (or subnational jurisdiction) that has the authority, decision-making power, and accountability for setting those standards. The professions are free to function like cartels in limiting access to their ranks. But the regulatory bodies they are subject to are creatures of the state.

Federalism Drives the Policy Bus ... Still

Given all of the literature on "governance" flowing from the different traditions identified by Hooghe and Marks, we expected to find multiple examples of the state being pulled apart, its authority being challenged, and its governance authority shifting to non-state entities. Instead, the energy, dynamism, creativity, and in many instances restrictions, of governance in North America continue to flow from the interaction of federal and sub-federal competition over policy. Two final examples, one each from Canada and the United States, make this abundantly clear: internal efforts to integrate Canada, and the advent of the Department of Homeland Security in the United States.

Canada and Internal Integration

An important interpretation of Canadian history centres on efforts to integrate itself economically and politically from east to west. Constitutionally powerful provinces with exclusive jurisdiction over policy areas of interest to the entire country (notably, natural resources) inherently set up conflict and competition among themselves and with Ottawa.

The territorial divisions of labour in Canada's federation continue to generate political and economic cleavages that have necessitated significant negotiations between the two levels of government aimed at reducing the impact of territorial barriers to trade and investment *within*. Canada currently lacks a single securities regulator. It also confronts a litany of inter-provincial barriers to trade and labour mobility in everything from beer and wine to whether electricians, plumbers, and carpenters who have apprenticed in one province can freely work in other provinces.

A number of important steps have been taken over the last several decades to mitigate some of these territorially based cleavages, starting in 1995 with the Agreement on Internal Trade (AIT). Interestingly, the text of the AIT reads remarkably like NAFTA concluded just a year earlier and had exactly the same purpose; reduction in territorially based barriers to trade. Yet the AIT was just as shallow as NAFTA institutionally, preserving too many provincial prerogatives in areas like labour mobility. In 2007, Alberta and British Columbia went further and concluded the awkwardly named Trade, Investment, Labour Mobility Agreement (TILMA) that sought to liberalize labour markets between the two provinces by, among other things, eliminating barriers to tradespeople working in each jurisdiction. The TILMA was joined by Saskatchewan in 2010 and renamed the New West Partnership (NWP), which Manitoba joined in 2017.

The members of the NWP claim to have formed the largest free trade zone in Canada – a remarkable claim for sub-federal governments of a sovereign state, and evidence of the continued importance of federalism as a form territorially anchored governance in North America. In July 2017 the entire country took another crack at improving the old AIT with the implementation of the Canadian Free Trade Agreement. After over 150 years of experimentation, competition, and conflict flowing from federalism in Canada, territorially based governance is very much alive and well.

Homeland Security and American Federalism

Federalism in the United States has evolved very differently than in Canada. Whereas Canada began in 1867 as a relatively centralized federation and became one of the most decentralized in the world, the United States began in 1776 as a loosely aligned collection of thirteen colonies that didn't work very well together until after 1789 and the ratification of the U.S. Constitution. Written explicitly to overcome a number of collective action problems through more centralized power in Washington, U.S. constitutional jurisprudence has routinely ruled in favour of federal action over state governments, particularly where the regulation of interstate commerce is concerned, and especially in times of national crisis.

Since its creation by Congress in 2003 as part of the broad national response to the 9/11 terrorist attacks on the United States, the Department of Homeland Security (DHS) has experienced well-documented growing pains.[43] A number of them were to be expected, given that creating DHS involved the largest reorganization of the federal government since the Second World War, cobbled together twenty-two agencies (each with its own culture) under a single bureaucratic roof, and created one of the largest single departments in the federal government at nearly 230,000 employees. Importantly, DHS has become the primary and dominant point of contact with the U.S. government for nearly everything that crosses North America's international borders.

However, DHS is also having a profound impact on the contours of American federalism. A major set of findings flowing from the 9/11 Commission Report to Congress in 2004 focused on the "failure" of government agencies to properly share information with each other.[44] In the restructuring that created DHS, how best to knock down bureaucratic silos and facilitate information sharing became a central issue. In many minds, creating a leviathan-like bureaucracy was antithetical to this task. Homeland Security has certainly become a leviathan, but alongside those qualities has emerged an impressive and powerful nationwide network of information gathering with significant impacts on federalism.[45]

For 2019, Congress proposed to appropriate $55.6 billion to DHS, which included significant source funding for a host of programs targeted at state and local governments. Among these sources of funding were:

- $538 million for the State Homeland Security Grant Program – an increase of $31 million above fiscal year 2018;
- $661 million for the Urban Area Security Initiative – an increase of $31 million above fiscal year 2018 – including $50 million for the Nonprofit Security Grant Program;

- $700 million for firefighter assistance grants;
- $249 million for pre-disaster mitigation grants; and
- $262.5 million for flood mapping.[46]

Federal spending powers in the context of federalism are the source of much conflict and controversy in nearly all federal systems. However, rapid assembly of DHS and its considerable discretionary budgetary authority to incentivize state and local governments into spending on homeland security infrastructure and preparedness, and cooperating extensively with federal authorities is dramatic.

Yet some states complain about the inadequacy of federal resources. Indeed, the scale of DHS and its congressional mandate have given the agency considerable regulatory and standards-setting power to which states and municipalities are often forced into compliance.[47] One by-product of DHS spending in the states has been the creation of more than fifty "fusion centers," mostly housed within the infrastructure of existing state and local first responders, all aimed at coalescing, sharing, and redirecting information among agencies at all levels of government throughout the country.[48] There is some debate about how and when the U.S. Supreme Court may yet limit federal intrusion into the states.[49] Moreover, the implications of the growing influence – or effectiveness – of DHS on federalism have yet to fully come into view. However, the capacity of DHS to spend and regulate defence of the homeland – particularly where states and local governments respond to unfunded mandates – will continue to shape the evolution of what is already North America's most important territorially based governance body.[50]

Conclusion

The overriding point of this chapter is a reminder that federalism remains a dominant framework of analysis for understanding patterns of governance in North America. Although the focus here has been on Canada and the United States, all three countries (Mexico included) have unique experiences with the policy competition, innovation, and experimentation incentivized by federal systems. Federalism routinely generates innovative policy proposals later adopted by the rest of the country, the best of which are sometimes considered across international borders. Whether it's universal health care in Saskatchewan, auto-emissions standards in California, regional climate change initiatives like the Regional Greenhouse Gas Initiative (RGGI), the grafting of Massachusetts health-care reforms (so-called RomneyCare) onto nationwide efforts (Obamacare), or Colorado's 2014 legalization of cannabis, which was studied by Canada in advance of its own nationwide legalization in 2018, each is ultimately the by-product of these "laboratories of democracy."

Yet this same dynamism also yields an understanding of the limitations federalism continues to place before policymakers wishing to understand the barriers to transcending the territoriality of governance. The territorial prerogatives of sub-federal governments sometimes incentivize ruinous competition (tax incentives for firms) and frequently, as in the case of Canada's internal trade, limit efforts to make the country more cohesive and efficient.

Finally, this chapter acknowledges the innovative work of scholars (including those in this volume) attempting to describe governance phenomena shaping the political economy of North America. However, this chapter suggests the exercise of some caution in grafting onto North America too much of the unique qualities of Europe's post-war governance. Europe has been wrestling with a complex mix of territorial and aterritorial governance, as some state functions are centralized in Brussels while localizing and outsourcing others. In North America, the ongoing experiment with federalism means more haphazard, experimental, and uneven forms of governance, most of which remain firmly anchored territorially.

NOTES

1 Stephen Clarkson, *Does North America Exist? Governing the Continent after NAFTA and 9/11* (Toronto: University of Toronto Press, 2008).
2 Robert Pastor, *The North American Idea: A Vision for a Continental Future* (New York: Oxford University Press, 2011).
3 Greg Anderson, "NAFTA on the Brain: Why Creeping Integration Has Always Worked Better," *American Review of Canadian Studies* 42, no. 4 (2012): 450–9.
4 Robert Pastor, *Toward a North American Community: Lessons from the Old World for the New* (Washington, DC: Institute for International Economics, 2001); see also Wendy Dobson, *Shaping the Future of the North American Economic Space: A Framework for Action*, Commentary no. 162 (Toronto: C.D. Howe Institute, 2002); Danielle Goldfarb, "The Road to a Canada-U.S. Customs Union," C.D. Howe Institute Commentary no. 184 (Toronto: C.D. Howe Institute, 2003); Goldfarb, "Beyond Labels: Comparing Proposals for Closer Canada-U.S. Economic Relations," Backgrounder No. 76 (Toronto: C.D. Howe Institute, 2003).

5 Gary Hufbauer and Jeffrey Schott, *NAFTA Revisited: Achievements and Challenges* (Washington, DC: Institute for International Economics, 2005).

6 Institute of Governance, Ottawa, "Defining Governance," https://iog.ca /what-is-governance/. Accessed, August 3, 2018.

7 See Greg Anderson, "Securitization and Sovereignty in Post-9/11 North America," *Review of International Political Economy* 19, no. 5 (December 2012): 711–41.

8 See respectively Thomas Friedman, *The Lexus and the Olive Tree* (New York: Farrar, Straus and Giroux, 1999); Friedman, *The World Is Flat: A Brief History of the Twenty-First Century* (New York: Farrar, Straus and Giroux, 2005).

9 Anderson, "Securitization and Sovereignty in Post-9/11 North America."

10 Dani Rodrik, "Sense and Nonsense in the Globalization Debate," *Foreign Policy* 107 (Summer 1997): 19–37.

11 Irfan Nooruddin and Joel Simmons, "Openness, Uncertainty, and Social Spending: Implications for the Globalization-Welfare State Debate," *International Studies Quarterly* 53 (2009): 841–66; Paul De Grauwe and Magdalena Polan. "Globalisation and Social Spending," CESifo Working Paper no. 585, March 2003; Brian Burgoon, "Globalization and Welfare Compensation: Disentangling the Ties That Bind," *International Organization* 55, no. 3 (Summer 2001): 509–51; Nita Rudra and Jennifer Tobin, "When Does Globalization Help the Poor?," *Annual Review of Political Science* 20 (2017): 287–307.

12 Bayless Manning, 'The Congress, the Executive and Intermestic Affairs: Three Proposals," *Foreign Affairs* 55, no. 2 (1977): 306–24; James Rosenau, "The Governance of Fragmegration: Neither a World Republic Nor a Global Interstate System," paper presented at the Congress of the International Political Science Association, Quebec City, 2000; Barry Wellman and Keith Hampton, "Living Networked On and Offline," *Contemporary Sociology* 28, no. 6 (1999): 648–54.

13 Liesbet Hooghe and Gary Marks, "Unraveling the Central State, but How? Types of Multi-Level Governance," *American Political Science Review* 97, no. 2 (May 2003): 223–43.

14 Hooghe and Marks, "Unraveling the Central State," 236.

15 Hooghe and Marks, "Unraveling the Central State," 236.

16 Hooghe and Marks, "Unraveling the Central State," 237.

17 Hooghe and Marks, "Unraveling the Central State," 238.

18 Hooghe and Marks, "Unraveling the Central State," 241.

19 Hooghe and Marks, "Unraveling the Central State," 241.

20 Greg Anderson, "Expanding the Partnership? States and Provinces in U.S.-Canada Relations," *Canadian-American Public Policy* 80 (2012): 1–44; Christopher J. Kukucha, *The Provinces and Canadian Foreign Trade Policy* (Vancouver: UBC Press, 2008).

21 Daniel Elazar, "The United States and the European Union: Models for Their Epochs," in *The Federal Vision: Legitimacy and Levels of Governance in*

the United States and the European Union, ed. Kalypso Nicolaidis and Robert Howse (London: Oxford University Press, 2001), 2.

22 Elazar, "The United States and the European Union," 2.

23 Elazar, "The United States and the European Union," 4.

24 Elazar, "The United States and the European Union," 5.

25 Elazar, "The United States and the European Union," 19.

26 *New State Ice Co. v Liebmann*, 285 US 262 (1932); see also G. Alan Tarr, "Laboratories of Democracy? Brandeis, Federalism, and Scientific Management," *Publius* 31, no. 1 (Winter 2001): 37–46.

27 Susan Strange, *The Retreat of the State* (Cambridge: Cambridge University Press, 1996).

28 Anne-Marie Slaughter, "Global Government Networks, Global Information Agencies, and Disaggregated Democracy," *Michigan Journal of International Law* 24 (2002–3): 1041–75; Eric Posner, "International Law and the Disaggregated State," *Florida State University Law Review* 32 (2004–5): 797–842.

29 William Riker, "Six Books in Search of a Subject or Does Federalism Exist and Does It Matter?," *Comparative Politics* 2, no. 1 (October 1969): 135–46.

30 Riker, "Six Books in Search of a Subject," 146.

31 Michael Burgess, *Comparative Federalism: Theory and Practice* (New York: Routledge, 2006), 76.

32 Burgess, *Comparative Federalism*, 265.

33 Burgess, *Comparative Federalism*, 260.

34 Lawrence Officer and Lawrence Smith, "The Canadian-American Reciprocity Treaty of 1855 to 1866," *Journal of Economic History* 28, no. 4 (1968): 598–623; Frederick Haynes, "The Reciprocity Treaty with Canada of 1854," *Publications of the American Economic Association* 7, no. 6 (1892): 7–70; Richard Johnston and Mike Percy, "Reciprocity, Imperial Sentiment, and Party Politics in the 1911 Election," *Canadian Journal of Political Science* 13, no. 4 (December 1980): 711–29.

35 Elazar, "The United States and the European Union," 2.

36 Technically the American Recovery and Reinvestment Act of 2009 or Public Law 111-5.

37 See World Trade Organization, Government Procurement Agreement, Appendices and Annexes to the GPA. Accessed 17 August 2020. https://www.wto.org/english/tratop_e/gproc_e/gp_app_agree_e.htm.

38 See U.S.-Canada Agreement on Government Procurement, 16 February 2010. Accessed 15 July 2018. https://ustr.gov/issue-areas/government-procurement/us-canada-agreement-government-procurement.

39 Greg Anderson, "Can Someone Please Settle This Dispute: Canadian Softwood Lumber and the Dispute Settlement Mechanisms of the NAFTA and WTO," *World Economy* 29, no. 5 (June 2006): 585–610; Anderson, "The

Reluctance of Hegemons: Comparing the Regionalization Strategies of a Crouching Cowboy and a Hidden Dragon," In *China and the Politics of Regionalization,* ed. Emilian Kavalski, 91–107 (Surrey, UK: Ashgate Publishers, 2009).

40 Engineers Canada, "Agreements on International Mobility," https://engineerscanada.ca/become-an-engineer/international-mobility-of-engineers/mutual-recognition-agreements.

41 Canadian Architectural Licencing Authorities, "Updates re. the NCARB/CALA Mutual Recognition Agreement and Acceptance of the Architect Registration Exam (ARE) (Ottawa: 21 January 2015). http://saskarchitects.com/wp-content/uploads/2015/03/January-2015-Bulletin-NCARB-CALA-Mutal-Recognition-Agreement-MRA.pdf.

42 Chartered Professional Accountants Canada "Canadian CPA Bodies Sign Mutual Recognition Agreement with U.S., Mexican Counterparts" (Ottawa: 1 November 2017). https://www.cpacanada.ca/en/connecting-and-news/news/media-centre/2017/november/mra-with-us-and-mexico.

43 Donald F. Kettl, *System under Stress: Homeland Security and American Politics* (Washington, DC: Sage, 2013). The period between the shambolic federal response to Hurricane Katrina (New Orleans) in 2005 through Hurricane Harvey (Houston) in 2017 was seen as a period of significant improvement in the federal response to large-scale disasters. However, Hurricane Irma, which devastated Puerto Rico, has once again put the federal response capacities under severe scrutiny.

44 Thomas Kean and Lee Hamilton, *The 9/11 Commission Report: Final Report of the National Commission on Terrorist Attacks Upon the United States* (Washington, DC: Government Printing Office, 2011), esp. 399–428.

45 Carmine Scavo, Richard C. Kearney, and Richard J. Kilroy Jr, "Challenges to Federalism: Homeland Security and Disaster Response," *Publius: The Journal of Federalism* 38, no. 1 (2007): 81–110.

46 S.3109, Department of Homeland Security Appropriations Act, 115th Congress, 2nd Session, 2018.

47 Patrick Roberts, "Dispersed Federalism as a New Regional Governance for Homeland Security," *Publius: The Journal of Federalism* 38, no. 3 (2009): 416–43.

48 Torin Monahan and Neal A. Palmer, "The Emerging Politics of DHS Fusion Centers," *Security Dialogue* 40, no. 6 (2009): 617–36.

49 Ilya Somin, "Federalism and the Roberts Court," *Publius: The Journal of Federalism* 46, no. 3 (2016): 441–62.

50 Peter Eisinger, "Imperfect Federalism: The Intergovernmental Partnership for Homeland Security," *Public Administration Review* 66, no. 4 (2006): 537–45.

3 Days of Future Past: Evaluating Canadian Foreign Trade Policies

CHRISTOPHER J. KUKUCHA

A major purpose of this volume is to evaluate the interaction of domestic and international policies on territorial boundaries created by treaties and state recognition in a sovereign Westphalian international system. In doing so, it considers the possibility that bordering is linked to actors and interactions where fluid networks have a history and culture that shift concepts of "space" beyond geography. This chapter will support and challenge these observations by highlighting the relevance of cross-cutting territorial and non-territorial considerations in the negotiation and practice of Canada's foreign trade policy.

Canadian governments, for example, have consistently pursued "defensive" interests in trade relations tied to sensitive sectors of Canada's economy, adopting negotiation strategies, border measures, custom administration procedures, and industrial policies. Other factors, however, point to aterritorial influences in this policy area. One case is an evolving historic, legal, and regulatory governance structure, resulting in a liberal rules-based trading system, consisting of institutions and regimes intruding into domestic jurisdiction with increasing emphasis on negative lists and international dispute panels limiting Canadian autonomy. Economic and market considerations result in similar outcomes as the result of established supply chains and entrenched path-dependent export and import markets. A series of sectoral and societal actors provide further pressure from industry associations, think tanks, labour, and other "critical" civil society groups. Further, normative considerations also appear to have an impact, at least rhetorically, as evident with Canada's current Progressive Trade Agenda (PTA). Finally, there is evidence to suggest that ideational factors shape Canada's foreign trade relations ranging from Keynesian and neoliberal economic influences to currently constructed American narratives embracing realpolitik, weaponized uncertainty, and permanent destabilization. Not surprisingly, all of these considerations have varying degrees of influence, depending on the time and space one examines over several decades.

The Relevance of Borders and Governments

First, it is necessary to highlight the ways in which borders, governments, and Westphalian sovereignty remain relevant in Canada's foreign trade policy. Following the Second World War, Keynesian principles of state intervention were widely, if imperfectly, adopted to manage and rebuild the capitalist economy through multilateral trade and financial institutions known collectively as the Bretton Woods system. The General Agreement on Tariffs and Trade (GATT), through several rounds of negotiations, successfully and significantly lowered barriers on a wide range of goods between members. These "first generation" GATT agreements, however, focused only on market access and tariff removal. As the GATT evolved, however, it also began to expand its agenda to include intrusive "second generation" trade rules touching on domestic jurisdiction. Not surprisingly, this slowed GATT negotiations. The regime also lost a level of legitimacy when the United States abandoned the gold standard and imposed a 10 per cent surcharge on imports in the 1970s. Twenty years later, the expanded membership of the European Union (EU) and its establishment of a monetary union reinforced these trends. Later problems included the rise of China, the 2008 financial collapse, and growing poverty in many parts of the world. As a result, the GATT regime and the subsequent World Trade Organization (WTO) struggled to transfer Western capitalist rules and norms at a pace similar to the early phases of the Bretton Woods system.[1]

In response to these developments, member states began to seek smaller forums of cooperation. Canada, for example, pursued bilateral trade agreements outside the WTO with its closest traditional trade partners. Initiatives included the Canada–United States Free Trade Agreement (CUFTA), the North American Free Trade Agreement (NAFTA), and the Canada–European Union Comprehensive Economic and Trade Agreement (CETA), all of which focused on tariffs but also on complex second-generation "behind the border" issues. Despite the evolution of trade negotiations, Canada maintained the same strategy throughout, prioritizing borders and self-interest in specific sectors. As Michael Hart has argued, Canadians earned a well-deserved reputation as "chisellers," pursuing offensive goals and trade liberalization where there was comparative advantage and defensive priorities where there were none.[2] NAFTA, for example, contained a wide range of exclusions on government procurement, energy-related services, subsidies, and financial services. It also largely excluded significant elements of agriculture and failed to address the contentious issue of monopolistic provincial liquor boards.[3] Other bilateral agreements followed a similar pattern, creating a range of agreements all adopting benchmark language that reflected Canada's offensive and defensive priorities. Though the list of issues grew ever longer, Canada's position as a "chiseller" remained consistent in broader subsequent negotiations, such as CETA, the Canada-Korea Agreement (CKFTA), and the Trans-Pacific Partnership (TPP).[4]

Border measures are another example of the ongoing relevance of territorial considerations for Canada and other states. Border measures include those taken by governments to limit the flow of goods in specific sectors. These barriers usually take the form of countervailing duties, anti-dumping measures, or quotas, and are justified as means to respond to the perceived unfair trade practices of other states. The United States, for example, has historically adopted border measures against Canada to counter what it argues are unfairly subsidized exports in the softwood lumber industry. More recently, President Donald Trump used questionable "national security" concerns to justify tariffs on aluminium and steel for a wide range of countries, including Canada, reminding Prime Minister Justin Trudeau, incorrectly, that "you guys burned down the White House" in the War of 1812.[5] Canada, however, worried that other states might dump aluminium and steel into the Canadian market as a result of U.S. tariffs, also adopted border measures.

Customs administration also demonstrates the ongoing relevance of borders and governments. Customs officials at border crossings enforce tariff schedule guidelines for commercial and non-commercial goods. In the Canadian case, a large number of goods travel duty-free across the borders of its trading partners, but tariffs and market access barriers continue in sectors such as alcohol, where provincial liquor boards control supply and pricing, and dairy products, which can have tariffs at or above 200 per cent. Customs officials, however, can also limit the cross-border movement of goods on the basis of a wide range of health and safety guidelines, as well as rules of origin criteria. The movement of workers across borders is also closely regulated by labour mobility provisions, especially for individuals seeking long-term employment in specific jurisdictions. To alleviate some of these issues, cargo preclearance initiatives have been adopted, especially in the automotive sector. These inspections can occur in numerous locations, but it is no coincidence that many custom brokers and clearance warehouses continue to be located in facilities at or near existing borders.[6]

Industrial policies, which are designed to enhance business competitiveness while serving a social purpose, usually providing jobs, generating new revenues, or encouraging research and development, further reinforce the importance of borders and governments.[7] Canada has a long history of industrial policies, beginning in the nineteenth century with Sir John A. Macdonald's National Policy. Other historic examples include the development of post-war nuclear energy and uranium exports, research and development in the aeronautics sector (such as the Avro Arrow), oil refineries and exploration in Atlantic Canada, the National Energy Program (NEP), and the Foreign Investment Review Agency (FIRA).[8] Although the popularity of industrial policies faded after the 1970s as most developed capitalist states adopted neoliberal economic priorities, it remained an important Canadian practice, especially in the aeronautics industry, most notably for Quebec-based Bombardier. In 2008 Ottawa provided the firm with a $350 million assistance program (in the form of a low-interest loan), which was followed by the Quebec government's $1 billion equity investment in Bombardier's C Series airliner in 2016. Although numerous governments provide assistance to domestic airline producers, including Boeing in the United States and Embraer in Brazil, foreign competitors argue that federal and provincial investments by Canadian governments provide an unfair advantage to Bombardier, allowing it to sell its planes for less than market value, resulting in demands for high protective tariffs in retaliation. The resulting international trade disputes are often based on some merit, but the political costs of removing this support often limits their impact on Canadian foreign trade policy.[9]

Emerging Governance Structures: Historical, Legal, and Regulatory

Despite the ongoing relevance of borders and governments, there is also evidence of aterritorial forms of influence. One example is historical, legal, and regulatory structures of governance. As noted, the post-war Bretton Woods regime saw many participants pursue offensive and defensive interests in negotiations based on perceived economic self-interest. In a broader sense, however, the shift from first-generation to second-generation agreements marked a significant change in the management of global trade. Tariff-based relationships guided the actions of states for centuries, but the emergence of a rules-based system, especially one focusing on "behind the border" issues, was an important evolution of the broader regime. Although this shift did not exclude the negotiation of tariffs and other border measures, it did ensure a new regime where these issues were negotiated as part of a broader rules-based agenda in more comprehensive international agreements. This process started slowly with the negotiation of a series of GATT codes on anti-dumping and subsidies, but by the 1990s the Uruguay Round produced new rules-based language in areas such as services, procurement, and technical barriers. NAFTA and WTO also began to include the use of "negative" lists in some areas of the agreement, sometimes exclusively but typically in the form of "hybrid" language combining positive and negative lists, such as the WTO's General Agreement on Trade in Services (GATS). Negative lists call for the non-discriminate movement of all goods and services, unless specifically *excluded* in a list of reservations, annexes, or appendices. Conversely, positive lists extend coverage only for issues voluntarily *included* by signatories. Together, these developments created a new form of legal and regulatory governance, evolving incrementally over several decades, for international trade.

For Canada, this resulted in a series of bilateral and multilateral rules-based agreements, such as the CUSFTA, NAFTA, and the WTO, that established benchmark rules and norms that increasingly intruded into provincial and territorial areas of jurisdiction, which became the foundation for future international negotiations.[10] The CKFTA provides a good example of the incremental evolution of the broader trade regime in a Canadian context. First, the CKFTA lowered a range of tariffs, adopting levels consistent with schedules for most other countries enjoying Most-Favoured Nation (MFN) status with Canada. The agreement also reflected "offensive" Canadian interests on rules-based issues, such as rules of origin, but in these cases the agreement largely transferred current language from the previous U.S. Korea Free Trade Agreement (KORUS) and the EU-Korea FTA. In this case, Canada benefited from adopting these benchmarks as it helped harmonize rules of origin for Canadian trade in Europe, the United States, and South Korea, which had positive implications for access to regional supply chains, especially in the automotive sector. The key, however, was that these norms were advanced as part of the overall regime, as opposed to specific Canadian self-interest. Having said that, the CKFTA also reflected Canadian "defensive" interests where tariff reductions in sensitive sectors such as dairy, poultry, and alcohol did not deviate significantly from previous agreements. In terms of alcohol, which is a processed agricultural product, provincial purchasing, pricing, and distribution practices were excluded and protection for the terms "Canadian whisky" and "Canadian rye whisky" provided. As always, Canada also excluded natural resources, "public services" such as health, public education, and other social services. Finally, provincial, territorial, and municipal procurement projects were not included in the Canada-Korea agreement.[11] Once again, however, these Canadian interests were already embedded in the regime as the result of previous benchmark language in other agreements.

Services, however, are more difficult to evaluate, given the differences between NAFTA and WTO benchmarks. The GATS adopted a hybrid but predominantly positive list for services, whereas NAFTA implemented a negative list approach, albeit with numerous exclusions in Annex 1 (non-conforming measures) and Annex 2 (future measures). The CKFTA is interesting because it reflects a NAFTA negative list approach to services, but its annexes do not extend obligations well beyond terms under GATS and other U.S. and European agreements. Although negative lists are important indicators of liberalization, it is crucial to focus on actual commitments in specific annexes. As such, CKFTA is considered to be a "GATS-plus" approach to services. This pattern is also evident in other issue areas, such as technical barriers and sanitary and phyto-sanitary standards, which again reflects regime norms.[12]

CETA, however, was an anomaly, given the expansive range of behind-the-border issues under negotiation, especially in sensitive sectors such as procurement. As with the CKFTA, Canada pursued a wide range of offensive tariff and market access issues, resulting in Canada and the EU agreeing to eliminate over 98.5 per cent of tariff lines, although, for the most part, tariffs were already low between all parties. Other offensive market access priorities, however, are more difficult to evaluate, given trade-offs that occurred during negotiations. In several cases, Canada was willing to drop asks on some trade rules in exchange for offensive market access. Rules of origin, for example, were a challenge, as Canada and the EU had different standards in a number of sectors. In cases such as automobiles (Ontario) and fish (Atlantic Canada and BC), Canadian products had difficulty meeting EU guidelines. The compromise was to grant "rules of origin derogations" allowing access for goods to a certain level that would not have to comply with EU regulations. For fish, this included important coastal exports of fish fillets, lobsters, salmon, herring, and sardines. Up to 100,000 Canadian passenger vehicles also received derogation for EU entry. The result, therefore, was improved market access, although to a lesser degree of incremental change under existing regime norms.[13]

Procurement, however, represented more significant liberalization, as it moved away from other rules-based benchmarks, especially as CETA allowed unprecedented access to EU suppliers for federal, provincial, Crown corporations, utilities, and MASH (municipalities, academic institutions, school boards, and publicly funded health and social services entities) procurement projects. Not only were provinces included, but also the "depth" of liberalization was significantly different, with access to utilities procurement, although not absolute, as well as provincial and territorial mass transit contracts. Sub-federal willingness to include these sectors was based on a perceived comparative advantage for some provinces, especially with hydroelectric power in Quebec, BC, and Manitoba. The significance of these developments was mitigated, however, by exclusions related to health care, security and policing, and public-private partnerships related to services and utilities. In most cases, procurement bidding thresholds also did not exceed existing commitments in the WTO's General Procurement Agreement (GPA) and Buy American Agreements.[14,15]

Canadian defensive interests related to market access in CETA, again reflecting broader regime norms, are also evident. CETA does allow higher imports of EU cheese (accounting for approximately 2 per cent of Canada's milk production), but other technical language does not dramatically extend agricultural obligations beyond WTO and NAFTA commitments, with the outright exclusion of poultry and eggs. The practices of provincial liquor boards related to listing, quotas, and purchasing, previously the focus of a number of Canada-EU disputes, also remains largely intact. The accord also adopts similar WTO language on technical barriers and sanitary and phyto-sanitary measures, including the Canada-EU Veterinary Agreement, both of which are sensitive to domestic areas of jurisdiction.[16] For services, CETA adopts a negative list, but its annexes do not move well beyond current "GATS-plus" models, such as CKFTA. This includes labour mobility and existing limitations on licensing and qualification requirements, although the agreement does point to the possibility of future liberalization in specific occupations where both parties have similar objectives.[17]

International trade disputes and rulings are also another example of emerging historical, legal, and regulatory governance posing some challenges to state autonomy. In some cases, these rulings establish quasi-precedents that guide governments for negotiated solutions, such as the series of NAFTA and WTO panel rulings that created parameters for previous Canada-U.S. Softwood Lumber Agreements (SLA 1996 and SLA 2006). Other disputes and rulings highlight Ottawa's difficulties in ensuring sub-federal compliance with regime norms. One of the earliest challenges centred on alcohol. Historically, Canadian provinces maintained monopolies or near-monopolies on the sale of alcoholic beverages. Provincial liquor control boards carefully regulated the sale and importation of all alcohol, with special emphasis on wine and beer, in order to protect local producers and raise government revenue. As such, foreign imports were subject to a wide range of measures, including high federal tariffs but also price mark-ups and quotas. The dispute between Canada and the European community (EC) was tied to a 1979 statement of intent signed by Canadian provinces designed to limit regulation of the sale of alcohol. The EC, however, questioned Ontario's marketing practices, especially on its differential mark-ups for domestic and imported beverages, and its listing, handling, and distribution practices. The matter was eventually submitted to a GATT dispute panel, which sided with the EC in October 1987. Subsequent discussions on alcohol were held with Europe, using these rulings as guidelines related to provincial practices over the next two decades.

International trade disputes have also targeted provincial procurement practices. In 2009 Ontario passed its Green Energy and Green Economy Act, which was tied to the province's feed-in tariff (FIT) program requiring a percentage of Ontario goods or labour for any awarded contracts. The following year Japan launched a WTO challenge, arguing that domestic content rules violated sections of the previous GATT framework as well as the WTO's Agreement on Subsidies and Countervailing Measures (SCM) and its Agreement on Trade Related Investment Measures (TRIMs). The European Union and the United States soon joined consultations on this challenge, and the EU launched its own formal complaint against Ontario's feed-in tariff (FIT) program in 2011. In December 2012 the WTO replied to both complaints in a single ruling, finding Ontario's practices in violation of Article 2.1 of the TRIMs and Article III.4 of the GATT (1994). Subsequent Canadian appeals were rejected, and after several mutually agreed upon extensions Canada announced it had complied with the WTO's findings in June 2014. Specifically, Ontario ended local requirements on large renewal energy procurement contracts and lowered similar provisions on smaller wind and solar electricity contracts awarded as part of the province's ongoing FIT program. Japan and the EU continue to monitor Ontario's procurement practices in this sector.[18]

Provincial policies have also been targeted under the investment provisions of NAFTA's chapter 11. In 2010 the international forest company Abitibi-Bowater launched a $500 million challenge related to Newfoundland and Labrador's decision to expropriate the company's water and timber rights and its hydroelectric holdings in the province, without compensation, after announcing its decision to close its final mill in Grand Falls–Windsor. In an effort to avoid a lengthy chapter 11 dispute, the federal government agreed to pay a $130 million settlement to Abitibi. This award has not been repaid by the government of

Newfoundland and Labrador. The regulatory powers of Nova Scotia were also challenged under NAFTA chapter 11 in *Bilcon* as the result of an environmental review related to a basalt quarry and marine terminal project in Whites Point, Nova Scotia, with the Delaware company claiming $500 million in damages. The panel also recently ruled against Canada in this dispute, although Bilcon was awarded only $7 million, largely for procedural costs, as opposed to its original argument based on projected lost profits from the project.[19]

Five of the other eight NAFTA chapter 11 challenges also focus on provincial practices. *Lone Pine Resources* involves an American company that was blocked from exploring oil and gas projects in Quebec as a result of Bill 18, which suspended all exploration in that sector in 2011. In two separate challenges from American companies, Windstream Energy and Mesa Power Group, challenged the provisions of Ontario's FIT program on wind projects in the province. In British Columbia, Mercer International claims that the BC Utilities Commission, the province's Ministry of Energy and Mines, and the Crown corporation BC Hydro, have all failed to develop a consistent policy of treatment for pulp mills and other facilities with self-generated power capacity. Finally, Mobil Investments Inc. has challenged the regulation of the oil and gas sector in Newfoundland and Labrador. These cases are in addition to previous challenges, such as Centurion Health, that centred on provincial regulatory provisions in BC that delayed the opening of a private surgical services facility, focusing on cosmetic, reconstructive, and general surgery in Vancouver. Although the U.S. complainant failed to pursue Centurion Health, all other claims remain active. Whether provinces fail to repay the federal government for any awarded damages, as in previous disputes, remains to be seen.[20]

These GATT, NAFTA, and WTO trade disputes and rulings are important for three reasons. First, they demonstrate Ottawa's commitment to the evolving norms of this regime in the dispute process but also in the central government's deference to panel decisions, for the most part, often in the face of sub-federal resistance. Second, as noted, these disputes and rulings have, at times, established guidelines for future government policy, at both levels of government, and set guidelines for negotiated settlements between governments, as in the case of the Canada-U.S. SLAs. Finally, international trade disputes and rulings can also serve as a "preventative regulatory regime" causing governments to reconsider policy options. During the 2003 provincial election in New Brunswick, for example, the high cost of automobile insurance was a key campaign issue, but Premier Bernard Lord's Conservative government abandoned it, fearful of the province's potential exposure under NAFTA's chapter 11. Ontario backed away from legislation mandating blank cigarette packaging in 1995 for similar reasons.[21]

The USMCA eliminated NAFTA's chapter 11 investment arbitration provisions between the United States and Canada, with provisions for a three-year phase-out of current obligations upon ratification. Investment disputes involving Canada and Mexico will be regulated under the TPP (renamed as CP-TPP) beginning in 2020. The United States, however, did view a USMCA side letter on alcohol as a victory in opening of provincial liquor markets. In the letter, British Columbia agreed to end discriminatory shelving practices for American wines in grocery stores. This side letter followed a broader Australian WTO dispute, yet undecided, targeting similar policies in BC, Ontario, Quebec, and Nova Scotia. Although the side letter and WTO dispute have the potential to alter established regime norms regarding Canada's sub-federal practices in this sector, it should be noted that both focus on distribution in private grocery stores, and not larger public provincial liquor outlets.[22]

Economic and Market Pressures

Economic and market pressures are other aterritorial forces that constrain government policy. One example is global value chains (GVCs). Over several decades, manufacturing has come to be dominated by GVCs, where parts are secured from a variety of national sources and production is spread across several states. GVCs also highlight the important and expanding linkages between trade, foreign direct investment, and international ownership. Many Canadian firms, especially those in the aerospace and automotive parts industries, are part of foreign multinationals and are deeply embedded in their global value chains. These transnational linkages, however, are still subject to some aspects of government control, usually defined in international agreements. Policymakers, for instance, still impose tariffs on intermediate goods included in final manufactured products. Governments also have numerous regulatory, tax, labour, and environmental controls that can affect the activities and profits of companies involved in supply chains.[23]

GVCs, however, also impose real limits on government policy options. As supply chains become longer and more complex, with inputs from an ever-increasing number of sources, it is difficult for policymakers to develop comprehensive economic and job-creation programs as rules and regulations in other jurisdictions are not always compatible. Further, as goods travel back and forth across borders as both inputs and final products, Canadian firms can be hampered by a lack of clear standards and certification requirements, as well as inconsistent customs procedures at borders. Seeking greater profitability, some firms will also relocate parts of their GVC overseas, reducing domestic jobs and investment.[24] Finally, Canada's longstanding "mature" supply chain network in the United States reduces incentives to participate in other production processes. Established supply chains can easily be lengthened or expanded, but are usually difficult to drastically alter, replace, or relocate.[25] Again, current protectionist American trade policies could challenge this conventional wisdom.

Canada's path-dependent export and import markets are another economic and market factor limiting government policy options. Specifically, these pressures are not "border dependent" and instead are driven by market considerations. During the last decade, Canada has shipped an average of 75.9 per cent of its exports to the American market, with highs of 78.9 per cent in 2007 and 76.2 per cent in 2016. There is no questioning the importance of the United States as Canada's top-ranked trading partner. Not surprisingly, similar consistency at much lower levels characterizes trade with Canada's other major export markets. China and the United Kingdom, for example, consistently rank second and third as export markets. In 2016, China received 4 per cent of Canadian exports, slightly ahead of the United Kingdom, which took in 3.3 per cent. China's share of imports from Canada also slowly increased from 2 per cent in 2007 to 3.7 per cent in 2011, while the United Kingdom increased its share from 2.8 per cent to 4.2 per cent, with shipments to Japan close behind. Mexico finished off the top five importers from Canada, with rates fluctuating minimally from 1.1 per cent in 2007 to 1.4 per cent in 2016.[26]

Canada's import figures exhibit a similar consistency. The United States is again Canada's top source of imports, averaging 51.9 per cent over the past decade. For the most part, this figure has fluctuated only slightly, from a low of 49.5 per cent in 2011 to a high of 54.3 per cent in 2014. China ranks second as a source for Canadian imports, incrementally increasing its share from 9.4 per cent in 2007, to 10.7 per cent in 2011, to 12 per cent in 2016. Shipments from Mexico, currently Canada's third major source for imports, follow a similar pattern, growing from 4.2 per cent in 2007 to 5.5 per cent in 2011 to 6.2 per cent in 2016. Fourth-ranked Germany registered similar but slightly lower totals. Japan rounds out the top five, with its share of imports slowly declining from 3.7 per cent in 2007 to 2.9 per cent in 2016.[27]

National numbers tell only part of the story. There are similar statistical patterns in Canada's provincial and regional economies. In British Columbia, diversifying trade with the Pacific Rim is a consistent political and economic objective. Yet there is no evidence to suggest that this has dramatically altered sectoral priorities. British Columbia, for example, has developed new "niche" markets in Asia's technology sector, but these industries are barely reflected in provincial trade statistics, which remain dominated by softwood lumber and liquefied petroleum. Similarly, Alberta has consistently relied on energy and agriculture, with international trade dependent on access to the North American market. The province has also attempted to increase trade with the Pacific Rim, but with limited success.[28]

Ontario has a more diversified economy than either British Columbia or Alberta. In addition to sectoral trade involving automobiles, transportation equipment, and telecommunications, the province has an extensive export profile that reflects its strong manufacturing base and engagement with numerous supply chains. Nonetheless, it still depends on the United States market. So too does Quebec, whose provincial economy is neither as diversified nor as dependent on one or two primary sectors. The province exports energy as well as a broad range of pulp and paper products. Quebec also has companies involved in the transportation and aeronautics sector, software services, industrial equipment, specialty food products, consulting engineering, and other professional services. At the same time, however, it faces pressure from the United States for greater liberalization in protected areas, especially the dairy and poultry sectors.[29]

In sum, Canada has distinct and long-standing economic priorities tied to path-dependent import and export relationships, which reinforce the incremental and inconsistent tendencies of Canadian foreign trade policy. Energy providers in Alberta want increased trade liberalization and greater market access to the United States. Ontario's automotive sector desires freer trade, but only with strict rules of origin. The aeronautics industry in Quebec demands protectionism, while softwood lumber is so contentious it is purposely omitted from foreign trade agreements and left to separate bilateral arrangements. Moreover, sectoral interests are not monoliths, and different producers within sectors often have divergent goals. The result is an exceptionally complicated policy process with numerous economic pressures and state and nongovernmental actors, often existing across borders, competing for influence.[30]

Sectoral, Societal, Normative, and Ideational Considerations

Additional aterritorial bordering processes include sectoral and societal factors, as well as normative and ideational considerations. On a sectoral basis, much of this activity is driven by Canada's corporate and business sector, usually through organizations such as the Canadian Cattlemen's Association (CCA), or such "umbrella groups" as the Business Council of Canada or the Canadian Chamber of Commerce. As Don Abelson noted over twenty years ago, these groups are linked with important "advocacy think tanks," prioritizing free markets, limited regulation, and other pro-business issues, best represented in Canada by the Fraser Institute and the C.D. Howe Institute.[31] There are deep linkages and networks between sectoral, umbrella, and think tank organizations, which enable ongoing and organized conversations that reinforce narratives and the basic principles of neoliberal capitalist priorities.[32] The role of diaspora communities in Canada has also influenced Canadian foreign trade relations, especially in the pursuit of specific trade agreements such as the Canada-Ukraine Free Trade Agreement (CUFTA).[33]

In contrast, "critical" societal actors (CSA) in this policy area are fewer and less organized. Historic trade unions and nationalist groups have increasingly given way to left-liberal think tanks, such as the Canadian Centre for Policy Alternatives (CCPA) or the Council of Canadians (COC), which first mobilized around CUFTA and NAFTA in the late 1980s and 1990s. The Pro-Canada Network, a coalition of these and other CSAs, was also created during this period, influencing the Canadian political discourse and the platforms of the New Democratic and Liberal parties. The COC and CCPA played a crucial role too in establishing the Trade Justice Network (TJN) in 2010, when several environmental, social justice, and Indigenous groups met in Ottawa to discuss the lack of dialogue and transparency associated with the CETA negotiations. The TJN would later expand to include several Canadian municipalities and like-minded societal groups in Europe, starting with a combined meeting in Brussels in the summer of 2010 after the fourth round of CETA talks.[34] Their efforts subsequently shifted to the renegotiation of NAFTA. Organized labour continues to play a prominent role in NAFTA 2.0 talks, with the high profile played by Unifor Canada's President Jerry Dias, including direct talks with Canada's chief negotiator, Steve Verheul.[35] Finally, recent student protests across Canada and the emergence of the Occupy Movement further focused attention on neoliberal inequalities, not all related to international trade.[36]

Normative considerations transcending territorial borders have also influenced Canada's foreign trade relations, albeit inconsistently and with an identifiable gap between rhetoric and reality. The original NAFTA, for example, contained side deals on both labour and the environment, which were copied and modified in several subsequent negotiations, when they were incorporated into these agreements as important provisions. The Canada-Chile trade agreement of 1997 was also the first to adopt a chapter on gender rights. Further, the current Trudeau government has promoted a "progressive trade agenda" in public statements and government documents. Although it is not explicitly defined as government policy, most accept that Canada's PTA approach includes a process for building consensus on the importance of international trade, but also one that highlights human security, class differences, the environment, and the empowerment of other marginalized groups. To this point, Canada proposed provisions on Indigenous peoples and gender in the early stages of the re-negotiation of NAFTA. Critics of these efforts, however, point out that recent and historical attempts to include normative considerations in Canada's trade relations typically follow an identified incremental pattern, as Ottawa frequently retreats from progressive language in the face of other competing commitments in multilateral and bilateral trade agreements. Modifications to benchmark language are similarly infrequent with numerous exclusions, and dispute-settlement mechanisms are either absent or lack efficiency and meaningful penalties. As such, it is important to acknowledge the aterritorial nature of these normative considerations but to also understand that governments, including Canada, maintain considerable autonomy in this issue area.

Finally, it is important to highlight the impact of aterritorial ideational considerations, especially on the construction and dismantling of the global rules-based trading system. As noted, in the post-war period principles of Keynesian liberal economics guided Western allies as they constructed the GATT system. Policymakers, fearing a return to interwar protectionism, used Keynesian ideas of regulation and state intervention to promote the freer flow of goods between member states and in turn bind these economies and governments with interdependent trading relationships. Neoliberal economics, which also gained intellectual attention during this period, began influencing government policies of deregulation and privatization in Canada, the United States, and Great Britain during the 1980s and 1990s. Associated with this was a reworking of the GATT trading system moving beyond tariffs and border measures to include rules-based regulation in areas such as procurement and services in a series of multilateral and regional agreements, such as the WTO and NAFTA.

Today, however, ideational considerations threaten to transform and possibly dismantle these commitments. The Trump administration, for example, has constructed new ideational narratives in its quest for protectionism. As Meredith Crowley and Dan Ciuriak have pointed out, the United States has consciously adopted an approach to trade that "weaponizes" uncertainty.[37] By seeking confrontation with other states and governments, including traditional allies, the goal is to create instability in other markets so companies will move jobs and investment back to the United States, using uncertainty as a non-tariff barrier not subject to the WTO or NAFTA.[38] As an extension, Jeffrey Goldberg has also cited this strategy as part of a broader Trump Doctrine on foreign policy, based on the realpolitik notion that the United States has a shifting series of alliances, suiting short-term American needs, as opposed to established liberal international regimes and institutions. In this quest, "permanent destabilization" creates a relative advantage for the United States, with both allies and its adversaries.[39] Ratification of the USMCA agreement may reduce such uncertainties, subject to ongoing effects of U.S. domestic political tensions.

Similar narratives challenging the established liberal trading system are echoed by China with its long history of large state-owned enterprises and questionable commitments to regime-based intellectual property rights. China also has a tendency to weaponize trade policy, especially when tied to broader regional or bilateral security and political goals. Recent discriminatory actions aimed at Canada, for example, are directly linked to Ottawa's decision to act on American extradition requests for Huawei Chief Financial Officer Meng Wanzhou as the result of alleged violations of U.S. trading sanctions against Iran. Following Meng's 2018 arrest in Vancouver, China imposed unilateral restrictions on imports of Canadian canola and pork. Further trade actions are also expected if Canada joins the United States and Australia in restricting Huawei products in the construction of a national 5G wireless network.

Conclusion

This chapter has examined territorial and aterritorial forces that affect Canadian foreign trade policy. Not surprisingly, evidence suggests that borders and governments are still relevant in Canada's global trade relations, especially in defensive trade interests, border measures, custom administration, and industrial policy. It is also clear, however, that non-territorial considerations are also important. First, a series of historic, legal, and regulatory governance structures have evolved over three generations to create a liberal rules-based trading system. These institutions and regimes consist of tariffs and border measures but also increasingly intrusive rules touching on domestic jurisdiction, as well as negative lists, and international dispute panels and rulings that limited Canada's autonomy. Economic and market pressures also create limitations in this policy area as the result of established supply chains and entrenched path-dependent export and import markets. A series of sectoral and societal actors do likewise, as the result of pressure from industry associations, think tanks, labour, and other "critical" civil society groups, which at times have influenced Canada to pursue normative considerations, such as Canada's current PTA, although admittedly often with a gap between rhetoric and reality. Finally, ideational factors have shaped Canada's approach to the creation and potential dismantling of the liberal rules-based trading order including Keynesian and neoliberal economic influences, as well as current U.S. and Chinese narratives prioritizing realpolitik and weaponized trade policy prioritizing uncertainty and destabilization.

NOTES

1 Christopher J. Kukucha, "Multilateralism and Canadian Foreign Trade Policy: A Long View," in *Seeking Order in Anarchy: Multilateralism as State Strategy*, ed. Robert W. Murray, 177–96 (Edmonton: University of Alberta Press, 2016).

2 Michael Hart, *Fifty Years of Canadian Tradecraft: Canada at the GATT 1947–1997* (Ottawa: Centre for Trade Policy and Law, 1998), 67.

3 Michael Hart, *Trade: Why Bother* (Ottawa: Centre for Trade Policy and Law, 1992).

4 Christopher J. Kukucha, "Canada's Incremental Foreign Trade Policy," in *The Harper Era in Canadian Foreign Policy: Parliament, Politics, and Canada's Global Posture*, ed. Adam Chapnick and Christopher J. Kukucha, 195–209 (Vancouver: UBC Press, 2016).

5 Daniel Victor, "A Fiery Affront, Back in 1812: But You Can't Blame Canada," *New York Times*, 6 June 2018, A17.

6 Eric Kulisch, "Canada, U.S. Push Plan for Getting Goods over Border Faster," *Automotive News Canada*, 24 February 2018, http://canada .autonews.com/article/20180224/CANADA01/302249991/canada-u.s. -push-plan-for-getting-goods-over-border-faster.

7 Anil Hira, "Understanding Industrial Policy," in *International Political Economy*, ed. Greg Anderson and Christopher J. Kukucha (Don Mills, ON: Oxford University Press, 2016), 343.

8 Michael Howlett, Alex Netherton, and M. Ramesh, *The Political Economy of Canada: An Introduction*, 2nd ed. (Don Mills ON: Oxford University Press, 1999), 298–300.

9 Patrick Leblond, "Boeing-Bombardier Dispute Is Trumpism at Its Worst," *Globe and Mail*, 28 September 2017, B4.

10 Kukucha, "Multilateralism and Canadian Foreign Trade Policy."

11   Canada, *Canada-Korea Free Trade Agreement: Creating Jobs and Opportunities for Canadians, Final Agreement Summary* (Ottawa: Public Works and Government Services Canada, 2014), 27–45.

12   World Trade Organization, "Overview of Services in RTAs: Positive or Negative List?" Asia-Pacific Economic Cooperation Workshop on Scheduling Services and Investment Commitments in FTAs, Singapore, 28–9 October 2014, http://mddb.apec.org/Documents/2014/CTI/WKSP5/14_cti_wksp5_010.pdf; Andrew Moroz, "Navigating the Maze: Canada, Rules of Origin and the Trans-Pacific Partnership (and Two Tales of Supply Chains)," in *Redesigning Canadian Trade Policies for New Global Realities*, ed. Stephen Tapp, Ari Van Assche, and Robert Wolfe, 423–46 (Montreal and Kingston: McGill-Queen's University Press, 2017).

13   Christopher J. Kukucha, "Federalism and Liberalization: Evaluating the Impact of American and Canadian Sub-Federal Governments on the Negotiation of International Trade Agreements," *International Negotiation* 22, no. 2 (2017): 259–84.

14   Sub-federal governments were added to Annex II of Canada's commitments to the WTO's Revised Agreement on Government Procurement in 2014.

15   David A. Collins, "Globalized Localism: Canada's Procurement Commitments under the CETA," 23 February 2015, http://papers.ssrn.com/sol3/papers.cfm?abstract_id=2568629.

16   Canada, Public Works and Government Services Canada, *Opening New Markets in Europe, Creating Jobs and Opportunities for Canadians* (Ottawa: 2013), 7–13.

17   European Commission, *CETA: Summary of the Final Negotiating Results*, February 2016, http://trade.ec.europa.eu/doclib/docs/2014/december/tradoc_152982.pdf.

18   Thomas J. Timmins, Wendy J. Wagner, and Neeta Sahadev, "The WTO Decision: What it Means for Ontario FIT 1.0 and 2.0 Projects," Mondaq, 3 June 2013, https://www.gowlings.com/KnowledgeCentre/article.asp?pubID=2910; World Trade Organization, "DS426: Canada: Measures Related to the Feed-in Tariff Program," implementation notified by respondent on 5 June 2014, https://www.wto.org/english/tratop_e/dispu_e/cases_e/ds426_e.htm.

19   Meinhard Doelle, "The Bilcon NAFTA Arbitration: The Damages Ruling," Dalhousie University Blogs, 1 March 2019, https://blogs.dal.ca/melaw/2019/03/01/the-bilcon-nafta-arbitration-the-damages-ruling/.

20   Canada, Global Affairs Canada, "NAFTA – Chapter 11 – Investment: Cases Filed against the Government of Canada," http://www.international.gc.ca/trade-agreements-accords-commerciaux/topics-domaines/disp-diff/gov.aspx?lang=eng.

21   Christopher J. Kukucha, *The Provinces and Canadian Foreign Trade Policy* (Vancouver: UBC Press, 2008), 193–7; Physicians for Smoke-Free Canada, *The Plot against Plain Packaging* (Ottawa: April 2008), http://www.smoke-free.ca/pdf_1/plotagainstplainpackaging-apr1%27.pdf.

22   Jesse Chase-Lubitz, "U.S. Hopes USMCA Wine-Win Will Break Open Canadian Market," Politico, 15 April 2019, https://www.politico.com/story/2019/04/15/usmca-wine-canada-1298447.

23   Emily Blanchard, "Leveraging Global Supply Chains in Canadian Trade Policy," in *Redesigning Canadian Trade Policies for New Global Realities*, ed. Stephen Tapp, Ari Van Assche, and Robert Wolfe, 209–28 (Montreal: Institute for Research on Public Policy, 2017).

24   Koen De Backer and Sébastien Miroudot, "New International Evidence on Canada's Participation in Global Value Chains," in *Redesigning Canadian Trade Policies for New Global Realities*, ed. Stephen Tapp, Ari Van Assche and Robert Wolfe (Montreal: Institute for Research on Public Policy, 2017).

25   Daniel Koldyk, Louis Quinn, and Todd Evans, "Chasing the Chain: Canada's Pursuit of Global Value Chains," in *Redesigning Canadian Trade Policies for New Global Realities*, ed. Stephen Tapp, Ari Van Assche, and Robert Wolfe (Montreal: Institute for Research on Public Policy, 2017), 283.

26   Canada, Industry Canada, "Trade Data Online: Trade by Product (HS) – HS Codes," http://strategis.gc.ca/sc_mrkti/tdst/tdo/tdo.php?lang=30&headFootDir=/sc_mrkti/tdst/headfoot&productType=HS6&cacheTime=962115865#tag.

27   Canada, Industry Canada, "Trade Data Online."

28   Kukucha, *The Provinces and Canadian Foreign Trade Policy*, 19–22.

29   Kukucha, *The Provinces and Canadian Foreign Trade Policy*, 15–19.

30   Kukucha, *The Provinces and Canadian Foreign Trade Policy*, 34–7.

31   Donald Abelson, "Environmental Lobbying and Political Posturing: The Role of Environmental Groups in Ontario's Debate over NAFTA," *Canadian Public Administration* 38, no. 3 (1995): 352–81; Abelson, "From Policy Research to Political Advocacy: The Changing Role of Think Tanks in American Politics," *Canadian Review of American Studies* 25, no. 1 (1995): 93–126.

32   William K. Carroll and Murray Shaw, "Consolidating a Neoliberal Policy Bloc in Canada, 1976–1996," *Canadian Public Policy* 27, no. 2 (2001): 1–23.

33   David Carment and Joe Landry, "Civil Society and Canadian Foreign Policy," In *Readings in Canadian Foreign Policy*, 3rd ed., ed. Duane Bratt and Christopher J. Kukucha, 277–89 (Don Mills, ON: Oxford University Press, 2015).

34   Stuart Trew, "Correcting the Democratic Deficit in the CETA Negotiations: Civil Society Engagement in the Provinces, Municipalities, and Europe," *International Journal* 68, no. 4 (2013): 568–75.

35 Laura Stone, "Unifor's Dias Emerges as Key Voice on NAFTA," *Globe and Mail*, 26 September 2017, A3.
36 Laura C. Macdonald and Jeffrey Ayres, "Civil Society and International Political Economy," in *International Political Economy*, ed. Greg Anderson and Christopher J. Kukucha, 329–42 (Don Mills, ON: Oxford University Press, 2016).
37 Meredith Crowley and Dan Ciuriak, *Weaponizing Uncertainty* (Toronto: C.D. Howe Institute, 2018).
38 Barrie McKenna, "Look Out, Canada: Trump May Be about to Get Serious about Protectionism," *Globe and Mail*, 12 January 2018, https://www.theglobeandmail.com/report-on-business/rob-commentary/look-out-canada-trump-may-be-about-to-get-serious-about-protectionism/article37593133/.
39 Jeffrey Goldberg, "A Senior White House Official Defines the Trump Doctrine: 'We're America, Bitch,'" *Atlantic*, 11 June 2018, https://www.theatlantic.com/politics/archive/2018/06/a-senior-white-house-official-defines-the-trump-doctrine-were-america-bitch/562511/.

4 International Regulatory Cooperation and Multilevel Governance: Motives, Methods, and Outcomes

GEOFFREY HALE

The spread of international regulatory cooperation has been significant in economic globalization since the late 1980s. International regulatory cooperation involves numerous processes that range from information sharing among governments, through the promotion of regulatory best practices and mutual recognition of regulatory goals and processes, to negotiated agreements and joint institutions to manage regulatory interdependence across national borders.[1] Such processes may be by-products of deliberate, centralized coordination through international organizations, reflecting the development of formal intergovernmental measures by specialized bodies, characterized by Slaughter as "vertical regulatory coordination." Alternatively, they may rely on private governance networks for developing international standards and other forms of private law.[2]

However, the prevalent forms of international regulatory cooperation in the twenty-first century include "horizontal regulatory cooperation" or "soft law" processes between specialized government agencies responsible for different regulatory fields in different countries.[3] Such measures preserve the forms and often the substance of domestic sovereignty, including dependence on domestic legislative authority and delegated administrative authority to specialized government departments or agencies subject to varying degrees of judicial oversight. However, these choices also reflect broader policy decisions, especially the facilitation of international trade and investment and the strengthening of international supply chains.[4] As such, they may respond to economic interests seeking to navigate technical and procedural differences in doing business across national borders, or balance governmental, economic, and societal interests, whether domestically or internationally focused.

In the context of this volume, international regulatory cooperation (IRC) is motivated by inherently aterritorial factors. However, it is also constrained by joint-decision problems in international relations, territorially based political constraints based on constitutional law, structural restrictions of national regulatory systems, and the realities of multilevel governance – the "interacting, reinforcing and colliding rule making and governance" processes at "international, federal, provincial and city/local community levels."[5]

This chapter explores the development of and constraints on IRC from a Canadian perspective since the mid-1990s. It provides a conceptual framework for regulatory cooperation, comparing the objectives of larger and smaller countries. It considers three broad approaches to regulatory cooperation: those involving centralized and/or horizontal coordination across governments, departmental or agency-led responses to periodic crises or focusing events, and incremental, agency-driven initiatives that reflect the technical and administrative priorities of sectoral regulators. It concludes by examining the prospects for ongoing regulatory cooperation within North America.

Its core hypothesis is that IRC in Canada and the United States is contingent on its relative priority within each country's domestic politics, reflecting both horizontal and policy- or sector-specific regulatory contexts. The priority given to IRC in specific policy fields reflects the intensity and persistence of political will, particularly if generated by external shocks and policymakers' recognition of the "intervulnerability" of national markets, regulatory systems, and wider security issues.[6] However, absent such crises, it is largely contingent on the extent and limits of enabling provisions both within international agreements and domestic horizontal and sector-specific legislation that governs regulatory systems. IRC is also subject to systemic constraints on the unilateral delegation of sovereignty, particularly domestic legal and legislative accountability, by cooperating states. Economic interdependence and normative considerations of efficiency and economic interoperability are important but secondary factors, which are often subordinated to broader political priorities of national governments and regulatory agencies.

International Regulatory Cooperation: Theories, Motives, Methods

The impetus for IRC derives from three longer-term trends, beginning with domestic U.S., British, and Canadian regulatory reform movements since the 1970s. Such initiatives focused primarily on economic aspects of regulation – particularly those governing market entry (and exit), price and/or output regulation, and detailed "command-and-control" approaches to regulation perceived to be overly restrictive of consumer choice and indifferent to relative cost effectiveness or implications for innovation.[7] Second, the growing internationalization of business activity reinforced the desire of regulators in such countries, particularly the United States, to promote information sharing, ultimately leading to the codification in multiple countries of "best regulatory practices" to combine protective elements of regulation with greater economic efficiency and growth.[8] Parallel North American and global trade negotiations subsequently sought to establish standards for and limits on technical (regulatory) barriers to trade (TBTs) and to encourage scientific and evidentiary bases for national health, safety, and environmental regulations while reducing discrimination against international trade and production.[9] Third, business groups also sought to persuade governments to pursue improved cross-border and broader IRC to facilitate trade and supply-chain efficiency arising from growing disaggregation and decentralization of production and distribution, sometimes described as "trade in tasks."[10]

After NAFTA's ratification, North American governments pursued bilateral and trilateral efforts to facilitate greater regulatory cooperation. Trade experts had previously recognized that non-tariff barriers, including technical differences in domestic regulations, product standards, related certification processes, and licensing requirements, were becoming the most significant challenge in facilitating trade and integrating production, distribution, and service networks across global regions.[11] NAFTA provided for the creation of about thirty working groups to identify and address coordination challenges within specific regulatory sectors. However, although some groups made modest progress, particularly in the deeply integrated chemicals sector,[12] most were relatively ineffective, if not stillborn, as noted by Kerr and Hobbs later in this volume.

In Canada, the External Advisory Committee on Smart Regulation (2004) emphasized the IRC's importance in complementing domestic efforts to promote "high levels of protection and economic competitiveness" – a position subsequently adopted as government-wide policy, albeit with diverse applications across departments.[13] Similarly, various U.S. federal agencies had pursued IRC initiatives. The Obama administration formalized these processes through Executive Order 13609 (2012), formally encouraging IRC in ways consistent with domestic legislative obligations and government-wide regulatory planning processes.[14] Subsequent U.S. trade initiatives, including the United States–Mexico–Canada Agreement (USMCA), have incorporated commitments to "good regulatory practices" consistent with these objectives within trade agreements.[15] Similar IRC initiatives may be seen within the European Union and ongoing OECD activities.

However, the existence of parallel national horizontal policy initiatives does not necessarily mandate responses by specific regulatory authorities. Scholars suggest that there are significant structural and cultural barriers to operational cooperation in reviewing and coordinating regulatory activities, even those in the United States and Canada that share similar business cultures. Donati argues that Canada's regulatory culture is characterized by broader public (or, at least, stakeholder) participation at early stages of the policy process aimed at problem definition and the exploration of regulatory options, but relatively closed processes later on[16] – although other observers demur, suggesting that relative openness is very much a function of the cultures of particular departments and agencies and their related policy communities.[17]

Conversely, U.S. regulatory processes tend to be characterized by limited "outsider" access at early stages of regulatory processes. However, the U.S. Administrative Procedure Act (APA) imposes extensive requirements to facilitate public participation and transparency in communications between regulators, major stakeholders, and broader publics.[18] Without specific legislated exemptions, communications from foreign governments are subject to such requirements, creating significant IRC challenges once regulatory processes enter the public domain. As a result, it is far easier for governments to cooperate on emerging and prospective areas of regulation than on coordinating existing regulatory processes – much to the frustration of business groups, which must function within these systems, although bilateral efforts through the U.S.-Canada Regulatory Cooperation Council (RCC), discussed below, have contributed to incremental improvements.[19]

Hoekman identifies several other systemic obstacles to regulatory cooperation, including mandate and coordination gaps among national agencies (including trade negotiators), information gaps among governments on technical and procedural requirements of particular regulatory systems, and coordination gaps between governments and businesses in participating countries.[20] Given differences in the relative size and interdependence of U.S. and Canadian economies, Canadian government officials and regulators must typically initiate efforts at relationship-building and cooperation with American counterparts (and those in other major trading partners).[21] However, internal incentives to invest the effort necessary for such cooperation depend heavily on the relative salience of cross-border relations for officials in such agencies and the extent of routine transgovernmental networks.

When specific economic sectors – most notably, Canada's automotive sector – depend overwhelmingly on a single export market, generally the United States, Canadian officials may align sectoral regulatory systems with changes to U.S. regulations, as with changes to Corporate Average Fuel Economy (CAFÉ) and designated emissions standards in recent years. However, such extensive harmonization requires regular adjustments by the smaller country to regulatory changes in the larger country, limiting the potential for regulatory innovation in the absence of a shared crisis.

Canadian governments are less likely to harmonize their regulatory systems or processes to U.S. or other international standards when faced by significant national or regional differences in industry structures or operating conditions. For example, major differences in the relative sizes and distribution of publicly traded companies in both countries effectively precluded Canada's wholesale adoption of corporate disclosure and accounting standards imposed by the U.S. Sarbanes-Oxley Act of 2003, despite widespread cross-listing of major corporations on Canadian and U.S. stock exchanges.[22]

This example illustrates two other major constraints on IRC. First, the greater the dependence of Canadian firms on domestic markets or on diverse export markets, the more likely governments are to pursue more flexible forms of regulatory cooperation. Second, the regional diversity of the Canadian economy often reinforces tendencies towards some degree of regulatory decentralization *within* Canada, despite ongoing efforts at interprovincial coordination, in which multinational firms pursue aterritorial regulatory standards but the prevalence of domestically focused firms contributes to the persistence of territorially based federal or provincial regulation.

Even with the best intentions, ongoing cross-border cooperation among regulators depends on shared regulatory objectives, similar methodologies and standards, comparable macro- and micro-regulatory processes, ongoing communication necessary for mutual trust, and substantial effort by both parties to maintain cooperation and effective parallelism. The following sections explore three different approaches to the pursuit of such objectives.

Centralized/Horizontal Regulatory Cooperation

Centralized approaches to international regulatory cooperation are the product of three broad forces. Overcoming institutional inertia required to reform regulatory systems requires a combination of intellectual leadership capable of demonstrating broader public benefits from such changes (especially relative to their prospective costs), strong support from societal and economic stakeholders, and authoritative political and bureaucratic leadership.[23]

As noted above, IRC's societal impetus has come from informal coalitions of internationally focused businesses whose competitiveness and capacity for growth depend on the efficiency of cross-border production and distribution networks. Post-2000 Canadian business pressures resulted from post-NAFTA economic integration, growing competitive pressures from inside *and* outside North America, and post-9/11 border thickening in response to U.S. security and other regulatory concerns.

Within government, federal central agency officials have led by creating institutional opportunities for promoting IRC, including advisory bodies such as the "Smart Regulation" Committee referenced earlier, the trilateral North American Competitiveness Council, and the Competition Policy Review Panel appointed in 2007 in response to a rising wave of foreign, particularly non-U.S. based, takeovers of major Canadian firms.[24] Senior political leaders and government officials in all three North American countries have embraced these proposals, if sometimes selectively and with different priorities.

These pressures contributed to successive separate, politically driven initiatives under the George W. Bush and Obama administrations. The trilateral Security and Prosperity Partnership (SPP, 2006–9) provided high-level political and bureaucratic leadership to "reboot" NAFTA's regulatory working groups, integrating border security measures with sectoral regulatory cooperation aimed at recovering momentum towards closer economic integration lost after 9/11. However, SPP's relative lack of transparency attracted strong, if ideologically varied nationalist opposition in all three countries. Its momentum bogged down in the absence of clear priorities among its roughly 300 projects. Another key constraint was political leaders' reluctance to pursue measures that required legislative approval, reflecting divided or minority government in each country.[25] Business stakeholder support was largely offset by critical interest group opposition, which undercut governments' efforts to give SPP a low public profile. Its largely incremental, technocratic processes proved politically self-defeating, being too bureaucratically focused to offer tangible expectations of substantial benefits to wider publics or political leaders, especially after the onset of financial crisis and recession.

SPP's failure led the Obama administration to open separate IRC processes with Mexico and Canada after 2010. The U.S.-Canada Regulatory Cooperation Council (RCC) initially enjoyed strong bureaucratic champions in both countries, presidential and prime ministerial support, and enough (twenty-nine) practical projects proposed by sectoral stakeholders and accepted by line agencies to offer prospects of substantial progress.[26] Coordination was centred initially in the U.S. Office of Information and Regulatory Affairs (OIRA) in the Office of Management and Budget (OMB), the Regulatory Cooperation Secretariat of Canada's Privy Council Office (PCO), and subsequently the Regulatory Affairs Secretariat of Canada's Treasury Board Secretariat – the central agency responsible for coordinating Ottawa's regulatory processes.

Line agencies developed functional work plans with cross-border counterparts to test specific proposals for change, conduct detailed stakeholder consultations (with public transparency), and provide more-or-less annual progress reports to national leaders – although it is unclear how much progress extended beyond "low-hanging fruit." Complementary bilateral initiatives were carried out on border-related issues (the "Beyond the Border" process) on both U.S. borders, as discussed elsewhere in this volume. Ongoing department-level consultations sought to "identify opportunities in the short, medium and long term for alignment" of emerging areas of regulation and, "ultimately, [to] evolve and implement regulatory systems in partnership."[27] Such consultations have normally avoided subjects associated with cross-border disputes, focusing instead on prospective and emerging regulatory issues and coordination of technical procedures in which each country's objectives are already comparable. However, Carberry suggests that the RCC process tended to place greater emphasis on moving towards the "synchronizing" regulatory *processes*, particularly ongoing internal reviews by line agencies, than on wholesale changes to regulatory systems: "Regulatory cooperation targets synchronized mutual improvements when things need updating or a new issue is faced. It's about aligning regulation, renewal and modernization schedules. Substantial results can be realized when it's time to improve and modernize some complex regulations."[28]

The first Joint Action Plan[29] of December 2011 spelled out four sets of sectoral initiatives in Agri-Food (particularly food safety), Transportation, Health and Personal Care Products (including workplace chemicals), and Environmental policy fields, along with working groups on small business and nano-materials. A second Joint Forward Plan in 2014 summarized (mainly procedural) outcomes of the first RCC workplan, lessons learned, and twenty-four sector-specific initiatives, expanding the original cluster of departments to include the U.S. Department of Energy and Natural Resources Canada, among others.[30] Table 4.1

Table 4.1. Regulatory Cooperation Partnerships since 2014

U.S. Agency	Canadian Counterpart	Topic(s)
Department of Agriculture	Canadian Food Inspection Agency	Meat inspection/ certification, animal health, plant health
Coast Guard (Homeland Security) & Environmental Protection Agency	Transport Canada	Marine safety and security
Consumer Products Safety Commission	Health Canada	Toy safety
Department of Energy	Natural Resources Canada	Energy efficiency standards, natural gas transportation standards
Environmental Protection Agency	Transport Canada	Locomotive emission standards
	Environment Canada	Vehicle and engine emissions
	Environment Canada and Health Canada	Chemicals management
	Pest Management Regulatory Agency	Crop protection products (pesticides)*
Food and Drug Administration	Canadian Food Inspection Agency	Food safety
	Health Canada	Pharmaceutical and biological products, over-the-counter products, medical devices,* veterinary drugs*
National Oceanic and Atmospheric Administration (Commerce)	Fisheries and Oceans Canada	Aquaculture
Occupational Health and Safety Administration (Labor)	Health Canada	Workplace chemicals, 2020: Workplace hazards*
Department of Transportation	Transport Canada	Connected vehicles
Pipelines and Hazardous Materials Safety Administration (Transportation)	Transport Canada	Transportation of dangerous goods

(Continued)

Table 4.1. (*Continued*)

U.S. Agency	Canadian Counterpart	Topic(s)
National Traffic Safety Administration (Transportation)	Transport Canada	Motor vehicle safety standards
Federal Railroad Administration (Transportation)	Transport Canada	Rail safety
Pipelines and Hazardous Materials Safety Administration (Transportation)	Natural Resources Canada	Explosives classification

* Continuing work: 2019–20 RCC Work Plan
Source: Canada-U.S. Regulatory Cooperation Council Joint Forward Report 2014;
U.S-Canada Regulatory Cooperation Council Work Plans 2019–20.

summarizes the bilateral department partnerships and proposed topics identified in this report. In June 2018, the two central agencies signed a memorandum-of-understanding (MOU) formalizing the basis for on-going bilateral cooperation.[31]

These developments demonstrate the numerous agencies involved in regulatory cooperation, the parallels and asymmetries in responsibilities between the two governments on various issues, and the highly technical nature of most discussions. Bilateral discussions continue under the Trump administration, notwithstanding broader political tensions. Although several sectors, notably food and rail safety, have seen progress on the development of mutual recognition processes and outcomes, these outcomes reflect either highly motivated agency leadership, particularly in the United States, or responses to significant external shocks in both countries, as discussed in the next section.

Carberry and others note that centralized horizontal regulatory coordination remains highly vulnerable to political drift and the deflection of political attention to more politically urgent or rewarding measures. Continued progress requires ongoing political buy-in on IRC's benefits (relative to the time and effort needed to achieve results) and minimizing political risks from motivated domestic stakeholders. The intensely "micro," technical character of resulting initiatives may reduce domestic political vulnerabilities. However, it also limits prospective political and economic benefits and the continued investment of political capital by senior political leaders. Notwithstanding these constraints, a side benefit of strengthening cross-border linkages and regulations is that it enables senior officials to deal with the second major mode of regulatory cooperation: agency-led crisis responses.

Agency-Led Crisis Responses

Major incidents or "focusing events" that draw public and political attention to regulatory gaps or inadequacies relating to cross-border movements of goods and people can provide a major impetus for cross-border or wider IRC. This subsection examines three sets of cross-border regulatory challenges that have prompted crisis responses from Canadian regulatory authorities, with varying degrees of interaction and cooperation with counterparts in the United States and other major advanced economies: managing the inter-vulnerability of electricity networks with extensive cross-border linkages, responding to successive disruptions in financial and capital markets both before and after the global financial crisis of 2007–9, and regulators' reactions to the Lac Mégantic rail tragedy of 2013 and subsequent derailments of trains carrying flammable products and other hazardous materials.

Electricity Policies

Overall levels of cross-border electricity trade in recent years have been comparable to interprovincial trade in Canada, although trade volumes and balances vary widely across provinces and from year to year.[32] As a result, the territorial dimension of electricity regulation enforced by provincial ownership of and jurisdiction over natural resources must accommodate aterritorial pressures to the extent that provincial electricity markets are interdependent with those of neighbouring U.S. jurisdictions.

Cross-border cooperation in electricity sector governance intensified during the 1990s following steps towards economic deregulation and growing interdependence of state and regional electricity markets by the U.S. Federal Energy Regulatory Commission (FERC). Provincial electricity regulators in Canada took steps to facilitate access by U.S. suppliers and maintain reciprocal access to U.S. regional markets. These actions resulted in substantial growth in Canadian electricity exports after 1995. Canadian diplomats subsequently lobbied Congress to maintain an open door to imports of Canadian hydropower by "zeroing out" such imports in calculating mandates for federal Renewable Portfolio Standards – removing Canadian hydro imports from both the total supply of generation from "renewable" sources and the total electricity supply used to calculate standards – thereby avoiding the creation of a major non-tariff barrier, although some state-level restrictions persist.[33]

The 1965 northeast blackout had prompted utilities to cooperate in developing cross-border electricity reliability standards. However, a major 2003 blackout that disrupted power supplies across much of eastern and central North America prompted calls for tighter U.S. federal regulation, necessitating new forms of cooperation to manage the operation of interconnected electricity transmission networks between Canada and the United States (and parts of Mexico). Quebec was spared the blackout, ironically as the result of preventive measures taken by American utilities and regulators due to differences in standard and practices.

Initial responses to create a North American Electricity Reliability Council (now "Corporation") to coordinate the activities of regional reliability organizations mandated by FERC and spilling over into Canada were further formalized through parallel legislation in Congress and Canadian provinces by 2008.[34] However, state and provincial organizations continue to have extensive autonomy in other areas of electricity regulation, including the design of state electricity markets (including permitting of power generation and transmission facilities within the state) and related issues of land use and environmental oversight. These realities are reflected in ongoing interest group litigation-driven regulatory bottlenecks in expanding interstate transmission networks, as with Massachusetts's difficulties in securing adequate Quebec electricity supplies to replace baseload power from the closure of a major nuclear plant in 2019.[35]

Similar politicization of provincial electricity policies has become increasingly visible in Canada, often resulting from rising internal costs and public concerns over the responsiveness of external political actors to internal provincial interests. These tensions have been reflected in Newfoundland and Labrador's longstanding dispute with Hydro Quebec over "wheeling" power exports through that province to neighbouring jurisdictions, and New Brunswick voters' angry backlash against NB Power's proposed merger with Quebec Hydro in 2009–10, among others. Provinces have never overcome their differences to negotiate an energy chapter within the 1994 Agreement on Internal Trade, although the so-called Canadian Free Trade Agreement of 2017 has adopted a "negative list" approach, which allows for extensive provincial carve-outs for procurement and other energy trade–related matters.[36] Gattinger explores competing governmental, industry, and societal priorities at greater length in chapter 17.

As a result of these conflicts in both countries, cross-border cooperation over electricity generation and transmission is largely contingent on the presence of and prospects for mutual commercial advantage between provincial and nearby state utilities, except for reliability standards noted above. Wider efforts at regulatory cooperation have generally proven no match for the offsetting politicization of local and regional electricity systems, reinforced by ongoing debates over mitigation of and adaptation to climate change.

Financial Services and Corporate Governance Regulations

Canada's financial sector is heavily interconnected with those of the United States and other major countries. Interconnections take three major forms. First, major financial institutions have long conducted business in foreign countries, bringing them under multiple national (and subnational) regulatory authorities responsible for banking, insurance, securities underwriting and distribution, and other forms of financial intermediation.[37] Second, these networks have supported the rapid growth of direct and portfolio investments, inward and outward, since the 1990s, contributing both to significant interdependence and regulatory overlap. Canadian businesses operating in the United States are subject to its financial, securities, and corporate governance rules (both federal and state), including financial reporting standards. Further complications flow from the growing international activity of independent (including state-sponsored and broader public sector) pension plans such as Quebec's Caisse de Dépôt and the Ontario Teachers' Pension Plan. Third, these factors reinforce extensive relations among financial sector (including securities) regulators and self-regulatory institutions (such as stock markets), which may result in some policy emulation and, more rarely, negotiation.

Great Britain's "big bang" regulatory reforms of 1985, which allowed competition and mergers across four traditional subsectors or "pillars" – personal and commercial banking, trust companies, insurance companies, and investment dealers – rapidly spread to Canada after Quebec and Ontario opened their markets to foreign securities dealers. Subsequent federal policy changes allowed Canada's major chartered banks to acquire most major Canadian securities dealers and trust companies well before Congress acted to repeal Depression-era Glass-Steagall barriers between commercial and investment banking in the late 1990s and liberalize interstate banking. However, this centralization of ownership has been offset by emergence of major new market actors, technologies, and financing techniques.

However, Ottawa's decisions to block two proposed major bank mergers in 1998, maintain barriers to mergers between large banks and insurance companies, and loosen restrictions on foreign ownership maintained Canadian financial institutions' domination of domestic banking and insurance markets. The Chrétien and Martin governments also encouraged major Canadian firms to expand their international operations, while progressively removing restrictions on international investments by pension funds by 2005, including major arm's-length federal and provincial investment management funds.[38]

These developments suggest aterritoriality resulting from selective policy adaptation and limited reciprocity within predominantly national (territorial) regulatory systems, and efforts by financial sector firms to take advantage of international opportunities. International developments may influence the evolution of Canadian financial sector regulations, and Canadian regulators have been active participants in international regulatory networks for many years. However, ultimate policy decisions appear to be shaped primarily by national (and sometimes subnational) policy criteria.[39]

Prior to 2000, Canada had paralleled many U.S. accounting and corporate governance standards while working within the International Accounting Standards Board (IASB). Canadian-based firms have been the largest source of U.S. corporate cross-listings on stock exchanges for many years, with comparable aggregate U.S. and Canadian trading volumes for these firms in the early 2000s.[40] This process was facilitated by U.S. recognition of Canadian listing requirements for publicly traded firms in its Multi-Jurisdictional Disclosure System as early as 1991.[41] A series of financial scandals among major U.S.-based firms led to intensified regulation and subsequent enforcement under the Sarbanes-Oxley Act of 2003. Sarbanes-Oxley (SOX) applied to the 181 Canadian-based firms – including most of Canada's largest publicly traded firms – whose securities were cross-listed on U.S. and Canadian (or other) markets in that year. (The comparable figure in late 2018 was 165, including 77 listings on the New York Exchange.)[42] One such investigation, while ultimately inconclusive, led to the 2009 demise of Nortel, once Canada's largest firm by market capitalization.[43]

Federal and provincial regulators and academic observers debated intensely whether to mandate highly prescriptive SOX-style accounting (and corporate governance) regulations or adopt more "principles-based" accounting standards resembling those in Great Britain through the Canadian Public Accountability Board. Ultimately, federal officials opted for the latter course, reflecting in part Canada's tiered and regionalized markets for publicly traded companies, with cross-listed firms allowed to follow the more intrusive (and administratively costly) U.S. standards.[44] Given the much larger number of small-cap firms on Canadian markets, Canadian regulators also allowed firms to disclose reasons for compliance or non-compliance with eighteen corporate governance "best practices" by 2004, rather than automatically following U.S. rules.[45] These developments have been reinforced – and sometimes anticipated – by intensive scrutiny by activist investors, often from outside Canada. For example, in 2011, short-sellers rather than regulators exposed fraudulent practices by Sino-Forest, a Chinese-based firm that had become Canada's largest forest company by market capitalization ($25 billion), ultimately resulting in bankruptcy and (belated) regulatory sanctions.[46]

More recently, international nongovernmental organizations such as Transparency International and major media outlets have criticized Canadian governments and financial institutions for weak regulations and internal controls over money-laundering and disclosure of beneficial ownership of corporations and trusts.[47] Money-laundering is closely associated with the activities of domestic and international organized crime, tax evasion through shell companies, and "grand" corruption involving both domestic and international transactions. Regulators and law enforcement agencies have identified money-laundering through financial institutions, the real estate sector, and provincially owned casinos, especially in British Columbia, as major areas of vulnerability requiring action from senior governments.[48] Canada has introduced amendments to legislation, partly in response to a 2014 statement of "high-level principles" from G-20 leaders. However, much will depend on implementation, sector compliance, and enforcement, particularly in financial institutions' far-flung international operations, which are subject to widely varied national regulatory standards and cultures.[49]

Notwithstanding longstanding decentralization of Canadian securities regulation, IRC in capital markets has long taken place through organizations such as the International Organization of Securities Commissions (IOSCO), which seeks to develop and promote "internationally recognized and consistent standards of regulation, oversight and enforcement" among its members in 115 jurisdictions (including Alberta, British Columbia, Ontario, and Quebec), while assisting members to identify and correct operational deficiencies through periodic audits.[50] Although primarily a "soft law" organization," its advisory function is analogous to that of the International Monetary Fund on macro-economic policies. Efforts by Ottawa and Ontario to create a national securities regulator have been slowed by persistent opposition from some provinces, especially Quebec, and adverse court rulings in 2011 and 2017[51] – again pointing to the significance of federalism in shaping the terms of regulatory cooperation. However, a Supreme Court ruling in late 2018 removed one major hurdle for eventual implementation of a jointly governed Cooperative Capital Markets Regulatory System, with support from Ottawa, five provinces, and Yukon.[52]

The 2007–9 international financial crisis prompted substantially increased cooperation among national regulators following bank bailouts in the United States, Great Britain, and some European countries. The evaporation of liquidity in the $32 billion asset-based commercial paper (ABCP) market in 2007, coinciding with the U.S. sub-prime mortgage crisis, required a complex mix of regulatory and market-driven actions and court protection, although ABCP investors were ultimately able to recover their funds.[53] The Bank of Canada provided markets with additional liquidity during the financial crisis and has subsequently maintained interest rates at relatively simulative levels without necessarily paralleling actions by the U.S. Federal Reserve, anticipating similar measures during the 2020 COVID pandemic. Canadian officials cooperated with bank regulators and central bankers through G-20 working groups, the Financial Stability Board (FSB), and the Basel Committee on Banking Supervision to coordinate broader policy objectives and, sometimes, more detailed measures during and after the financial crisis. Resulting requirements such as higher capital ratios and stress tests, with details determined by national regulators subject to external "peer review," sought to ensure major financial actors were capable of withstanding future crises.[54] These developments suggest growing aterritoriality, but still contingent on the support of national governments.

Longworth suggests that Canada "found itself between hawks and doves"[55] in these processes. Canadian political leaders and officials supported provisions requiring major banks to maintain contingent debt capital that could be converted to equity should they require additional equity in future crises as an alternative to government bailouts. However, they worked successfully to block higher international taxes on transactions or bank profits, arguing that countries with effective regulatory systems should not burden their financial sectors to support countries whose regulatory inadequacies had contributed to past bailouts.[56] However, as persistently low global interest rates, reinforced by quantitative easing (sizeable purchases of financial assets by U.S. and European central banks), contributed to huge increases in Canadian mortgage and consumer debt – a clear example of aterritorial policy effects – federal regulators incrementally tightened domestic rules governing mortgage eligibility and reduced federal guarantees on such lending to offset risks of rising debt levels to financial stability and the security of Canadian households. British Columbia and Ontario also introduced added tax measures on non-resident homebuyers to limit effects of foreign, largely non-institutional capital inflows on house prices.[57]

In summary, Canada's financial sector regulatory systems are heavily influenced by both formal and informal capital flows, and challenges of balancing the pursuit of market openness with effective, responsible, corporate governance. Canadian federal and provincial regulators have long been heavily engaged in IRC. However, their ability to influence international trends and maintain domestic ("territorial") operational autonomy through soft-law regimes appears significantly greater in multilateral settings. Fundamental differences with the United States in regulatory and market structures result in a mixture of emulation, parallelism, and territorial autonomy, particularly in sectors with more decentralized governance structures induced by Canada's decentralized federal system. Crises, actual or imminent, may create opportunities for greater cooperation. However, the reality of actual cooperation is generally contingent on whether such crises are centred in U.S. markets (in which case American unilateralism is the norm) or more widely distributed.

Rail Transportation Safety

The interdependence of transportation subsectors within and beyond North America is discussed extensively in chapters 8 and 9. Regulatory and related political bottlenecks to the expansion of domestic and cross-border pipelines in each country discussed by Monica Gattinger in chapter 17 resulted in growing amounts of crude oil and other hazardous materials being shipped by rail until the collapse of global oil prices at the outset of the 2020 COVID pandemic. The 2013 Lac Mégantic rail tragedy, in which the derailment and explosion of a runaway train carrying highly flammable crude oil from North Dakota killed forty-seven people in a small town near the Quebec-Maine border, and several other major train derailments in both countries intensified public and regulatory attention to the growth of hazardous materials shipments across long distances and national borders.

The resulting enquiry pointed to numerous inadequacies in rail safety standards and financial safeguards to ensure smaller railways could assume liability for accidents following the sector's reorganization in the 1990s and 2000s. Canadian regulators responded to these events with emergency regulations that addressed staffing, tank car design, maintenance of rail lines, monitoring of train operations, information sharing on hazardous materials shipments with municipalities and emergency responders, related response capacities, and insurance requirements for carriers and shippers.[58] However, it was vital for Canadian officials to network with U.S. counterparts to ensure reasonable alignment with regulatory changes to be introduced by the latter, including timeframes for implementation that could accommodate the production of adequate amounts of new rolling stock for railways in each country. These conversations initially took place outside the RCC framework noted above, but were facilitated by it.

Each country's transportation safety board was in regular contact to share information on derailment investigations, leading to an unusual joint statement in January 2014 calling for specific regulatory changes to rail car standards, railway risk analysis processes, and emergency response capacities in both countries.[59] In July 2014, the U.S. Department of Transportation (DOT) followed the Canadian initiative to remove or phase out older railcars and impose lower speed limits on shipments, especially in urban areas.[60] Both governments announced updated regulations in May 2015,[61] although subsequent implementation appears to taken place faster in Canada.

In this case, Canada's regulatory structure, with its provisions for emergency regulations, provided federal officials with a "first-mover advantage" in responding to crises, while engaging the slow-moving U.S. system. Although cross-cutting interests were present in both countries, well-established professional relationships enabled effective transgovernmental coordination and the adaptation of regulatory processes to subsequent events. Regulatory coordination is closest in areas governing standards for rolling stock, whose economic utility depends on cross-border movements. Regulatory divergence continues in other areas, ranging from rules that govern disclosure of hazardous materials shipments in each country to the pace of phasing out "less crash-resistant" tank cars, which has been accelerated in Canada.

With persistent pipeline bottlenecks resulting in growing shipments of oil (and other hazardous materials) by rail in both countries until 2020, concerns over public safety and the interoperability of railroads necessitate ongoing cross-border regulatory cooperation on rail safety. Policy shocks and opportunities provided by the Regulatory Cooperation Council have contributed to the acceleration of such processes and deeper institutionalization than might have occurred otherwise.

Agency-Led Incremental Responses

Much of functional IRC activity reflects agency-led incremental initiatives or responses, which may reflect the internal leadership priorities or respond to external policy stimuli. Internal incentives to expand IRC activities reflect the extent to which regulatory challenges facing executive departments or regulatory agencies arise from substantial or systemic changes in their operating environments originating outside national jurisdictions involving both potential benefits and risks. Senior regulatory officials tend to be highly sensitive to their legislative mandates – whether as external requirements that limit their exercise of discretion (and shift the potential burden of accommodation to foreign governments) or mandates that shape the context for exercising such discretion.[62]

Airline security issues raised in response to the 9/11 terrorist attacks created the necessity for shared protocols and technical standards to enable effective screening and communication among border agencies, airlines, airports, and public agencies issuing personal identification, building on deeply institutionalized global networks, as discussed in chapter 8. Significant increases in international financial flows were accompanied by growing levels of trade, investment, and migration. However, they also require ongoing processes to facilitate cooperation among financial regulators, law enforcement agencies, and financial institutions to identify and correct large- and smaller-scale illicit financial transfers – whether to detect and curtail corruption, tax evasion, proceeds or financing of other criminal activities. Growing levels of international trade raise questions of the capacity of government agencies to monitor the quality of or validate foreign production and inspection standards for food products, pharmaceuticals, and animal feed, and to identify and curtail the production and/or importation of other counterfeit goods. The development and spread of new technologies bring with them the need to integrate such technologies into the dealings of businesses, individuals, and governments – often involving movement or transactions across national borders.

Carberry and others have suggested that the coordination of such standards-setting may be easier in areas of emerging standards, as opposed to existing ones – although the accommodation (and, sometimes, promotion) of particular domestic interests is far from unknown in such settings.[63] Cross-border cooperation of business groups and industry associations may be useful in identifying regulatory gaps or anomalies, along with possible changes to regulations or related operational processes of governments (and businesses). However, identifying viable approaches involves extensive information sharing between and among national agencies on the objectives, processes, and methods of regulation (including perceptions of risks to be managed, and evaluations of existing processes to address those risks).

Movement towards harmonization is more common in areas in which Canadian exporters depend on U.S. markets for a disproportionate share of their overall production or revenues, especially when major policy and operational objectives are shared between Canadian and U.S. regulators, as with the automotive and chemical (including household products) sectors.[64] However, the identification of opportunities for and barriers to closer harmonization often depends on the design and outcome of pilot projects to identify both common and substantially different national operating processes. Such processes have been used to test proposed changes to both passenger and cargo pre-clearance (e.g., "inspected once, cleared twice") initiatives in recent years, with decidedly mixed results. In some cases, the technical legal wording mandating varied approaches may be within the jurisdiction of the line department or agency. In others, it may reflect broader interpretations of administrative or constitutional law, which limit agency discretion or require explicit (and perhaps conditional) legislative authorization.

However, U.S. regulators may take the initiative in IRC processes as part of a broader global strategy to manage regulatory risks. The U.S. Food and Drug Administration (FDA) has extensive jurisdiction over imports of processed foods and pharmaceutical products. As with regulators in Canada and other agricultural exporting countries, it maintains extensive relations with the *Codex Alimentarius* Commission, the U.N. Food and Agriculture Organization (FAO), the World Health Organization, and other international bodies to develop broader food safety standards and evaluate scientific testing methodologies. To facilitate IRC discussions with other countries, the FDA has secured legislative authorization to engage in confidential information-sharing agreements with foreign governments and their regulatory agencies, thus enabling closer levels of technical cooperation – a significant exception to normal APA public disclosure requirements. In recent years, FDA has developed formal processes, with transparent criteria, to enable the negotiation of systems recognition agreements on food safety standards with other countries that confirm the comparability of food safety system objectives, processes, and outcomes. Sands describes this IRC approach as "mutual recognition of functional equivalence" or MuRFE.[65] Health Canada and the Canadian Food Inspection Agency concluded mutual recognition agreements with FDA in 2016, although these agreements do not apply to fresh food products under the jurisdiction of the U.S.

Department of Agriculture.[66] However, most such IRC arrangements in other areas are significantly narrower in scope.

The RCC process discussed above has provided a broader "umbrella" framework for bilateral IRC discussions between Canadian and U.S. government agencies, including commitments to advance notice of regulatory initiatives or the review of existing regulations, which enables relevant agencies to anticipate and review potential policy changes within the context of their own responsibilities. However, whatever encouragement may be provided by central agencies, U.S. IRC processes tend to be more decentralized than in Canada and highly dependent both on agency-specific legislative contexts and the priorities of agency leaders.[67] As a result, Canadian regulators need to build and maintain ongoing relationships with U.S. counterparts to discuss the interactions of their regulatory systems and be prepared to engage on issues of mutual interest and concern as they arise. Ultimately, however, the extent of bilateral regulatory cooperation depends either on the willingness of U.S. officials to accommodate Canadian interests, usually as a result of implications for significant U.S. domestic interests, or Canadian officials' willingness to accommodate U.S. policy shifts as a result of disproportionate sectoral dependence on U.S. markets in particular areas.

Conclusion

These findings suggest that Canadian governments engage in extensive bilateral regulatory cooperation at the level of information exchange and regulatory parallelism to U.S. and broader international initiatives, while leaving room for the accommodation of legal and structural economic differences with major trading partners. However, these differences also contribute to differentiated patterns of integration across economic and policy sectors – particularly in areas with extensive provincial jurisdiction.

The relevance of international (particularly U.S.) regulatory influence on Canada's regulatory systems largely depends on the degree to which specific economic or industry sectors depend on U.S. markets, are mainly domestically focused, or have more diversified markets. The automotive sector, with its dependence on foreign investment and primarily U.S. markets, particularly for assembly plants, falls into the first group. There has been considerable parallelism in export-oriented agri-food sector regulations for many years, reflecting wide variations in market diversity and competitive efforts at market branding in different subsectors, but also similarities of food safety risks and histories of cooperation in broader international regulatory contexts. Canada's financial, rail, and air transport sectors have fallen into the third category for many years, as strategic sectors with very different histories of market regulation, sensitivities to domestic regional interests, and types of international exposure.

The structures of Canadian (and U.S.) federalism also play significant roles in levels of regulatory cooperation. Shared federal oversight is usually a necessary but rarely sufficient condition for extensive bilateral regulatory cooperation, although the dependence of major provincial electric utilities on cross-border export strategies led to the accommodation of FERC's market liberalization strategy. Measures to address shared risks predating other regulatory reforms encouraged cooperation among utilities on reliability standards. The greater the role of provincial or state governments in sector-specific regulatory activity, the less scope is available to central governments to pursue regulatory cooperation except in specific areas of uncontested federal authority. Such constraints may be mitigated when central and sub-central governments have shared objectives and philosophies of regulation, or even when there are such communities of interest among subnational governments, as seen with selective cooperation on climate change. However, such measures are subject to political defection – particularly when economic disruptions or broader societal divisions undermine the basis for domestic, let alone cross-border consensus, as seen in efforts to price greenhouse gases, expand pipelines, or (in Canada) centralize oversight of securities regulations. The greater the number of policy goals to be integrated in specific policy settings, the more difficult it becomes to maintain sustained intergovernmental or cross-border cooperation.

Fortunately for both governments and societal stakeholders, regulatory cooperation is not synonymous with harmonization. Rather, it can accommodate varying degrees of cooperation, mutual accommodation, and parallel actions adaptable to persistent economic and social differences between the two countries. As such, regulatory cooperation is inherently contingent, opportunistic, and dependent on the balance of internationally and domestically focused interests in specific policy and economic sectors and subsectors.

NOTES

1 Organisation for Economic Co-operation and Development (OECD), *International Regulatory Cooperation: Addressing Global Challenges* (Paris: April 2013); OECD, *International Regulatory Cooperation and Trade* (Paris: May 2017).

2 Anne-Marie Slaughter, *A New World Order* (Princeton, NJ: Princeton University Press, 2004); Tim Büthe and Walter Mattli, *The New Global Rulers: The Privatization of Regulation in the World Economy* (Princeton, NJ: Princeton University Press, 2011).

3 Miles Kahler and David A. Lake, eds., *Governance in a Global Economy* (Princeton, NJ: Princeton University Press, 2003); Slaughter, *New World Order*; John J. Kirton and Michael J. Trebilcock, eds., *Hard Choices, Soft Law: Voluntary Governance in Global Trade, Environment and Social Governance* (Aldershot, UK: Ashgate, 2004); John J. Kirton and Jenilee Guebert, "Soft Law, Regulatory Coordination and Convergence in North America," in *Borders and Bridges: Canada's Policy Relations in North America*, ed. Monica Gattinger and Geoffrey Hale, 59–76 (Toronto: Oxford University Press, 2010).

4 Michael Hart, "Risks and Rewards: The Case for Accelerating Canada-U.S. Regulatory Reform," in *Rules, Rules, Rules, Rules: Multilevel Regulatory Governance*, ed. G. Bruce Doern and Robert Johnson, 27–51 (Toronto: University of Toronto Press, 2006); Robert Carberry, "Making a Good Thing Better: Finishing What Was Started and Leveraging NAFTA to Advance Canada-U.S. Regulatory Cooperation" (Washington, DC: Wilson Center, March 2018).

5 G. Bruce Doern and Robert Johnson, "Multilevel Regulatory Governance: Concepts, Context and Key Issues," in *Rules, Rules, Rules, Rules: Multilevel Regulatory Governance* (Toronto: University of Toronto Press, 2006), 3.

6 Charles Doran, *Forgotten Partnership: U.S.-Canada Relations Today* (Baltimore, MD: Johns Hopkins University Press, 1984), 8, 53–66.

7 Andrew Downer Crain, "Ford, Carter and Deregulation in the 1970s," *Journal of Telecommunication and High Technology Law* 5 (2017): 413–47; Economic Council of Canada, *Reforming Regulation: Final Report* (Ottawa: Supply and Services Canada, 1981); Walter Block and George Lermer, eds., *Breaking the Shackles: The Economics of Deregulation – A Comparison of U.S. and Canadian Experience* (Vancouver: Fraser Institute, 1991).

8 Rosemary Foot, S. Neil MacFarlane, and Michael Mastanduno, eds., *U.S. Hegemony and International Organizations: The United States and International Organizations* (New York: Oxford University Press, 2003).

9 Bernard Hoekman and Michel Kostecki, *The Political Economy of the World Trade Organization*, 3rd ed. (New York: Oxford University Press, 2011), 236–59; Hoekman, "International Regulatory Cooperation in a Supply Chain World," in *Redesigning Canadian Trade Policies for a Supply Chain World*, ed. Stephen Tapp, Ari Van Assche, and Robert Wolfe, 365–94 (Montreal: Institute for Research in Public Policy and McGill-Queen's University Press, 2017).

10 Gene M. Grossman and Esteban Rossi-Hansberg, "Trading Tasks: A Simple Theory of Offshoring," *American Economic Review* 98, no. 5 (2008): 1978–97; Tapp, Van Assche, and Wolfe, *Redesigning Canadian Trade Policies*.

11 Hoekman, "International Regulatory Cooperation," 366.

12 Kirton and Guebert, "Soft Law, Regulatory Coordination and Convergence."

13 Canada, External Advisory Committee on Smart Regulation, *A Regulatory Strategy for Canada* (Ottawa: September 2004), 18; Canada, Treasury Board Secretariat, *Guidelines on International Regulatory Obligations and Cooperation* (Ottawa: 17 August 2007).

14 U.S. Federal Register, "Executive Order 13609: Promoting International Regulatory Cooperation" (Washington, DC: 4 May 2012); 26413–15.

15 "Chapter 28 – Good Regulatory Practices," United States–Mexico–Canada Trade Agreement (Washington, DC: Office of the U.S. Trade Representative, 1 October 2018).

16 Eugene Donati, "Opposed Triangles: Policy-making and Regulation in Canada and the United States," *Policy Options* (April 2001): 44–9.

17 W.T. Stanbury, "Corporate Power and Political Influence," in *Mergers, Corporate Concentration and Power in Canada*, ed. R.S. Khemani and W.T. Stanbury, 417–31 (Halifax: Institute for Research in Public Policy, 1988); William Coleman and Grace Skogstad, *Public Policy and Policy Communities* (Toronto: Copp Clark, 1992); Geoffrey Hale, *Uneasy Partnership: The Politics of Business and Government in Canada*, 2nd ed. (Toronto: University of Toronto Press, 2018), 122–9.

18 Donati, "Opposed Triangles"; Susan E. Dudley and Jerry Brito, *Regulation: A Primer* (Washington, DC: Mercatus Center, George Mason University and Center for Regulatory Studies, George Washington University, 2012), 35–40.

19 Confidential Interviews, Government of Canada, national Canadian and U.S. business associations 2012–15; Simon Lester and Inu Manak, "A Framework for Rethinking NAFTA for the 21st Century: Policies, Institutions, and Regionalism." *CTEI Working Paper 2017–10* (Geneva: Centre for Trade and Economic Integration, 2017), 7–10; Geoffrey Hale, "Regulatory Cooperation in North America: Diplomacy Navigating Asymmetries," *American Review of Canadian Studies* (2019): 123–49.

20 Hoekman, "International Regulatory Cooperation," 368.

21 Jeff Heynen and John Higginbotham, Advancing Canadian Interests in the United States: A Practical Guide for Public Officials (Ottawa: School of Public Service, 2004); Geoffrey Hale, *So Near and Yet So Far: The Public and Hidden Worlds of Canada-U.S. Relations* (Vancouver: UBC Press, 2012).

22 Cécile Carpentier and Jean-Marc Suret, "The Canadian and American Financial Systems: Competition and Regulation," *Canadian Public Policy* 29, no. 4 (December 2003): 431–47.

23 James Q. Wilson, "The Politics of Regulation," in *The Politics of Regulation*, ed. James Q. Wilson, 357–94 (New York: Basic Books, 1980); Milton and Rose Friedman. *The Tyranny of the Status Quo* (San Diego: Harcourt, Brace,

Jovanovich, 1984); Peter A. Hall, ed., *The Political Power of Economic Ideas* (Princeton: Princeton University Press, 1989).

24 Competition Policy Review Panel, *Compete to Win: Final Report* (Ottawa: Industry Canada, June 2008); North American Competitiveness Council, "Meeting the Global Challenge: Private Sector Priorities for the Security and Prosperity Partnership of North America (Washington, Mexico City, Ottawa: August 2008).

25 Greg Anderson and Christopher Sands, *Negotiating North America: The Security and Prosperity Partnership* (Washington, DC: Hudson Institute, 2007); Alicja Gluszek, "The Security and Prosperity Partnership and the Perils of North American Regionalism," *NortéAmerica* 9, no. 1 (January 2014), 7–54.

26 Carberry, "Making a Good Thing Better," 2.

27 Carberry, "Making a Good Thing Better," 3–4.

28 Carberry, "Making a Good Thing Better," 5.

29 Canada, Privy Council Office and United States, Office of Information and Regulatory Affairs, "Joint Action Plan for the Canada-U.S. Regulatory Cooperation Council" (Ottawa and Washington: December 2011);

30 Canada, Privy Council Office and United States, Office of Information and Regulatory Affairs, "Canada-U.S. Regulatory Cooperation Council Joint Forward Plan 2014" (Ottawa and Washington: August 2014)

31 Canada. Treasury Board Secretariat and United States, Office of Information and Regulatory Affairs, "Memorandum of Understanding between the Treasury Board of Canada Secretariat and the United States Office of Information and Regulatory Affairs Regarding the Canada-United States Regulatory Cooperation Council" (Ottawa and Washington: 4 June 2018).

32 Canada. Natural Resources Canada. *Energy Fact Book* (annual) (Ottawa: 2015–17); Hale, "Cross-Border Energy Infrastructure and the Politics of Intermesticity," in *Canada among Natons 2018–2019*, ed. David Carment, Inger Weibust, and Christopher Sands, 163–92 (New York: Palgrave Macmillan, 2019).

33 Monica Gattinger, "Canada-United States Energy Relations: Test-Bed for North American Policy-Making?" *Canadian-American Public Policy* 77 (September 2011): 8–12.

34 Gattinger, "Canada-United States Energy Relations," 17–21.

35 Hale, "Cross-Border Energy Infrastructure."

36 Douglas M. Brown, *Market Rules: Economic Union Reform and Intergovernmental Policy-Making in Canada* (Montreal and Kingston: McGill-Queen's University Press, 2002), 164–7; Internal Trade Secretariat, "Canadian Free Trade Agreement: Consolidated Version" (Winnipeg: 2017).

37 Christopher Kobach and Joe Martin, *From Wall Street to Bay Street: The Origins and Evolution of American and Canadian Finance* (Toronto: University of Toronto Press, 2018).

38 Canso Investment Council, "Canadian Financial Markets and the Foreign Property Rule" (Richmond Hill, ON: 2005); Canada. Department of Finance, "Pension Plan Investment in Canada: The 30 Percent Rule" (Ottawa: June 2016).

39 Organisation for Economic Co-operation and Development, "Regulation of Insurance Industry and Pension Fund Investment, OECD Report to G-20 Finance Ministers and Central Bank Governors" (Paris: September 2015); Tony Porter, "Canada, the FSB, and the International Institutional Response to the Current Crisis," in *Crisis and Reform: Canada and the International Financial System: Canada among Nations 2014*, ed. Rohinton Medhora and Dane Rowlands, 71–6 (Waterloo, ON: Centre for International Governance Innovation, 2014),

40 Eric Chouinard and Chris D'Souza, "The Rationale for Cross-Border Listings," *Bank of Canada Review* (Winter 2003–4): 23–30; Michael Halling, Marco Pagano, Otto Randl, and Josef Zechner, "Where Is the Market? Evidence from Cross-Listings in the United States," *Review of Financial Studies* 21, no. 2 (2008): 735–7.

41 British Columbia Securities Commission, "72–701 Guide for Use of the Multijurisdictional Disclosure System by Canadian Issuers in the U.S. Market (Previously NIN#91/22)," (Victoria: June 1, 2001).

42 Chouinard and D'Souza, "The Rationale for Cross-Border Listings," 24–5; "The complete list of Canada stocks listed on U.S. markets," TopForeignStocks.com.

43 James E. Bagnall, *100 Days: The Rush to Judgment That Killed Nortel* (Ottawa: Ottawa Citizen, 2013).

44 This discussion reflects the author's extended conversations with provincial regulators in 2004–6. See also Geoffrey Hale and Christopher Kukucha, "Investment, Trade and Growth: Multilevel Regulatory Regimes in Canada," in *Rules, Rules, Rules, Rules: Multilevel Regulatory Governance*, eds. G. Bruce Doern and Robert Johnson (Toronto: University of Toronto Press, 2006), 190–205.

45 Nancy Hoi Bertrand and Pamela Hughes, "Canadian Inter-Listed Companies: Navigating the maze of governance requirements," *Ivey Business Journal* (September–October 2004).

46 Mark MacKinnon and Andy Hoffman, "Inside the Sino-Forest Storm," *Globe and Mail*, 18 June 2011, B1; Peter Foster, "Short-Sellers Beat Regulators on Sino-Forest," *National Post*, 31 August 2018, FP13.

47 Transparency International, "Canada; Beneficial Ownership Transparency," (Berlin: 2015); Alistair MacDonald, Paul Vieira, and Vipal Monga, "The Money-Laundering Hub on the U.S. Border? It's Canada," *Wall Street Journal*, 9 August 2018; Denis Meunier, "Hidden Beneficial Ownership and Control: Canada as a Pawn in the Global Game of Money

Laundering," *Commentary # 519* (Toronto: C.D. Howe Institute, September 2018).

48 Dean Beeby, "Ottawa's Secret Report on Money-Laundering Points Fingers at Canada's Banks," CBC News, 5 April 2018; Gary Mason, "B.C. Became a Gangster's Paradise: Who Will Take the Blame?" *Globe and Mail*, 29 June 2018.

49 Alistair MacDonald, Rita Trichur, and Will Connors, "Money-Laundering Fears Fuel an RBC Retreat," *Wall Street Journal*, 3 February 2015; Beeby, "Ottawa's Secret Report."

50 International Organizations of Securities Commissions. www.iosco.org /about.

51 Reference re: Securities Act [2011] 3 SCR 837; John Tuzyk and Liam Churchill, "National Securities Regulator on the Ropes? Quebec Court of Appeals Rules Proposed Cooperative System Unconstitutional," *Blake's Business Class*, 15 May 2017; John Tuzyk and Liam Churchill, "Supreme Court Hearing Leave Cooperative Capital Markets Regulatory System in Limbo" (Toronto: Blake, Cassells and Graydon, 28 March 2018).

52 Barbara Shecter, "Bid to Create National Markets Watchdog Advances with Supreme Court Ruling," *Financial Post*, 10 November 2018, FP1.

53 Jacquie McNish and Tara Perkins, "ABCP Investors Face New Hurdle," *Globe and Mail*, 17 March 2008, B1; Paul Halpern et al, *Back from the Brink* (Toronto: Rotman-UTP Publishing, 2016).

54 David Longworth, "The Global Financial Crisis and Financial Regulation: Canada and the World," in *Crisis and Reform: Canada and the International Financial System: Canada among Nations 2014*, ed. Rohinton Medhora and Dane Rowlands, 87–102 (Waterloo, ON: Centre for International Governance Innovation, 2014).

55 Longworth, "Global Financial Crisis," 97.

56 Paul Vieira, "Canada Wins in Bid to Rebuff Bank Tax," *National Post*, 24 April 2010, A14.

57 Janet McFarland, "Canada's Cooling Experiment," *Globe and Mail*, 2 August 2017, A8–9.

58 Geoffrey Morgan, "Ottawa Beefs Up Crude-by-Rail Rules," *Financial Post*, 21 February 2015, FP2.

59 Kim Mackrael and Jacquie McNish, "U.S. Canada Issue Rail Safety Warnings," *Globe and Mail*, 24 January 2014, A3.

60 Jim Snyder, "U.S. Aiming to Phase Out Older Oil Tanker Rail Cars," *Financial Post*, 24 July 2014, FP10.

61 Kim Mackrael, "Canada, U.S. Unveil Tougher Rail Safety Standards for Transporting Oil," *Globe and Mail*, 1 May 2018.

62 For example, see Eugene Bardach and Robert A. Kagan, *Going by the Book* (New York: Transaction, 2002).

63 Olivier Falck, Christian Gollier ,and Ludger Woessmann, eds., *Industrial Policy for National Champions* (Cambridge, MA: MIT Press, 2011).

64 Porter, *The Day after NAFTA*, 8; Kirton and Guebert, "Soft Law, Regulatory Coordination and Convergence."

65 Christopher Sands, "The Case for a 'Mutual Recognition' Deal with the U.S." *Policy Options*, 12 May 2017.

66 United States. Food and Drug Administration, Canada Food Inspection Agency and Health Canada. "FDA - CFIA and Health Canada, Food Safety Systems Recognition Arrangement" (Washington and Ottawa: May 2016).

67 Hale, "Regulatory Cooperation in North America."

5 Who Is Us? The Shifting Sands of Foreign Direct Investment Policies

GREG ANDERSON

One of the more remarkable consequences of the end of the Cold War was the scholarly interest in reconsidering some of the most fundamental aspects of international relations: sovereignty, the coherence of the state, and the disruptive rise of a host of non-state actors poised to re-order conceptions of global politics. The oft-disruptive effects of these forces on domestic economic, social, and political relationships have triggered numerous cross-cutting pressures on many governance stakeholders, notably different elements within governments themselves. In some contexts, stakeholders have sought assistance from governments in adapting to these pressures. In some cases, that assistance comes in the form of territorially derived protective actions, such as trade actions, as outlined by Chris Kukucha in chapter 3, designed to shield those interests from internal and external risk. However, in other areas, we also see the evolution of aterritorial or extraterritorial "governance" arrangements prompted by a mix of market forces, governmental actions, elements of transnational civil society, or the norm-creating function of international institutions.

The range of "governance" mechanisms drawing the attention of scholars often defies simple categorization or description, making assessment of their impact on traditional definitions of sovereignty and borders challenging. Indeed, there is a "moving target" quality to fully understanding how everything from the treatification of international law or pooling of sovereignty through state action compares with both the porousness of borders and the apparent "retreat of the state" in many areas.[1] It is unsurprising, then, that a murk of scholarly terminology has evolved as a way to think about the varieties of "governance" we observe around us, among them multilevel governance, multidimensional governance, subsidiarity, intermesticity, the perforated state, and the disaggregated state.[2] Among the most uncomfortable of coined terms, but perhaps also the most appropriate to the task, is James Rosenau's notion of "fragmegration," capturing the two strongest and broadest pressures on the modern state: integration and fragmentation.[3]

There are few areas of global politics that epitomize the tensions between integration and fragmentation, those between territorial or aterritorial governance, or the tensions between the state, non-state, and market-driven forces more than the governance of foreign direct investment (FDI) flows. Indeed, the steady expansion of private capital flows throughout the post-war period has incentivized and necessitated important innovations to international regimes governing those flows. In important and recently controversial ways, governance of international capital flows has become endemic to the broader debate in the politics of the global economy about the "retreat of the state" as the global economy has confronted the state with seemingly difficult trade-offs between openness and protection. Capital flows are simultaneously sought and vilified, reshaping debates about FDI's benefits and generating anxieties about sovereignty and self-determination. At its core, the governance of FDI flows fundamentally challenges the meaning of borders for demarcating the sources of our prosperity. Indeed, as former U.S. Labor Secretary Robert Reich argued many years ago, FDI forces everyone to confront the very notion of "Who is us?"[4]

Global capital flows are shaped by a complex mix of domestic, regional, and multilateral governance regimes. It is a set of regimes initially driven in equal parts by state and private, market-oriented actors. However, the debate about these regimes and norms has been joined by intense, critical scrutiny of them by a host of civil society actors. The focus of this chapter is on two categories of capital flow governance in the North American context; the design and functionality of Chapter 11 of the North American Free Trade Agreement (NAFTA) in the wider context of Investor State Dispute Settlement (ISDS) processes, and the separate domestic investment review regimes of Canada and the United States anchored in national security considerations. In both examples we see significant ebb and flow in the location of governance, each of which challenges our understanding of the territoriality of regimes governing investment. Interestingly, in both domains of investment governance, we observe in the late 1980s and early 1990s the ascendancy of market-driven rules and deference to the decision-making of private actors relative to governments. However, blowback from a different set of non-state entities (civil society organizations) and the growing prominence of state-run enterprises have helped push the pendulum back towards a reassertion of territoriality and state prerogatives over investment governance.

In many respects, the NAFTA Chapter 11 experience became a global catalyst for reassertion of territorially defined, state-driven investment governance as concerns mounted with each Chapter 11 case regarding the implications of private, transnational, or aterritorial regime formation, which critics assert accelerated the retreat of the state from key areas of policymaking. Similarly, the end of the Cold War also marked a period of relaxed concerns about the nexus of foreign investment and national security to one more recently driven by sensitivities over even minor investments that parallels the advent of contemporary electoral populism.

The plan of this chapter is as follows. Part 1 will briefly review the importance of FDI flows to global politics and the rationale behind the attraction of capital flows as part of domestic economic growth. Part 2 will outline the sources of controversy and the impact of FDI on the territoriality of governance. Part 3 will then explore the parallel evolution investment review regimes focused on national security that have periodically blocked specific investments into Canada and the United States.

On the surface, these two regimes seem categorically distinct. Embedding ISDS in trade agreements like NAFTA in the 1990s represented the post-war ascendancy of private capital as principal architect of global investment regimes, while investment review loomed in the background as the remnant of a heavy-handed national security state. Yet we also observe the evolution of these regimes as paralleling one another temporally. Both are the by-product of a broader ebb and flow of post-war efforts to regulate capital flows. Both are in the midst of a significant reconsideration.

Foreign Direct Investment: The Good, the Bad, and the Ugly

The BITs and Pieces of Investment Governance

One central issue plaguing international commercial relations is that the private interests at the heart of international flows of goods, services, and capital have traditionally lacked any "personality" within customary international law. Without international "personality," private investors have had difficulties binding themselves contractually to sovereign hosts in ways that secure their market access for private investments, as would be the case in their home markets – the principle of just compensation for expropriation of property is applied very differently across countries. The absence of such rules has long been thought to be an important disincentive on capital flows from developed to developing countries and a partial explanation for the historical tendency for foreign direct investment (FDI) to flow so predominantly between developed economies with more consistency where expropriation is concerned.

The main mechanism for mitigating these problems in the post-war era has been the emergence of bilateral investment treaties (BITs).[5] The use of BITs between state parties to define the treatment of private investment, including rules for dispute settlement and compensation, has offered private interests a form of "personality" within international law through which they can defend their interests vis-à-vis host governments.

Put differently, BITs represent the ascendancy of rules designed to extend a state-centric, territorially defined international legal regime to incorporate non-state, private actors like firms. Importantly, the binding dispute-settlement mechanisms contained within these regimes amount to an aterritorial legal architecture binding on host states and accessible only to foreign, private holders of capital.

While there is some debate about how to fully maximize the benefits of FDI,[6] one historical challenge for developing states is actually getting FDI to flow in their direction at all. The Organisation for Economic Co-operation and Development (OECD) reports that in 2007 FDI outflows from OECD countries reached a record US$1.82 trillion in value, with outflows from the United States alone amounting to US$333 billion.[7] However, the disparity in flows of FDI between rich and poor countries is as stark as ever. As reported by the OECD, there is a strong correlation between FDI outflows from rich countries and FDI inflows to poor countries. In 2007, developing countries matched the record growth in FDI outflows from the OECD by attracting record inflows amounting to US$471 billion. However, the BRIC countries (Brazil, Russia, India, and China) accounted for 50–60 per cent of all developing country inflows.[8] In 2014, OECD countries still accounted for 40 per cent of FDI inflows and 70 per cent of all outflows, in spite of serious economic crises among them.[9] Not much has changed. According to the United Nations, there were nearly US$3 trillion worth of foreign direct investment flows globally in 2015. Unfortunately, the lion's share of those flows was (and remains) between wealthy OECD countries, accounting for more than 70 per cent of all outflows and more than 50 per cent of all inflows (see table 5.1).

Moreover, if we take away the large flows into and out of China, investment flows into the developing world represent a paltry 19 per cent of all global inflows. Economic theory suggests that capital ought to naturally flow from regions in which capital is abundant (rich,

Table 5.1. Global Flows and U.S., EU Shares

	2015 ($U.S. millions)
Global outflows	1,474,424
Global inflows	1,762,155
High-income OECD outflows	1,098,527 (74.5% of total)
High-income OECD inflows	698,064 (55% of total)
U.S. outflows	316,549
EU (28) outflows	487,150
U.S. inflows	379,894
EU (28) inflows	439,457
U.S. + EU (28) outflows	803,699 (61% of global flows)
U.S. + EU (28) inflows	819,351 (54% of global flows)
Least Developed Country outflows (China)	701,090 (47% of global total)
Least Developed Country inflows (China)	335,121 (19% of global total)

Source: United Nations Conference on Trade and Development, UNCTADstat.

industrialized, OECD countries), and therefore inexpensive, to those regions in which it is scarce (poor, developing countries) and therefore expensive. The reasons for this discrepancy are multi-fold but include things such as poor infrastructure, lack of market proximity, access to a skilled labour force, and weak domestic legal protections for property – especially property held by foreigners.

North America's 500-Pound Investment Gorilla

At first blush, North America would not be the first place where the introduction of a regime of aterritorial investment protection rules would need to be implemented. Indeed, the original BIT between Germany and Pakistan in 1959 reflects the overwhelming pattern of post-war investment protection agreements: dyads of developed and developing countries. According to UNCTAD, there are nearly 3,000 BITs and 350 treaties with investment provisions within, the overwhelming majority of which are between dyads of developed and developing countries.[10] The reason for this recurring pattern of dyads flowed from concerns about comparatively weak property-rights protections in developing countries and the lack of legal recourse for foreign holders of capital in the event of expropriation or nationalization by the host state.

In North America, the main source of comparable concern was Mexico, notably its nationalization of PEMEX in 1938. However, investment was also an irritant between Canada and the United States in the 1970s

Table 5.2. North American FDI Inflows ($U.S. Millions)

	1985		2006		2015		2016	
	Inflows	% NA	Inflows	% NA	Inflows	% NA	Inflows	% NA
United States	20,490	86.0	237,136	74.4	348,402	82.3	391,104	86.6
Canada	1,372	5.8	60,293	18.9	41,512	9.8	33,721	7.4
Mexico	1,983	8.3	21,147	6.6	33,181	7.8	26,738	5.9
Total	23,845		318,576		423,095		451,563	

Table 5.3. North American FDI Outflows ($U.S. Millions)

	1985		2006		2015		2016	
	Outflows	% NA	Outflows	% NA	Outflows	% NA	Outflows	% NA
United States	13,388	76.6	224,220	83.9	303,177	79.5	299,003	82.0
Canada	3,862	22.1	46,213	17.2	67,036	17.5	66,402	18.2
Mexico	222	1.2	5,758	2.1	10,733	2.8	−787	0
Total	17,472		267,191		380,946		364,618	

and 1980s as Ottawa implemented a series of nationalist energy policies favouring domestic over foreign (mostly U.S.) investment in the energy sector.[11] However, when Canada and Mexico went looking for new sources of foreign capital in the late 1980s and early 1990s, the United States was the most obvious and proximate source.[12]

Indeed, part of the political sales pitch made to Canadian and Mexican voters around the NAFTA was that a stable, binding set of investment rules would allow Canada and Mexico to attract a greater share of North America's investment. Firms hitherto wary of Canadian or Mexican tendencies towards economic nationalism could be assured their investments would not be targets of regulatory harassment because of their foreign status ("national treatment"). In the Canadian case especially, it was hoped that a highly educated workforce, quality health care, and first-world infrastructure would draw far more capital inflows.

However, as the tables and graphs below depict, the impact of those efforts is hard to discern. For example, the United States in 1985 accounted for 86 per cent of all inflows into North America (see table 5.2). In 2016, that proportion was essentially unchanged (86.6 per cent in 2016), albeit at a significantly higher share of GDP. Some of the persistent gap can be correlated with the similarly persistent gap in GDP growth over the last thirty years (see tables 5.2 and 5.3; figures 5.1 and 5.2), but the fact is that

Table 5.4. North American Gross Domestic Product ($U.S. Billions)

	1985		2006		2015		2016	
	GDP	% NA	GDP	% NA	GDP	% NA	GDP	% NA
United States	4,347	88.0	13,163	86.0	17,946	86.9	18,624	87
Canada	364	5.8	1,271	8.3	1,553	7.5	1,529	7
Mexico	184	3.7	839	5.4	1,151	5.5	1,046	5
Total	4,895		15,273		20,650		21,199	

Figure 5.1. Gross Domestic Product: Total and per Capita, Current and Constant (2010) Prices, Annual, 1970–2016. Measure (U.S. Dollars at Current Prices in Millions)

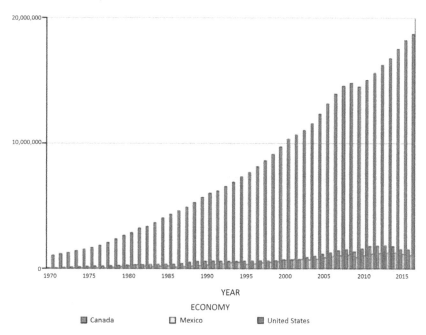

Source: UNCTADStat, United National Conference on Trade and Development.

investment rules don't appear to have obviously redistributed the shares of investment capital flowing into each of the three countries.

The uneven results flowing from investment protection rules were enjoined by a broader collection of anxieties about the origins and policy impact of flows that were stimulated. In 1990, Robert Reich asked which of the following two firms was more important to the economic health of the country:

Figure 5.2. Foreign Direct Investment: Inward and Outward Flows and Stock,
Annual, 1970–2016. Direction (Inward, (Measure (U.S. Dollars at Current
Prices in Millions, Mode (Flow)

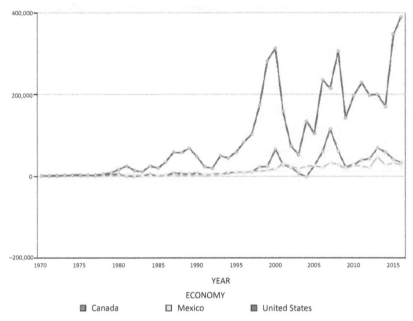

Source: UNCTADStat, United Nations Conference on Trade and Development.

Corporation A is headquartered north of Toronto. Most of its top managers are citizens of Canada. All of its directors are Canadians, and Canadian investors hold a majority of its shares. But, most of Corporation A's
employees are non-Canadians. Indeed, the company undertakes much of
its R&D and product design, and most of its complex manufacturing outside the borders of Canada in Asia, Latin America, and Europe. Within the
Canadian market, an increasing amount of the company's product comes
from its laboratories and factories abroad.

    Or

Corporation B is headquartered in another industrialized nation. Most
of its top managers and directors are citizens of that nation, and citizens
of that nation hold a majority of its shares. But most of Corporation B's
employees are Canadians. Indeed, Corporation B undertakes much of its
R&D and new product design in Canada. And it does most of its manufacturing in Canada. The company exports an increasing proportion of

its Canadian-based production, some of it even back to the nation where Corporation B is headquartered.[13]

One's answer highlights many of the tensions of territorial and aterritorial governance of foreign direct investment as it has evolved in the post-war period.

The Evolution of Private Investment Regimes

A Lot Happened in 1994

For most of the post-war period, BITs generated virtually no controversy. Indeed, in 2015, *Germany and the United States had more than three hundred BITs with developing countries.* However, two things happened in 1994 that would bring investment rules out of relative obscurity, making them a focus of broader opposition to economic liberalization. In January 1994, the North American Free Trade Agreement (NAFTA) entered into force and incorporated what was essentially U.S. BIT model language in the text: Chapter 11.[14] Across the Atlantic, the European Energy Charter was expanded, renamed the Energy Charter Treaty[15] in late 1994 and, like the NAFTA, infused with ISDS provisions. Because both were agreements that included developed state parties between whom large amounts of capital flowed, they also made it far more likely those same developed states would be subject to ISDS disputes.

It didn't take long. In 1998, The Loewen Group, a Canadian funeral services firm, challenged an adverse Mississippi court ruling under the terms of NAFTA Chapter 11, upending the political dynamics of investment protection rules.[16] Never before had such rules been used by a *firm from a developed country* to challenge treatment of an investment in *another developed country.* Such cases under NAFTA began to pile up. As of mid-2019, there have been fifty-five Chapter 11 cases alleging discriminatory treatment at the hands of a NAFTA government. Interestingly, only seventeen of the fifty-five have been filed against Mexico, the rest against the United States (seventeen) and Canada (twenty-one).

It wasn't supposed to be this way. Yet controversy flowed as the NAFTA experience with Chapter 11 led critics to conclude investment rules were being used to undermine (not enhance) the rule of law by giving foreign firms an exclusive pathway to challenge the state's sovereign power to regulate in the public interest. The perception that ISDS was being used to attack the state's regulatory power was compounded in late 1999 when Methanex Corp., a Canadian petrochemical firm, challenged a California regulation banning a fuel additive proven to be

Figure 5.3. Number of Investment Dispute Cases per Year

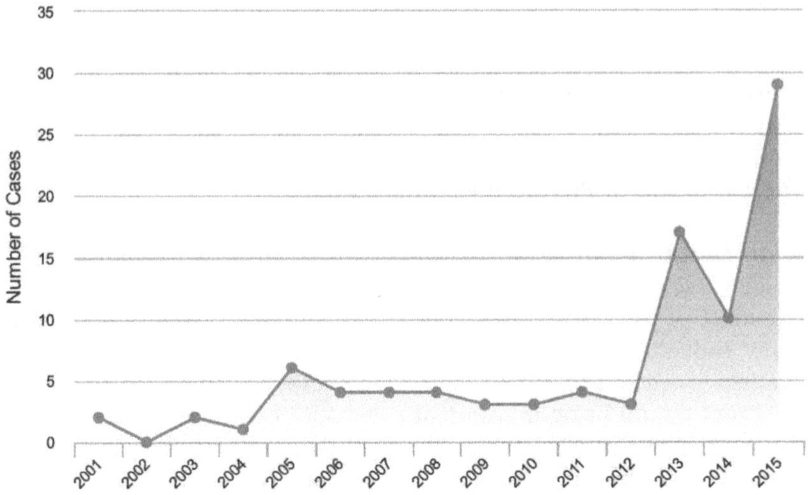

Source: Dr. Dorte Fouquet, Becker Buttner Held Consulting AG, "Current Arbitration Cases under the Energy Charter Treaty," Vienna Forum on European Energy Law, April 15, 2016.

toxic to groundwater supplies.[17] For critics, Chapter 11 had provided a set of aterritorial legal tools for foreign firms to challenge a host of regulatory measures, including those protecting the environment.

Across the Atlantic, the European Energy Charter was expanded, renamed the Energy Charter Treaty in late 1994, and, like NAFTA, infused with ISDS provisions. In North America, Mexico was the assumed target of investment protection, in Europe it was Russia or former Soviet republics. Until the mid-2000s, that's how things unfolded. Yet, in recent years, Energy Charter Treaty ISDS cases have spiked (figure 5.3), with firms based primarily in developed countries (figure 5.4) alleging property rights violations in similarly developed countries with histories of stable property rights (figure 5.5).

Fallout from Vattenfall

Spain has regrettably been the single biggest target of ISDS suits under the Energy Charter Treaty flowing from changes to that country's renewable energy markets. However, it was the Swedish energy company Vattenfall's two ISDS claims against Germany in 2009 and 2012 that set off alarm bells all over Europe. In 2009, Vattenfall alleged

Figure 5.4. Investment Dispute Cases Heatmap. Frequency of Investor-State Dispute Cases per Country

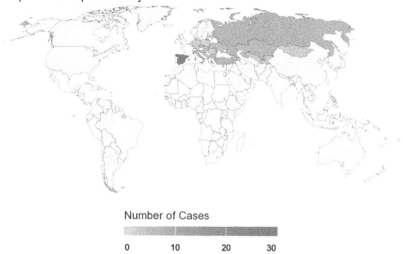

Number of Cases

0   10   20   30

Source: International Energy Charter.

Germany was unfairly and arbitrarily phasing out certain kinds of coal-fired power generation in which the firm had taken a large stake.[18] More controversial was Vattenfall's 2012 decision to sue Germany over its decision to shutter all of its nuclear power facilities in the wake of the Fukushima nuclear disaster.[19] The political consequences of the Vattenfall cases could not have been worse and nearly derailed the Comprehensive Economic and Trade Agreement (CETA) negotiations with Canada.[20] The emerging case pattern reinforced the growing sense that governments had negotiated their way into a powerful aterritorial legal regime running parallel to domestic law, but inaccessible to domestic stakeholders. And, to top it all off, it was almost entirely unaccountable to voters.

Vattenfall forced all parties to the CETA to rapidly rethink their positions[21] and resulted in a hastily cobbled together public consultation in Europe.[22] The Canadians were not the only ones on European minds. On the horizon were the proposed Trans-Atlantic Trade and Investment Partnership (TTIP) negotiations with the Americans, the first such negotiations over investment since the failed Multilateral Agreement on Investment in 1998.[23]

Across the pond, the Methanex challenge rattled nerves in all three NAFTA governments,[24] prompting them to issue an "interpretation"

Figure 5.5. International Energy Charter Case Distribution

Spain 47	Albania 3
Italy 12	Macedonia 1
Germany 3	Bulgaria 5
Czech Republic 6	Romania 4
Poland 1	Moldova 2
Slovakia 1	Turkey 6
Russia 6	Georgia 3
Ukraine 4	Azerbaijan 2
Romania 4	Kazakhstan 5
Hungary 5	Uzbekistan 1
Slovenia 1	Tajikistan 1
Croatia 2	Kyrgyzstan 1
Bosnia Herzegovina 2	Mongolia 3
Latvia 1	European Union 1

Total: 133 Cases

Source: International Energy Charter, Statistics of Energy Charter Cases, Updated 2020. https://www.energycharter.org/fileadmin/DocumentsMedia/News/Statistics_Cases_under_the_Energy_Charter_Treaty_as_of_1_June_2020.pdf.

of elements of the Chapter 11 text in July 2001.[25] That same year, the United States launched its first formal review of its BIT model since the 1994 model BIT was originally embedded in the NAFTA.

When unveiled in 2004, the new U.S. model BIT included substantial changes to stem criticism of NAFTA Chapter 11, effectively reasserting the kind of territoriality in investment governance from an earlier time. Among the changes was explicit language limiting the capacity of firms to challenge the state's right to regulate in the public interest, clearer definitions of "expropriation," and the standards of treatment that firms could expect.[26]

Similarly, the Vattenfall cases against Germany, which imperilled the CETA talks with Canada, prompted important reforms that have become the baseline for European negotiators in subsequent trade talks.[27] The hastily organized European review generated recommendations similar to those in the evolving U.S. model BIT asserting the state's right to regulate in the public interest, and enhanced restrictions on the scope for private firms to claim expropriation. Importantly, and different from the U.S. model, the EU would promote the establishment of a permanent "investment court,"[28] an especially significant proposal that likely saved the CETA politically.

Aterritorial Governance of Private Capital: It Was a Good Run

Interestingly, Europe and the United States have taken slightly different approaches to reasserting territoriality into the governance of FDI. Europe has proposed the creation of a multilateral investment court that would institutionalize most governance of FDI in a multilateral institution.[29] Europe's proposal would, in effect, claw back nearly all the aterritorial evolution of post-war investment rules by anchoring them in an institution that is the creature of member states. The institution itself would be aterritorial in that it would independently adjudicate, but it would do so on the basis of rules set by states, the contours of which would undoubtedly privilege sovereign authority. The ill-fated Multilateral Agreement on Investment from the late 1990s, the last major multilateral effort to establish global investment rules, never proposed a formalized institution such as a court.[30]

The United States, on the other hand, has similarly sought to restrain the growing aterritoriality of investment protection regimes, but mostly through the use of linguistic changes aimed at limiting the scope of application. Indeed, with the exception of the recently renegotiated NAFTA, or USMCA, the United States has continued to advocate for strong investment protections that include ISDS dispute-settlement mechanisms.[31] While the USMCA eliminates the application of aterritorial investment rules between Canada and the United States and severely limits the use of ISDS to specific areas of U.S.-Mexico FDI, the most recent revision to the U.S. model BIT in 2012 preserves the status quo of aterritoriality for disputes between private actors and the state. The Trump administration's apparent preference for eliminating investment protections is, of course, at odds with much of post-war U.S. foreign policy and may not last beyond the forty-fifth president. On the other hand, the evolving set of doubts about investment protections flowing out of agreements in the 1990s suggest that a more strongly territorial reset of these regimes is here to stay.

The Territorial State, National Security, and Investment

Perhaps more than any other dimension of the global economy, foreign direct investment blurs territorial boundaries directly via cross-border flows of capital; everything from portfolio capital infusions, to investments in new physical assets such as green- or brown-field developments, mergers, and acquisitions among firms prompt the reorientation of supply and production chains.

The changes arising from private flows of capital have periodically revealed an uncomfortable politics because of the tensions inherent in securing the benefits of FDI for the state while maintaining traditional sovereign control. Indeed, FDI is constantly pushing us to redefine "Who is us?"; which is more important to Canada – a Toronto-headquartered firm with most of its employees abroad, or a Tokyo-headquartered firm employing thousands of Canadians in Ontario?

Almost parallel to the pursuit of investment governance in trade agreements, highly statist investment-review regimes occupy a similarly awkward space in threading some of the same needles. Capital inflows have historically prompted doubts about who the beneficiaries of that capital are or whether the benefits outweigh the costs. Does foreign capital mean foreign ownership? If so, what are the implications for domestic labour? Foreign capital and know-how may be exactly what is needed to revitalize lagging or inefficient industries. But what if those new capital infusions are in strategically important or sensitive sectors? How does that change perceptions of sovereign control?

In both Canada and the United States, investment review regimes have emerged, ebbed, and flowed, in virtual lockstep with concerns about implications of foreign ownership for sensitive or strategically important sectors. As noted by Robert Pastor back in 1980, "Each antiforeign wave washed ashore a flotsam of restrictive and exclusionary legislation which receded slightly or became buried and forgotten in the sand only to advance again with a new wave."[32] Indeed, the origins of investment review in both Canada and the United States are rooted in some of the standard anxieties about growing economic openness to the rest of the world. Indeed, national security reviews have become a highly territorial, last-line-of-defence check on the liberalization of capital flows.

In 1973, the Canadian Parliament legislated into existence the Foreign Investment Review Act (FIRA), the purpose of which was written into the Act itself:

> The extent to which control of Canadian industry, trade and commerce has become acquired by persons other than Canadians and the effect thereof on the ability of Canadians to maintain effective control over their economic environment is a matter of national concern, and ... it is therefore expedient to establish a means by which measures may be taken ...to ensure that, in so far as is practicable after the enactment of this Act, control of Canadian business enterprises may be acquired by persons other than Canadians ... only if it has been assessed that the acquisition of control of those enterprises or the establishment of those new businesses, as the

case may be, by those persons is or is likely to be of significant benefit to Canada having regard to all of the factors to be taken into account under this Act for that purpose.[33]

In May 1975, President Ford signed an executive order creating the Committee on Foreign Investment in the United States (CFIUS) with a slightly different purpose, the protection of national security: "Section 1. Policy. International investment in the United States promotes economic growth, productivity, competitiveness, and job creation. It is the policy of the United States to support unequivocally such investment, consistent with the protection of the national security."[34]

Rationales for review are couched in terms ranging from cultural sensitivity, to economic autonomy, and even actual national security concerns, but both FIRA and CFIUS have the same goal – asserting territorially defined control over foreign capital. Moreover, the immediate impetus for their emergence in the 1970s was also the same – a dramatic spike in inflows of capital from the likes of a rapidly modernizing Japan, state-run oil firms, or in the case of Canada, a single source like the United States.[35]

Since their creation in the 1970s, FIRA and CFIUS have undergone a number of revisions – some relaxing the rules, others toughening them – not coincidently tailored to the politics of the moment. For example, in 1984, FIRA received a semantic change to the less defensive-sounding Investment Canada Act (ICA), in part because of accumulating evidence FIRA was discouraging inward (mostly U.S.) FDI, and generating conflict with the United States.[36] It was an important change by Ottawa that many credit with facilitating the conclusion of free trade with the United States in 1987, but one that temporarily set aside Canadian concerns about the implications of FDI for sovereignty.[37]

Following its creation in 1975 by executive order, CFIUS was formalized by Congress in 1976 but faded from view as the perceived threat from petrodollars receded. There was never a name change, but according to Jackson, between 1975 and 1980, CFIUS met only ten times and never seemed sure about the circumstances under which it should intervene.[38] Like FIRA and its successor the ICA, CFIUS has been formalized and amended several times, notably by the so-called Exon-Florio Amendment to the 1988 Omnibus Trade and Competitiveness Act, and by the so-called Byrd Amendment to the 1992 Defense Authorization Act.[39] Each of these amendments served to toughen and expand the scope of CFIUS, and did so in response to the perception of specific "threats": high-profile Japanese acquisitions or the growth of state-owned-enterprises.[40]

The Big Stick of Investment Review

One of the most interesting things about investment review as wielded by states is how seldom it is actually used. In addition to infrequent meetings, Jackson also notes that CFIUS reviews have resulted in the rejection of just *four* proposed investments.[41] Moreover, Jackson also reports that in 2008–15, there were just 925 proposed mergers and acquisitions reported to CFIUS. That sounds like a lot, but for perspective, in 2015 alone there were 1,800 foreign investment transactions in the United States.[42] In other words, only a fraction of proposed investments ever receive CFIUS attention. Under the ICA since 1985, Canada has reviewed more than 1,700 applications, approving all but 9. In 2015, the Harper government amended the ICA to include a "net benefit" determination as part of the assessment, again to pacify public concerns about high-profile acquisition proposals in the Canadian energy sector – this time from Asia. Yet, according to Industry Canada's annual investment report, there were 737 investment filings in 2016–17, with 22 assessed under the "net benefit" standard. All were approved.[43]

Nevertheless, approval or disapproval is hardly the whole story where this particular brand of investment governance is concerned. Indeed, ICA and CFIUS loom large as territorially based checks or gatekeepers on foreign direct investment flows because of the "chilling" effect the very threat of review has had on investments from specific sources or into particular sectors.

From the mid-2000s onward, FDI flows from state-run enterprises have been the focus of considerable concern. Much like the capital flows from oil-rich states that precipitated the creation of investment review in the United States in particular, the recent focus of concern has been on forms of state capitalism. One of the most important contemporary challenges in calibrating investment review has been discerning the strategic intent of firms based in non-market economies like China.[44] What are the implications of investments from firms whose senior management may be deeply tied to a state's political bureaucracy? Will capital flows from such firms be motivated primarily by commercial considerations or are there more nefarious geostrategic considerations flowing from the firm's interconnectedness with the home state? In both Canada and the United States, uncertainty about the implications of "foreign ownership" have become hotly contested once again. And, much as high-profile Japanese acquisitions in the 1980s sounded alarm bells, similarly proposed acquisitions of assets by state-linked Chinese firms are again ringing them.

The Impetus for Reform: The State vs State-Owned Enterprises

Doubts about the wisdom of investment protections embedded into trade agreements over the last two decades have paralleled a general decline in the pace of liberalization of investment rules. According to the OECD, investment regimes around the world have become more restrictive in this period.[45] It is a less liberal approach to investment that includes a variety of measures intended to discourage foreign capital inflows. Such measures include British Columbia's real estate tax on second homes, aimed at curbing money laundering through the real estate market.[46] Similarly, the activities of state-led investment review regimes have also been on the rise, thus contributing to a more restrictive regulatory environment for capital.

The contemporary politics of foreign investment review have added worries about the intent of firms from non-market economies – particularly those in which the state has deep or controlling interests – to the long-standing anxieties that typically come with private capital flows. What are the implications of merger or controlling stake in a domestic firm if all corporate activity is still subject to domestic law? Is the relative inactivity or inaction under the ICA or CFIUS a sign that none of this is very important? Are these regimes merely window-dressing to demonstrate action is possible? Or does the presence of these regimes result in transparent and stable transactions?

Investment Canada

The most notorious recent example of Ottawa intervening to "review" a foreign acquisition of a Canadian firm was the 2012 acquisition of Calgary's Nexen Energy by China's National Overseas Oil Company (CNOOC) for C\$15 billion. The Harper government ultimately approved the acquisition. However, concerns have been raised about whether CNOOC has lived up to the conditions imposed on the deal by Ottawa, all of which were symbolically inflamed in January 2019 when CNOCC announced the Nexen name would disappear and become a full subsidiary of CNOOC.[47]

The Nexen purchase prompted a significant rethink by the Harper government about such transactions. Indeed, at the same time the government approved the acquisition, the door on similar transactions in the future was closed and new, lower-value thresholds for triggering ICA reviews were announced.[48] The Harper government's amendments to the ICA also included new distinctions and financial thresholds for review, depending on whether the investment is from a World

Trade Organization member state – and presumably a market-based economy. However, state-owned enterprises, many of which emanate from WTO member states like China, are subject to much lower triggering thresholds: C$389 million for SOEs vs C$1.5 billion for non-SOEs.[49]

Since 2007, implementation guidance has ensured the net-benefit assessment will "include whether the non-Canadian adheres to Canadian standards of corporate governance (including, for example, commitments to transparency and disclosure, independent members of the board of directors, independent audit committees and equitable treatment of shareholders), and to Canadian laws and practices, including adherence to free market principles."[50]

In 2013, the ICA expanded the definition of SOE to include individuals acting under the direction of a foreign government and individuals/entities directly or indirectly influenced by a foreign government.[51] Both the "net-benefit" test and national security reviews are governed by a vague set of criteria, leaving considerable discretion to those driving the review.[52] Importantly, "under the Act, the burden of proof is on foreign investors to demonstrate to the satisfaction of the Minister that proposed investments are likely to be of net benefit to Canada."[53] Where "national security reviews are concerned, investors are encouraged to reach out to the Minister of Innovation, Science and Economic Development at the earliest stages of development to *discuss* their investment prior to formal review" (italics mine).[54]

In 2018, Ottawa invoked "national security" as the justification for blocking the proposed $1.5 billion acquisition of Canadian construction giant Aecon by CCC International Holdings, an arm of the state-owned China Communications Construction Company.[55] For critics, the rejection of a significant proposed acquisition by a state-run enterprise represented an overdue shift in posture by Ottawa away from pliantly seeking Chinese capital, regardless of its source.[56] Yet the arbitrary qualities of territorially based investment review can also be seen in the rejection on *national security grounds* of the C$30 million acquisition of a *fire suppression systems firm* by a Chinese company because it was located just two miles from the headquarters of the Canadian Space Agency.[57]

CFIUS

Much as the Nexen case sounded alarm bells in Canada, the proposed 2006 acquisition of key American port operations by DP World (Dubai Ports World) arguably rang even more, given its temporal proximity to the 9/11 terrorist attacks.[58] That DP World was based in the Middle East and was a state-owned enterprise only exacerbated broader fears about critical infrastructure security that generated twenty-five legislative measures in the 109th Congress (2005–7) alone addressing foreign investment.[59] On top of standard suspicions about foreign ownership making the United States more vulnerable to financial decisions made abroad, the connection between DP World and the United Arab Emirates – it is somewhere in the Middle East, after all – only intensified the scrutiny. Was the proposed DP World acquisition a strictly commercial venture? Was there a nefarious geostrategic rationale behind the proposal? Indeed, acquisition of U.S. port operations was, in fact, indirect ,since the real target of DP World was a British firm, Peninsular and Oriental Steam Navigation Company (P&O). Would DP World have drawn so much attention if it was headquartered in Germany or France rather than the UAE?

Starting in 1992 with the Byrd Amendment noted earlier, the CFIUS was tasked with sorting through the connections between firms proposing U.S. investments and their home state governments.[60] Curiously, as the DP World acquisition was raising hackles, France's Alcatel was in the midst of its own CFIUS review of a proposed takeover of Lucent Technologies. The Bush administration's approval Alcatel's takeover came with a number of conditions, including allowing CFIUS to reopen the deal at any time to review its compatibility with national security.[61] In 2013–15, proposed acquisitions of major U.S. telecoms firms did not seem to bother the CFIUS. Indeed, in 2013, the acquisition of Sprint-Nextel by Japan's SoftBank was approved. In 2014, the committee approved the acquisition of part of Alcatel-Lucent's business by China Huaxin Post and Telecommunications Co., and in 2015 cleared Nokia's acquisition of the remainder of Alcatel-Lucent, making it a Finnish company.

DP World's acquisition of P&O was eventually finalized, but only after DP World divested itself of the combined firm's U.S. port holdings. Interestingly, the CFIUS was poised to approve the acquisition until news of the proposed transaction reached the U.S. Congress, which took measures to block the deal.

It all points to a certain inconsistency or incongruence in the application of national security reviews. U.S. Treasury data from 2014 to 2016 depicted in table 5.6 show how seldom CFIUS completes investigations with recommendations against the investment – just two rejections between 2014 and 2016.

Spurred by the anxieties around the BP World acquisition, Congress went further and distilled twenty-five recently proposed investment review measures into the Foreign Investment and National Security Act of 2007 (FINSA). The FINSA redefined the composition of the CFIUS,

Table 5.6. Covered Transactions, Withdrawals, and Presidential Decisions, 2014–2016

Year	Number of notices	Investigations	Presidential decisions	Notices rejected	Total notices withdrawn	Notices withdrawn and resubmitted
2014	147	51	0	1	12	7
2015	143	66	0	1	13	9
2016	172	79	1	0	27	15

Source: United States Treasury, Committee on Foreign Investment in the U.S., "Reports and Tables," https://www.treasury.gov/resource-center/international/foreign-investment /Pages/cfius-reports.aspx.

assigned it new review responsibilities, and expanded the criteria for reviewing proposed takeovers by SOEs.[62] Much as Canada had done in 2012 by amending the Investment Canada Act, FINSA established a number of new, but vague, criteria for sorting through the often-murky relationship between a firm's financial activities and its home state government. The FINSA revised formal timelines for review: a thirty-day national security review followed by a subsequent forty-five-day investigation to address any concerns after which the president formally has fifteen days to make a determination.[63]

However, it is the informal review process that has drawn attention and demonstrates most clearly how territorially defined governance regimes are reasserting power over global capital flows. Under guidelines set by FINSA, virtually all proposed investments into the United States must be "informally" vetted by the CFIUS. An informal vetting such as this ostensibly permits any national security objections to be identified and dealt with prior to the scrutiny, and potential embarrassment for the state-owned enterprise, entailed by a formal review. Most importantly, it is at the informal review stage that the state can quietly tell the firm their investment is not welcome. As the data in table 5.6 depict, the period 2014–16 saw 52 (of 462 total) notifications to the CFIUS "withdrawn" from consideration. Critics have complained that the informal process is opaque and contains no real criteria for evaluation, yet places the onus on firms to satisfy a moving checklist of demands by the state. Indeed, failure to engage with the informal reporting and review process virtually ensures the proposed investment will be denied.[64]

As the stock of Chinese direct investment in the United States has grown, so too have calls for additional CFIUS reform.[65] Critics of CFIUS argue the committee has been ill-equipped to review greater volumes

Table 5.7. Covered Transactions by Acquirer Home Country or Economy, 2013–2015, Top 5

	2013	2014	2015	Total
China	21	24	29	74
Canada	12	15	22	49
United Kingdom	7	21	19	47
Japan	18	10	12	40
France	7	6	8	21

Source: United States Treasury, Committee on Foreign Investment in the U.S., *Annual Report to Congress, 2015*, 16, https://www.treasury.gov/resource-center/international /foreign-investment/Pages/cfius-reports.aspx.

of investment flows for national security concerns (see table 5.7). Moreover, it is not simply the volume or value of the investments that is of concern, but investments in specific types of technologies that CFIUS is ill-suited to evaluate.[66]

In 2017, several congressional measures were proposed to strengthen CFIUS once again, the most prominent of which is the Foreign Investment Risk Review Modernization Act (FIRRMA), later codified in the 2019 defence authorization bill.[67] Among the reforms under FIRRMA is the requirement that CFIUS review minority stakes (not just full acquisitions) in high-tech firms as well as whether U.S. FDI receives reciprocal treatment in countries like China, where critics allege compulsory local partnership and technology transfer requirements are frequently imposed.[68]

Critics worry that FIRRMA also leaves the mandate and scope of CFIUS activities too broad and the evaluation criteria for review too ill-defined to generate the kind of certainty firms require when making investment decisions. Moreover, the failure to limit the scope of national security reviews has suddenly become especially problematic because of the Trump administration's sweeping use of "national security" as the reason for 2018's steel and aluminium tariffs.[69] Indeed, now that Trump has opened the door to using national security as the basis for trade restrictions in virtually any sector, restricting the scope of national security reviews may become even more difficult.

At a minimum, national security reviews of FDI amount to an evolving, fluid regime of deeply territorial governance that most often approves and seldom rejects proposed investments. More frequent, however, is the mere spectre of review that both Canada and the United States periodically assert with changes to the Investment Canada Act or CFIUS.

Conclusions

This chapter began by arguing that foreign direct investment was a barometer for the interconnectivity of the global economy in the post-war era. Moreover, efforts to alternately restrict, liberalize, or otherwise govern those flows indicated many of the changes affecting how we defined the significance of borders in the global economy. The many benefits of foreign direct investment – capital formation, skills transfer, competition, higher wages – had always contended with the trade-offs – too few local spin-offs, not enough skills transfer, high cost to local firms, and lost policy autonomy. The post-war era has seen a roughly tandem ebb and flow of two important kinds of policy regimes affecting private capital flows: transnational investment rules (ISDS), and state-based screening in the name of national autonomy (culture, national security, critical infrastructure).

The early post–Cold War period witnessed the retreat of the state from active roles in the governance of FDI flows. In the case of investment rules (ISDS), the state negotiated the creation of sets of rules within trade agreements binding the state to a set of protocols around the regulation of private property. However, as the case history of these provisions unfolded in the 1990s and 2000s, it became more evident that those rules had created a de facto legal regime accessible only to multinational firms operating in host countries. In both Europe and North America, firms began using binding, aterritorial legal regimes to challenge many of the traditional sovereign prerogatives of the state itself. That, in turn, generated a backlash against the aterritoriality of governance many saw as unaccountable and undemocratic. It is a backlash that has cast a cloud over the future of ISDS, with most proposals for new rules containing significant reassertions of territorially based governance and, in the case of the new successor to the NAFTA, the USMCA, the elimination of those rules altogether.

The same basic story can be told of state-based review of investment for national security (or other) purposes. These review processes have experienced important periods of activity and dormancy that have risen and fallen with the perceived threat posed by significant flows of foreign direct investment from perceived adversaries – economic, geopolitical, or military. In the 1970s and 1980s, it was OPEC states and Japan whose acquisition of high-profile assets raised public anxieties about sovereignty. In the 1990s, investment review went into hiding, more or less giving way to broader forces of liberalization in trade and investment that were broadly challenging the territoriality of sovereign prerogatives.

Yet, with the post–Cold War expansion of the WTO and participation of non-market economies, many of which are increasingly thought of as geo-strategic rivals, statist review of foreign direct investment is once again becoming more vigorous and evolving heavily in the direction of territorially based forms of strictures that are reasserting the functionality and sovereign prerogatives of the borders from whence those rules emanate.

In short, where foreign direct investment is concerned, borders have come roaring back to life. Moreover, where foreign direct investment was provocatively posing the question "Who is us?," the contemporary reassertion of territoriality into investment governance suggests the state will be answering that question more often.

NOTES

1 Susan Strange, *The Retreat of the State* (Cambridge: Cambridge University Press, 1996).

2 Michael Burgess, *Comparative Federalism: Theory and Practice* (New York: Routledge, 2006); Anne-Marie Slaughter, "Global Government Networks, Global Information Agencies, and Disaggregated Democracy," *Michigan Journal of International Law* 24 (2002–3): 1041–75; David A. Lake, "The New Sovereignty in International Relations," *International Studies Review* 5, no. 3 (2003): 303–23; Stephen Krasner, "Sharing Sovereignty: New Institutions for Collapsed and Failing States," *International Security* 29, no. 2 (2004): 85–120.

3 James Rosenau, "The Governance of Fragmegration: Neither a World Republic Nor a Global Interstate System," *Studia Diplomatica* 53, no. 5 (2000): 15–40.

4 Robert Reich, "Who Is Us?," *Harvard Business Review* 59 (January/February 1990): 53–64.

5 Jeswald Salacuse, "BIT by BIT: The Growth of Bilateral Investment Treaties and Their Impact on Foreign Investment in Developing Countries," *International Lawyer* 24, no. 3 (1990): 655–75.

6 See Edward M. Graham, *Fighting the Wrong Enemy: Antiglobal Activists and Multinational Enterprises* (Washington, DC: Institute for International Economics, 2000), 3–7; Deborah L. Swenson, "Why Do Developing Countries Sign BITs," *U.C. Davis Journal of International Law & Policy* 12 (2005–6): 131–55; Jeswald Salacuse and Nicholas Sullivan, "Do BITs Really Work? An Evaluation of Bilateral Investment Treaties and Their Grand Bargain," *Harvard Journal of International Law* 46 no. 1 (Winter 2005): 67–130.

7 *OECD Investment News* 7, June 2008.

8 *OECD Investment News* 7, June 2008; World Bank, "FDI Trends," *Public Policy for the Private Sector*, note no. 273, September 2004.

9 Organisation for Economic Co-operation and Development, "FDI in Figures," April 2015.

10 UNCTAD, Investment Policy Hub, "Most Recent IIAs," http://investmentpolicyhub.unctad.org/IIA.

11 Errol P. Mendes, "The Canadian National Energy Program: An Example of Assertion of Economic Sovereignty or Creeping Expropriation in International Law," *Vanderbilt Journal of Transnational Law* 14, no. 3 (Summer 1981): 475–507; George Glover, "Canada's Foreign Investment Review Act," *Business Lawyer* 29, no. 3 (1974): 805–22; Steven Globerman, "Canada's Foreign Investment Review Agency and the Direct Investment Process in Canada," *Canadian Public Administration* 27, no. 3 (1984): 313–28; Warren Grover, "The Investment Canada Act," *Canadian Business Law Journal* 10 (1985): 475–82.

12 Carlos Salinas De Gortari, *Mexico: The Policy and Politics of Modernization* (Mexico City: Plaza Y Janes, 2002), 37–47, 394– 91; Michael Hart, *A Trading Nation: Canadian Trade Policy from Colonialism to Globalization* (Vancouver, UBC Press, 2002): 298–304.

13 Adapted from Reich, "Who Is Us?"

14 As of 1 July 2020, the NAFTA was superseded by the new United States-Canada-Mexico Agreement. However, the text of the original NAFTA can be found at the Organization of American States electronic archive: http://www.sice.oas.org/Trade/NAFTA/NAFTATCE.ASP.

15 International Energy Charter, "The Energy Charter Treaty," https://energycharter.org/process/energy-charter-treaty-1994/energy-charter-treaty/

16 See *Loewen Group v United States*, http://investmentpolicyhub.unctad.org/ISDS/Details/24.

17 See *Methanex v United States*, https://www.italaw.com/cases/683.

18 https://www.iisd.org/library/background-paper-vattenfall-v-germany-arbitration.

19 Nathalie Bernasconi-Osterwalder, "Background Paper on Vattenfall v. Germany Arbitration," International Institute for Sustainable Development, July 2009, https://www.iisd.org/library/state-play-vattenfall-v-germany-ii-leaving-german-public-dark.

20 Shehab Khan, "CETA: Belgian State Rejects Controversial EU-Canada Trade Deal," *Independent*, 15 October 2016, https://www.independent.co.uk/news/world/europe/ceta-belgian-state-wallonia-eu-canada-deal-comprehensive-economic-and-trade-agreement-brexit-a7363386.html.

21 European Commission, "Online Public Consultation on Investment Protection and Investor-to-State Dispute Settlement (ISDS) in the Transatlantic

Trade and Investment Partnership Agreement (TTIP)" (Brussels: 13 January 2015), http://trade.ec.europa.eu/doclib/docs/2015/january /tradoc_153044.pdf.

22 European Commission, "Online Public Consultation"; Laura Puccio and Roderick Harte, "From Arbitration to the Investment Court System: The Evolution of the CETA Rules" (Brussels: European Parliamentary Research Service, June 2017), 1–32; Janyce McGregor, "EU Quietly Asks Canada to Reword Trade Deal's Thorny Investment Clause," CBC, 21 January 2016, http://www.cbc.ca/news/politics/canada-europe-trade -isds-ceta-1.3412943.

23 See Graham, *Fighting the Wrong Enemy*; Stephen J. Canner, "The Multilateral Agreement on Investment," *Cornell International Law Journal* 31, no. 3 (1998): 657–81.

24 *Methanex v United States*.

25 See Office of the United States Trade Representative, "Joint Statement by the NFTA Free Trade Commission: Building on a North American Partnership" (Washington, DC: 31 July 2001), https://ustr.gov/about-us /policy-offices/press-office/press-releases/archives/2001/july/joint -statement-nafta-free-trade-commission-; Greg Anderson, "How Did Investor-State Dispute Settlement Get a Bad Rap? Blame It on NAFTA, of Course," *World Economy* 40, no. 12 (2017): 2954–5.

26 See Kenneth Vandevelde, "A Comparison of the 2004 and 1994 US Model BITs: Rebalancing Investor and Host Country Interests," *Yearbook on International Investment Law and Policy* 2009 (2008): 283–316; see also Mark Kantor, "The New Draft Model U.S. BIT: Noteworthy Developments," *Journal of International Arbitration* 21, no. 4 (2004): 383–96.

27 European Commission, "Online Public Consultation."

28 European Commission, "Investment in TTIP and Beyond: – The Path for Reform," Concept Paper, May 2015, http://trade.ec.europa.eu/doclib/ docs/2015/may/tradoc_153408.PDF; see also European Commission, "EU Finalises Proposal for Investment Protection and Court System for TTIP," news release, 12 November 2015.

29 See European Commission, "Recommendation for a Council Decision, Authorising the Opening of Negotiations for a Convention Establishing a Multilateral Court for the Settlement of Investment Disputes," 13 September 2017. See also the Comprehensive Economic and Trade Agreement (CETA) Article 8.29; see also Department of State, *Report of the Advisory Committee on International Economic Policy Regarding the Draft Model Bilateral Investment Treaty*, 11 February 2004; and Department of State, *Report of the Subcommittee on Investment of the Advisory Committee on International Economic Policy (ACIEP) Regarding the Model Bilateral Investment Treaty*, September 2009.

30 Graham, *Fighting the Wrong Enemy*.
31 One exception to this was the 2004 U.S.-Australia FTA, which did not contain ISDS provisions on investment. This was mostly at the insistence of Australia. However, ISDS as applicable to both countries returned in the context of the 2016 Trans Pacific Partnership (TPP), which the Trump Administration promptly withdrew from in January 2017.
32 See Robert Pastor, *Congress and the Politics of U.S. Foreign Economic Policy* (Berkeley: University of California Press, 1980), 220.
33 Canada, Parliament, Foreign Investment Review Act §3 1 (1), 12 December 1973; see Glover, "Canada's Foreign Investment Review Act," 805–22; Globerman, "Canada's Foreign Investment Review Agency," 313–28.
34 President Ford, Executive Order 11858, 7 May 1975.
35 C.S. Eliot Kang, "US Politics and Greater Regulation of Inward Foreign Direct Investment," *International Organization* 51, no. 2 (1997): 301–33; Globerman, "Canada's Foreign Investment Review Agency"; Earl Fry, *The Politics of International Investment*, (New York: McGraw-Hill, 1983), 17; Pastor, *Congress and the Politics*, 220–43.
36 Mendes, "Canadian National Energy Program," 475–507.
37 Michael Hart, Bill Dymond, and Colin Robertson, *Decision at Midnight: Inside the Canada-U.S. Free Trade Negotiations* (Vancouver: UBC Press, 1994), 16–18, 222; Fry, *Politics of International Investment*, 77–106; Grover, "Investment Canada Act," 475–82.
38 James Jackson, "The Committee on Foreign Investment in the United States (CFIUS)" (Washington, DC: Congressional Research Service, 2018), 3–4.
39 Kang, "U.S. Politics and Greater Regulation," 303–5; Jackson, "Committee on Foreign Investment," 5–8.
40 Jackson, "Committee on Foreign Investment," 4–6; See also Laura D'Andrea Tyson, *Who's Bashing Whom? Trade Conflict in High-Technology Industries* (Washington, DC: Peterson Institute for International Economics, 1993).
41 Jackson, "Committee on Foreign Investment," 1.
42 Jackson, "Committee on Foreign Investment," 21.
43 Investment Canada Act, 2016–17 (Ottawa: Innovation, Science and Economic Development Canada, 2017), 3.
44 China Institute, University of Alberta, "State-Owned Enterprises in the Chinese Economy Today: Role, Reform, and Evolution" (Edmonton: University of Alberta, 2018).
45 Stephen Thomsen and Fernando Mistura, "Is Investment Protectionism on the Rise?," OECD Investment Division, 6 March 2017.
46 Maureen Maloney, Tsur Somerville, and Brigitte Unger, *Combatting Money Laundering in BC Real Estate* (Victoria: Ministry of Finance, Government of

British Columbia, 31 March 2019); see also Denis Meunier, *Hidden Beneficial Ownership and Control: Canada as a Pawn in the Global Game of Money Laundering*, C.D. Howe Commentary No. 519 (Toronto: C.D. Howe Institute, September 2018).

47 Jameson Berkow, "'A Parade of Broken Promises': How CNOOC Stumbled with Its Nexen Takeover." BNN Bloomberg, 15 September 2017, https://www.bnnbloomberg.ca/a-parade-of-broken-promises-how-cnooc-stumbled-with-its-nexen-takeover-1.857533; Deborah Jaremko, "Nexen Name to Disappear as Subsidiary Absorbed into CNOOC International," JWNEnergy.com, 16 January 2019, https://www.jwnenergy.com/article/2019/1/nexen-name-disappear-subsidiary-absorbed-cnooc-international/.

48 See *Globe and Mail*, "Harper Approves Nexen, Progress Foreign Takeovers," 7 December 2012; *Toronto Star*, "Ottawa Toughens Rules for Oil-sands Buyouts by Foreign State-Owned Firms," 7 December 2012; Canada, Investment Canada Act, RSC 1985, c 28 (1st Supp.) Part IV.

49 See Investment Canada Act, "Thresholds for Review," https://www.ic.gc.ca/eic/site/ica-lic.nsf/eng/h_lk00050.html.

50 Investment Canada Act, "Guidelines," https://www.ic.gc.ca/eic/site/ica-lic.nsf/eng/lk00064.html#p2.

51 Canada, Investment Canada Act, RSC 1985, c 28 (1st Supp.) 3(c). http://laws-lois.justice.gc.ca/eng/acts/I-21.8/page-1.html#h-3.

52 Ibid., 20(a-f) and 25.1.

53 Investment Canada Act, "Guidelines."

54 Investment Canada Act, "Guidelines on the National Security Review of Investments," https://www.ic.gc.ca/eic/site/ica-lic.nsf/eng/lk81190.html.

55 Robert Fife and Steven Chase, "Trudeau Cabinet Blocks Chinese Takeover of Aecon over National Security Concerns," *Globe and Mail*, 23 May 2018, https://www.theglobeandmail.com/politics/article-ottawa-blocks-chinese-takeover-of-aecon-over-national-security/.

56 Terry Glavin, "Blocking China's Takeover of Aecon Was Trudeau's Only Defensible Option," *Maclean's*, 24 May 2018. https://www.macleans.ca/opinion/blocking-chinas-takeover-of-aecon-was-trudeaus-only-defensible-option/.

57 *Globe and Mail*, "Ottawa's 'National Security Review' a Warning to Foreign Investors," 1 July 2015.

58 Jackson, "Committee on Foreign Investment," 1.

59 Jackson, "Committee on Foreign Investment," 2.

60 Jackson, "Committee on Foreign Investment," 8.

61 Jackson, "Committee on Foreign Investment," 8.

62 Jackson, "Committee on Foreign Investment," 9–10.

63 Jackson, "Committee on Foreign Investment," 12–14.

64 Kang, "U.S. Politics and Greater Regulation," 303.
65 Matthew Goodman and David Parker, "The China Challenge and CFIUS Reform," *Global Economics Forum* 6, no. 3 (Washington, DC: Center for Strategic and International Studies, 31 March 2017).
66 David Francis, "The Lights Are On at the Committee on Foreign Investment in the United States, but Nobody's Home," *Foreign Policy*, 22 June 2017.
67 Formally, S. 2098, "To modernize and strengthen the Committee on Foreign Investment in the United States to more effectively guard against the risk to the national security of the United States posed by certain types of foreign investment, and for other purposes," 115th Congress, 2nd Session, 8 November 2017; Public Law 115-232, Title XVII, "John S. McCain National Defense Authorization Act for Fiscal Year 2019," 115th Congress, Second Session, 13 August 2018.
68 Martin Chorzempa, "Confronting China through CFIUS Reform: Improved but Still Problematic," Petersen Institute for International Economics, 13 June 2018. https://piie.com/blogs/trade-investment-policy-watch/confronting-china-through-cfius-reform-improved-still?utm_source=update-newsletter&utm_medium=email&utm_campaign=2018-06-18.
69 Chorzempa, "Confronting China." See also Chad Bown, Cathleen Cimino-Isaacs, and Melina Kolb, "Will Trump Invoke National Security to Start a Trade War?" (Washington, DC: Petersen Institute for International Economics, 5 July 2017). https://piie.com/blogs/trade-investment-policy-watch/will-trump-invoke-national-security-start-trade-war.

PART TWO

Market Movements, Human Flows, and Canada's Multidimensional Borders

6 Cross-Border Movements and Governance: A Multidimensional Shifting Landscape

GEOFFREY HALE

A fundamental characteristic of the modern national state involves the exercise of "control" over populations within territorial boundaries, subject to conditions and limits defined by domestic legal/ constitutional systems, international agreements, and technological and administrative capacity. Concepts of "borders" and practices for managing and regulating cross-border movements have evolved to reflect their enormous diversity, whether of goods, services, people, or capital, across regions, economic sectors, and modes of transportation and communication, including the internet.

Although Canada and other countries continue to administer controls over movements of people and goods at physical borders or "frontiers," their relative effectiveness frequently depends on aterritorial measures. Facilitative measures may include the cooperation of other governments and businesses engaged in legal forms of cross-border transportation and communications. Such cooperation frequently depends on reciprocity or mutual accommodation. Without it, the sheer volume of market and human flows can contribute to unintended consequences: what Stephen Flynn has characterized as the "open border paradox," in which facilitating legal trade and travel can enable growing volumes of illicit goods (e.g., guns, drugs, and counterfeit or unsafe goods), irregular migrants and human trafficking, and the "hardened border paradox," in which efforts to reinforce border security and other controls expand incentives to evade them.[1] National governments' relative capacities to project "bordering" policies beyond their physical borders with the help of technological innovation often influence the structure and effectiveness of other states' border-management objectives and policies, whether to enable or constrain them – or both.

Numerous studies have attempted to disaggregate the interaction between trade patterns, demographic trends, and security metrics and bordering policies.[2] Some studies point to significant effects of post-9/11 security policies, including identification requirements, within and beyond North America on trade and travel flows. Subsequent developments – at least before the 2020 COVID pandemic – appear to have been driven primarily by economic factors, not least fluctuations in exchange rates.[3]

However, domestic, North American, and wider international political and regulatory factors remain significant to movements of goods, services, and capital, along with variations in flows of people resulting from international travel, tourism, and migration. Political and social disruptions in Africa, the Near East, South Asia, and parts of Latin America have unleashed a wave of "forced displacements" unmatched since the end of the Second World War – almost 80 million people according to United Nations reports.[4] Canada is not immune from such pressures, whether in humanitarian terms or stresses on administrative and policy systems. Major shifts in U.S. trade policies, combined with continuing structural changes in the global economy, have posed major challenges to the global trading system and to North American trade and investment regimes central to Canada's economic policies and the business strategies of Canadian businesses and investors since the 1980s.[5] These realities remain a challenge to border management in Canada (and elsewhere) across a wide range of policy sectors.

This chapter provides an overview of the implications of evolving border policies and practices for patterns of trade and travel between Canada and the United States and in a wider international context, including uncertainties over implementation of changes to North American trade rules contained in the U.S.-Mexico-Canada Agreement (USMCA) and the longer-term effects of the 2020 global pandemic on many international supply chains. It summarizes shifts in travel and freight movements since the 1990s, reflecting differences in economic activity across regions, exchange rate shifts, and adaptation to post-2001 security measures. It outlines the realities of major trade corridors embodied in the Canadian federal government's pursuit of regionally varied Gateway and "Beyond-the-Border" strategies. Finally it summarizes current debates on rapidly expanding forms of cross-border trade including e-commerce and digital trade.

Shifting Border Regimes: 1995–2020

Canada's cross-border policy environment has experienced three tectonic shifts since NAFTA's ratification in 1993. Border policies during the 1990s typically involved incremental adjustments to sharp market-driven increases in flows of people, goods, and capital across national borders as businesses and citizens adjusted to growing North American economic integration.

The North American border policy environment after 2001 was framed by political, psychological, and administrative responses to the 9/11 terrorist attacks, persistent extra-legal immigration to the United States, and rapid growth of Asia-Pacific trade. These factors reinforced the politicization of U.S. border policies, despite efforts at intergovernmental cooperation under the Smart Border Accord of 2001 and its U.S.-Mexican counterpart, and parallel reorganizations of numerous "legacy" agencies into the U.S. Department of Homeland Security (DHS) and Public Safety Canada in 2003.[6] Symmetries and asymmetries of risk and border-management policies depended on the willingness and capacity of "homeland" and border security officials in each country to pursue shared approaches to threat assessments – which did not take place until 2009–10[7] – along with reciprocity or mutual recognition of specific security arrangements. Such arrangements took varied forms in different travel and shipping modes, reflecting their combinations of risk assessments, industry structures and practices, and domestic legal and constitutional arrangements.

Canada aligned its maritime security regime with American and other major trading partners through the Container Security Initiative (CSI), a U.S. initiative internationalized through the International Maritime Organization (IMO), to facilitate pre-screening of shipping containers before their arrival in North America.[8] Container traffic passing through Canadian (and North American) Pacific Coast ports doubled between 1990 and 2000 and increased another 50 per cent through 2007, reflecting rapid growth in Asia-Pacific trade.[9] But such arrangements did not replace U.S. border screening for multimodal rail or truck shipments.

Canadian and U.S. governments introduced parallel, variably integrated trusted trader (C-TPAT, PIP, FAST) and traveller (NEXUS, Global Entry) programs, using "risk management" approaches in hopes of limiting cross-border trade and travel disruptions – particularly for automotive and other sectors with deeply integrated cross-border supply chains.[10] However, sustained collaboration was constrained by subsequent political developments in each country, especially congressional unilateralism and Mexico's post-2006 narco-insurgency.[11]

Economic imperatives from the 2008–9 recession initially drove policy developments under the Obama administration, leading to greater intergovernmental collaboration in separate processes with Mexico and Canada. The Beyond-the-Border (BTB) initiative[12] facilitated some aspects of cross-border travel, including improved coordination of screening systems and trusted traveller programs. The two countries signed an expanded bilateral pre-clearance agreement in 2015,[13] ratified in 2017. BTB contributed to the streamlining of customs and inspection processes for some freight (particularly air freight) shipments. These initiatives remain a work-in-progress.

The third tectonic shift reflects growing domestic protectionist pressures leading to President Trump's election in 2016, subsequent renegotiation of NAFTA subject to threats of unilateral American withdrawal, and the aggressive use of U.S. trade laws to manage or restrict trade. The Trump administration has made unprecedented use of substantial national security tariffs from multiple countries (including Canada and Mexico) under Section 232 of the ironically named Trade Expansion Act of 1962, prompting proportionate retaliation by both countries before a 2019 trilateral agreement governing steel and aluminium trade. In turn, Canada imposed safeguards on third-country steel imports to contain potential import surges and avoid further U.S. protection against "rerouted" steel imports – although safeguard exemptions became a key factor enabling industry commitment in 2018 to build Canada's first liquid natural gas export terminal in Kitimat, BC.[14] Further resort to Section 232 tariffs on Canadian aluminium imports in 2020 prompted retaliatory Canadian tariffs. President Trump ostentatiously threatened to impose similar tariffs against Canadian automotive exports during USMCA negotiations. Although U.S. courts have overruled selected tariff measures on newsprint and aircraft, unilateral U.S. actions against Canadian exports made preservation of arm's-length dispute resolution among the Trudeau government's top priorities in these negotiations.

Notwithstanding NAFTA's benefits, border effects on trade remained significant, even before the conclusion of USMCA negotiations.[15] House and Herrera note that about 50 per cent of non-energy-related U.S. imports from Canada and 40 per cent from Mexico in 2016 entered under WTO tariff rates rather than administratively cumbersome NAFTA rules-of-origin.[16] Uncertainties from U.S. trade policies, noted by Kukucha in chapter 3, have reshaped incentives for locating future private sector investments across North America. USMCA's rules of origin, including a 75 per cent North American content requirement, new rules governing wage rates, steel and aluminium content, and potential import quotas on cross-border automotive trade increase prospects of significant border thickening. Other USMCA provisions require Canada to strengthen its administrative capacity to intercept shipments containing counterfeit and pirated goods "in-transit" through Canada between other markets.[17]

NAFTA renegotiations also addressed numerous horizontal regulatory and technical border management issues reflecting broader "modernization" agendas associated with other trade agreements, including shared commitments to "good regulatory practices,"[18] anti-corruption measures,[19] disciplines on operations of state-owned enterprises, principles governing sanitary and phyto-sanitary (SPS) regulations and technical barriers to trade (TBTs), updated rules governing digital trade, customs rules affecting trade facilitation, and *de minimis* rules governing application of tariffs and domestic consumption taxes to cross-border purchases, including e-commerce.[20] Although most of these measures govern domestic legislation, their practical impact will depend largely on the technical wording of legislation implementing USMCA.

An emerging challenge to cross-border travel involves growing de-alignment of Canadian and U.S. laws (and of U.S. federal and state laws) governing possession and use of cannabis products, particularly marijuana. Implementing a high-profile promise from the 2015 election, the Trudeau government legalized recreational use of marijuana in 2018, although its distribution and sale are to be regulated by provinces. Eleven U.S. states, including five border states (Washington, Alaska, Michigan, Vermont, and Maine) have legalized recreational marijuana use at time of writing, while thirty-three others permit medical use.[21] By contrast, U.S. federal laws currently prohibit marijuana importation and allow border officials to prohibit future entry of non-Americans found in possession of the drug or who admit to marijuana use. These developments point to the need for ongoing information campaigns by both governments to limit risks of legal action against border crossers from both countries.[22]

The immediacy of these issues receded during the 2020 COVID pandemic, with its widespread disruption of domestic and international supply chains. Ottawa and Washington negotiated a consensual closure of borders to non-essential travel while allowing transport-related and other essential workers continued passage. While some of these restrictions have been relaxed at time of writing, subject to evolving public health (including quarantine) rules, prospects for and timing of a return to pre-pandemic conditions remain highly uncertain.

Identification, Information Sharing, and Border Management Regimes

The cumulative effects of post-9/11 security concerns and U.S. domestic debates over enforcement of immigration laws have significantly affected rules and incentives for cross-border mobility in both countries. Before 2001, there were few impediments, apart from prior criminal convictions, for cross-border travel by U.S. or Canadian citizens, in sharp contrast with the U.S.-Mexican border, which became quasi-militarized after the mid-1990s.[23] Initial congressional efforts to track entry and exit of non-U.S. nationals in response to growing extra-legal migration from Mexico and Central America were stymied by domestic business and public opposition.[24] Post-9/11 preoccupations with border security and illegal immigration led Congress to legislate "secure identification" (ID) requirements as a condition of cross-border travel, domestic air travel, and entry to federal facilities through the Western Hemisphere Travel Initiative (WHTI) in 2004. However, implementing the 9/11 Commission's proposals for comprehensive entry-exit tracking systems proved impossible without Canadian cooperation, which depended tacitly on U.S. cooperation on other border issues.[25]

Congress also attempted to standardize security criteria and ID verification rules for issuing drivers' licences (and qualifying for domestic programs such as Social Security). DHS standardized electronic (e-) passports and biometric identifiers as an international standard for air travel through the International Civil Aviation Organization (ICAO) in 2006, but with the pace of implementation subject to national discretion. These measures inevitably spilled over into Canadian regimes governing passports and other forms of identification, providing an "aterritorial" policy framework that has both shaped and responded to the interaction of national legislative and regulatory regimes. After extensive consultation shaped by its domestic cost-recovery mandate, Passport Canada introduced a ten-year e-passport in 2013, phasing-out existing five-year passports.[26]

In addition, both countries now maintain extensive biometric databases for visa applicants, applicants for permanent residence and refugee status,[27] and members of "trusted traveller" programs. U.S. law now requires extensive inter-agency information sharing of biometric information for security and law enforcement purposes – a legacy of the 9-11 Commission's concerns with security siloes. However, legislative and regulatory provisions for protection of data privacy and verification of individual records lag those enabling biometric data collection. A 2017 executive order has mandated 100 per cent facial identification verification for all international travellers, including U.S. citizens, as part of a comprehensive exit verification policy, beginning with the twenty largest U.S. airports.[28] These policies have raised significant concerns about mission creep, sharing of personal information between and outside government agencies, including their private contractors, and security of public and private sector databases, given persistent cyber-security concerns.[29] A 2018 report suggests that these concerns remain major constraints on Canada's expanded use of biometric technologies, although NEXUS cards issued from 2020 will incorporate facial recognition technology.[30]

Neither Canada nor the United States have embraced the continental European tradition of comprehensive national identity cards, reflecting shared traditions of limited government and suspicion of concentrated power. Both countries divide responsibility for issuing primary identification documents between central and subnational governments. Provincial and state governments are responsible for birth certificates and driver's licences, the former linked to jurisdiction over maintenance of vital statistics. Passports and naturalization documents are issued by central governments. All U.S. and Canadian states, provinces, and territories are members of the American Association of Motor Vehicle Administrators (AAMVA), a "soft-law," non-regulatory organization that recommends standards for driver's licences and fosters cooperation among licensing agencies through regional councils. The Association of American Railroads' (AAR) Technical Services group performs a similar function in setting technical standards for U.S. and Canadian railroads, although the U.S. Federal Railroad Administration and Transport Canada retain overall regulatory oversight.[31]

WHTI requirements challenged traditions of open travel between the two countries while enforcing a "one border policy" on northern and southern U.S. borders. It provoked intense pushback by business stakeholders, border communities, and northern border state members of Congress, resulting in delayed implementation and DHS acceptance of alternatives such as "enhanced" driver's licences in five U.S. states (Washington, Vermont, New York, Michigan, and later Minnesota) and four (later two) Canadian provinces (BC, Manitoba, Quebec until 2014, and Ontario until 2019) with sizeable border communities.[32] Efforts by successive administrations to impose national licence-issuing standards under the 2005 REAL-ID Act faced significant state resistance for both civil liberties and fiscal reasons, including objections to unfunded mandates.[33] Following multiple delays, only one state (Oklahoma, vs. nineteen in 2018) remained in non-compliance (with "extensions") in mid-2020, risking eventual federal sanctions (including non-recognition of licences for boarding commercial airplanes or entry to federal facilities), although some observers estimate that two-thirds of outstanding U.S. driver licences fail to meet REAL-ID standards.[34] By contrast, passports are encouraged but not required for Canadian citizens re-entering Canada. Other government-issued photo ID may be used for domestic air travel.

Canadians adjusted more rapidly to enhanced security measures, as noted in the comparison in table 6.1 of national rates of passport

Table 6.1. Passport Ownership: United States and Canada (% of Population)

	United States	Canada
2018	42.0	64.7
2016	40.8	64.0
2012	36.2	67.0
2009	31.8	53.0
2005	21.9	35.8
2002	19.2	31.3

Table 6.2. Passport Ownership: Canadian Provinces and Territories (% of Population)

	2012	2015		2012	2015
Ontario	72.4	64.8	New Brunswick	54.1	53.2
British Columbia	71.8	64.4	Saskatchewan	54.9	51.6
Alberta	67.0	59.2	Prince Edward Island	49.2	49.2
Yukon	65.8	62.4	Nova Scotia	50.0	48.9
Canada	67.0	60.0	Newfoundland/Labrador	44.0	43.6
Quebec	59.1	55.2	Northwest Territories	45.2	43.1
Manitoba	59.6	55.1	Nunavut	17.7	17.5

Sources: Bureau of Consular Services, U.S. Department of State, 2018; Passport Canada, annual reports 2002–3 to 2012–13 (Ottawa: Immigration, Refugees, and Citizenship Canada); *Passport Program Annual Report: 2014–2015* (Ottawa: 2016); *Passport Program Annual Report: 2017–2018* (Ottawa: 2020).

possession in the U.S. and Canada. But Anderson et al. have shown that post-9/11 border security measures effectively "nullified" benefits of rising Canada-U.S. exchange rates on cross-border travel, particularly same-day trips, by Canadians, while reinforcing their negative effects for American land-based travellers.[35] However, cross-border air travel has increased significantly in both directions since 2000, notwithstanding initial post-9/11 security effects.

Responses to WHTI illustrate the asymmetrical nature of bilateral relations. Canadian rates of passport ownership (65 per cent in 2018) remain substantially greater than U.S. rates (65 per cent in 2019), with significant regional differences in both countries. Unlike the United States, Canada's passport program has been subject to full cost-recovery through user fees since the 1990s, paralleling its airport operations regime.[36]

The Smart Border Accord provided for collaborative design and administration of "trusted traveller" programs by U.S. and Canadian border security agencies to facilitate cross-border travel by pre-identified

Table 6.3. NEXUS-Enabled Trips at Major Border Crossings, Airports, 2014–2015 (%)

Land		Air/ Rail	
Whirlpool Niagara Falls	100.00	Pearson International Terminal 1 (Toronto)	7.00
Boundary Bay (BC)	42.64	Calgary International Airport	7.00
Douglas (BC) (I-5)	38.08	Billy Bishop Airport, Toronto	6.10
Sarnia Blue-Water Bridge (ON) (I-94)	18.01	VIA Traffic Ops	4.90
Windsor Ambassador Bridge (ON)	16.14	Ottawa Macdonald-Cartier International Airport	4.80
Fort Erie Bridge (I-190-QEW) (ON)	15.20	Edmonton International Airport	4.60
Huntingdon (Quebec)	12.96	Trudeau International Airport, Montreal	4.20
Windsor-Detroit Tunnel	11.86	Winnipeg James Armstrong Richardson International Airport	3.20
Queenston Bridge (ON) (I-190)	5.46	Pearson International Terminal 3 (Toronto)	3.10
Lacolle (Quebec) (A-15/I-87)	5.42	Halifax International Airport	2.70

*NEXUS-enabled crossings at major travel corridors in New Brunswick, Manitoba, Alberta, and the crossing between Fort Frances, ON, and International Falls, MN, are below 1 per cent.
Source: Canada Border Services Agency, "Evaluation of the Trusted Traveler Programs (Air, Land, Marine)" (Ottawa: November 2016), exhibits 11, 16.

"low-risk" travellers. Personal interviews, background checks, and fingerprint records are required for enrollees, updated every five years. NEXUS program membership expanded rapidly after separate land, sea, and air modes were integrated in 2007, from 78,000 in FY 2005 to 285,000 in 2009, 1.2 million in 2015, and 1.4 million in 2017.

Canadians accounted for 82 per cent of enrolment.[37] NEXUS-enabled trips increased by 115 per cent between fiscal years 2011 and 2015, with 83 per cent of crossings at highway ports-of-entry. Parents with NEXUS cards may enrol their children (under age eighteen) without charge.

However, NEXUS use appears to be largely localized at land crossings, as noted in table 6.3. NEXUS trips in the West Coast's "Cascade Corridor" increased from 12 .per cent of crossings in 2010 to 36 per cent in 2015,[38] with somewhat smaller volumes in southwestern Ontario and Quebec. Border agency staffing of NEXUS or "Ready" lanes for enhanced-ID often appears poorly correlated with related traffic volumes.[39]

NEXUS use at major Canadian airports, while modest, is more evenly spread. Since 2014, the U.S. Global Entry program has cross-enrolled

NEXUS members, enabling more efficient screening in major U.S. airports through the TSA Pre-Check program. Similar measures are available to NEXUS members travelling through major Canadian airports.

Entry-Exit

The post-2011 Beyond-the-Border (BTB) process led to Canadian accommodation of longstanding U.S. calls for reciprocal record sharing of border entry data from one country to document exit from the other. Entry-exit was intended to address visa compliance, especially "overstays" by non-nationals, and other legal requirements such as removal of persons denied asylum or refugee status, following a major overhaul of Canada's heavily backlogged refugee adjudication system in 2009–11. Entry-exit's first stage was implemented for third-country nationals and (lawful) permanent residents (non-nationals) of each country in mid-2013.[40] Information-sharing provisions on U.S. citizens entering Canada and Canadian citizens returning from the United States took effect in 2017 and 2019 respectively, following legislative delays in Canada.[41]

This legislation has also affected the taxable status of travellers in each country – directly affecting cross-border commuters, professional athletes and entertainers, vacationing Canadians "snowbirds," long-distance truck drivers, and others travelling extensively for business in the United States; eligibility for social benefits (most provinces require annual residence of seven months to qualify for health insurance); immigration enforcement; and tracking of criminal suspects and fugitives. U.S. tax law determines residency based either on 183 days' physical presence per calendar year, or a cumulative 183 days over three years calculated on a weighted scale.[42]

Irregular Migration

Canada's legal structure governing asylum-seekers is governed by a combination of international commitments under the 1951 Refugee Convention, domestic court rulings, and the 2004 Safe Third Country Agreement with the United States. The Refugee Convention and a subsequent 1967 United Nations protocol commit signatories not to force persons to return to countries where they face serious threats to their life or freedom. Since legislative changes in 1989 prompted by the Supreme Court's 1985 *Singh* decision, Canadian law has required independent adjudication of refugee claims by the federal Immigration Review Board (IRB) following oral hearings, not just administrative reviews by government officials.[43] Decisions of the IRB's Refugee Appeal Division may be appealed to the Federal Court. Asylum claims are administered separately from humanitarian resettlement of refugees screened by Canadian immigration officials and/or the UN High Commission on Refugees (UNHCR). The Safe Third Country Agreement (STCA) of 2004 enabled Canada to manage flows of asylum-seekers more efficiently (and limit "asylum-shopping") by requiring persons entering Canada at land border crossings to seek asylum in the first "safe country" in which they arrive, unless meeting specified exceptions. The United States is the only "safe third country" currently listed under the agreement.[44]

Annual asylum claim volumes vary widely, depending on international conditions affecting migrant flows and federal visa requirements for visitors and applicants from countries with significant numbers of unsuccessful asylum claimants (as with Mexico, Romania, and the Czech Republic in 2009) – sometimes described as "push and pull" factors in migration flows. Media reports indicate significant annual variations in rejection rates for visitor visas – from 17 per cent in 2013 to 26 per cent in 2017, with some observers questioning federal officials' reasoning in their use of administrative discretion.[45] Other key factors include federal resource allocations and tribunal processes, which may contribute to or diminish relative efficiency. IRB backlogs peaked at 57,000 claims in 2002, 62,000 (2009), and 75,000 (2019), a more than two-year caseload, leading to federal administrative and legislative changes (see figure 6.1).[46]

Legislative changes in 2010 and 2012 streamlined internal IRB processes and limited subsequent appeals, reducing its "legacy" backlog of refugee claimants from 59,000 in 2010 to 5,800 by 2016.[47] A recent review of the federal asylum system notes that it "is characterized by the complex interaction of the IRB, IRCC, CBSA and the Federal Court," each of which has "independent organizational planning and accountability systems" so that "a horizontal system is being managed vertically" across varied silos.[48]

However, events in and beyond North America have triggered rapid increases in asylum claims since early 2017, resulting from exploitation of a major SCTA loophole. U.S. law provides Temporary Protected Status (TPS) to foreign nationals from designated countries "due to conditions in the country that temporarily prevent the country's nationals from returning safely," such as "armed conflicts or environmental disasters" or "in certain circumstances, where the country is unable to handle the return of its nationals adequately," but generally without eligibility for permanent residence.[49] TPS status may be extended at the discretion of the secretary of homeland security.

In 2017, the Trump administration announced plans to withdraw TPS status from Haitians, followed by subsequent plans to end TPS

Figure 6.1. Historical View of the Activities of the Refugee Protection Division (Input, Output, Resources, and Backlog)

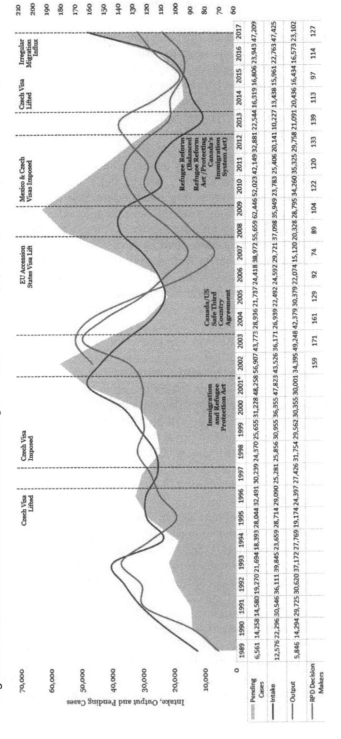

	1989	1990	1991	1992	1993	1994	1995	1996	1997	1998	1999	2000	2001*	2002	2003	2004	2005	2006	2007	2008	2009	2010	2011	2012	2013	2014	2015	2016	2017
Pending Cases	6,561	14,258	14,580	19,270	21,694	18,393	28,044	32,491	30,239	24,370	25,655	31,228	48,258	56,907	43,773	28,936	21,737	24,418	38,972	55,659	62,446	52,023	42,149	32,881	22,544	16,319	16,806	33,943	47,209
Intake	12,576	22,296	30,546	36,111	39,845	23,659	28,714	29,090	25,281	25,856	30,955	36,355	47,823	43,526	36,171	26,939	22,492	24,592	29,721	37,098	35,949	23,783	25,406	20,141	10,227	13,438	15,961	22,763	47,425
Output	5,846	14,294	29,725	30,620	37,172	27,769	19,174	24,397	27,426	31,754	29,562	30,355	30,001	34,395	49,248	42,379	30,379	22,074	15,120	20,328	28,795	34,260	35,325	29,758	21,091	20,436	16,434	16,573	23,102
RPD Decision Makers														159	171	161	129	92	74	89	104	122	120	133	139	113	97	114	127

for more than 400,000 nationals of several other countries,[50] despite continuing political instability. Although litigation has blocked implementation of these measures through mid-2020, these announcements triggered an initial upsurge of migrants to Canada, with most crossing beyond regular ports of entry to claim asylum, taking advantage of the SCTA loophole noted above.[51]

The largest number of irregular migrants – about 30 per cent in 2017–19[52] – have come from Nigeria, using visitor visas to enter the United States in transit to Canada. Information (and sometimes misinformation) conveyed by social media has reinforced these trends.[53] Canadian border agencies have worked quietly with U.S. State Department officials to limit this backdoor means of entry.[54] However, the refugee adjudication system is increasingly backlogged, stretching emergency housing systems and social services in Montreal and Toronto to their limits.[55] Federal funding increases announced in 2019 are expected to stabilize but not significantly reduce waiting periods.[56] The Trudeau government has sought to renegotiate SCTA to apply to refugee claims made by all entrants from the United States, not just those entering at ports of entry. However, the Trump administration has shown little interest in its reopening.

A 2020 Federal Court ruling challenged the STCA's constitutionality, pointing to the substantial erosion of due process rules for asylum seekers under the Trump administration.[57] The ultimate outcome of this case may hinge on the degree of policy continuity following the 2020 presidential election.

Trends in Cross-Border Trade and Travel: Shifting and Varied

The study of cross-border trade and travel movements is complicated by major differences in "northbound" and "southbound" traffic – travel flows are often "east-west" on the Ontario-Michigan, New Brunswick–Maine, and BC/Yukon-Alaska borders – physical and temporal modes of travel (highway, air, rail, same-day, multi-day), and trade (truck, rail, air, marine freight, and e-commerce). Key data sources for this study are from Statistics Canada, subject to periodic methodological changes, and the U.S. Bureau of Transportation Statistics for southbound traffic and trade flows, organized by clustered ports of entry.

The value and mode of cross-border shipments vary with changes in trade patterns, reflecting shifts in exchange rates, major commodity prices, and regional economic activity. Truck freight has traditionally accounted for the largest share of cross-border trade – 57.7 per cent in 2017 (see table 6.4). Rail shipments – primarily bulk commodities and

Table 6.4. Canada-U.S. Cross-Border Trade by Mode of Shipment: 2014–17 (Billions of Current U.S. Dollars; Imports Plus Exports)

	2014	%	2015	%	2016	%	2017	%
Total	658.20	100.00	575.20	100.00	544.00	100.00	582.40	100.00
Truck	354.00	53.80	335.20	58.30	327.20	60.10	336.10	57.70
Rail	104.20	15.80	90.50	15.70	88.40	16.20	94.20	16.20
Pipeline	89.10	13.50	53.30	9.30	45.80	8.40	61.60	10.60
Vessel	38.50	5.90	28.00	4.90	18.00	3.30	22.90	3.90
Air	28.40	4.30	26.80	4.70	26.10	4.80	27.10	4.70
Mail	0.05	0.01	0.04	0.04	0.04	0.01	0.04	0.01
Other		6.70		7.20		7.10		6.90

Source: U.S. Bureau of Transportation Statistics, 2018; https://transborder.bts.gov; accessed 20 July 2018; author's calculations.

Table 6.5. Changes in Cross-Border Truck Traffic, by Country of Origin, 2000–2019 (%)

2000 = 100	2000	2003	2007	2009	2013	2019
Canadian trucks	100.0	99.0	94.4	72.6	76.9	95.0
U.S. trucks	100.0	92.0	87.6	69.7	82.8	48.8

Source: Statistics Canada, table 24-10-0002-01 (formerly CANSIM table 427-0002) (Ottawa: 11 August 2020); author's calculations.

containerized goods – rank second, with pipeline shipments third, both reflecting significant fluctuations in commodity prices. These statistics do not capture the growing market for electronic (e-)commerce.

Canada's Atlantic and Pacific ports have grown rapidly since 1990, reflecting overall trade growth, particularly bulk commodities and container traffic. The latter grew 206 per cent between 1990 and 2007 and 281 per cent through 2016 – 968 per cent among Pacific ports. The port of Metro Vancouver, which passed Montreal as Canada's largest port in 1999, now processes more than twice as many containers annually as Canada's leading Atlantic coast port. Prince Rupert has emerged as a major trans-shipment point for goods travelling between northeast Asia and the U.S. Midwest, which accounts for two-thirds of its eastbound traffic.[58] International Falls, MN, has surpassed Sarnia–Port Huron as the busiest rail border crossing since 2011.

About half of Canada's U.S. exports and 73 per cent of U.S. exports to Canada in 2015 were shipped by truck.[59] Table 6.5 indicates that total truck crossings have fluctuated sharply with increased security measures, exchange rate shifts, and economic activity after 9/11.[60] Four major crossings in southern Ontario – the Ambassador Bridge between

Table 6.6. Major Truck Trade Corridors – Total Cross-Border Truck Traffic (%)

Detroit, Ambassador Bridge	I-75	28%	Sweetgrass (MT/AB)	I-15	2%
Buffalo, Peace Bridge	I-190	17%	Derby Line (VT/QC)	I-89	2%
Port Huron-Bluewater Bridge	I-69/94	13%	Jackman (ME/QC)	US-201	2%
Blaine (WA)	I-5	6%	Houlton (ME/QC)	I-95	2%
Pembina (ND)	I-29	4%	Highgate Springs (VT)	I-91	2%
Alexandria Bay (NY)	I-81	4%	Portal (ND/SK)	US-52	2%
Sumas (WA)		3%			

Source: Maoh, Shakil, and Anderson (2014, 14); United States, Bureau of Transportation Statistics; author's calculations.

Detroit and Windsor, the Niagara frontier's Peace and Queenston Bridges, and Sarnia–Port Huron's Bluewater Bridge – have accounted for about 58 per cent of truck crossings in recent years. Four corridors in BC's Lower Mainland, Manitoba (Emerson/Pembina), and Eastern Ontario (Lansdowne/Alexandria Bay) accounted for 16 per cent. Six regional corridors provided a further 12 per cent (see table 6.5, figure 6.2).

Fluctuations in truck crossings reflect overall levels of economic activity – as demonstrated by the sharp drop in cross-border trade volumes in 2007–9, its sectoral distribution, exchange rate shifts, and regulatory changes in each country. U.S. truck crossings dropped 41 per cent between 2013 and 2019 to less than half of pre-2000 levels, while Canadian truck crossings increased 23.5 per cent during the same period,

Cross-border and wider international travel has also increased substantially since the 1990s, but with significant fluctuations in the volume of Canadian, U.S., and other foreign residents returning to or entering Canada. Tables 6.5 and 6.7 summarize shifts in total truck traffic and auto travel volumes across Canada's borders since the mid-1990s.

Total volumes of travellers declined about 20 per cent in 2001–3. Economic recovery and exchange rate parity contributed to 24.5 per cent growth in overall travel volumes between 2009 and 2013 – 43.2 per cent for southbound Canadians, before subsiding with the loonie's post-2013 depreciation, with travel to other countries growing even faster. By contrast, the number of Americans travelling to Canada dropped 52.1 per cent in 2001–9, reflecting increased security requirements (including passports), falling U.S. exchange rates, and the effects of the 2008–9 recession, before recovering modestly after 2014.

Variations by mode of travel have been equally significant. After a sharp drop in cross-border travel during the 2009 recession, Canadian travel to the United States increased sharply in 2010–13, reaching

Figure 6.2. Top Ports for U.S. Trade with Canada and Mexico, 2016 (Billions of Dollars)

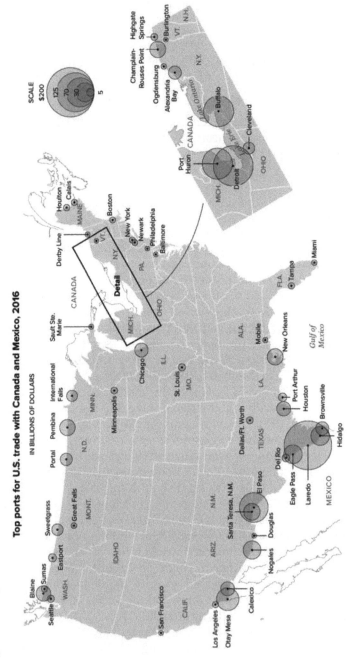

Top ports for U.S. trade with Canada and Mexico, 2016

Table 6.7. Northbound Auto Travel – Canada and Selected Provinces, by Vehicle Origin (2000 = 100)

	2003	2007	2013	2019
Canadian	83.9	94.9	133.4	96.3
American	81.7	54.5	43.0	49.3
BC Canadian	75.7	85.1	**200.3**	142.3
BC US	88.6	64.3	64.5	**82.7**
ON Canadian	87.2	99.5	121.3	90.7
ON US	*77.2*	*50.0*	*37.1*	*41.4*
QC Canadian	79.7	*84.5*	109.0	78.0
QC US	101.4	68.0	57.4	68.3
NB Canadian	86.5	95.0	*88.9*	*56.3*
NB US	93.1	66.0	42.6	*40.5*

Note: Bold: highest among provinces; bold italic: lowest among provinces
Source: Statistics Canada, table 24-10-0002-01 (formerly CANSIM 427-0002); author's calculations.

52.3 per cent above 2001 levels: 46.6 per cent by car and 70.6 per cent by air. The sharp drop in the Canada-U.S. exchange rate after 2013 affected cross-border car travel (dropping 31.3 per cent in 2013–16 before stabilizing), much more than air travel (down 3.2 per cent). However, wider international travel volumes increased 25.0 per cent during this period. Travel to Canada by third-country residents increased 19.0 per cent; American visitors to Canada increased 16.7 per cent.[61]

Table 6.7 compares the evolution in cross-border travel patterns by Canadians and Americans. The decline in American cross-border automotive traffic after 9/11 was reinforced by declining U.S. exchange rates in 2003–8. The normalization of border security measures (and the higher Canadian dollar) encouraged rapid increases in Canadian cross-border traffic in 2009–13, while U.S. traffic remained stagnant. Conversely, the falling loonie reduced Canadian automotive (and truck) traffic to 2008 levels after 2013, only modestly offset by U.S.-based cross-border travel. These data suggest a structural shift in cross-border travel patterns by Americans, with Canadian traffic more sensitive to exchange rate shifts.

Temporal and regional patterns of cross-border travel have also diverged substantially. In 2000, Americans were almost as likely as Canadians to engage in same-day travel, typically for shopping, recreation, or, in border communities, work or studies. Same-day trips dropped from 77.6 to 58.1 per cent of total American visits between 2000 and 2016 – most sharply in 2003–8. Same-day crossings by Canadians dropped from 81.3 to 72.3 per cent of total travel during this period – stabilizing after 2009.

However, these figures disguise major regional differences across border regions noted in Table 6.7. Cross-border travel in the BC Lower Mainland doubled in 2009–13, before subsiding in response to exchange rate shifts. All three Prairie provinces experienced post-2009 growth in cross-border travel until falling exchange rates and a prolonged economic slowdown reversed this trend after 2018. By contrast, the post-2009 recovery in cross-border travel has been less pronounced and post-2014 decline more noticeable in Quebec and, especially, New Brunswick, where income and passport possession levels are lower than in most other provinces bordering the United States (see table 6.2), and ease of crossing the border is sometimes linked to language issues.[62]

The 2020 global coronavirus pandemic, which resulted in border restrictions for all but "non-essential traffic," with provisions for the preservation of cross-border supply chains,[63] reduced the volume of border crossings in both directions by more than 90 per cent. The extent and duration of these restrictions and their broader social and economic effects remain to be determined at time of writing.

E-Commerce and Digital Trade: Licit and Illicit

The rapid growth of electronic (e-)commerce, the purchase of goods and services over the internet, has largely replaced traditional debates over cross-border shopping as an ongoing source of domestic and cross-border contention. Estimates of e-commerce trade vary substantially but are growing rapidly. A 2019 J.P. Morgan report estimates that Canadian e-commerce sales reached $61 billion in 2018, about 5 per cent of consumer spending, up from 2.5 per cent in the first half of 2017 and 1.4 per cent in 2012, with cross-border sales accounting for an estimated 16.7 per cent of sales volumes.[64] Canada Post reports that international tracked packet shipments – typically e-commerce purchases under 1 kilogram – increased 20.9 per cent in 2018 alone.[65] These trends accelerated during the 2020 pandemic, with its widespread substitution of online shopping for traditional retail activity. These trends have also expanded opportunities for illicit trade, as discussed below.

U.S. authorities distinguish between aspects of e-commerce involving physical delivery of products ordered through the internet, and digital trade: the delivery of products or services through digital (or electronic) networks.[66] The latter includes the exchange, transmission, and storage of data, including books, movies, and music, on computer servers. Different forms of e-commerce include business-to-consumer (B2C), business-to-business (B2B), and hybrid purchases in which consumers pick up electronically ordered goods at brick-and-mortar stores.

Levels of cross-border shopping (and same-day travel) have fluctuated with bilateral exchange rates, peaking in 1991, and enjoying a modest rebound in 2010–14,[67] but subsiding since 2015. With traditions of cross-border smuggling to avoid excise taxes dating to the eighteenth century, Canadian border agencies have long focused on border tax collection as a major priority. However, seeking to harmonize cross-border tax exemptions, if not restrictions on other products, the Harper government reduced same-day travellers' exemption to $200 in 2012 (excluding alcohol and tobacco), while shortening the $800 tax and duty-free threshold from seven days to forty-eight hours.

USMCA provides for increased *de minimis* thresholds (DMTs) on cross-border e-commerce transactions from $20 to $40 for sales tax purposes and $150 for customs duties, although a 2017 study suggests that sales tax enforcement is spotty, at best, for low-value postal shipments, compared with those shipped through international courier firms – with revenue effects estimated at $1.3 billion annually.[68] By contrast, the U.S. e-commerce customs threshold increased from $200 to $800 in 2016. However, Australia eliminated its previously sizeable DMT ($A1,000) in 2017.

Similarly, reports suggested that postal shipments face far lower levels of inspection for criminal law enforcement purposes than courier shipments – raising concerns over rising volumes of international drug shipments through the mails, particularly fentanyl and other shipments from China, associated with the opioid epidemic, which killed an estimated 4,460 Canadians and 67,367 Americans in 2018.[69] More recently, CBSA has adopted risk-based screening of mail packets integrated with both destination data on drug-related deaths and countries of origin (e.g., China).[70]

Consumers will benefit from higher DMT thresholds, but retail interests lobbied for lower thresholds as the result of probable effects of tax differentials on sale and market share. A 2017 study suggests that Canadian online prices average about 12 per cent above comparable U.S. prices before taxes and shipping costs, although differentials vary widely across product categories. Explanations include higher labour, transportation, and regulatory costs in Canada, and lower bargaining power with foreign suppliers as the result of smaller market sizes.[71] The impact of USMCA changes on retail competition and government revenues remains somewhat speculative, depending on market factors and the efficiency of tax collection and enforcement on offshore vendors.

Quebec and Saskatchewan imposed sales taxes on online transactions on out-of-province firms in their 2018 budgets, requiring sales tax registration for all online firms marketing into the province, although enforcement prospects remain unclear.[72] The OECD has issued guidelines for destination-based taxation of international e-commerce transactions, and several countries have extended domestic value-added taxes (VATs) to cross-border transactions.[73] The European Union is planning to phase in a new VAT regime (with no DMTs) to capture offshore e-commerce revenue in 2019–21.[74] Canada's auditor general has strongly criticized major revenue losses to governments and competitive disadvantages for Canadian businesses from Ottawa's current failure to tax cross-border e-commerce sales.[75] France introduced an interim tax of 3 per cent of French revenues of major global internet platforms in 2019, pending international agreement on broader OECD proposals to introduce new corporate income allocation rules for such firms,[76] only to suspend the tax to forestall retaliatory U.S. tariffs.[77] During the 2019 federal election, the federal Liberals and other parties proposed an income tax on "targeted advertising services and digital intermediation services" sold in Canada by foreign technology firms with more than $1 billion in domestic sales not currently subject to its tax laws. However, the Trudeau government has deferred introducing such a tax until it can do so as part of a wider "aterritorial" regime.[78]

Canadian and U.S. laws make customs duties and sales taxes from foreign vendors the legal responsibility of importers/purchasers. Value-added and provincial sales taxes on goods and services in Canada range from 5 to 15 per cent, while the United States has no national sales tax. In 2018, citing changing market conditions, the U.S. Supreme Court reversed an earlier precedent in *South Dakota v Wayfair*, allowing individual states to collect sales taxes from out-of-state e-vendors above modest thresholds, while leaving room for future congressional action.[79]

Other ongoing issues relating to digital trade include rate-setting mechanisms and anomalies for international postal rates, especially those involving developing countries, the interaction of domestic privacy laws with mobility of data through the internet, vendor liability for cyber-security violations, trade policy debates over geographic restrictions on the location of data servers (and their implications for law-enforcement practices and national security), capacity for enforcement of safety standards for e-commerce (particularly pharmaceutical) products sourced outside North America, and the enforcement of intellectual property rights – whether related to trade in counterfeit goods (a major U.S. concern), or Canadian cultural industries. These issues illustrate the complexity of globalization and the persistence of national regulatory contexts in dealing with expanding commercial and data flows, particularly in coming to grips with unanticipated effects of continuing technological change.

In 2018, the Trump administration gave notice of its planned withdrawal from the Universal Postal Union, which sets international postal rates and standards by consensus among its 192 members.[80] This decision followed several years of debate over "terminal" postal rates for international package deliveries, particularly from China, which subsidized developing countries' international e-commerce, while increasing costs for postal services in Canada, the United States, and elsewhere. A 2019 agreement allowed the United States to raise rates to cover its costs on such shipments, while enabling other large importers to phase in higher rates in 2021–5.[81]

Canada Post, which shares U.S. concerns, had previously agreed with China in 2017 to double terminal fees for its cheaper, slower "untracked packets" service in 2018–21.[82] Such solutions are likely to require integration of small parcel data in existing Advanced Customs Information processes currently applicable to larger shipments to receive expedited processing, and information-sharing requirements between postal services and national customs and law enforcement officials such as those legislated by Congress in 2018.

Conclusion

Border policies in Canada and the United States reflect different national, political, and institutional contexts and very different levels of interdependence between the two countries, along with the need for sustained cooperation on numerous issues ranging from facilitating borders for trusted travellers and traders to joint law-enforcement initiatives such as the post-9/11 growth of Integrated Border Enforcement (IBETs) and National Security Enforcement Teams (INSETs) involving law-enforcement agencies from both countries. Even so, major differences in each country's legal systems and policing cultures enforce predominantly territorial approaches to governance, with cooperation based on ongoing recognition of each country's jurisdictional boundaries.

Formal cooperation under the Smart Border Accord and subsequent Beyond the Borders (BTB) process has contributed to extensive, if sometimes selective cooperation on border and related issues, as discussed in chapter 10. However, political and administrative support for such processes depends on shifts in each country's domestic political climate, and U.S. officials' perception of shared policy and operational commitments. Both countries have cooperated in "pushing out" borders through a wide range of administrative measures, particularly for air travel, trusted travellers, and traders. Resulting cooperation in facilitating cross-border travel was far more extensive than cooperative enforcement of bordering policies on cross-border freight shipments under the Obama administration, even before extensive trade challenges launched by the Trump administration. Although information on cross-border infrastructure planning has improved since 2011, actual cooperation on particular projects between national governments remains strictly limited, as discussed in chapter 8, and subnational cooperation (and related interest group pressures) vary widely across states and provinces.

Migration-related border policies remain resolutely territorial, with limited exceptions, such as the Safe Third Country Agreement, reflecting continuing differences in emphases and relative effectiveness of national immigration policies and strong societal commitment to national sovereignty in both countries. Bilateral cooperation, reinforced by domestic policy shifts in Canada, has been central to the relative effectiveness of Canada's asylum system. However, subsequent developments demonstrate its practical limitations, reflecting both broader global trends and persistent differences in societal attitudes towards immigration (and trade) policies in each country, not just the Trump administration's disruptive policy innovations. As a result, Canadian governments must make substantial investments in border management and related "behind the border" support services and administrative capacity to manage these challenges, even while continuing to pursue cross-border cooperation in areas of shared interest.

The effects of electronic and digital commerce have created new and growing aterritorial pressures on Canadian borders and related tax-collection and law-enforcement policies. Cross-cutting domestic pressures for greater flexibility from consumers, stronger tax enforcement from domestic retailers, and varied levels of public resistance to an expanded tax base, along with the technical complexities associated with the rapid evolution and diversification of e-commerce, have led federal officials to pursue cooperative approaches with other OECD countries.

The complexity of border policies reflects the extent and diversity of Canada-U.S. economic and societal relations, along with the growing complexity of trade, travel, and migration issues with other countries. Effective border management requires cooperation between and among national governments to achieve efficient flows of legal trade, travel, and migration. However, ongoing differences in national legal systems and political priorities, whether associated with democratic or other forms of governance, will continue to shape the interaction among governments and their ability to manage such differences through reciprocity and mutual recognition of standards.

NOTES

The author thanks Tannis Schilk for her research assistance for this chapter, and Emmanuel Brunet-Jailly, Victor Konrad, and anonymous reviewers for their helpful comments.

1 Stephen E. Flynn, "The False Conundrum: Continental Integration vs. Homeland Security," in *The Re-Bordering North America? Integration and Exclusion in a New Security Environment*, ed. Peter Andreas and Thomas J. Bierstecker, 110–27 (New York: Routledge, 2003).

2 Steven Globerman and Paul Storer, "The Impacts of 9-11 on Canada-U.S. Trade," Research Report #1 (Bellingham, WA: Border Policy Research Institute, Western Washington University, July 2006); Alexander Moens and Nachum Gabler, "Measuring the Costs of the Canada-U.S. Border" (Vancouver: Fraser Institute, August 2012); Susan Bradbury, "Irritable Border Syndrome: The Impact of Security on Travel across the Canada-U.S. Border," *Canadian-American Public Policy* 79 (December 2012); William P. Anderson, Hanna F. Maoh, and Charles M. Burke, "Passenger Car Flows across the Canada-U.S. Border: The Effect of 9/11," *Transport Policy* 35 (2014): 50–6; W. Mark Brown, "How Much Thicker Is the Canada-U.S. Border? The Cost of Crossing the Border by Truck in the Pre- and Post-9/11 Eras," Economic Analysis Research Paper Series 11F0027M – No. 99 (Ottawa: Statistics Canada, July 2015).

3 Steven Globerman and Paul Storer, "An Assessment of Future Bilateral Trade Flows and Their Implications for U.S. Border Infrastructure Investment," Research Report #21 (Bellingham, WA: Border Policy Research Institute, Western Washington University, December 2014); Brown, "How Much Thicker."

4 United Nations High Commission for Refugees, "Figures at a Glance" (New York: 2020). https://www.unhcr.org/figures-at-a-glance.html.

5 Douglas Porter, ed., *The Day after NAFTA: Economic Impact Analysis* (Toronto: BMO Capital Markets, 17 November 2017), https://commercial.bmoharris.com/media/filer_public/c9/d3/c9d39192-7924-462f-a453-2bd510460a29/appmediahero_imagebmo_economics_special_report__the_day_after_nafta_-_icb.pdf; Chad P. Bown and Greg Autry, "Is the International Trading System Broken," *Economist*, 7 May 2018; Martin Hesse, "WTO Faces Existential Threat in Times of Trump," *Der Spiegel*, 30 June 2018.

6 Peter Andreas and Thomas J. Biersteker, eds., *The Rebordering of North America* (New York: Routledge, 2003); Edward Alden, *The Closing of the American Border* (New York: Norton, 2008).

7 United States, Department of Homeland Security and Public Safety Canada, *United States–Canada Joint Border Risk and Threat Assessment* (Washington, DC, and Ottawa: July 2010).

8 United States, Department of Homeland Security, *Container Security Initiative: In Summary* (Washington, DC: May 2011); Geoffrey Hale and Christina Marcotte, "Borders, Trade and Travel Facilitation," in *Borders and Bridges: Canada's Policy Relations in North America*, ed. Monica Gattinger and Geoffrey Hale, 100–19 (Toronto: University of Toronto Press, 2010).

9 Geoffrey Hale, *So Near Yet So Far: The Public and Hidden Worlds of Canada-U.S. Relations* (Vancouver: UBC Press, 2012), 259; American Association of Port Authorities, "North America: Container Port Traffic in TEUs" (Alexandria, VA: 2010).

10 Alden, *Closing of the American Border*; Hale, *So Near and Yet So Far*, 251–75.

11 Hale and Marcotte, "Borders, Trade and Travel Facilitation"; George W. Grayson, *Mexico: Narco-Violence and a Failed State?* (New York: Routledge, 2010); Richard C. Kilroy Jr., Abelardo Rodriguez Sumano, and Todd S. Hataley, *North American Regional Security: A Trilateral Approach?* (Boulder, CO: Lynne Rienner, 2013).

12 United States, White House and Canada, Privy Council Office, *Beyond the Border: A Shared Vision for Economic Security and Economic Competitiveness* (Washington, DC and Ottawa: December 2011).

13 United States, Department of Homeland Security and Canada, "United States and Canada Sign Preclearance Agreement" (Washington, DC: 16 March 2015), http://www.dhs.gov/news/2015/03/16/united-states-and-canada-sign-preclearance-agreement.

14 Robert Fife and Greg Keenan, "Ottawa Weighs Tactics to Forestall Dumping of Steel," *Globe and Mail*, 13 March 2018, A1; Naomi Powell, "Canada Fears Foreign Steel," *Financial Post*, 27 March 2018, FP1; Shawn McCarthy, Brent Jang, and Justine Hunter, "Ottawa Clears Way for Proposed LNG Terminal on B.C. Coast," *Globe and Mail*, 26 September 2018, A1.

15 W. Mark Brown, "Overcoming Distance, Overcoming Borders: Comparing North American Regional Trade," Economic Analysis Research Paper Series 11F0027 No. 008 (Ottawa: Statistics Canada, April 2003); Brown, "How Much Thicker."

16 Brett House and Juan Manuel Herrera. "NAFTA: Uncertainty to Persist throughout 2018" (Toronto: Scotiabank Global Economics, 14 December 2017), 1.

17 Clifford Sosnow and Peter E. Kirby, "The USMCA: A First Look at Key Contentious Issues," *International Trade and Customs Law Bulletin* (Toronto: Fasken, 3 October 2018).

18 Jeff Weiss, "Good Regulatory Practice in the United States," presentation to High-Level Symposium to Enhance Regulator Expertise on Technical Barriers to Trade, Mexico City (Washington, DC: Department of Commerce, 9 February 2016); United States, Office of United States Trade Representative

("USTR"), *Summary of Objectives for the NAFTA Renegotiation* (Washington, DC: 17 July 2017).

19 Matthew Jenkins, "Anti-Corruption and Transparency Measures in Trade Agreements" (Berlin: Transparency International, 2017).

20 Colin Robertson, "NAFTA: A Primer for the Montreal Round" (Calgary: Canadian Global Affairs Institute, January 2018), 10.

21 Governing the Future of States and Localities, "State Marijuana Laws in 2019 Map," https://www.governing.com/gov-data/safety-justice /state-marijuana-laws-map-medical-recreational.html.

22 Laurie Trautman, "Cross-Border Collaboration in the Cascadia Region," paper presented at the Association for Borderland Studies World Conference, Vienna, 10 July 2018; Lornet Turnbull and Katie Zezima, "Legal Pot in Canada Will Mean Legal Peril at the Border," *Washington Post*, 13 July 2018, A3.

23 Andreas and Biersteker, *Rebordering of North America*; Alden, *Closing of the American Border*.

24 Christopher Sands, "Fading Power or Rising Power: 11 September and Lessons from the Section 110 Experience," in *Readings in Canadian Foreign Policy*, ed. Duane Bratt and Christopher J. Kukucha, 249–64 (Toronto: Oxford University Press, 2007).

25 United States, National Commission on Terrorist Attacks on the United States, *The 9/11 Commission Report* (Washington, DC: 2004), 387–9; Geoffrey Hale, "People, Politics, and Passports: Contesting Security, Trade, and Travel on the US-Canadian Border," *Geopolitics* 16, no. 1 (2011): 27–69.

26 Connie Hache, "Financing Public Goods and Services through Taxation or User Fees: A Matter of Public Choice?" (PhD diss., University of Ottawa, 2015).

27 Green and Spiegel LLP, "Then and Now: Canada's Collection of Biometric Data," 10 July 2018. https://www.gands.com/knowledge-centre/blog -post/insights/2018/07/10/then-and-now-canada-s-collection-of -biometric-data.

28 Executive Order 13780: Executive Order Protecting the Nation from Foreign States, 6 March 2017; Davey Alba, "The U.S. Government Will Be Scanning Your Face at 20 Top Airports, Documents Show," *Buzzfeed News*, 11 March 2019.

29 Drew Harwell, "Hack Pulls the Veil Off Secret Eyes at Border," *Washington Post*, 24 June 2019, A1.

30 Public Policy Forum, "The Next-Level Border: Advancing Technology and Advancing Trade" (Ottawa: 30 July 2018), 3; Luana Pascu, "Nexus Kiosks at Canadian Airports Upgraded with Facial Recognition," *Biometric Update. com*, 29 October 2019.

31 Association of American Railroads, "Welcome to AAR's Technical Services," https://www.aar.com/standards/index.html.

32 Sands, "Fading Power or Rising Power"; Hale, "Politics, People and Passports."

33 Mark Sanford, "REAL-ID Side Effects," *Washington Times*, 14 April 2014; National Conference of State Legislatures, "The Real ID: State Legislative Activity in Opposition to the REAL ID Act" (Washington, DC: January 2014).

34 United States. Department of Homeland Security, "REAL-ID," https://www.dhs.gov/real-id; Katia Hetter, "Most Americans Will Need a New ID to Fly, Starting in October," CNN News, 22 February 2020. https://www.cnn.com/travel/article/real-id-us-travel-requirements-electronic-filing/index.html.

35 Anderson, Maoh, and Burke, "Passenger Car Flows," 53–5.

36 Passport Canada, "International Comparison of Passport-Issuing Authorities," (Ottawa: March 2012).

37 Canadian government fiscal years end March 31. Canada Border Services Agency, "Evaluation of the Trusted Traveler Programs (Air, Land, Marine)" (Ottawa: November 2016), Exhibit 9.

38 International Mobility and Transportation Corridor Program, "Cascade Gateway Monthly Volumes" (Bellingham, WA: April 2016).

39 For example, see Buffalo and Fort Erie Public Bridge Authority, "Buffalo-Niagara International Bridge Officials Call for NEXUS Reform to Improve Border Travel, Security" (Buffalo–Fort Erie: 23 January 2017).

40 Canada Border Services Agency. *Annex Regarding the Sharing of Biographic Entry Data to the 2003 Statement of Mutual Understanding on Information Sharing* (Ottawa: Public Safety Canada, 2012).

41 Jim Bronskill, "Canada Border Services Agency Sharing Information on American Border Crossings with Homeland Security," *Toronto Star*, 31 August 2017; Canada, *Bill C-21: An Act to amend the Customs Act*, First Session, 42nd Parliament, 64–67 Elizabeth II, 2015–2018.

42 United States. Internal Revenue Service, "Substantial Presence Test," (Washington, DC: 3 August 2017).

43 Neil Yeates, *Report of the Independent Review of the Immigration Review Board: A Systems Management Approach* (Ottawa: Immigration, Refugees and Citizenship Canada, April 2018), 7.

44 Immigration, Refugees and Citizenship Canada, *The Safe Third Country Agreement* (Ottawa: 23 June 2016).

45 Geoffrey York and Michelle Zilio, "Access Denied: Canada's Refusal Rate for Visitor Visas Soars," *Globe and Mail*, 9 July 2018, A1.

46 Yeates, *Report of the Independent Review*, 10–11; Maura Forrest, "Asylum-Claim Backlog Could Hit 100,000 by 2021," *National Post*, 29 May 2019, A4.

47 Yeates, *Report of the Independent Review,* 13–14.

48 Yeates, *Report of the Independent Review.*

49 U.S. Citizenship and Immigration Services, "Temporary Protected Status," https://www.uscis.gov/humanitarian/temporary-protected-status.

50 Zuzana Cepla, "Fact Sheet: Temporary Protected Status" (Washington, DC: National Immigration Forum, 5 April 2019). https://immigrationforum .org/article/fact-sheet-temporary-protected-status/; Congressional Research Service, *Temporary Protected Status: Overview and Current Research Issues, RS-20844* (Washington, DC: Library of Congress, 1 April 2020). https://fas.org/sgp/crs/homesec/RS20844.pdf.

51 Richard Gonsales, "Trump Administration Ends Temporary Protected Status for Hondurans," National Public Radio, 4 May 2018, https://www .npr.org/sections/thetwo-way/2018/05/04/608654408/trump -administration-ends-temporary-protected-status-for-hondurans; Alan Freeman, "Fearing Their Legal Status Will End, Asylum Seekers in U.S. Flee to Canada," *Washington Post,* 4 August 2017, A8; see also U.S. Citizenship and Immigration Services, "Temporary Protected Status."

52 Immigration and Refugee Board Canada, "Refugee Protection Claims Made by Irregular Border Crossers," https://irb-cisr.gc.ca/en/statistics /Pages/irregular-border-crossers-countries.aspx.

53 Wael Nasser, "Irregular Border Crossings and Asylum in Canada: Study on the Irregular Migration from Nigeria to Canada" (Honours thesis, University of Lethbridge, 2018).

54 Michelle Zilio, "Illegal Border Crossings from U.S. Increase by Nearly 23 Percent from June to July," *Globe and Mail,* 15 August 2018; Christian Leuprecht, "Renewing the Social Contract: Sustainable Policy Approaches to 'Irregular Migration'" (Ottawa: Macdonald-Laurier Institute, 2019).

55 Tristin Hopper, "Canada Wants to Stem Migrant Tide at Border," *National Post,* 2 May 2018, A1; Tavia Grant, "Wave of Asylum Seekers Floods Toronto's Shelters," *Globe and Mail,* 21 June 2018, A1.

56 Forrest, "Asylum-Claim Backlog."

57 Nicholas Keung, "U.S.-Canada Asylum Pact Violates Charter, Court Rules," *Toronto Star,* 23 July 2020, A1.

58 Eric Atkins, "With Two New Global Deals, Canada's Container Growth Is Forecast to Surpass the U.S.," *Globe and Mail,* 22 February 2018, B1.

59 Hanna F. Maoh, Shakil A. Khan, and William P. Anderson, "Truck Movements across the Canada-U.S. border: The Effects of 9/11 and Other Factors," *Journal of Transport Geography* 53 (2016): 12.

60 Maoh, Khan, and Anderson, "Truck Movements across the Canada-U.S. Border," 13; U.S. Bureau of Transportation Statistics, "Border Crossing Data," author's calculations.

61 Geoffrey Hale, "Market Flows, Human Flows and the Canada-U.S. Border" (paper presented to Association for Canadian Studies in the United States, Las Vegas, 15 October 2015); Canada, Statistics Canada, "International Travellers Entering or Returning to Canada, by Type of Transport," CANSIM Table 427-0001 (Ottawa: 17 October 2017).

62 Élisabeth Vallet and Andréanne Bissonnette, "The Quebec/United States Border: Language, an Asset or Liability to Border Crossing?" (paper presented to Western Social Science Association, San Antonio, TX, 5 April 2018).

63 United States, Department of Homeland Security, "19 CFR Chapter 1 – Notification of Temporary Travel Restrictions Applicable to Land Ports of Entry and Ferries Service between the United States and Canada" (Washington, DC: 20 March 2020).

64 J.P. Morgan, "E-commerce Payment Trends" (New York: 2019), https:// www.jpmorgan.com/merchant-services/insights/reports/canada; Adrienne Warren, "NAFTA: Raising Canada's Duty-Free Threshold on E-Commerce" (Toronto: Scotiabank Global Economics, 25 August 2017), 1–3.

65 Canada Post Corp., *2018 Annual Report* (Ottawa: April 2019), 14.

66 United States, International Trade Commission, *Digital Trade in the U.S. and Global Economies, Part I*, USITC Publication 4415 (Washington, DC: July 2013).

67 Derek Burleton and Admir Kolaj, *Canada-U.S. Border Spending: A Reversal of Fortunes* (Toronto: TD Economics, 8 February 2016).

68 Bruno Basalisco, Jimmy Gardebrink, Martina Facino, and Henrik Okholm, "E-Commerce Imports into Canada: Sales Tax and Customs Treatment" (Copenhagen: Copenhagen Economics, March 2017).

69 Basalisco et al., "E-Commerce Imports into Canada"; Arthur Herman, "Crisis in the Mail: Fixing a Broken International Package System" (Washington, DC: Hudson Institute, April 2017); Anthony Zurcher, "Opioid Addiction and Death Mail-Ordered to Your Door," BBC.com, 22 February 2018; Centers for Disease Control and Prevention, "Drug Overdose Deaths" (Atlanta: 19 March 2020); Claire Brownell, "For Fentanyl Importers, Canada Post Is the Shipping Method of Choice," *Maclean's*, 7 March 2019; Public Health Agency of Canada, *National Report: Apparent Opioid-Related Deaths in Canada* (Ottawa: June 2019).

70 Mike Leahy, Canadian Border Services Agency, presentation to Pacific NorthWest Economic Region conference, Saskatoon, SK, 22 July 2019.

71 PriceWaterhouseCoopers LLP, *Rise in Canada's De Minimis Threshold: Economic Impact Assessment* (Toronto: Retail Council of Canada, December 2017), 21–5.

72 Les Perreaux, "Quebec Pre-Election Budget Contains Massive Boost in Spending," *Globe and Mail*, 28 March 2018; CBC News, "Netflix Now Charging Sask. Customers PST for Streaming Services," 19 January 2019.

73 OECD, *Addressing the Challenges of the Digital Economy: Action 1 – 2015 Final Report*, OECD Base Erosion and Profit Shifting Project (Paris: October 2015); John McCarthy, "Digital Tax Trends: International Plans to Tax the Digital Economy," www.taxamo.com, 26 April 2018.

74 PricewaterhouseCoopers, *Rise in Canada's De Minimis Threshold*, 4; European Commission, "Modernizing VAT for Cross-Border e-Commerce" (Brussels: 11 December 2018).

75 Canada, Office of the Auditor General, "Report 3: Taxation of E-Commerce" (Ottawa: May 2019).

76 Jeffrey Trossman and Jeffrey Shafer, "The Big Shake-Up: Making Sense of the OECD Digital Tax Proposals," E-Brief #297 (Toronto: C.D. Howe Institute, 12 November 2019). https://www.cdhowe.org/sites/default/files /attachments/research_papers/mixed/e-brief%20297.pdf.

77 Keith Bradsher, "France Says Tax Deal with U.S. Is Closer," *New York Times*, 22 January 2020, B6.

78 Sean Silcoff, "Liberals Vow to Tax Foreign Tech Giants on Digital Ads, Services," *Globe and Mail*, 30 September 2019, B1; Janyce McGregor, "As U.S. Threatens Retaliation on Digital Taxes, Canada Waits for OECD Talks," CBC.ca, 22 January 2020, https://www.cbc.ca/news/politics /davos-digital-tax-wednesday-1.5436372.

79 Rosalie Wyonch, "Bits, Bytes and Taxes: VAT and the Digital Economy in Canada," Commentary #487 (Toronto: C.D. Howe Institute, August 2017); United States, Bureau of Customs and Border Protection, "Internet Purchases" (Washington, DC: 31 January 2018); *South Dakota v Wayfair*, [2018] 585 US; Robert Barnes, "Supreme Court Rules That States May Require Online Retailers to Collect Sales Taxes," *Washington Post*, 22 June 2018.

80 Herman, "Crisis in the Mail"; United States, Government Accountability Office, "International Mail: Information on Changes and Alternatives to the Terminal Dues System" (Washington, DC: October 2017); Eliot Kim, "Withdrawal from the Universal Postal Union: A Guide for the Perplexed," *Lawfare*, 31 October 2018.

81 Abigail Abrams, "U.S. Avoids Postal 'Brexit' as Universal Postal Union Reaches a Deal," *Time*, 26 September 2019.

82 Paul Steidler, "Canada's Strong Stand to Fix International Postal System," (Arlington, VA: Lexington Institute, 15 February 2019); Bill Curry, "Canada Post Phasing Out Controversial Shipping Discount for Chinese Goods," *Globe and Mail*, 9 November 2017.

7 Reforming High-Skilled Temporary Worker Programs in Canada and the United States: Sticks and Carrots

MEREDITH B. LILLY

The workplace of "the fourth industrial revolution"[1] suggests that future competition for skilled labour will be fierce and global. With aging populations throughout most OECD countries, the shortage of high-skilled workers is recognized as a serious problem. High old-age dependency ratios (number of persons aged sixty-five or over per hundred working-aged persons) in countries such as Canada and the United States will continue to challenge national economies to sustain adequate and appropriately skilled labour forces, which will in turn reduce the capacity of governments to finance the social programs on which many citizens – especially retirees – depend.[2]

Immigration systems in Canada and the United States play a vital role in addressing such labour shortages. Both countries have long histories of attracting immigrants of all skill levels and have benefited from geographic isolation and controlled immigration systems that allow their governments to select immigrants according to various criteria. Canada, in particular, has a merit-based immigration system that, although not without its challenges, is admired internationally.

Nevertheless, current debates about immigration are fuelled by populist anti-migration sentiment, particularly in the United States and parts of Western Europe.[3] Constant effort and attention is focused on whether immigration systems are admitting the right number and "type" of immigrants to bolster labour market productivity and economic growth, and by extension, to maintain popular support. The corollary implies that failure to carefully select and admit migrants into North America according to such economic goals then leads to populist anti-immigration sentiment more generally.[4]

Lost in such debates are several difficult realities facing North America. As the result of low fertility rates and population aging, net migration will account for upwards of 80 per cent of population growth in the world's high-income countries leading into 2050.[5] While technological disruption and artificial intelligence may lower the demand for low-skilled work in the future, competition for high-skilled immigrants is expected to increase in Western economies.[6] At the same time, rapid development in China and India (two top source countries for immigration to both Canada and the United States) as well as other fast-growing Asian economies will inevitably mean those countries will offer more opportunities to their own highly educated citizens in the coming decades, thereby reducing incentives to migrate elsewhere. By comparison, relatively low levels of economic growth are expected for Canada and the United States over the long term. Since upwardly mobile, educated migrants tend to be attracted to high-growth economies, Canada and the United States may face difficulty attracting such migrants in the future.[7]

In this context, major employers and investors increasingly regard the capacity to move skilled professionals around the globe as essential to business success.[8] An increasing body of evidence suggests that the ability to hire and move workers from overseas with ease is becoming a factor in decision-making by global companies when establishing new locations.[9] Despite facing similar challenges regarding skilled labour shortages, Canadian and U.S. governments are following divergent paths on temporary labour mobility for skilled professionals. Canada has increasingly liberalized its temporary entry policies for skilled workers, while the United States is revisiting its visa policies with a view to greater restrictions.[10]

This chapter explores the changing visa policies and rules surrounding temporary skilled labour mobility in Canada and the United States. It begins with a short introduction to the evolution of temporary foreign worker programs in each of Canada and the United States. It then outlines the two countries' major high-skilled labour mobility programs: the H-1B visa in the United States and Canada's high-wage stream of the Temporary Foreign Worker (TFW) Program. It analyses the problems identified with programs in both countries, and subsequent reforms undertaken by Canada between 2014 and 2017. It suggests that Canada's TFW-reform efforts were successful because of their two-part approach: restricting access to the high-skilled TFW program by closing loopholes and avenues for abuse, and simultaneously streamlining entry for high-skilled workers in high-demand occupations. Finally, it outlines lessons learned by Canada during this process with a view to offering policy options for U.S. reforms.

High-Skilled Temporary Worker Programs in the United States and Canada

Throughout their histories, Canada and the United States have relied on temporary foreign labour to fill job shortages. Most temporary foreign workers (also called migrant workers) in both countries have been concentrated in low-income sectors and low-skilled jobs, especially in agriculture.[11] Much of the public and academic discourse has also focused on low-skilled migrant workers[12] and whether flows of such workers into North America exacerbate job competition and contribute to declining wages and working conditions for citizens.[13]

The United States first created temporary entry visas for high-skilled workers via the 1952 Immigration Act.[14] This category of migrants remained very small, accounting for less than 3 per cent of overall immigration flows.[15] The United States further expanded visa availability to high-skilled workers via the 1965 Immigration and Nationality Act. But even at that time, policymakers were concerned about the potential for foreign workers to displace Americans and required economic needs tests to demonstrate that American workers were not being displaced, nor were their wages adversely affected. According to Usdansky and Espenshade,[16] admissions remained low, with only 8 per cent of all U.S. immigrants arriving on employment-based visas in 1980.

Yet, other forms of immigration rose dramatically during the 1970s and 1980s – including major increases in illegal migration. This spurred calls for reforms to the broader immigration system, focused largely on stemming the flow of family-reunification immigration upon which the U.S. system is predicated. The 1990 Immigration Act introduced many changes, including the creation of a series of employment-based visas. At that time, the H-1B visa was created, and set an annual cap of 65,000 professionals to be admitted with this visa. H-1B workers required a bachelor's degree, their stay in the United States was limited to six years, and they were to be paid prevailing wages.[17]

In the late 1990s, Congress expanded the 65,000 cap to allow 115,000 H-1B visa holders to enter the United States annually in 1999 and 2000, rising to 195,000 annually for fiscal years 2001–3. This temporary increase was meant to enable the United States to prepare for Y2K computer problems anticipated at the turn of the twenty-first century.[18] However, the cap reverted to 65,000 after 2003.

In 2004, Congress passed the H-1B Visa Reform Act to renew a number of sunsetting program safeguards. These included ensuring that American workers would not be displaced by H-1B workers, that employers would make efforts to hire Americans first, and that federal authorities could investigate complaints and enforce the legislation. New restrictions were also placed on the program, including higher processing fees for applications and requiring H-1B workers to be paid prevailing wages for jobs.[19] An additional 20,000 H-1B visas were allocated to graduates from American universities for a combined annual total of 85,000.[20]

Between 2007 and 2017, upwards of 250,000 H-1B applications were approved annually, with an average of 147,000 visas actually issued.[21] This discrepancy arises from the fact that the same worker can be represented by multiple applications, or employers may apply for spots never filled with workers. Many visas are issued above the 85,000 cap because many employer types (such as universities and non-profits) are exempt, as are H-1B visa holders who renew their visa or change employers.[22]

Canada's temporary foreign worker program is situated within a broader policy context focused on economic migration – a direction that began in the 1960s and has been implemented systematically since the 1990s.[23] Specifically, Canada has set annual immigration targets with a deliberate focus on economic factors, using a points-based system to privilege skilled and educated foreign nationals who were expected to succeed in the Canadian labour force. Canada's approach is well documented in the literature and is admired internationally, yet it is also subject to criticism. Concerns have been raised regarding the validity of methods used to select immigrants, the government's capacity to predict the skills required to succeed in the labour force, and the broader ethics of favouring high-skilled immigrants over others.[24] Although family reunification remains an established mechanism for immigration to Canada, economic migration is now entrenched as the primary pathway for entry. Canada's high public support for immigration is also strongly linked to this economic policy orientation.[25]

Canada's experience with temporary high-skilled labour migration is recent, with numbers of high-skilled workers remaining very low until the twenty-first century. Although the Temporary Foreign Worker (TFW) Program has been in existence since 1973, most arrivals were concentrated in low-skilled, caregiving, and agricultural streams. In the past decade, roughly one-quarter of all TFWs admitted to Canada have been high-skilled workers.[26] Before outlining the specific changes made to Canada's temporary foreign worker program, it is important to also understand the complicating influence of international labour mobility agreements on the movement of high-skilled labour to both Canada and the United States. International treaties such as the General Agreement on Trade in Services (GATS) govern the global labour movement of business visitors, investors, intra-corporate transferees at senior management levels, and some skilled professionals.[27] Both Canada and the United States, as GATS signatories, have committed to allowing limited movement of professionals into their countries through parallel temporary entry tracks, without the need for economic needs tests required by temporary foreign worker programs. Of course, Canadian and American firms also benefit from GATS provisions that allow them to transfer their own national employees internationally to other signatory countries.

In addition, labour mobility provisions in bilateral free-trade agreements (FTAs) signed by the United States and Canada allow for temporary entry by high-skilled professionals on a reciprocal basis. NAFTA, as the principal FTA responsible for the presence of foreign professional workers in both countries, covers temporary entry of sixty-three categories of professionals. Canada has now also signed additional labour mobility provisions for professionals under FTAs with the European Union (CETA) and the Comprehensive and Progressive Trans-Pacific Partnership (CPTPP).[28]

Each country handles oversight and management of workers arriving in North America under these international agreements differently. The United States uses H-1B and L-1 intra-corporate transferee visas for individuals entering under GATS, while separate purpose-created visas such as the NAFTA TN visa cover entries of individuals arriving in the United States under specific FTAs.[29] For example, in Canada, the temporary foreign worker program operates separately and is overseen by Employment and Social Development Canada. Immigration, Refugees and Citizenship Canada oversees management of workers arriving in Canada under international agreements under the International Mobility Program.[30]

Table 7.1 outlines high-skilled worker temporary admissions to Canada and the United States under each of these programs from 2010 to 2015. The two countries' statistics are not directly comparable for several reasons. For example, Canadian data reflect the number of high-skilled workers who *entered* the country, while U.S. data reflect the number of *visas issued* to H-1B and L-1 workers. In addition, each country accounts for temporary admissions differently, reflecting slightly different underlying data. For instance, Canada's "significant benefit" stream falls under the Immigration Mobility Program, and entrants are exempt from economic needs tests.[31] The category reflects an approximate 50:50 ratio of intra-corporate transferees (comparable to the U.S. L-1 visa) and those deemed by Canadian officials to offer significant social, cultural, or economic benefits to the country.

Table 7.1. Canada and U.S. High-Skilled Temporary Worker Admissions by Major Programs

	2010	2011	2012	2013	2014	2015
Canadian high-skilled worker entries						
High-wage TFWs*	36,057	42,741	63,108	46,396	23,813	21,960
Needs test exempt – significant benefit to Canada**	16,065	18,822	20,502	30,490	34,437	24,060
Trade and international agreement	19,550	20,624	22,052	22,151	21,567	19,707
Canada total	**71,672**	**82,187**	**105,662**	**99,037**	**79,817**	**65,727**
U.S. non-immigrant visas issued*						
H-1B visas	117,409	129,134	135,530	153,223	161,369	172,748
L-1 intra-corporate transferees	74,719	70,728	62,430	66,700	71,513	78,537
U.S. total	**192,128**	**199,862**	**197,960**	**219,923**	**232,882**	**251,285**

Source data: * Canada, Office of the Auditor General, "Report 5 – Temporary Foreign Worker Program – Canada, Employment and Social Development Canada," 2017 *Spring Reports of the Auditor General of Canada to the Parliament of Canada*, http://www.oag-bvg.gc.ca/internet/English/parl_oag_201705_05_e_42227.html.
** Canada, Immigration, Refugees and Citizenship Canada, *Express Entry Year End Report 2016*.
*** United States, Department of State, Bureau of Consular Affairs, "Nonimmigrant Visa Issuances by Visa Class and by Nationality, FY1997–2017 NIV Detail Table," Nonimmigrant Visa Statistics, https://travel.state.gov/content/travel/en/legal/visa-law0/visa-statistics/nonimmigrant-visa-statistics.html.

There are a number of other temporary paths to entry into Canada and the United States not reflected in the table, including spousal admission programs, streams for university professors, and reciprocal work arrangements to allow foreign students to work while studying. Nevertheless, it reflects the primary categories of migration used to capture temporary foreign workers entering Canada and the United States primarily for their advanced skills, rather than their student status or familial relationships.

Concerns Emerge about Misuse of Temporary Entry Programs

Many have suggested that temporary entry by high-skilled workers has not been controversial as the result of a perceived shortage of skilled workers in many sectors and the positive benefits for the host country.[32]

However, over the past twenty years, global outsourcing firms have emerged in both countries to enable companies to lower costs, leading some to question the validity of reported labour shortages.[33] In the U.S. information and technology (ICT) sector, Matloff[34] has documented how foreign high-skilled workers granted temporary entry to the United States were paid less and were less qualified than Americans. More recently, Bound et al.[35] have found evidence to support Matloff's findings, suggesting foreign workers in the sector reduce opportunities for American workers. Hira[36] and Martinez[37] have also documented how the U.S. H-1B visa program has been used to outsource American ICT jobs and drive down sector wages. Although the H-1B program explicitly forbids hiring foreign workers to displace domestic employees, some employers have used loopholes related to wages and hiring criteria to undermine its integrity.

Repeated congressional efforts to address abuses have failed to deliver results. In fact, aside from modest changes in 2004, the U.S. program has not been overhauled since the H-1B's creation in the 1990 Immigration Act.[38] Meanwhile, the numbers of ICT workers entering the United States via the H-1B program has only increased.

Canadian studies have not investigated these relationships for high-skilled workers specifically. However, several scandals emerged in 2012 and 2013 whereby skilled Canadians were either bypassed for work opportunities in favour of foreign hires or were terminated and replaced with foreign workers by outsourcing firms who were violating the terms of Canada's TFW program.[39] Concern that Canada's TFW program had become a convenient workaround for the country's backlogged immigration system led to immediate and widespread reforms under the Conservative government led by Stephen Harper. Changes to the TFW program coincided with a major overhaul of Canada's immigration system led by Immigration Minister Jason Kenney, from one largely reliant on long wait lists for approved applicants to a new nimble "just-in-time" system focused on changing labour market demands.[40] The next section details reforms to Canada's TFW program between 2014 and 2017 and how they intersected with Kenney's changes to the broader treatment of skilled workers by the immigration system.

Canada's Reforms to the Temporary Foreign Worker Program

Canada introduced major reforms in 2014 to address abuses of the program and ensure that Canadian workers were not being displaced by foreign workers hired to permanently outsource Canadian jobs.[41] Figure 7.1 presents the major reforms undertaken. The rationale for

Figure 7.1. Canadian Temporary Foreign Worker Program Reforms for High-Skilled Workers (2013–2015)

Canadian Temporary Foreign Worker Program Reforms for High-Skilled Workers (2013–2015)

TFW Program Reorganized to Restrict Entry to Essential Workers When No Canadians Are Available (Oversight: ESDC)

Tighten TFW Program

- Pay prevailing wages
- More comprehensive and rigorous labour market impact assessment
- Longer duration and broader domestic recruitment requirements
- Prohibit dislocation of local workforce, increase penalties for violation including permanent bans on employers
- Increase monitoring and compliance
- Eliminate language requirements other than English/French

Streamline TFW Program

- Prioritize processing for highest paid, most skilled and shortest duration applications to program
- Restrict renewals
- Create transition plans to hire local workers
- More resources devoted to use of more and better labour market information on shortages
- Increase application fees to support monitoring and enforcement

International Mobility Program to Facilitate Entry (Oversight: IRCC)

Canadian Interests and International Agreements

- Provisions in the General Agreement on Trade in Services
- Provisions in Canadian trade agreements: NAFTA; CPTPP; CETA; and others
- Based on reciprocity with other highly developed countries
- Limited other categories of workers to reflect Canadian interests
- Not dependent on employer demands or labour market shortages
- New monitoring and compliance system for prevailing wages and Cdn worker dislocations
- Entry permits are open, not linked to employer

Additional Reforms to Facilitate Entry (2015–2017)

Global Talent Stream Pilot (2017–2019)

For Referred Employers

10-day approval process for pre-approved occupation lists and specialized roles

Express Entry System Launched (2015)

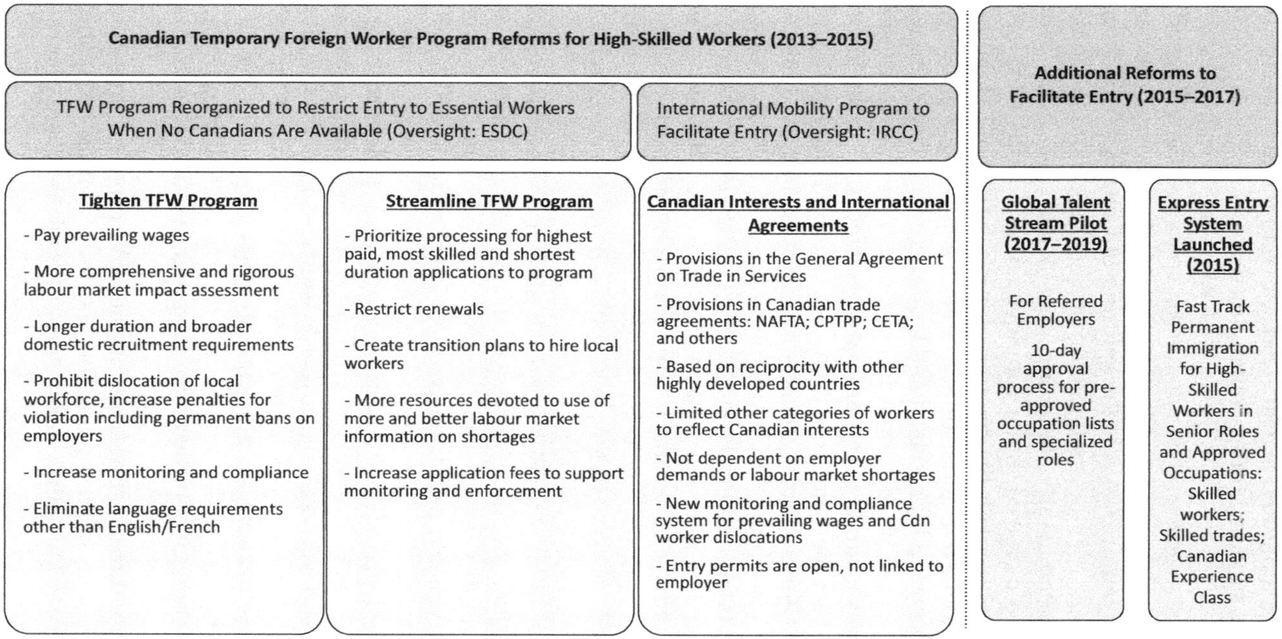

Fast Track Permanent Immigration for High-Skilled Workers in Senior Roles and Approved Occupations: Skilled workers; Skilled trades; Canadian Experience Class

each individual change is documented extensively in government reports[42] and not replicated here. Instead, the common threads that tie the reforms into a broader strategy are outlined below.

1. *The broader societal, economic, and employment influences driving low-skilled, low-wage migration into Canada are categorically different from those related to high-skilled and high-wage work.* The most significant reform undertaken in 2014 was the splitting of Canada's TFW program into two streams: (1) the continuation of a smaller, more tailored TFW program predominantly targeting low-skilled/low-wage workers that would be managed under Employment and Skills Development Canada (ESDC), and (2) the creation of the International Mobility Program to be managed under Immigration Canada (IRCC) that would streamline entry for high-skilled and high-demand occupations as well as those entering Canada under international agreements. The majority of concerns about the program have been focused on the low-wage streams, and on the work conditions of child- and elder-care workers and seasonal agricultural workers primarily from Mexico, Latin America, and the Caribbean.[43] By separating Canada's TFW program from the International Mobility Program, the government ensured greater scrutiny over the temporary migration of low-skill/low-wage workers into Canada. This reform is consistent with research evidence by Hugo[44] and Panizzon,[45] who have found that labour mobility trends of low-skilled workers differ considerably from those of high-skilled workers.

Under the reformed TFW program, stricter measures were introduced to reduce the attractiveness of the program for hiring Canadians, to ensure that labour shortages were genuine and there were no Canadians available to fill the positions, and to increase monitoring and enforcement. The TFW program application process became more expensive and placed the burden on employers to demonstrate local labour shortages before foreign workers could be hired. Government monitoring and enforcement standards were also increased, with more checks and balances to ensure that foreign workers hired through the program were not abused and that employers were not using the program to disadvantage Canadian workers. Strict penalties were also introduced for employers who abuse the program, including permanent bans for repeat offenders.

At the same time, a sub-stream of the TFW program known as the "highest-demand, highest-paid, and shortest-duration occupations"[46] was also created to facilitate quick entry of workers where domestic labour shortages had already been clearly established. In this case, skilled trades and other high-demand occupations, workers being offered positions in the top 10 per cent of wages earned by Canadians, and where workers were expected to stay in Canada for less than six months would all be offered a ten-day service standard for application review and approval. Overall, changes to the TFW program were intended to ensure that employers access it only as a last resort. At the same time, it was important that the program did not become so restrictive that entry to Canada became a barrier to business operations.

2. *The newly created International Mobility Program facilitates faster entry of high-skilled, high-wage workers based on Canada's national interests and reciprocal trade agreements with other countries.* Using the International Mobility Program, the federal government can unilaterally, or in cooperation with provinces, facilitate quick entry of high-demand occupation workers into Canada. In such cases, demonstrated labour shortages have already been established and labour market assessments would duplicate information already gathered about labour market needs. For example, the Canada–British Columbia Immigration Agreement is used to streamline entry of ICT workers into that province.[47]

In addition, the International Mobility Program also covers foreign professionals entering Canada under FTA provisions enshrined in NAFTA, CETA, and the CPTPP. These workers can move between member states to work for a limited period without being required to satisfy economic needs tests required under the TFW program.[48] Such workers gain entry to Canada's labour markets irrespective of the country's labour force needs. Each year, approximately twenty thousand workers enter Canada via international trade agreements; of those, 90 per cent enter under NAFTA provisions.

Although such arrangements could be viewed as a major gap in Canada's TFW reforms, the use of international agreements to facilitate entry to Canada has generally escaped criticism. This may be because Canadians seeking to work in the United States have benefited disproportionately from NAFTA's provisions: ESDC[49] estimates that nearly forty thousand Canadians reside in the United States via the agreement. Still, with new trade agreements in place with Europe and the Asia-Pacific via the CPTPP, Canada must anticipate that more foreign workers will enter Canada under their provisions in coming years. Careful monitoring can ensure that such high-skilled migratory streams continue to benefit Canadian workers and society overall.

3. *More paths to permanent residency reduce incentives to abuse Canada's temporary entry programs.* Changes to Canada's temporary entry programs were complemented by the creation of additional fast-track permanent immigration programs for high-skilled workers in high

demand. Led by Immigration Minister Jason Kenney,[50] Canada's new Express Entry System was launched in 2015, facilitating permanent residency for high-skilled workers via three streams: (1) the Federal Skilled Worker Program, (2) the Federal Skilled Trade Program, and (3) the Canadian Experience Class.[51] The program enjoys multi-partisan support and successfully transitioned from its 2015 launch under the previous Conservative government to the Liberal government under Justin Trudeau. Since its establishment, the program has invited just over thirty thousand highly ranked workers to move to Canada permanently in 2015 and again in 2016.[52]

With a rapid and adaptive program that offers many more avenues for highly skilled workers to immigrate permanently, Canada's Express Entry System reduces incentives to misuse the parallel temporary entry system. In addition, significant latitude allows Canada's federal immigration minister to adjust the program's targeted admission numbers and criteria directly using regulatory instruments (Ministerial Instructions) rather than requiring legislative changes subject to Parliament's approval.[53]

Impact of Canada's Reforms

Canada's 2016 Auditor General's report evaluated the reforms to Canada's TFW program. Much of the report's findings focused on the low-wage stream of the program, rather than the high-wage stream. The report found that, while reforms reduced the numbers of temporary foreign workers in Canada, the global financial crisis and subsequent economic slowdown in Canada also played a role.[54] For example, many skilled trades workers entered Canada to work in Alberta's oil sands during the commodity boom. A major slowdown of the sector led to a drop in high-skilled TFWs even before government reforms were implemented. This trend can be seen in the top line of table 7.1, which clearly demonstrates the drop in high-wage temporary foreign workers admitted to Canada beginning in 2013, after peaking in 2012. Other streams of temporary entry for both countries have generally increased year-over-year.

The report also found that while the reforms helped address some abuses of the TFW program, the implementing department (ESDC) relied too heavily on employer attestations rather than independent assessments of labour market needs. Similarly, the auditor general found that ESDC did not have sufficient monitoring and evaluation in place to assess whether the reforms had unintended consequences on the Canadian labour market.

The incapacity of ESDC to measure the impact of the changes also meant that the government had no evidence to respond to criticism from employer groups and experts who believed the Harper government's reforms were overly restrictive at best, and totally unnecessary at worst.[55] Some Canadian employer groups and industry stakeholders indicated that the reforms had severely reduced the ability of employers to fill labour shortages quickly.[56] The Harper government was largely unmoved by such protests, prioritizing system integrity and domestic political concerns over those expressed by employers. Further, any potential retaliatory impacts on the ability of Canadian firms to send their employees overseas via intra-corporate transferee rules following reforms were largely ignored.

In response to the outcry by Canadian employer groups over the difficulties they faced bringing foreign workers to Canada, the Liberal government introduced additional reforms in 2017 to loosen restrictions on the low-skilled streams of the TFW program and the Live-In Caregiver program.[57] For high-skilled workers, the Trudeau government introduced a new two-year pilot program under the TFW program in 2017 called the Global Talent Stream (See figure 7.1). It allows qualified employers to apply for and receive work permits in just two weeks for "unique and specialized" workers and those filling established occupational shortages in computer and ICT sectors. Employers must be referred to the program by approved municipal, regional, provincial, and federal partner organizations. As of July 2019, there were just over forty approved organizations, which were closely connected to federal officials who administered the program.[58] Additional program eligibility criteria included specialized training, five or more years of specialized experience, and offering salaries over C$80,000.[59]

To summarize, although there were some problems with the roll-out of reforms as identified by the auditor general and some employer groups, the overall lack of controversy about the changes was remarkable. It is suggested that this is because Canada's reforms were designed to augment both the "stick" and "carrot" approaches to managing entry of temporary workers into Canada. Greater restrictions (sticks) were placed on use of the TFW program and were combined with stronger monitoring and enforcement to reduce the attractiveness of the program overall, and to penalize employers who abused the program. Simultaneously, the government paired the new restrictions with efforts to facilitate the flow of high-demand workers into Canada (carrots), thereby accelerating processing time and the response of programs to labour market gaps. Implications of Canada's reforms for the United States are discussed further below.

Table 7.2. Major Problems Identified with H-1B Program and Possible Reforms

Problem	Canadian 2014 reform	Potential U.S. reform
Wages too low	Pay prevailing wage	Pay prevailing wage and prioritize; set new wage floors
Education/skills not specialized enough	Standardize skill levels – approved occupation lists	Minimum bachelor's degree, prioritize
Outsourcing firms dislocating local workforce	Enforce and increase penalties	Enforce and increase penalties
Program not focused on real labour shortages	Labour Market Impact Assessment/labour info	Improve use of labour info
No requirements to make efforts to hire local workers or advertise jobs	30-day advertising, increased domestic recruiting	Advertising and local hiring
Use of temporary streams for permanent labour	Restrict renewals, transition plans for local labour	Reduce length of visas and single renewal
No formal audit or compliance program, insufficient penalties	Audit and compliance, strict enforcement, penalties, blacklist	Audit and compliance, enforcement and penalties, blacklist
Presence in host country tied to employer	No solution	Possible de-linking

Canadian Sources: Canada, Employment and Social Development Canada, "Overhauling the Temporary Foreign Worker Program." U.S. Sources: U.S. Congress, Senate, S. 180; U.S. Congress, House, H.R. 1303; U.S. Congress, House, H.R. 170; U.S. Congress, House, H.R. 2233.

American Attempts at Reform

While Canada moved quite swiftly to reform its TFW program when concerns were raised, the United States has faced serious difficulties in making meaningful changes to its own H-1B visa program. Over the past decade, U.S. lawmakers have identified issues they would like to address.[60] A review of congressional bills tabled to reform the H-1B visa reveals that many of the challenges and proposed solutions resemble those already addressed by Canada. Table 7.2 summarizes the problems and solutions identified by four major H-1B reform bills before the 115th Congress[61] and compares them to the measures undertaken by Canada. These bills enjoy bipartisan support: two reflect twin bills before both the Senate[62] and Congress[63] that outline comprehensive reforms; one offers moderate reforms focused primarily on ensuring that the H-1B program does not displace American workers;[64] and one is a narrow bill[65] focused on increasing the salary floor for H-1B positions from $60,000 to $100,000 (with an annual inflator).

The table suggests that – if Canada's reforms offer any guidelines for others – appropriate U.S. policy solutions have already been identified. Given the weaknesses in Canada's program identified by the auditor general following implementation of its reforms, the Canadian experience also underscores the need to support legislative reforms with resources to better measure and identify labour market gaps and enforce penalties.

Lessons from Canada for the United States

While Canada and the United States can find much common ground for the reform of their respective high-skilled temporary worker programs, there are several major differences. One may be the disproportionate U.S. emphasis on increasing restrictions on the H-1B program while ignoring arguments to facilitate entry for truly high-demand workers. Though some attempts have been made to increase the cap on H-1Bs and avenues for H-1B workers to remain in the United States (such as Republican Senator Orrin Hatch's "I-squared" bill of 2015[66]), these efforts have been overshadowed by greater focus on limitations and enforcement. As outlined above, Canada's reforms paired greater restrictions with the creation of additional streams to facilitate entry for specific categories of workers and perhaps points to one reason that Canada's reforms have rolled out without controversy.

Although the Trump administration may also have signalled a "stick-and-carrot" approach similar to Canada's for reforming the H-1B program, there have been few substantive changes to the program since. In April 2017 President Trump issued an executive order entitled "Buy American and Hire American," which contains two directives related to the H-1B program: (1) strengthen the immigration system to prioritize American workers for jobs, and (2) propose reforms to the H-1B visa program that prioritize the most high-skilled and highest-paid workers.[67] In 2018, reports emerged that more H-1B applications were denied than in the past, suggesting that the "stick" objective had been prioritized by U.S. officials.[68]

However, the second component of the executive order suggests there may also be openness to introducing some policy "carrots" to facilitate entry for high-skilled workers. For example, the lottery structure for the H-1B visa is deeply unpopular and many stakeholders have called for its elimination.[69] The United States could pursue a points-based H-1B visa system similar to Canada's Global Talent Stream whereby only the highest-skilled, highest-paid immigrants are offered H-1B spots. This would satisfy Trump's direction to U.S. officials to "suggest reforms to help ensure that H-1B visas are awarded to the most-skilled or highest-paid petition beneficiaries."[70] It would also discourage outsourcing firms who typically pay less and hire lower-skilled employees from accessing the visas in bulk. Many of the controversies surrounding abuse of the H-1B program, the reported dislocation of American workers, and the risk that the program drives down American wages have been concentrated in this category of employers.[71] Progress on the president's executive order has been incremental. In 2019, the Trump administration revised one minor aspect of the H-1B registration process to increase the likelihood that applicants with graduate degrees would be selected in the lottery over lower-skilled applicants.[72]

It is also important to generate buy-in from interested stakeholders in system changes. Employers who are genuinely focused on using temporary entry programs to address labour shortages are likely to support reforms that prioritize the highest-skilled and highest-paid workers and recognize the need to maintain political and public confidence in the H-1B program. For example, in Canada, specialized sectors such as the video gaming industry have worked constructively with government to find solutions that satisfy the sector's need to resolve labour shortages quickly by accessing foreign workers while also addressing the country's desire to prepare more Canadians to fill the positions over the long term.[73] Similarly, Microsoft founder Bill Gates has been particularly vocal on the need to reform the H-1B visa program in the United States.[74] Politicians of all stripes – in both countries – have generally responded well to this group of employers and the frustrations they experience accessing temporary entry programs.

The Canadian combination of *both* restrictive *and* facilitative reforms may offer an important path to break congressional deadlock in the United States over these issues. By working to open up quick entry pathways to qualified employers, the United States could gain greater industry support from some of the country's most high-profile ICT companies. Linking facilitative reforms to greater efforts to also strengthen enforcement and restrict access to jobs with demonstrated labour shortages could offer lawmakers with a total package that could potentially satisfy both sets of political interests. Nevertheless, U.S. polarization between Republicans and Democrats has made bipartisan cooperation on any initiatives almost impossible. Though Obama also faced great difficulty achieving bipartisan support for many of his initiatives, the Trump administration's blatant anti-immigration posture and the forced separation of families crossing the U.S. border illegally in 2018 make it difficult to imagine how the current administration could make the compromises necessary to generate sufficient bipartisan support for any meaningful reforms to the H-1B program.

Linking Temporary Foreign Work to Permanent Residence

Although this chapter has focused primarily on the short-term relocation of highly skilled foreign workers, the broader immigration context also plays a key role in both countries. Canada and the United States have very different immigration systems. In Canada, successive Conservative and Liberal governments have implemented deliberate economic migration strategies to attract more skilled workers to become citizens. Relatively speaking, foreign workers with advanced degrees in sectors with established labour shortages find it straightforward to move to Canada permanently. This has the effect of reducing both demand on the TFW program and the likelihood of gaming among would-be permanent immigrants.

By contrast, the U.S. immigration system is not merit-based, but is instead oriented to country-of-origin quotas and family reunification. This emphasis can result in very long wait lists for permanent residency in the United States, even for skilled workers. For example, Pierce and Gelatt[75] have outlined how many H-1B visa holders are Indian nationals who wait more than a decade for green cards. Unlike in Canada, there is also an absence of political consensus among U.S. lawmakers on whether pursuing an economic migration policy is even desirable, and the debate over whether to focus the American system more deliberately on economic migration remains highly controversial.[76]

The precarious situation faced by such foreign workers exacerbates concerns about the "indentured" nature of long-term H-1B work, which may place downward pressure on wages for the foreign nationals themselves as well as more broadly in the sectors that employ them. It is for this reason that some lawmakers advocate eliminating penalties for H-1B visa holders who leave their jobs with sponsoring employers.[77] Although Canada did not de-link foreign workers from employers in reforms to its own TFW program, the government announced in 2019 that it is considering this change.[78]

Broad societal support for immigration can also offer positive spillover effects on attitudes towards temporary foreign worker programs in the two countries. Historically, Canada has been insulated from large-scale flows of unregulated migrants at its borders. This has been partially the result of geography (a strong border to the south and oceans on all other sides) but also due to a carefully designed, merit-based system of immigration that helps maintain public confidence in the value of immigration to Canadian society.[79] Still, the uptick of refugee claimants crossing the Canada-U.S. border, particularly to Quebec, since 2017 has revealed Canada's incapacity to manage large flows of illegal migration across its southern border, and may test the political consensus in Canada about the benefits of immigration. Other countries such as Germany and the United Kingdom have experienced declines in public support for immigration following such movements of unauthorized migrants.[80] In the United States, a cumbersome immigration system and the presence of millions of unregulated workers have contributed to negative public perceptions about immigration generally.[81] Negative public opinion about immigration can then potentially undermine the ability of lawmakers to improve specific programs such as the H-1B visa system, since it is politically easier to dismiss all migration programs as inherently problematic, regardless of the policy reality.[82]

Looking Forward: Challenges for Bilateral Relations

The divergent policy paths followed by the United States and Canada towards temporary entry may also have consequences for bilateral relations. Through its reforms, Canada has become an easier country for high-skilled workers to access temporarily and permanently. This reality may be something of a two-edged sword. On the one hand, it may serve to reverse the "brain drain" of Canadians to the United States. However, on the other, it may incentivize companies and foreign workers whose ultimate goal is to move to the United States to instead use Canada as a temporary landing pad. Each issue is discussed in turn below.

Reversing the Brain Drain

An increasing body of evidence suggests that global firms in the digital sectors are making investment location decisions based on the ability to move high-skilled labour in and out of the proposed host country.[83] Even prior to Donald Trump's election to the U.S. presidency, several large American ICT firms established offices in Canada, primarily in Vancouver and Toronto. According to congressional testimony offered by Microsoft founder Bill Gates, cumbersome U.S. immigration policies influenced those decisions.[84] Clearly, immigration policies have not been the only factor in this northern migration of high-tech firms. More than a decade of investment by provincial and federal governments has helped create attractive climates for high-tech companies to flourish, and several Canadian universities have supplied high-quality gradu-ates to the American ICT sector for decades.[85]

Nevertheless, since Trump's election, anecdotal evidence suggests there has been an acceleration of movement north to Canada, particu-larly among foreign workers in the United States whose H-1B status has been threatened by the president.[86] Canada appears to be a net ben-eficiary of the perceived clamp-down and has sought to capitalize on the uncertainty in the United States through such initiatives as the "Go North" campaign in Silicon Valley. In addition to foreign workers in the United States on H-1B visas, some of the workers attracted north are Americans as well as Canadians returning home.[87] Canada has also experienced a major uptick in the number of international students ap-plying to study at Canadian universities since Trump's election.[88]

Canada as a Temporary Landing Pad for U.S. Immigration

The divergent policy paths towards temporary entry followed by the United States versus Canada may also have other unanticipated conse-quences. Since implementing its reforms, Canada offers a much more nimble and efficient suite of programs to relocate high-skilled workers temporarily. Canada's reforms have not gone unnoticed by global com-panies in the ICT sector.[89] This raises the concern that employers could exploit policy differences between the two countries in order to use Can-ada as a temporary way station for foreign workers ultimately destined for the United States. For example, the 1995 GATS agreement enables companies to move employees into both countries as intra-corporate transferees, allowing them to by-pass national economic needs tests. In order to qualify as an intra-corporate transferee, workers must be continuously employed by their firm for a minimum of one year prior to relocating.[90] Now that Canada has reformed its temporary worker programs and streamlined entry for high-skilled workers, companies have several avenues to hire new foreign workers into Canadian jobs quickly. After a year or more of training at the company in Canada, those employees can then move to the United States as intra-corporate transferees without resorting to the H-1B lottery system.

Thus far, there is a dearth of research evidence on this phenomenon. Nevertheless, the building blocks are now in place for companies to use Canada's programs to bypass the U.S. H-1B program. For example, in 2017, the Microsoft Centre of Excellence in Canada was opened in Van-couver after receiving an exemption from the federal government from the normal requirements of the TFW program.[91] The Microsoft Centre serves as an international training location for 400 new Canadian and international Microsoft hires who receive training for a year or more in Vancouver before being posted to jobs internationally. Conceivably, these employees could then be relocated to the United States or to any other GATS member country as intra-corporate transferees.

Even if companies do seek to use Canada as a temporary landing pad to train high-skilled workers for ultimate placement elsewhere, the degree to which this represents a public policy problem is debatable. For Canada, the outcome of Microsoft's co-location in Vancouver has been the establishment of an important high-tech hub with spin-off companies establishing nearby, and more than $90 million in direct annual investments in the city.[92] Microsoft could have established a similar centre in any number of countries, and Canadian governments at all levels worked flexibly and creatively to bring those benefits to Vancouver.

However, if Microsoft established its training centre in Canada primarily as a response to the Unite States' intransigent labour mobility policies, it represents significant economic and efficiency losses for the United States. Additional moves north by other companies could lead to bilateral tension between the two countries' governments.[93] It could also exacerbate the well-known problem of Canada's private sector being regarded as a gateway to the U.S. market, rather than a permanent location for long-term growth.

Of course, attractive policies for skilled migration represent only one factor in location decisions of global companies such as Microsoft and Amazon. Attractive tax treatment by national and subnational authorities, local factors such as climate, infrastructure, liveability, affordability, and proximity to an international airport have all been named as factors influencing location decisions.[94] U.S. corporate tax reform and a pro-business environment in the United States have already had a negative impact on Canada's ability to attract foreign investment.[95] Further research in this area is recommended.

Conclusion

Certainty of supply of skilled labour is becoming an increasingly important factor in the modern globalized workplace. As trade in services represents the lion's share of expected growth in employment and global trade in the coming years, countries are rethinking their international commitments to labour mobility. Global high-tech companies in the United States that have come to rely heavily on trained foreign labour are increasingly looking north to Canada to establish satellite campuses where open immigration policies allow them to quickly hire and train foreign workers.

This chapter has evaluated reforms to high-skilled migration programs in Canada that have enabled such movements of workers. Thus far, such policies have remained relatively uncontroversial in Canada, largely as the result of its more open approach to immigration and the economic success of high-skilled immigrants in the country. The chapter also reveals that Canada can offer workable solutions for the United States to address problems in its own high-skilled migration programs. However, political appetite to overhaul the program in a meaningful and forward-focused fashion may be the real challenge for the United States, a domestic challenge that only Americans can confront.

NOTES

The author gratefully acknowledges research funding support from the Social Sciences and Humanities Research Council.

1 Klaus Schwab, "The Fourth Industrial Revolution: What It Means and How to Respond," *Foreign Affairs,* 12 December 2015. https://www.foreignaffairs.com/articles/2015-12-12/fourth-industrial-revolution.

2 United Nations, *International Migration Report 2015: Highlights* (New York: United Nations Department of Economic and Social Affairs, Population Division, 2016).

3 Ian Bremmer, *Us vs. Them: The Failure of Globalism* (New York: Portfolio/Penguin, 2018).

4 Jens Hainmueller, Michael J. Hiscox, and Yotam Margalit, "Do Concerns about Labour Market Competition Shape Attitudes toward Immigration? New Evidence," *Journal of International Economics* 97, no. 1 (2015): 193–207.

5 United Nations, *International Migration Report.*

6 Barry Edmonston, "Canada's Immigration Trends and Patterns," *Canadian Studies in Population* 43, no. 1–2 (2016): 78–166.

7 Meredith B. Lilly, "How Demographic Transition Can Help Predict Canada–US Trade Relations in 2042," *International Journal of Canadian Studies* 55 (2017): 67–76, https://doi.org/10.3138/ijcs.55.07.

8 Katherine Richardson, "Attracting and Retaining Foreign Highly Skilled Staff in Times of Global Crisis: A Case Study of Vancouver, British Columbia's Biotechnology Sector," *Population, Space and Place* 22, no. 5 (2016): 428–40.

9 Richard Florida, Kathrine Richardson, and Kevin Stolarick, "Locating for Potential: An Empirical Study of Company X's Innovation Centre in Vancouver, British Columbia," Martin Prosperity Institute Working Paper Series: Ontario in the Creative Age, REF. 2009-WPONT-020, October 2009, http://martinprosperity.org/media/pdfs/Locating_for_Potential

-Richardson.pdf; Information Technology Association of Canada, "The Importance of Global Workers in Canada's ICT and Digital Media Industries," January 2014, http://theesa.ca/wp-content/uploads/2015/08/ITAC-white-paper.pdf.

10 Meredith B. Lilly, "Advancing Labour Mobility in Trade Agreements," *Journal of International Trade Law and Policy* 18, no. 2 (2019): 58–73.

11 Margaret L. Usdansky and Thomas J. Espenshade. "The H-1B Visa Debate in Historical Perspective: The Evolution of U.S. Policy toward Foreign Born Workers," UC San Diego Working Papers, 2000, 3, https://escholarship.org/uc/item/8qf435d5.

12 Graeme Hugo, "Demographic Change and International Labour Mobility in Asia-Pacific – Implications for Business and Regional Economic Integration: Synthesis," in *Labour Mobility in the Asia-Pacific Region: Dynamics, Issues and a New APEC Agenda*, ed. Graeme Hugo and Soogli Young, 1–62 (Singapore: Institute of Southeast Asian Studies, 2008); Marion Panizzon, "Standing Together Apart: Bilateral Migration Agreements and the Temporary Movement of Persons under 'Mode 4' of GATS," *Centre on Migration, Policy and Society*, Working Paper No. 77 (University of Oxford, 2010), https://www.wti.org/media/filer_public/6b/e7/6be710fa-b343-447a-998e-c90aeac5ea83/wp1077_marion_panizzon_2.pdf.

13 Dominique Gross, "Temporary Foreign Workers in Canada: Are They Really Filling Labour Shortages?" C.D. Howe Institute Commentary No. 407 (April 2014), https://www.cdhowe.org/sites/default/files/attachments/research_papers/mixed//commentary_407.pdf.

14 Julie R. Watts, "The H-1B Visa: Free Market Solutions for Business and Labor," *Population Research and Policy Review* 20 (2001): 143–56; Grace Martinez, "Legal Immigrants Displacing American Workers: How U.S. Corporations Are Exploiting H-1B Visas to the Detriment of Americans," *UMKC Law Review* 86, no. 1 (2017): 209–36.

15 Usdansky and Espenshade, "H-1B Visa Debate," 6.

16 Usdansky and Espenshade, "H-1B Visa Debate," 7.

17 Usdansky and Espenshade, "H-1B Visa Debate"; Watts, "H-1B Visa"; Martinez, "Legal Immigrants Displacing American Workers."

18 Usdansky and Espenshade, "H-1B Visa Debate"; Watts, "H-1B Visa."

19 United States, Department of Labor, "Fact Sheet #62A: Changes Made by the H-1B Visa Reform Act of 2004," last modified July 2008, https://www.dol.gov/whd/regs/compliance/FactSheet62/whdfs62A.pdf.

20 Chad Sparber, "Choosing Skilled Foreign-Born Workers: Evaluating Alternative Methods for Allocating H-1B Work Permits," *Industrial Relations* 57, no. 1 (2018): 3–34.

21 United States, Citizenship and Immigration Services, "Number of H-1B Petition Filings, Applications and Approvals, Country, Age, Occupation,

Industry, Annual Compensation ($), and Education, FY2007–FY2017," US Citizenship and Immigration Services Claim 3, 15 November 2017.

22 Sarah Pierce and Julia Gelatt, "Evolution of the H-1B: Latest Trends in a Program on the Brink of Reform," Migration Policy Institute, March 2018.

23 Ana M. Ferrer, Garnett Picot, and William Craig Riddell, "New Directions in Immigration Policy: Canada's Evolving Approach to the Selection of Economic Immigrants," *International Migration Review* 48, no. 3 (1 September 2014): 846–67, https://doi.org/10.1111/imre.12121.

24 Ather H. Akbari and Martha MacDonald, "Immigration Policy in Australia, Canada, New Zealand, and the United States: An Overview of Recent Trends," *International Migration Review* 48, no. 3 (2014): 801–22, https://doi.org/10.1111/imre.12128; Harriet Duleep and Mark Regets, "U.S. Immigration Policy at a Crossroads: Should the U.S. Continue Its Family-Friendly Policy?," *International Migration Review* 48, no. 3 (2014): 823–45, https://doi.org/10.1111/imre.12122; Ferrer, Picot, and Riddell, "New Directions in Immigration Policy"; Gary P. Freeman, "Can Liberal States Control Unwanted Migration?" *ANNALS of the American Academy of Political and Social Science* 534, no. 1 (1994): 17–30, https://doi.org/10.1177/0002716294534001002.

25 Ferrer, Picot, and Riddell, "New Directions in Immigration Policy."

26 Gross, "Temporary Foreign Workers in Canada"; Canada, Employment and Social Development Canada, "Overhauling the Temporary Foreign Worker Program," 2014, https://www.canada.ca/content/dam/canada/employment-social-development/migration/documents/assets/port-folio/docs/en/foreign_workers/employers/overhauling_TFW.pdf; Canada, Parliament, House of Commons, Standing Committee on Human Resources, Skills and Social Development and the Status of Persons with Disabilities (HUMA), *Temporary Foreign Worker Program*, 42nd Parliament, 1st Session, September 2016, https://www.ourcommons.ca/Content/Committee/421/HUMA/Reports/RP8374415/humarp04/humarp04-e.pdf.

27 Hugo, "Demographic Change"; Julia Nielsen, "Labor Mobility in Regional Trade Agreements," in *Moving People to Deliver Services*, ed. Aaditya Mattoo and Antonia Carzaniga, 93–112 (Washington, DC: World Bank, 2003); Laura R. Dawson, "Labour Mobility and the WTO: The Limits of GATS Mode 4," *International Migration* 51, no. 1 (2013): 1–23.

28 Lilly, "Advancing Labour Mobility in Trade Agreements."

29 Lilly, "Advancing Labour Mobility in Trade Agreements."

30 Canada, Employment and Social Development Canada, "Overhauling the Temporary Foreign Workers Program," 2014, https://www.canada.ca/content/dam/canada/employment-social-development/migration/documents/assets/portfolio/docs/en/foreign_workers/employers/overhauling_TFW.pdf.

31 Canada, Immigration, Refugees and Citizenship Canada, "International Mobility Program: Canadian Interests – Significant Benefit General Guidelines [R205(a) – C10]," last modified 18 September 2014, https://www.canada.ca/en/immigration-refugees-citizenship/corporate/publications-manuals/operational-bulletins-manuals/temporary-residents/foreign-workers/exemption-codes/canadian-interests-significant-benefit-general-guidelines-r205-c10.html.
32 Edmonston, "Canada's Immigration Trends and Patterns"; Magnus Lofstrom and Joseph Hayes, "H-1Bs: How Do They Stack Up to US Born Workers?" IZA Discussion Paper no. 6259, December 2011, http://ftp.iza.org/dp6259.pdf; Kenneth Geisler, "Fissures in the Valley: Searching for a Remedy for U.S. Tech Workers Indirectly Displaced by H-1B Visa Outsourcing Firms," *Washington University Law Review* 95, no. 2 (2017): 465–506.
33 Martinez, "Legal Immigrants Displacing American Workers."
34 Norman Matloff, "Immigration and the Tech Industry: As a Labour Shortage Remedy, for Innovation, or for Cost Savings," *Migration Letters* 10, no. 2 (2013): 211–28.
35 John Bound, Gaurav Khanna, and Nicholas Morales, "Understanding the Economic Impact of the H-1B Program on the U.S.," NBER Working Paper no. 23153, February 2017, http://www.nber.org/papers/w23153.
36 Ronil Hira, "Immigration Reforms Needed to Protect Skilled American Workers." Testimony in a hearing before the U.S. Senate Judiciary Committee, 17 March 2015, https://www.judiciary.senate.gov/imo/media/doc/Hira%20Testimony.pdf; Ron Hira, "Candidates' Plans to Change Controversial H-1B Guestworker Program Highlight Need for an Overhaul," *Conversation*, 1 March 2016, https://theconversation.com/candidates-plans-to-change-controversial-h-1b-guestworker-program-highlight-need-for-an-overhaul-55482.
37 Martinez, "Legal Immigrants Displacing American Workers."
38 Martinez, "Legal Immigrants Displacing American Workers."
39 Canada, Employment and Social Development Canada, "Overhauling the Temporary Foreign Worker Program"; CBC, "B.C. Mine's Temporary Foreign Workers Case in Federal Court," CBC News, 3 April 2013, http://www.cbc.ca/news/canada/british-columbia/b-c-mine-s-temporary-foreign-workers-case-in-federal-court-1.1374502; CBC, "RBC Publicly Apologizes to Employees Affected by Outsourcing Arrangement," CBC News, 6 April 2013, http://business.financialpost.com/news/fp-street/rbc-apology-outsourcing.
40 Canada, Immigration, Refugees and Citizenship Canada, "Speaking Notes for Jason Kenney, P.C., M.P., Minister of Citizenship, Immigration and Multiculturalism," 2 November 2012, https://www.canada.ca/en/immigration-refugees-citizenship/news/archives/speeches-2012/jason-kenney-minister-2012-11-02.html.

41 Canada, Employment and Social Development Canada, "Overhauling the Temporary Foreign Worker Program."

42 Canada, Employment and Social Development Canada, "Overhauling the Temporary Foreign Worker Program."

43 Christine Knott, "Contentious Mobilities and Cheap(er) Labour: Temporary Foreign Workers in a New Brunswick Seafood Processing Community," *Canadian Journal of Sociology* 41, no. 3 (2016): 375–98; Ethel Tungohan, "From Encountering Confederate Flags to Finding Refuge in Spaces of Solidarity: Filipino Temporary Foreign Workers' Experiences of the Public in Alberta," *Space and Polity* 21, no. 11 (2017): 11–26; Arthur Sweetman and Carsey Warman, "Canada's Temporary Foreign Workers Programs," *Canadian Issues* (Spring 2010): 19–24; Hugo, "Demographic Change."

44 Hugo, "Demographic Change."

45 Panizzon, "Standing Together Apart," xlvii.

46 Panizzon, "Standing Together Apart"; Canada, Employment and Social Development Canada, "Overhauling the Temporary Foreign Worker Program."

47 Canada, Immigration, Refugees and Citizenship Canada, "International Mobility Program: Federal-Provincial or Territorial Agreements [R204(c)] (LMIA exemption code T13)," last modified 13 October 2017, https://www.canada.ca/en/immigration-refugees-citizenship/corporate/publications-manuals/operational-bulletins-manuals/temporary-residents/foreign-workers/exemption-codes/federal-provincial-territorial-agreements-r204-lmia-exemption-code-t13.html.

48 Lilly, "Advancing Labour Mobility in Trade Agreements."

49 Canada, Employment and Social Development Canada, "Overhauling the Temporary Foreign Worker Program."

50 Canada, Immigration, Refugees and Citizenship Canada, "Speaking Notes for the Honourable Jason Kenney."

51 Canada, Immigration, Refugees and Citizenship Canada, "Who Can Apply: Canadian Experience Class (Express Entry)," last modified 14 October 2017, https://www.canada.ca/en/immigration-refugees-citizenship/services/immigrate-canada/express-entry/become-candidate/eligibility/canadian-experience-class.html?_ga=2.254487425.880335319.1512612908-900327754.1512612908.

52 Canada, Immigration, Refugees and Citizenship Canada, *Express Entry Year End Report 2015*, last modified 15 March 2017, https://www.canada.ca/en/immigration-refugees-citizenship/corporate/publications-manuals/express-entry-year-end-report-2015.html.

53 Canada, Immigration, Refugees and Citizenship Canada, "Who Can Apply."

54 Office of the Auditor General of Canada, *Report 5: Temporary Foreign Worker Program – Employment and Social Development Canada, 2017*, http://www.oag-bvg.gc.ca/internet/English/parl_oag_201705_05_e_42227.html.

55 Philip Cross, "Liberals Expand TFW Program as Proof of Mess Arrives," *Financial Post*, 15 June 2017, http://business.financialpost.com/opinion /philip-cross-the-liberals-expand-the-temporary-foreign-workers -program-just-as-proof-arrives-of-what-a-mess-its-become.

56 Information Technology Association of Canada, "Importance of Global Workers"; Sarah Anson-Cartwright, "Immigration for a Competitive Canada: Why Highly Skilled International Talent is at Risk," Canadian Chamber of Commerce, 14 January 2016, http://www.chamber.ca/media /blog/160114-immigration-for-a-competitive-canada/.

57 Canada, Parliament, House of Commons, Standing Committee on Human Resources, Skills and Social Development, *Temporary Foreign Worker Program*.

58 Canada, Employment and Social Development Canada, "Program Requirements for the Global Talent Stream," https://www.canada.ca/en /employment-social-development/services/foreign-workers/global -talent/requirements.html.

59 Canada, Department of Finance, *Budget 2017: Building a Strong Middle Class* (Ottawa, 2017), https://www.budget.gc.ca/2017/docs/plan/budget- 2017-en.pdf; Canada, Employment and Social Development Canada, "Program Requirements for the Global Talent Stream."

60 Martinez, "Legal Immigrants Displacing American Workers."

61 Many of the bills tabled in the 115th Congress have been tabled in previous sessions as well.

62 U.S. Congress, Senate, *H-1B and L-1 Visa Reform Act of 2017*, S. 180, 115th Cong., 1st sess., introduced in Senate 20 January 2017, https://www .congress.gov/115/bills/s180/BILLS-115s180is.pdf.

63 U.S. Congress, House, *H-1B and L-1 Visa Reform Act of 2017*, H.R. 1303, 115th Cong., 1st sess., introduced in House 2 March 2017, https://www .congress.gov/115/bills/hr1303/BILLS-115hr1303ih.pdf.

64 U.S. Congress, House, *American Jobs First Act of 2017*, H.R. 2233, 115th Cong., 1st sess., introduced in House 28 April 2017, https://www.con- gress.gov/115/bills/hr2233/BILLS-115hr2233ih.pdf.

65 U.S. Congress, House, *Protect and Grow American Jobs Act*, H.R. 170, 115th Cong., 1st sess., introduced in House 3 January 2017, https://www .congress.gov/115/bills/hr170/BILLS-115hr170ih.pdf.

66 U.S. Congress, Senate, *I-Squared Act of 2015*, S. 153, 114th Cong., 1st sess., introduced in Senate 13 January 2015.

67 White House. "Presidential Executive Order on Buy American and Hire American," 18 April 2017, https://www.whitehouse.gov/presidential -actions/presidential-executive-order-buy-american-hire-american/.

68 Stuart Anderson, "New Evidence USCIS Policies Increased Denials of H-1B Visas," *Forbes*, 25 July 2018, https://www.forbes.com/sites

/stuartanderson/2018/07/25/new-evidence-uscis-policies-increased
-denials-of-h-1b-visas/#5687d8105a9f.

69 United States, Congress, House, Committee on Science and Technology, *Competitiveness and Innovation on the Committee's 50th Anniversary with Bill Gates, Chairman of Microsoft,* 11th Cong., 2nd sess., 2008, serial no. 110-84.

70 White House, "Presidential Executive Order on Buy American and Hire American."

71 Hira, "Immigration Reforms Needed to Protect Skilled American Workers."

72 United States, Department of Homeland Security, "Registration Requirement for Petitioners Seeking to File H-1B Petitions on Behalf of Cap-Subject Aliens" (Federal Register, 31 January 2019), https://www
.federalregister.gov/documents/2019/01/31/2019-00302/registration
-requirement-for-petitioners-seeking-to-file-h-1b-petitions-on-behalf
-of-cap-subject.

73 Information Technology Association of Canada, "Importance of Global Workers."

74 United States, Congress, House, Committee on Science and Technology, *Competitiveness and Innovation.*

75 Pierce and Gelatt, "Evolution of the H-1B."

76 Akbari and MacDonald, "Immigration Policy in Australia, Canada, New Zealand, and the United States"; Duleep and Regets, "U.S. Immigration Policy at a Crossroads."

77 U.S. Congress, House, H.R. 2233.

78 Her Majesty the Queen in Right of Canada, "Notice to Interested Parties — Introducing Occupation-Specific Work Permits under the Temporary Foreign Worker Program," *Canada Gazette Part 1* 153 25, no. 25 (June 22, 2019): 2951–7.

79 Canada, Immigration, Refugees and Citizenship Canada, *2017 Consultations on Immigration Levels, Settlement and Integration – Final Report,* last modified 1 November 2017, https://www.canada.ca/en/immigration
-refugees-citizenship/corporate/transparency/consultations/2017
-consultations-immigration-levels-settlement-integration-final-report
.html.

80 Bremmer, *Us vs Them,* 24.

81 Bremmer, *Us vs Them,* 25.

82 Hainmueller, Hiscox and Margalit, "Do Concerns about Labour Market Competition Shape Attitudes toward Immigration?"

83 Richardson, "Attracting and Retaining Foreign Highly Skilled Staff," 433; Sheng-Jun Xu, "Skilled Labor Supply and Corporate Investment: Evidence from the H-1B Visa Program," SSRN, 1 March 2017, http://dx.doi.org
/10.2139/ssrn.2877241.

84 United States, Congress, House, Committee on Science and Technology, *Competitiveness and Innovation*.

85 Mike McDerment, "Canada Can't Fall Behind in the Global Race for Tech Talent," *Globe and Mail*, 26 June 2017, https://www.theglobeandmail.com/report-on-business/rob-commentary/canada-cant-fall-behind-in-the-global-race-for-tech-talent/article35459472/; Bourree Lam, "Canada Wants Silicon Valley's Tech Employees," *Atlantic*, 9 May 2017, https://www.theatlantic.com/business/archive/2017/05/canada-tech/525930/.

86 Lam, "Canada Wants Silicon Valley's Tech Employees."

87 Lam, "Canada Wants Silicon Valley's Tech Employees"; Joel Rose, "Canada's Tech Firms Capitalize on Immigration Anxiety in the Age of Trump," NPR, 9 June 2017, https://www.npr.org/2017/06/09/532220824/canadas-tech-firms-capitalize-on-immigration-anxiety-in-the-age-of-trump.

88 Susan Svrluga, "'I'll Be in Canada': More Students Are Looking to Head North," *Washington Post*, 27 March 2017, https://www.washingtonpost.com/news/grade-point/wp/2017/03/27/ill-be-in-canada-more-students-are-looking-to-head-north/?noredirect=on&utm_term=.af606834265e.

89 U.S. Congress, House, Committee on Science and Technology, *Competitiveness and Innovation*.

90 Nielson, "Labor Mobility in Regional Trade Agreements."

91 Canada, Immigration, Refugees and Citizenship Canada, "International Mobility Program: Federal-Provincial or Territorial Agreements [R204(c)] (LMIA Exemption cCde T13)," last modified 13 October 2017, https://www.canada.ca/en/immigration-refugees-citizenship/corporate/publications-manuals/operational-bulletins-manuals/temporary-residents/foreign-workers/exemption-codes/federal-provincial-territorial-agreements-r204-lmia-exemption-code-t13.html.

92 Edoardo De Martin, "Opinion: Why Microsoft Chose Vancouver for Its Excellence Centre," *Vancouver Sun*, 17 June 2016, http://vancouversun.com/opinion/opinion-why-microsoft-chose-vancouver-for-its-excellence-centre.

93 U.S. Congress, House, Committee on Science and Technology, *Competitiveness and Innovation*.

94 Amazon, "Amazon HQ2 Request for Proposals," 2017, https://images-na.ssl-images-amazon.com/images/G/01/Anything/test/images/usa/RFP_3._V516043504_.pdf.

95 Bank of Canada, "Canada's Economic Expansion: A Progress Report," remarks by Timothy Lane, deputy governor of the Bank of Canada, 8 March 2018, https://www.bankofcanada.ca/wp-content/uploads/2018/03/remarks-080318a.pdf.

8 Managing Cross-Cutting Interdependencies: Canada's Cross-Border Transportation and Infrastructure Regimes in a North American and International Context

GEOFFREY HALE

The interaction of domestic, cross-border, and wider international transportation regimes with domestic and cross-border infrastructure regimes in all modes of transportation is critical in facilitating trade flows and related private sector investments. At the same time, responsible domestic governance requires integration of major policy concerns with significant, if not predominantly, domestic implications, creating an "intermestic"[1] context for large elements of transportation policy. Key domestic policy preoccupations include political responsiveness to frequently competing regional and local interests in the location and operation of transportation services and related infrastructure, public safety and environmental considerations, and the cost-effectiveness of (and processes for setting priorities in allocating) infrastructure investments, whether private, public, or a combination of both.

Canada's geography has shaped its economic history, the evolution of transportation policies, and contemporary movements of trade and travel, which provide market demand for transportation services and related infrastructure. Historically, these policies have reflected different degrees of regulatory parallelism, cooperation, or coordination informed by four broad realities:[2]

- the extent to which to which market movements in particular sectors and subsectors are subject to domestic and/or international governance and related processes for border management, including the relative symmetry or asymmetries of interdependence;[3]
- the relative institutional symmetry (or asymmetries) and concentration (or diffusion) of legislative and regulatory responsibilities within national (or subnational) governments, and regulatory and administrative agencies;[4]

- degrees of similarity or differentiation between national and subnational government objectives and priorities, including relative unity or diffusion of policy focus: the number and relative importance of policy priorities in the mandates of particular agencies;[5]
- degrees of similarity or differentiation between policy instruments and settings used to implement – and often coordinate – these objectives, and the strengths and weaknesses of mechanisms for intergovernmental coordination.[6]

Although transportation policy regimes evolve slowly in both Canada and the United States, significant changes in policies or in the allocation (and financing) of spending on transportation infrastructure are subject to intense contestation by domestic economic interests. These realities lend themselves to the diffusion of governance and regulatory responsibilities within national and subnational governments in both countries, contributing to higher and more persistent levels of policymaking territoriality in each country unless a common external threat renews incentives for closer cooperation. At the same time, interdependence among transportation networks, while varied across different modes, has contributed to the emergence of transgovernmental and transnational networks[7] to manage these issues in ways that often cultivate reciprocity across national borders.

This chapter explores the context for Canada's transportation and related infrastructure policies and the interaction of territorial institutions and extraterritorial forces in four key subsectors central to the structuring of cross-border transportation and related infrastructure regimes: trucking, railways, airlines, and the major nodes in which they intersect.

The Domestic and Intermestic Contexts for Canada's Transportation Policies

The literature on transportation systems emphasizes their quality as networks facilitating overlapping and competing objectives at multiple levels of analysis.[8] For example, intermodal shipments accounted for 23 per cent of the revenues of Canada's two largest railways in 2018.[9] Transportation hubs centred in major maritime ports – Vancouver, Montreal, Prince Rupert, Halifax – major inland ports, and Canada's eight largest international airports serve as key network nodes. Short-line railways and trucking firms are in direct competition for intermodal business. The dominant business model of international courier and freight-forwarding firms such as United Parcel Service (UPS), Federal Express, and increasingly, Canada Post, is more frequently premised on the growth of electronic (e-)commerce deliveries, raising questions of competitive and fiscal equity between domestic and international businesses, as discussed in chapter 6. Bottlenecks in Canadian and U.S. urban and port infrastructure (and other chokepoints in infrastructure networks), the frequent politicization of major infrastructure projects, and the periodic fragility of key network elements often dwarf border congestion as major sources of risk to the efficiency and resilience of cross-border market and travel movements.[10]

Doern, Prentice, and Coleman[11] note several key "aterritorial" factors contributing to interdependence, policy emulation, and parallelism in Canadian transportation and related infrastructure policies, including continuing effects of "economic deregulation and privatization dynamics" from the 1980s and 1990s, systemic and regulatory pressures arising from trade liberalization (and more recent trade disputes), multilateral and bilateral transportation agencies (e.g., IMO, ICAO, and International Transport Forum), post-9/11 security policies, and interest group pressures to maintain national cabotage restrictions across transportation modes. In many cases, these relationships substantially predate the closer economic integration of the 1990s and post-9/11 security preoccupations, and have helped to shape governmental and societal responses to them.[12] Similarly, the actions and responses of national governments and their respective agencies, and of the businesses and individuals engaged in cross-border economic activity reflect interactive dynamics subject to continuing change – whatever the relative stability of particular sectoral and subsectoral regimes.

These realities are balanced by several major "territorial" factors. Domestic transport and related infrastructure policies are relatively politicized in both countries, reflecting competition over the regional distribution of federal investments, ideological competition over investment priorities (e.g., roads vs. mass transit, integrative vs. competitive approaches to relationships between climate change and economic development), competing fiscal and internal policy priorities ("shovel-readiness" vs. cost-benefit analyses vs. systemic perspectives and priorities),[13] and NIMBYism, sometimes verging on "BANANA-ism."[14]

Transportation planning issues often overlap with jurisdictional complexities of federalism. Interprovincial (and interstate) transportation falls under federal jurisdiction in both countries. Ownership of border infrastructure is more complex, as discussed below. Highway infrastructure "behind borders" is usually owned by provinces (or territories) and states, with provinces accounting for 86 per cent of capital improvements on the "national" highway system of core, feeder, and northern routes in 2006–16[15] (see figure 8.1). By comparison, the U.S.

Figure 8.1. Canada's "National" Highway System

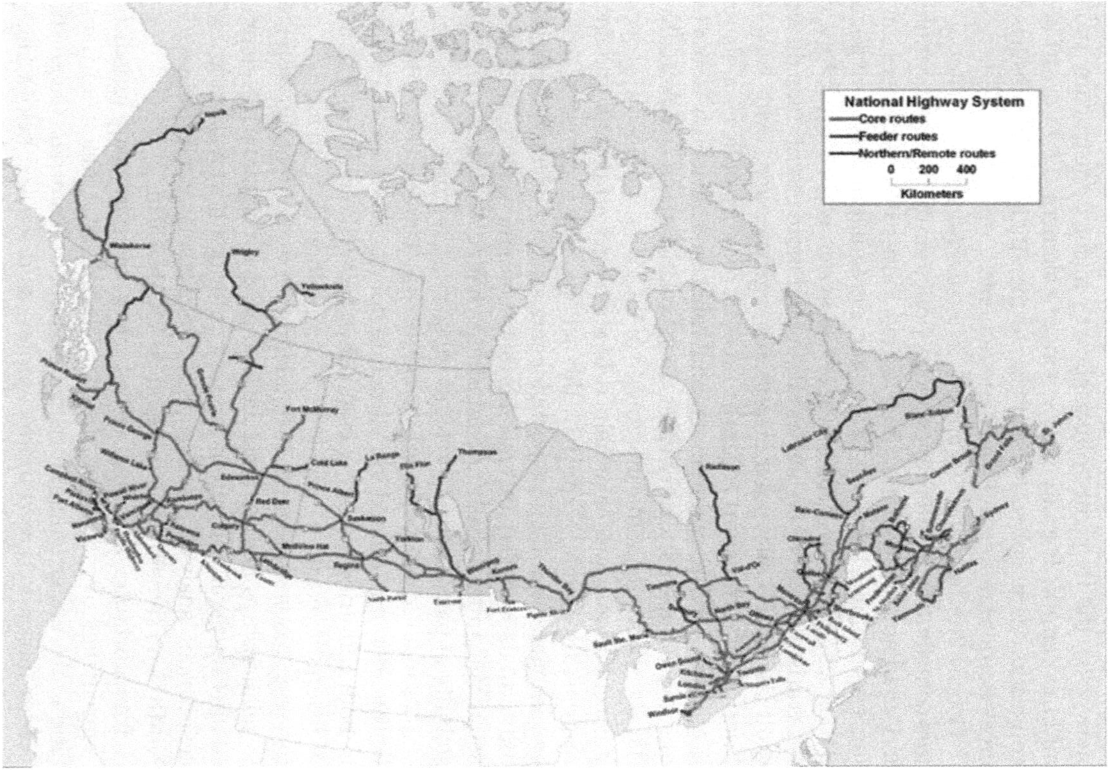

government accounted for 43 per cent of comparable capital spending in 2003–12, with wide variations across states[16] and intense competition within Congress over funding levels and distribution of highway and transit funding. Most freight railways are privately owned in both countries, although passenger rail is generally delivered by federal and subnational government agencies or corporations.

Infrastructure planning is frequently subject to the intensive politicization of budgetary priorities, along with pervasive interest group competition in fields such as grain transportation and land-use planning (the latter usually a mix of provincial/state and municipal responsibilities). A major federal report observes that "a key consideration [is often] to ensure that funds were dispersed on a 'fair share' basis across Canada" and that "the bottom-up approach to project identification left little room for the selection of projects of national scope and strategic importance."[17]

Jurisdiction is often widely shared, requiring cooperation and stakeholder engagement to achieve cross-border consensus. Government policies to attract private capital to complement public sector investments, with related tolls, user fees, competition, and freight rates, continue to evolve in both countries. To facilitate this coordination, the Canadian Transportation Act Review released in 2016 recommended the creation of intermodal and sector-specific strategies and investment plans, as well as processes for defining and organization infrastructure projects of long-term national importance in a Transportation Infrastructure Plan and Projects Pipeline.[18]

Such complications are particularly important in managing major urban and inter-urban transportation planning projects – such as the Canadian government's multi-year Gateway and corridor projects – and major cross-border initiatives such as the Cascades Corridor in the BC Lower Mainland–Puget Sound region or the Detroit River International Crossing project, which has been embroiled in years of acrimonious litigation and political manoeuvring.[19] This section summarizes regulatory contexts for two major elements of border infrastructure: port-of-entry facilities and related transportation infrastructure. It then explores key elements of domestic and cross-border governance of three major transportation subsectors: truck freight, rail freight, and airlines and their implications in balancing domestic and international policy priorities and pressures.

Border Infrastructure

Ownership and regulatory oversight of border infrastructure is heavily segmented. Land port-of-entry (POE) facilities along the Canada-

Table 8.1. Ownership of Cross-Border Bridges and Tunnels

	U.S.			Private*	Canada		
	Federal	State	Municipal		Federal	Provincial	Municipal
Ontario-MN	0	2	–	1	–	2	–
Ontario-MI**	0	2	1	3	3	–	1
Ontario-NY	1	5***	1	–	3	3	–
NB-Maine	0	10	–	–	–	10	–
Total	1	19	2	4	6	15	1

* The Ambassador Bridge (Detroit-Windsor) is privately owned by the Detroit International Bridge Co. (Moroun family) and its Canadian subsidiary. The Fort Frances–International Falls Bridge is jointly owned by Boise Inc. and Resolute Forest Products. Two rail tunnels crossing the St. Clair and Detroit Rivers are privately owned by Canadian National and OMERS Infrastructure Management respectively.
** The Windsor-Detroit Bridge Authority responsible for building and operating the new Gordie Howe International Bridge is a wholly owned Canadian government agency.
*** Ogdensburg Bridge and Port Authority is a New York State agency, which is the sole operator of the Ogdensburg-Prescott Bridge.

U.S. border are typically owned by federal agencies: the Canada Border Services Agency (CBSA) in Canada, the U.S. General Services Administration (GSA), which typically owns, but sometimes leases, larger POE facilities, and the Department of Homeland Security (DHS), which often owns smaller POEs. A handful of land POEs – Sweetgrass/Coutts (MT-AB) and three smaller crossings on the Yukon/Alaska ("Top of World Hwy.") and BC/Washington State (Osoyoos, Carson) borders – are jointly owned by the two governments. Ownership of cross-border bridges and tunnels in the Great Lakes–St. Lawrence and New Brunswick–Maine border regions is idiosyncratic (see table 8.1). Maritime ports are typically governed by federally chartered port authorities in both countries, while terminal facilities with ports often have multiple private owners. Joint governance authorities that operate most bridges in the Great Lakes–St. Lawrence Region usually operate on consensus, although Canadian and U.S. ambassadors had to intervene in a 2013 dispute between New York Governor Cuomo and Canadian members of the Peace Bridge Authority to resolve a dispute that could have resulted in the unilateral dissolution of the Authority.[20]

Port infrastructure is governed by eighteen port authorities, which function as commercial government business enterprises (but not agencies) of the federal government under the Canada Marine Act of 1998.[21] As such they must be fully self-funding. Port authorities frequently contract with private firms or other Crown corporations that manage specialized terminal facilities within their boundaries. However, the Vancouver-Fraser Port Authority has come to dwarf most other Canadian ports, accounting for 45 per cent of port traffic and two-thirds of Port Authority income in 2014–15 – although its traffic volumes are less than one-tenth of those of global mega-ports such as Shanghai or Singapore.[22] Port authorities may also manage other businesses, as with Ports Toronto with Billy Bishop International Airport on Toronto Island. Smaller ports are typically more specialized and may face financial challenges from market or infrastructure disruptions – as with the evolution of federally owned Ridley Terminals at the Port of Prince Rupert since the early 2000s (finally sold in 2019), or the chronically troubled Port of Churchill, Manitoba, and its owner, the Hudson's Bay Railroad, rescued from bankruptcy in 2018.[23]

Port authorities located within major urban areas – particularly Vancouver, Montreal, Halifax, Toronto, and Hamilton – face complex intergovernmental relations in their dealings with provincial, municipal, and sometimes First Nations governments, along with major stakeholders such as railroads. As federal agencies, port authorities are subject to the constitutional "duty to consult" with First Nations whose lands, treaty, and constitutional rights may be affected by shifting patterns of port operations and their environmental impact. At the same time, port authorities compete with one another and with U.S. and Mexican counterparts to provide efficient loading and off-loading capacity as part of global maritime networks dominated by a handful of major shipping companies. The challenges of managing intergovernmental and related public-private sector relations across the nodes and networks of Canada's major transportation corridors are critical to expanding and enhancing their capacity to integrate the management of trade corridors (including maritime ones) with regional and local economic development and quality of life (including environmental) issues across Canada. The federal government's Oceans Protection Plan, announced in 2016, reflects the need to balance regional and local ecosystem protection concerns affecting the interaction of shipping activities with other commercial, tourism, and environmental concerns.[24] As with airport governance, discussed below, Ottawa's need to raise funds to finance new national infrastructure projects has triggered debates over the desirability and feasibility of privatizing port facilities – although federal officials appear to have backed away from these initiatives for the foreseeable future.[25]

Intergovernmental cooperation on border infrastructure has been sporadic, although more systematic among Canadian federal, provincial, and municipal governments through the federal Gateways strategy since 2006. Although the Smart Border Accord helped to improve coordination of cross-border law enforcement after 2001, sizeable U.S.

investments in border infrastructure under the 2009 American Recovery and Reinvestment Act were unilateral and heavily influenced by congressional log-rolling. The Canada-U.S. Beyond-the-Border (BTB) process intended to improve coordination of border security and facilitation has sought to improve coordination of domestic border infrastructure investments after 2011, as seen in successive phases of the Border Infrastructure Investment Plan (BIIP).[26] These measures appear to have improved interdepartmental cooperation *within* central governments while expanding information sharing *between* governments.[27] However, effective cross-border coordination of infrastructure planning appears to be the exception rather than the rule, except in limited cases in which local stakeholders managed to institutionalize such cooperation, as in the BC Lower Mainland–Puget Sound region and along the Maine–New Brunswick border.

One significant challenge facing border agencies is the management of small and remote ports of entry, which account for almost half of land border crossings, each generally processing fewer than fifty cars and ten trucks daily. Key issues involve coordination of hours of service, in consultation with stakeholder communities, investigation of (and overcoming constraints on new investments in) joint location opportunities, potential for and limits on remote processing of travellers, and border and law enforcement agencies' capacity to provide physical back-up in locations at considerable distances from the nearest city or town. BTB pilot projects are investigating the viability of "pre-registration" of local travellers for remote processing.[28]

Trucking

The evolution of trucking regulation in Canada and the United States since the 1980s reflects elements of both territorial and aterritorial governance noted above, mediated by differences in relative centralization of federal regulatory powers. Economic deregulation – the elimination of market entry in intra- and interprovincial trucking and provincial price regulations – in Canada followed similar U.S. initiatives during the 1980s.[29] Palmer suggests federal (interprovincial) truck and rail rate regulation (excluding grain) was fairly modest after the Canadian Transportation Act of 1967, with significant differences in the former's restrictiveness across provinces. Competition from truckers helped to discipline rail freight rates in many subsectors.[30] Operational and safety regulations have been historically more decentralized in Canada, reflecting greater provincial regulatory discretion and enforcement capacity.

In 1980, the U.S. Motor Vehicle Act and Staggers Act effectively eliminated truck and rail rate regulation, while allowing confidential contracts. Canada Transportation Act changes in 1987 (and 1996) paralleled and extended these measures, contributing to significantly increased domestic and cross-border competition, although restrictions remained on foreign ownership of Canadian trucking firms. However, pressures from increased competition and trends towards greater traffic volumes and bigger loads have increased the prominence of safety and environmental regulations, including those related to dangerous goods, with persistent regulatory decentralization often creating internal and cross-border barriers to freight shipments. Rules governing hours of service, permitted truck configurations, and the enforcement of operational regulations have reflected varying degrees of federal-provincial and interprovincial cooperation (and negotiation) within Canada.

Several other "aterritorial" factors help to structure cross-border trucking markets. A 1991 bilateral MOU provided for mutual recognition of truck and bus driver licences. Canada and the United States subsequently achieved "common format and contents for North American driver log-books; reciprocity of medical standards; and substantial harmonization of regulations governing the transportation of hazardous material."[31] However, U.S. protectionist measures against Mexican trucks have remained a major irritant, and Quebec-based truckers report inconsistent recognition of French-language documents in some states. Bissonnette and Vallet note that English proficiency has become increasingly important for Quebec-based trucking firms hiring for cross-border routes since 2011 in response to more assertive patterns of regulatory enforcement of drivers with limited facility in English by U.S. border officers.[32]

Inter-jurisdictional compacts including the International Registration Program (IRP) and International Fuel Tax Agreement (IFTA) provide for prorated apportionment of registration fees and highway taxes in all provinces and forty-eight states based on distances driven in each jurisdiction.[33] After 9/11, U.S. and Canadian governments sought to coordinate trusted trader programs for cross-border shippers and truck drivers, although with continuing cross-border differences for the former. Since 2011, both countries have worked to facilitate cross-border truck freight with the roll-out of electronic manifest processes under Automated Commercial Information (Canadian) / Automated Commercial Environment (U.S.) programs, which provide advance notification of freight shipments and relevant customs documentation. But customs clearance remains unpredictable for small and occasional shippers. Some progress has been made towards enabling "in-bond" transit, allowing for uninterrupted shipments from two points in one country across the territory of the neighbouring country. Ongoing technological changes, including remote monitoring of vehicles through "Intelligent Transportation Systems," the development of autonomous (self-driving) vehicles, and related safety and cybersecurity protocols are ongoing areas of cross-border regulatory cooperation under the Canada-U.S. Regulatory Cooperation Council (RCC).

However, several territorial and "aterritorial" measures contribute to barriers to cross-border trade and "border thickening." Although Canada's western provinces have established common vehicle length, weight, and configuration standards for highway usage, continuing differences across states and provinces remain a longstanding regulatory obstacle for long-haul truck freight. These rules are particularly challenging for over-dimensional loads,[34] as symbolized by the six-year "megaloads" controversy surrounding shipment of oil sands construction equipment from South Korea to northern Alberta via Oregon, Idaho, and Montana.[35] New hours-of-service regulations introduced by the U.S. federal government in 2014–17 have created additional challenges for long-haul truckers, reflecting competing outlooks of major carriers (and their unions), owner-operators, and safety advocates. Trucking firms that have responded to the aging of the long-haul trucking workforce by recruiting drivers from outside North America face continuing challenges as a result of different immigration policies that place restrictions on citizens of third countries.

Finally, despite signing of a bilateral pre-clearance agreement in 2015, border agencies have been unable to navigate differences between national legal systems to secure pre-clearance of truck cargoes entering the United States at high-volume ports-of-entry – a major goal for Canadian exporters. The critical barrier appears to be constitutional limits on the extraterritorial application of U.S. laws to Canadian nationals detained at pre-clearance facilities used to administer U.S. customs laws, and vice-versa.

Railways

Railway regulation in North America remains predominantly national, although there has been significant policy parallelism since the 1980s. Most rail infrastructure is privately owned by individual railways in both countries. Canada restricts foreign ownership of railways, although permitting track-sharing agreements between Canadian and U.S. carriers. However, the Canadian rail system is integrated into a broader continental system spanning the United States and Mexico. Doern et al. observe

Table 8.2. Railway Industry Revenues – Canada

	Canadian National	Canadian Pacific
Transborder	34%	30%
• Southbound (CN 2016)	21%	24%
• Northbound (CN 2016)	12%	8%
Domestic Canada	17%	16%
Domestic United States	15%	15%
Mexico		1%
Global	34%	36%
Asia	na	32%
Europe	na	4%
Total revenue: 2018 (in $bn.)	$14.3	$7.3

Sources: Canadian National Railway, *Together into Our Next Century: 2019 Investor Fact Book* (Montreal: 2019), 2; Canadian Pacific Railway, 2019 *Investor Fact Book* (Montreal: 2019), 8.

that "the North American model of system interrelationships among capacity, congestion, system optimization and levels of service is the overriding feature of Canadian freight rail transportation."[36] System optimization refers to the interaction of shippers, carriers, and other transportation modes and nodes to maximize efficient operation consistent with public safety, system resilience, and environmental sustainability.

Canada's two largest railways, Canadian National and Canadian Pacific, have extensive cross-border and U.S. operations as two of North America's seven Class I railways.[37] Cross-border traffic and freight destined for markets outside North America account for about two-thirds of each carrier's revenues (see table 8.2). Railway shipments are the dominant mode of shipment for long-haul bulk commodities and many industrial goods. However, container traffic accounts for about one-quarter of system revenues.

Relaxation of rate regulation began with 1967 amendments to the Canada Transportation Act, except for grain transportation, which was heavily subsidized until the mid-1990s.[38] U.S. rate deregulation after 1980 reinforced this trend as part of a broader strategy to increase competition and provide revenues necessary for large-scale private investment to expand railway infrastructure.[39] Average Canadian freight rates dropped by one-third after inflation between 1988 and 2013 and have been among the lowest in the world in recent years.[40] Both major railways continue to make large-scale investments in capacity and safety improvements, although short-line railways, which are often provincially regulated, face significant cost pressures in meeting higher safety standards, making capital improvements, and adopting new technologies.[41]

These processes were complemented by the privatization of Canadian National in 1995, the bitter thirty-year political battle among farm groups that culminated in federal legislation ending the Canadian Wheat Board's grain marketing monopoly in 2012,[42] and other regulatory changes contributing to the diversification of Western Canadian agriculture. Transitional challenges following the Wheat Board's demise has contributed to considerable political competition among grain growers, other shippers, and railways to influence federal regulatory policies, reflected in the 2014 Fair Rates for Grain Farmers Act and 2017 amendments to the Canada Transportation Act.[43] Immigration rules require the switching of Canadian and U.S. crews at or near border crossings. These factors have reinforced the primarily domestic political focus of railway policies, while encouraging competition. Increased domestic and cross-border competition have resulted in significantly greater operating efficiencies, while provoking the (unprecedented) hostile takeover of Canadian Pacific in 2012 by an investor group seeking to parallel improvements at Canadian National.[44]

Integration is reinforced by significant parallels in railway regulation between the two countries, although Canada's regulatory regime remains somewhat more interventionist, particularly for grain transportation and accessing long-haul inter-switching for shippers.[45] The U.S. tax system has been more supportive of new investment through faster depreciation rates, even before its 2017 tax reform bill.[46]

The integration of rail networks and shipments across North America requires cooperation among national regulatory authorities, particularly for safety and climate change–related issues. However, many of the challenges required to network management are place-specific, requiring policy responses tailored to particular locales and networks involving domestic stakeholders: governments, shippers, and societal, together with railways themselves.

Both Harper and Trudeau governments have made the formalization of cooperation on new regulatory development, in consultation with shippers, railways, and other stakeholders, an RCC priority since 2011. Key issues range from rail safety and the regulation of dangerous goods shipments, particularly tank car specifications following the 2013 Lac Mégantic disaster, to locomotive emissions regulations.[47]

The greatest challenges to mobility are likely to be regionalized, particularly effective land-use management in and around railway corridors, maritime and inland port facilities; upgrading and replacement of critical infrastructure at potential chokepoints, while addressing effects

Table 8.3. Average Cars per Freight Train, Canada

1990	70.4
1995	62.9
2000	72.2
2005	79.3
2010	90.6
2015	109.6
2017	113.3

Source: Statistics Canada, "Railway Industry Summary Statistics on Freight and Passenger Transportation," table 23-10-0057-01, formerly CANSIM table 404-0016 (Ottawa: 27 July 2019).

of urban traffic congestion resulting from progressively longer trains (see table 8.3); interaction between and possible separation of freight and passenger rail systems, especially in urban areas; and adapting to production surges in particular industry sectors without disrupting overall networks.[48] Implementing these objectives while also enabling development of and cooperation between national transportation systems will remain a major challenge in coming years.

Air Transport

The rapid growth of cross-border and international air travel since the mid-1990s, combined with post-9/11 security measures, has transformed relationships between air transport and border management policies. Air travellers accounted for 37.5 per cent of persons, including returning Canadians, entering Canada in 2018, compared with 12.9 per cent in 1995. Canadians accounted for 65.7 per cent of international travellers in 2018, down since 2013, but up from 55.5 per cent in 1995. U.S. residents and those of other countries have made up variable but comparable shares of passenger arrivals during that period (see table 8.4). Most cross-border and international air travel takes place through four major gateways: Toronto-Pearson, Montreal-Trudeau, Vancouver, and Calgary International Airports (see table 8.5). Edmonton, Winnipeg, Halifax, and Quebec City generate smaller numbers of international passengers. Air freight, which accounted for $125.2 billion in 2016, 14.1 per cent of total Canadian trade (vs. 4.8 per cent of Canada-U.S. trade), is focused on high value commodities and industrial parts such as precious metals, minerals (gold, diamonds), pearls, pharmaceuticals, airplanes, helicopters and their parts.[49]

Table 8.4. International Air Travel Trends: Passenger Arrivals in Canada

		1995	2002	2006	2009	2013	2018
Canadians	Air US	3,939,922	4,750,236	5,600,452	6,156,868	8,102,658	9,634,508
	Air other	3,530,725	4,825,029	6,723,927	8,218,047	9,574,330	11,954,939
Americans	Air	3,028,671	4,228,173	4,175,598	3,472,848	3,995,719	5,471,508
Other	Air	2,962,692	3,818,688	3,848,260	4,052,534	3,817,601	5,780,203
Total		13,462,010	17,622,126	20,348,237	21,900,297	25,490,308	32,841,870
Air travellers as % of all persons entering Canada		12.9	17.7	22.8	27.7	25.9	37.5
Air travellers entering Canada (%)							
Canadians		55.5	54.3	60.6	65.6	69.3	65.7
Americans		22.5	24.0	20.5	15.9	15.7	16.7
Other		22.0	21.7	18.9	18.5	15.0	17.6

Source: Canada, Statistics Canada, "International Travelers Entering or Returning to Canada, by Type of Transport," CANSIM table #24-10-0041-01 (formerly CANSIM table 427-0001) (July 2019).

Table 8.5. Key Air Passenger Markets (Incoming/Outgoing), 2016 (%)

	United States	Europe	Asia	Other*
Toronto	44	23	9	29
Montreal	35	35	1	29
Vancouver	49	14	29	9
Calgary	65	18	2	15

* Include Mexico, Cuba, Dominican Republic, Australia, New Zealand
Source: Transport Canada, *Transportation in Canada: 2016* (Ottawa: 2017).

Table 8.6. Key Elements of Aterritorial, Territorial Governance Systems

Aterritorial	Territorial
Governance	
ICAO/ IATA	Transport Canada
Warsaw Convention	* market access (licensing/cabotage)
International passport regime	* safety, * consumer protection (provincial)
Open Skies (bilateral, reciprocal)	Airport security (CATSA)
Air travel security regimes	Airport ownership / fiscal obligations
	Air navigation (NavCan)
Immigration (reciprocal visa agreements)	
Electronic travel advisory requirements	
Markets	
International airline alliances	Foreign ownership restrictions
Ticketing/Code-sharing agreements	Domestic duopoly + regional carriers
Cross-border airport competition	Labour relations (federal)
	Local airport authorities (monopoly rents)

* Subset of Transport Canada

Air transport policies are complicated by four major interrelated objectives and sets of processes, particularly in the relationships between major domestic and international governance regimes outlined in table 8.6. First, Canadian policies have sought to balance public accessibility to safe, relatively competitive air travel with the economic viability of Canadian airlines since economic deregulation and Air Canada's phased privatization after 1987, as the new regime experienced extended transitional challenges. A 2005 U.S. government report notes that "the airline industry is characterized by high fixed costs, cyclical demand for its services, intense competition, and vulnerability to external shocks," including fluctuating fuel costs, interest rates, and geopolitical instability, making airlines "more prone to failure than many other businesses,"[50] including Canada's then two largest carriers: Canadian Airlines International (2000), and Air Canada (2003–4), and most major U.S. carriers.[51] These trends have fostered growing industry concentration and oligopolistic competition in both countries. In 2016, Air Canada, Westjet and their regional affiliates accounted for 92 per cent passenger "seat-kilometres,"[52] resulting in minimal competition on many intra-regional routes.[53] Air Canada accounted for 49 per cent of the transborder market in 2018, with an additional 10 per cent for its codeshare partner, United Airlines, while codeshare partners Westjet and Delta report a further 28 per cent share.[54] Onex Corp.'s acquisition of Westjet and Air Canada's proposed takeover of Air Transat in 2019–20 parallel trends towards industry consolidation in the United States and elsewhere, although the latter has faced prolonged scrutiny from both Canadian and European competition regulators.[55]

Second, domestic regimes are embedded within wider international governance and market regimes and networks characterized by national government intervention, reciprocal and sometimes multilateral cooperation on market access, security, the regulation of international migration, and persistent asymmetries between Canadian and U.S. markets. Regimes governing air travel have long been regulated through a combination of international (including bilateral) agreements and domestic legislation and regulatory instruments. The Warsaw Convention (1929) and subsequent Montreal Convention (1999) are international agreements, the latter negotiated through ICAO, that regulate liability for commercial carriage of persons, luggage, or goods.[56] They are part of extensive international standards that have evolved since the 1920s governing aspects of air navigation that range from personnel licensing through meteorological services, to aircraft operation, air traffic services, air accident inquiries, environmental protection, aircraft security, safety management, and transportation of dangerous goods by air.[57] Individual countries may adopt supplementary standards, which often evolve into recommended best practices under international "soft law," until adopted as international standards by ICAO.

ICAO also coordinates international standards for air traffic control (ATC), although the implementation of such standards remains under the control of national governments or their designated agencies. Canada privatized responsibility for air navigation in 1995, under the authority of Nav Canada, a private, non-profit agency managed by a stakeholder-dominated board, and funded through surcharges on airline tickets and revenues from external activities.

Air traffic control in industrial countries is frequently managed by government-owned, arm's-length corporations, although the U.S. Federal Aviation Administration combines traditional civil aviation regulatory functions with air navigation services, and Britain's NATS (formerly National Air Traffic Service) is a government-controlled public-private partnership that also manages Spain's ATC system while providing extensive international consulting services.[58] Such functions necessarily involve protocols for routine cooperation among national agencies in managing movements from the airspace of one country to another.

Nav Canada is also the lead investor in Aireon LLC, a U.S.-based firm leading the development of satellite air navigation system to enable real-time monitoring of flights over ocean regions.[59] This project, based on networks of quasi-public and private actors integrating new technologies with ground-based ATC systems subject to approval by national regulators, adds a new dimension to the intermestic governance of air travel. Aireon investors and clients include air navigation agencies in at least twenty-eight countries, along with major aircraft manufacturers and airlines. Anticipated benefits include more effective air traffic control at substantial distances from "national" airspace, accommodating growing volumes of commercial aircraft.[60]

Canada's airline regime, while increasingly market-based, remains perched between the predominantly market-driven U.S. and European Union regimes, and heavily subsidized state airlines in other global regions. Canada's relatively small, widely dispersed population, limited competition on many routes, and predominantly user-pay system for airport management and security have resulted in rather higher airfares than in more densely populated countries. Competition is also constrained by foreign ownership limits, although the use of variable voting shares has permitted some dilution of these limits.[61] The Transportation Modernization Act, authorized by Parliament in 2018, increased airlines' foreign ownership limits to 49 per cent, while retaining 25 per cent limits for any single international investor or combination of international air carriers.[62]

These distinctions, along with Canada's extensive geography and widely dispersed populations outside its three or four largest markets, complicate efforts to negotiate bilateral, reciprocal "Open Skies" agreements in the absence of WTO subsidy disciplines.[63] They also sustain preserving political incentives for cabotage restrictions preventing foreign-based carriers from carrying travellers (or freight) between different points within other countries' domestic markets, as noted in chapter 9. In responses, major Canadian carriers have developed international alliances with foreign-based airlines to facilitate network development through code-sharing on ticket sales, enabling reciprocal passenger access to one another's flights and airline hubs. However, the federal Competition Bureau effectively vetoed a proposal by Air Canada and United Airlines to integrate their cross-border flight schedules more closely in 2012 in response to potential effects on competition.[64]

Third, North American airport and airline security policies are characterized by mixtures of collaboration (on trusted traveller systems and security intelligence), cooperation (on passenger pre-screening and market alliances among air carriers), partial parallelism (on security and immigration policies), and competition (among airports, especially those in border regions). Post-9/11 measures for international screening and exchange of information on travellers built upon international regulatory architecture under the International Air Transport Association (IATA). These measures included the standardization of electronic and biometric passport requirements by 2006, although actual implementation of these measures by different countries remains a work in progress.

U.S. border agencies have pre-screened travellers to the United States passing through major Canadian airports since 1974, with pre-clearance extended to cross-border rail and ferry travellers under a revised agreement signed in 2015 and ratified in 2017. Both the United States and Canada have visa waiver programs for citizens from selected industrial (and, in Canada, Commonwealth) countries whose citizens are seen to pose limited risks of illegal migration, although the EU reportedly required Canada to eliminate visas for residents of the Czech Republic (2013) and Romania and Bulgaria (2017) before signing the 2016 Comprehensive Economic and Trade Agreement (CETA). Both Canada and the United States have introduced Electronic Travel Authorization (ETA, U.S. ESTA) systems to pre-screen travellers from other countries for security risks as part of broader North American perimeter security measures since 2014, while maintaining reciprocal exemptions for one another's citizens.[65]

In addition, third-country travellers holding a U.S. visa may transit through Toronto, Vancouver, and Calgary airports without requiring a Canadian visa. This provision enables airport authorities (and airlines) to pursue strategies to expand roles as airline "hubs" connecting international passengers in transit to third countries, especially the United States, following patterns of other major international hub airports.[66] Industry advocates also seek to position Canadian airports as international hubs for air cargo, typically involving shipments of higher value and accounting for about 30 per cent of Canada's trade outside North America in 2013.[67]

These measures provide a significant reciprocal or "aterritorial" framework that both shapes and responds to the interaction of national legislative and regulatory regimes governing the ownership and operations of airlines (including safety regimes), the administration of immigration and other security regimes, and the evolving use of airports as extraterritorial, not just conventional ports-of-entry. These measures were extended significantly through unilateral measures, complementary Canadian responses, and bilateral agreements under the Beyond-the-Border (BTB) initiative after 2011, which brought U.S. and Canadian security measures into closer alignment.

Both airport and airline industry agencies point to the necessity of significant domestic policy shifts to increase the efficiency and productivity of Canadian airports and airlines. However, there are significant divisions among key stakeholders over potential changes in policy priorities and related business models for airports and airport security– not least questions of ownership, regulatory structure and accountability, and who should pay for desired improvements.

Canada's user-pay system for airports and airport security, discussed below, finances all airport security costs from user fees levied on travellers, rather than by a mix of user fees and taxpayer subsidies, as in the United States. Airlines and consumer groups have been highly critical of the efficiency and cost-effectiveness of the Canadian Air Transport Security Agency (CATSA), noting that user fees revenues have consistently exceeded costs (and previous deficits) in recent years, with surplus revenues reverting to the federal government. The 2015 Canadian Air Transportation Review was highly critical of CATSA's relative inefficiency and limited use of innovative risk-management strategies such as those used in some European countries.[68] Suggested policy alternatives range from privatizing airport security to enhance efficiencies to reverting its core budgetary funding as a basic public service.[69] These challenges have grown with the devastating effect of the global COVID pandemic on air travel, forcing major airports to increase their improvement fees to offset revenue losses in other areas.[70]

Fourth, Canada's unique system of major urban airport authorities and non-profit air navigation anchors a user-pay fiscal regime, noted above, for the maintenance and upgrading of infrastructure, air navigation, and security-related costs, which reflects an uneasy and evolving balance of competing policy objectives. The federal government devolved responsibility for managing airports to non-profit airport authorities with independent borrowing powers and responsibility for financing airport improvements in 1995, while maintaining ownership of airport lands and drawing about $300 million annually in net revenues from long-term leases. This policy contrasts sharply with prevailing U.S. policies of government ownership and funding, and those many European countries, which have sold major airports to private, for-profit corporations. During periods of high exchange rates, major Canadian airports near the border, including Montreal, Toronto, Winnipeg, and Vancouver, experienced intense cross-border competition for passengers from regional U.S. airports, particularly those offering access to deep discount airlines and their networks.[71]

The divided ownership regime of national airports lacks effective accountability provisions for fee-setting and financial management by what are effectively local monopolies, while smaller national airports often lack the passenger volumes to meet government mandates. However, the airport and airline sectors are deeply divided over proposals for the privatization of airports, particularly without regulatory oversight of fee-setting practices. Montreal and Toronto appear open to securing institutional investment to finance their evolution into global transportation hubs serving international passengers in transit to U.S. and other markets. Other major airports seek to maximize local control to promote regional economic development. These debates suggest the need for flexibility in future policy developments.

Conclusion: Borders, Hubs, Networks, and the Persistence of Territoriality in an Interdependent World

Four broad trends have shaped economic interdependence and adaptation to broader international forces in the interaction of border management and transportation systems since 2000. First, evolving security concerns have made the integration of security considerations a precondition for efficient operation of international and cross-border transportation systems. Second, although nurturing bilateral trade relations and related investment opportunities remain critical issues for Canadian (and varied provincial) economic policies, continuing asymmetries of power and economic interdependence create important incentives for efforts to diversify and expand Canada's broader trade relationships in the Asia-Pacific, Europe, and Latin America, influencing both border and transportation policies. Third, these trends are reinforced by growing traffic at Canada's major maritime and airports, with their critical roles as national, North American, and international gateways and transit points, closely linked with other international and regional hubs, further integrating Canada within continental economic and travel networks. Fourth, persistent and pervasive U.S. domestic political conflicts require Canadian and Mexican governments to pursue cooperative economic and transgovernmental relations wherever possible, while maintaining a continuing "capacity for choice"[72] in domestic and wider international policy relations.

The realities of interdependence enhance the value of cooperation on border and national security issues. However, governments sensitive to the domestic political implications of such issues cannot "delegate" oversight over security and public safety without placing their own legal and democratic accountability (and electoral prospects) at risk. Intergovernmental cooperation may facilitate the "pushing out of borders" beyond physical frontiers. But the range of public safety risks to be managed and contained amid the growth of global trade, travel, and virtual networks necessitates the preservation of domestic policy capacity in managing borders and overseeing transportation networks even while investing in both global and continental "gateways" and "perimeters," and cooperating with neighbouring and allied governments. Strengthening domestic capacity while pursuing international cooperation have thus become what Sokolsky and Lagassé have called the "belt and suspenders" of managing borders effectively while promoting the efficient movement of people, goods, service, and capital within and across national borders.[73] If anything, this "both-and" imperative of bilateral policy relations has been reinforced by the domestically focused nationalist agenda embraced by the Trump administration and its often erratic policy style.

Border and transportation policy management may be characterized by higher constitutional symmetry than many other fields of Canada-U.S. policy relations, as noted in table 8.7. However, the diversity of functional relationships also lends itself to a sufficient diffusion of policy responsibilities within and across governments to complicate and decentralize policy cooperation. Traditional asymmetries of power and relative interdependence across these sectors generally ensure that U.S. policy formation remains mainly domestically oriented, effectively privileging functional cooperation in specific areas over system-wide cooperation on border management, border and transportation infrastructure planning, and assorted transportation policy subfields. In both rail and air transportation, volumes of Canadian trade and travel to and from wider international markets are now comparable to cross-border flows. However, the networked and intermodal nature of these movements has contributed to the continuing integration of these "outward" and "southward" aspects of Canadian public policies.

Bilateral movements of people and goods continue to predominate on Canada's land borders. But persistent regional variations and

Table 8.7. Charting Interdependence, Intermesticity and Policy Fields

	Land Borders		All Borders			
	Trucking	Railway	Travel/tourism	Infra-structure	Security + Administration	Air
Export Dependence	Medium-high Regional variations	Medium-high	Medium-high	High (Gateways) corridors	High (BTB)	Medium-high
Degree of market integration	Sectorally varied	High (including maritime border)	Regionally, sectorally varied			Medium-high
Institutional (a)symmetry (federalism)	Symmetrical/ diffuse	Symmetrical (federal + state-provincial for short lines)	Symmetrical/ diffuse	Mixed/diffuse	Symmetrical/federal	Symmetrical
Degree of governance internationalization	Asymmetrical * trusted traders * reciprocal agreement	Primarily domestic + BTB, RCC	Asymmetrical * trusted travellers * international (ICAO)	Mainly domestic (except border bridges)	Mainly domestic. some bilateral, international	Mixed
Style of policy relations	Parallelism + reciprocity	Parallelism + reciprocity	Parallelism + cooperation	Unilateral + cooperation	Parallelism + cooperation	Cooperation + parallel

differences require intergovernmental cooperation in promoting complementary regional policies in engaging the dynamics of globalization. Failure to do so risks a reversion to self-defeating, zero-sum policies of inter-regional and inter-sectoral competition, which have often spurred mutually reinforcing movements of defensive political nationalism and regionalism within Canada. The effective management of globalization and the cross-cutting pressures of North American integration thus remains the most effective safety valve for managing continuing regional and other domestic tensions in Canada.

NOTES

1 Bayless Manning, "The Congress, the Executive and Intermestic Affairs: Three Proposals," *Foreign Affairs* 55, no. 2 (1977): 306–24; Geoffrey Hale, *So Near Yet So Far: The Public and Hidden Worlds of Canada-U.S. Relations* (Vancouver: UBC Press, 2012), 4, 13.

2 Geoffrey Hale and Monica Gattinger, "Variable Geometry and Traffic Circles: Navigating Canada's Policy Relations in North America," in *Borders and Bridges: Canada's Policy Relations in North America*, ed. Monica Gattinger and Geoffrey Hale, 371–75 (Toronto: Oxford University Press, 2010); Bernard Hoekman, "International Regulatory Cooperation in a Supply Chain World," in *Redesigning Canadian Trade Policies for a Supply Chain World*, ed. Stephen Tapp, Ari VanAssche, and Robert Wolfe, 365–94 (Montreal: Institute for Research in Public Policy and McGill-Queen's University Press, 2017).

3 Douglas Porter, ed., *The Day after NAFTA: Economic Impact Analysis* (Toronto: BMO Capital Markets Economics, 20 November 2017); Dan Ciuriak et al., "The NAFTA Renegotiation: What If the U.S. Walks Away" (Toronto: C.D. Howe Institute, 28 November 2017).

4 Hoekman, "International Regulatory Cooperation," 368.

5 Kenneth Dyson and Angelos Sepos, "Differentiation as Design Principle and as Tool in the Political Management of European Integration," in *Which Europe? The Politics of Differentiated Integration*, ed. Kenneth Dyson and Angelos Sepos (Basingstoke, UK: Palgrave Macmillan, 2010), 3–4; Paul Sando, "Water and Political Relations between the Upper Plains States and the Prairie Provinces: What Works, What Doesn't, and What's All Wet," in *Beyond the Border: Tensions across the Forty-Ninth Parallel in the Great Plains and Prairies*, ed. Kyle Conway and Timothy Pasch, 133–50 (Montreal and Kingston: McGill-Queen's University Press, 2013).

6 Dyson and Sepos, "Differentiation and Design Principle," 4; Hoekman, "International Regulatory Cooperation"; Geoffrey Hale, "Regulatory

Cooperation in North America: Diplomacy Navigating Asymmetries," *American Review of Canadian Studies* 49, no. 1 (March 2019): 123–49.

7 Geoffrey Hale, "Transnationalism, Transgovernmentalism, and Canada-U.S. Relations in the 21st Century," *American Review of Canadian Studies* 43, no. 4 (December 2013): 494–511.

8 For example, Canada, Transportation Act Review, *Pathways: Connecting Canada's Transportation System to the World*, vol. 1 (Ottawa: Transport Canada, October 2015); G. Bruce Doern, John Coleman, and Barry Prentice, *Canadian Multimodal Transport Policy and Governance* (Montreal-Kingston: McGill-Queen's University Press, 2019).

9 Canadian National Railway. *Together into Our Next Century: Investor Fact Book: 2019* (Montreal: 2019): 2; Canadian Pacific Railway, *Investor Factbook 2019* (Montreal: 2019): 8.

10 Anne Goodchild, Steven Globerman, and Susan Albrecht, "Service Time Variability at the Blaine, Washington International Border Crossing and the Impact on Regional Supply Chains," Research Report #3 (Bellingham, WA: Border Policy Research Institute, June 2008). For examples of the effects of critical infrastructure disruption, see Geoffrey Hale and Cailin Bartlett, "Managing the Regulatory Tangle: Critical Infrastructure Security and Distributed Governance in Alberta's Major Traded Sectors," *Journal of Borderland Studies* 35, no. 1 (June 2018): 257–79, doi.org/10.1080/08865655.2017.1367710.

11 Doern, Coleman, and Prentice, *Canadian Multimodal Transport Policy and Governance*, 95–134.

12 Reginald C. Stuart, *Dispersed Relations: Americans and Canadians in Upper North America* (Washington and Baltimore: Wilson Center Press and Johns Hopkins University Press, 2007); Kyle Conway and Timothy Pasch, eds., *Beyond the Border: Tensions across the Forty-Ninth Parallel in the Great Plains and Prairies* (Montreal and Kingston: McGill-Queen's University Press, 2013).

13 Canada, Transportation Act Review, *Pathways: Connecting Canada's Transportation System to the World* (Ottawa: Transport Canada, October 2015), 21.

14 NIMBY: "Not in my back yard"; BANANA: "Build absolutely nothing anywhere near anything." Monica Gattinger, "Making a 'MESS' of Energy Policy," *American Review of Canadian Studies* 42, no. 4 (December 2012): 460–73.

15 Canada, Transport Canada, *Transportation in Canada: 2016* (Ottawa: 2017), 7.

16 United States, Federal Highway Administration, "Federal Share of State Highway Capital Expenditures, by State," table HF-202FS (Washington, DC: January 2016).

17 Canada, Transportation Act Review, *Pathways*, 1:21.

18 Canada, Transportation Act Review, *Pathways*, 1:23–7.

19 John Daly, "Double-crossing," *Globe and Mail*, 23 November 2017.

20 Paul Koring and Gloria Galloway, "A Symbol of Peace, a War of Words," *Globe and Mail*, 29 May 2013, A8–9; Tom Precious and Robert J. McCarthy, "Cuomo, Canadian Ambassador Tout Peace Bridge Deal," *Buffalo News*, 26 June 2013; Munroe Eagles, "At War over the Peace Bridge: A Case Study in the Vulnerability of Binational Institutions," *Journal of Borderland Studies*, May 2018, DOI:10.1080/088065655. 2018.1465354.

21 Canada, Public Services and Procurement Canada, "Enterprise Crown Corporations and Other Government Business Enterprises" (Ottawa: 2 June 2016); Michael Ircha, "Ports and Shipping: Opportunities and Challenges" (paper presented to "Borders in Globalization" conference, MacEachen Institute, Dalhousie University, Halifax, March 2018).

22 Theo van de Kletersteeg, "Federal Port Review: 2014–15," *Canadian Sailings*, 22 February 2017; Ircha, "Ports and Shipping."

23 Geoffrey Hale, *Uneasy Partnership: The Politics of Business and Government in Canada*, 2nd ed. (Toronto: University of Toronto Press, 2018), 252–4; Rob Drinkwater, "Churchill Rail Deal Boosts Hopes Shuttered Port Will Be Revitalized," *Globe and Mail*, 3 September 2018, B1.

24 Ircha, "Ports and Shipping"; Canada, Transport Canada, "Canada's Oceans Protection Plan: What It Means for Canada's Regions" (Ottawa: 7 November 2016).

25 Bill Curry, "Federal Government Postpones Study of Airport Privatization," *Globe and Mail*, 21 April 2018.

26 Canada, Transport Canada, *Border Infrastructure Investment Program 2.0 – Canada–United States* (Ottawa: December 2014); Canada, Transport Canada, *Border Infrastructure Investment Program 3.0: Canada–United States.* (Ottawa and Washington: August 2016).

27 United States, United States–Canada Transportation Border Working Group (TBWG), *Action Plan 2015–2017* (Washington, DC, and Ottawa: 2015).

28 Transport Canada, *Border Infrastructure Investment Program 3.0*, 116–20; confidential interview, U.S. and Canadian government agencies, 2015.

29 John P. Palmer, "Truck and Rail Shipping: The Deregulation Evolution," in *Breaking the Shackles: Deregulating Canadian Industry*, ed. Walter Block and George Lermer, 151–69 (Vancouver: Fraser Institute, 1991).

30 Palmer, "Truck and Rail Shipping," 152–4.

31 Mary R. Brooks, "Mapping the New North American Reality: The Road Sector," *Policy Options* (June 2004). https://policyoptions.irpp.org/wp-content/uploads/sites/2/assets/po/north-american-integration/brooks.pdf.

32 Élisabeth Vallet and Andréanne Bissonnette, "Frontière Québec-États-Unis: La langue, facilitareur ou obstructe au franchissement" (paper presented to Association of Borderland Studies conference, San Antonio, TX, 4 April 2018).

33 Ontario Ministry of Transportation, "International Registration Plan" (Toronto: October 2017); Ontario Ministry of Finance. "International Fuel Tax Agreement" (Toronto: November 2017).
34 Van Horne Institute, Prolog Canada Inc., and JRSB Logistics Consulting, "Overdimensional Loads: A Canadian Solution" (Calgary: Van Horne Institute, July 2015).
35 Nathan Vanderklippe, "Transportation Woes Threaten to Delay Imperial Oil Sands Project," *Globe and Mail*, 28 June 2011; Betsy Z. Russell, "441-Foot-Long Megaloads Bound for Montana through CdA," *Idaho Spokesman-Review*, Boise, ID, 20 December 2013; Rebecca Boone, "'Megaload' Settlement Bans New Big Truck Loads on Idaho Road," *Missoulian*, Missoula, MT, 27 January 2017.
36 Quoted in Canada, Transportation Act Review, *Pathways*, 123.
37 The others are Burlington Northern Santa Fe (BNSF), CSX Corp., Kansas City Southern, Norfolk Southern, and Union Pacific.
38 Doern, Coleman, and Prentice, *Canadian Multimodal Transport Policy and Governance*, 198ff.
39 CPCS, "Evolution of Canadian Railway Economic Regulation and Industry Performance under Commercial Freedom" (report prepared for Railway Association of Canada, Ottawa: 28 November 2014); Canada, Transportation Act Review, *Pathways*.
40 Railway Association of Canada, "Canadian Transportation Act Review" (Ottawa: 4 October 2014), 8; Canada, Transportation Act Review, *Pathways*, 118.
41 Canada, Transportation Act Review, *Pathways*, 120, 129–30.
42 Arthur Kroeger, *Retiring the Crow Rate: A Study in Political Management* (Edmonton: University of Alberta Press, 2009); Kristen Courtney et al., *Bill C-18: An Act to Reorganize the Canadian Wheat Board and make consequential amendments to certain other Acts* (Ottawa: Library of Parliament, December 2011).
43 Doern, Coleman, and Prentice, *Canadian Multimodal Transport Policy and Governance*; Levi Wood, Allan White, and James Clements, "Bill C-49" (presentations to 22nd Annual Fields on Wheels Conference, Transport Institute, University of Manitoba, December 2017).
44 Canada, Transportation Act Review, *Pathways*, 120; Terence Corcoran, "Historic Coup at CP Rail," *National Post*, 4 May 2012, FP11.
45 Railway Association of Canada, "Canadian Transportation Act Review," 10–14; Wood, White, and Clements, "Bill C-49."
46 Canada, Transportation Act Review, *Pathways*, 127–8.
47 Laureen Kinney, "Rail Safety in Canada since Lac Mégantic" (presentation to Oil by Rail Safety Symposium, Lakewood, WA. Seattle: Center for Regional Disaster Resilience, 27 April 2016; Canada, Treasury Board

Secretariat, *Canada–United States Regulatory Cooperation Council E-Newsletter* (Ottawa: March 2016).

48 Railway Association of Canada, "Canadian Transportation Act Review," 19, 27; Canada, Transportation Act Review, *Pathways*, 1:141–3.

49 Transport Canada, *Transportation in Canada: 2016*, 11, 31; Statistics Canada, "Merchandise Imports and Exports: Customs Basis," CANSIM table 228-0081 (Ottawa: December 2017).

50 United States, Government Accountability Office, *Commercial Aviation: Bankruptcy and Pension Problems Are Symptoms of Underlying Structural Issues*, GAO-05-945 (Washington, DC: September 2005).

51 Paul S. Dempsey, "Airline Bankruptcy: The Post-Deregulation Epidemic" (Montreal: Institute for Air and Space Law, McGill University, March 2011).

52 Transport Canada, *Transportation in Canada: 2016*, 14–15.

53 Konrad Yakabuski, "For Quebeckers, Airline Passenger Rights Bill Is a Flight of Fancy," *Globe and Mail*, 7 February 2018, B4.

54 Edward Russell, "Air Canada Continues to Eye Closer United Ties," FlightGlobal.com, 27 September 2018.

55 Eric Atkins, "Transat A.T. Inc. Delays Closing $720 Million Takeover by Air Canada," Globe and Mail, 28 July 2020.

56 Alfred Pelsser, *The Postal History of ICAO* (Montreal: ICAO, updated 30 November 2017).

57 Pelsser, *Postal History of ICAO*.

58 Dorothy Robyn, "Alternative Governance Models for the Air Traffic Control System: A User Cooperative versus a Government Corporation" (Washington, DC: Brookings Institution, 6 April 2015), https://www .brookings.edu/blog/fixgov/2015/04/06/alternative-governance -models-for-the-air-traffic-control-system-a-user-cooperative-versus-a -government-corporation/.

59 Eric Atkins, "NATS Takes Equity Stake in NavCanada's Space-Based Air Traffic System," *Globe and Mail*, 17 May 2018, B2.

60 Eric Atkins, "NavCanada's Ambitious Plan: A Satellite Network to Track Planes around the World," *Globe and Mail*, 25 October 2017; Aireon, "Inter-active Timeline," www.aireon.com.

61 Canada, Transportation Act Review, *Pathways*, 1:190, 195–8.

62 Karen McCrimmon, MP, "Speech on Bill C-49, Transportation Modernization Act," 42nd Parliament, 1st Session (Ottawa: House of Commons, 31 October 2017).

63 Canada, Transportation Act Review, *Pathways*, 1:198.

64 Russell, "Air Canada Continues to Eye Closer United Ties."

65 Hale, *So Near and Yet So Far*, 260–2; Canadian Press, Postmedia News, "Many Visitors to Canada Will Pay $7 Fee under Security Plan," *Ottawa*

Citizen, 21 June 2014; Canadian Press, "Canada-EU Visa Dispute Could Impact Free-Trade Deal Vote, Ambassador Says," *Toronto Star*, 12 April 2016.

66 Kristine Owram, "Toronto's Pearson Airport Wants to Be a Mega-Hub, but High Costs and Congestion Stand in the Way," *Financial Post*, 2 December 2016; Hillary Marshall, "Toronto Needs to Become North America's Next Airport Mega Hub," *Huffington Post*, 2 March 2017.

67 Gerry Bruno, "Connecting Canada: An Aviation Policy Agenda for Global Competitiveness and Economic Prosperity" (submission to Canadian Airports Council, Canadian Transportation Act Review, Ottawa, January 2015), 25–8.

68 David Gillen and William G. Morrison, "Aviation Security: Costing, Pricing, Finance and Performance," *Journal of Air Transport Management* 48 (September 2015): 1–12; Canada Transportation Act Review, *Pathways*, 1:201–2; Bill Curry, "National Airlines Council of Canada urges Ottawa on Decision of Privatizing Airport Security," *Globe and Mail*, 1 August 2018, A1.

69 Barry E. Prentice, "Canadian Airport Security: The Privatization of a Public Good," *Journal of Air Transport Management* 48 (September 2015): 52–9; Canada, Transportation Act Review, *Pathways*, 1:201–2.

70 Ryan Tumilty, "Canada's Airports Poised to Hike Fees in Wake of Pandemic," National Post, 13 August 2020, A5.

71 Mary-Jane Bennett, *Airport Policy in Canada* (Winnipeg: Frontier Institute for Public Policy, August 2012); Vijay Gill, *Driven Away: Why More Canadians Are Choosing Cross-Border Airports* (Ottawa: Conference Board of Canada, October 2012).

72 George Hoberg, ed., *Capacity for Choice: Canada in the New North America* (Toronto: University of Toronto Press, 2002).

73 Joel J. Sokolsky and Philippe Lagassé, "Suspenders and a Belt: Perimeter and Border Security in Canada-US relations," *Canadian Foreign Policy Journal* 12, no. 3 (2006): 15–29.

9 Dividing and Uniting Transportation Border Markets: The Role of Cabotage

BARRY E. PRENTICE AND JOHN COLEMAN

International borders are usually determined by geographical features or historical political developments and seldom by economic logic. As a result, borders often cut through and divide border regions into national enclaves. Border markets generally lack the economic vitality of regions more distant from the frontier, but as Walther observes they are vital components of the economy, despite being located at the margins of the state.[1] The economic truncation of border markets is often accentuated by restrictions on transportation. The territorial impediments to transport usually occur at the frontier, frequently manifested as costs from waiting and inspection. However, prohibitions on foreign transport services away from the border can limit the integration of transportation networks and further impede the efficient movement of passengers and goods within the border region.

"Free trade" in transportation services is known as cabotage. This chapter examines the implications for border regions of the restrictions on cabotage for different modes of transport. Rail, trucking, aviation, and marine modes are considered in the Canada-U.S. cross-border markets. The role of cabotage restrictions in constraining cross-border networks is examined through an economics lens. It demonstrates there is reduced prosperity of border regions, which we argue is caused at least partly by constrained transport options for people and especially freight.

Cabotage restrictions in North America are pervasive and often intentionally subtle, like other non-tariff barriers to trade, with the result that more than one regulatory instrument may restrict the practice. All kinds of different obstacles are used to prevent foreign competition in domestic markets, from vehicle specifications and customs rules, to restrictions on foreign labour. With land transport, the main target for protectionism is the crew, rather than the vehicle. Restrictions on cabotage are designed to protect employment for drivers and other crew members, and the disposition of railway rolling stock and truck power-units ("tractors") and trailers is almost immaterial. With air transport, things are less clear-cut, because passenger airlines tend to be viewed in more parochial terms. That is partly a hold-over from the early years of commercial aviation when countries showed off their progressive credentials (and national pride) by designating "flag carriers" to serve international routes, and government regulation was the accepted policy paradigm for all modes. It is also partly because, in an era of knowledge industries, people travelling on business are tantamount to "trade." That has steadily raised the contribution of airports and air travel to urban growth, and governments are generally loath to play laissez-faire about who serves them.

In this chapter, the modes of transport are considered individually, using examples of market truncation. We begin with a review of the literature on the economic advantages and disadvantages of border regions and an assessment of the evidence in North America.

Border Market Truncation

Hacker et al. examine the integration of the regional economies of Eastern Europe into the large European Union.[2] In their examination of economic development in the former political frontiers they review the theoretical concepts applicable to the border separation factor. Depending on location and ease of crossing, border regions can be viewed as backward and impaired, or positively positioned to obtain the benefits from transnational flows and global supply chains. Figure 9.1 presents the advantages and disadvantages of location that have been assigned to border regions.

It appears that the merits of a border location are related to their degree of openness. Walther describes this as "porosity" and Hacket et al. refer to it as "permeability."[3] If trade is free to move easily across the border with low transaction costs, then the region can take advantage of the business opportunities afforded by its peripheral location. The wider the economic differentials in the cross-border markets (resources, costs, prices, etc.), the greater these opportunities are. Of course, other factors such as ethno-linguistic networks and transportation infrastructure can also influence the formation of border effects, as observed by Schulze and Wolf in the Habsburg Empire prior to the First World War.[4] In North America, we could expect to see more permeability across U.S. borders with English-speaking regions of Canada than with Quebec or Mexico, where it is more difficult to develop communications and social networks.

Figure 9.1. Theoretical Economic Advantages and Disadvantages of Border Regions

Disadvantages of border markets
1 Peripheral location to the central economic regions of the country
2 Limited or truncated economic sphere
3 High trans-boundary transaction costs (trade barriers, infrastructure, regulations, etc.)
Advantages of border markets
1 Easier access to lower-cost cross-border resources
2 Better contact with new, international markets
3 Diversification of supply sources and directions of sale
4 Richer cultural environment (tourism, cross-border learning)

Adapted from R. Scott Hacker et al., eds., *Emerging Market Economies and European Economic Integration* (Northampton, MA: Edward Elgar Publishing, 2004).

Geographic barriers tend to funnel Canada-U.S. traffic flows through gateway locations at strategic bridges and border crossings. The physical porosity of these border points yields the advantages of contact and diversification. The regional markets at major pairs of border cities like Seattle-Vancouver, Detroit-Windsor, and Buffalo–Niagara Falls have well-integrated cross-border supply chains and serve major Canada-U.S. trade corridors. Other gateways along the Canada-U.S. frontier serve smaller corridors and local cross-border traffic. As illustrated in figure 9.2, a small number of border locations accounts for most of the traffic. The crossings at the Bluewater Bridge (Sarnia) and the Ambassador Bridge (Windsor) account for 45.2 per cent of all two-way trade by truck. The Peace Bridge and Queenston Bridge across the Niagara frontier account for 16 per cent. The Pacific Highway and Huntington in British Columbia represent 5.7 per cent of the bi-national trade volume. Despite the much larger population in Quebec, its two crossings (Lacolle and Rock Island) account for only 6.6 per cent. That may reflect the linguistic barrier. The three Prairie border crossings (Emerson, Coutts, and North Portal) represent 10.6 per cent of two-way traffic. All the other border crossings together represent only 15.9 per cent of Canada-U.S. truck-borne trade.

The economics of the Canada-U.S. border regions creates an interesting irony. Regions of the United States lying immediately south of the Canada-U.S. border represent some of the least economically attractive parts of their country. But in Canada, the areas lying within

Figure 9.2. Percentage Share of Two-Way Canada–U.S. Truck Crossing, 2017

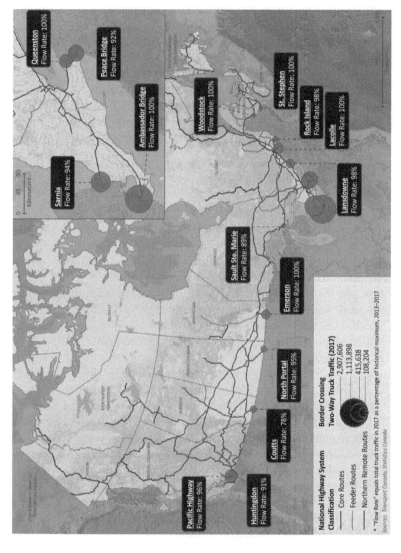

National Highway System
Classification
— Core Routes
— Feeder Routes
— Northern Remote Routes

Border Crossing
Two-Way Truck Traffic (2017)
2,907,606
1,113,898
415,638
108,204

* "Flow Rate" equals total truck traffic in 2017 as a percentage of historical maximum, 2013–2017

Sources: Transport Canada, Statistics Canada

Queenston
Flow Rate: 100%

Peace Bridge
Flow Rate: 92%

Ambassador Bridge
Flow Rate: 100%

Sarnia
Flow Rate: 94%

Woodstock
Flow Rate: 100%

St. Stephen
Flow Rate: 100%

Rock Island
Flow Rate: 98%

Lacolle
Flow Rate: 100%

Lansdowne
Flow Rate: 98%

Sault Ste. Marie
Flow Rate: 89%

Emerson
Flow Rate: 100%

North Portal
Flow Rate: 95%

Coutts
Flow Rate: 78%

Pacific Highway
Flow Rate: 96%

Huntingdon
Flow Rate: 91%

200 kilometres north of the very same border are the most economically desirable parts of *this* country. Clearly the colder climate has some effect on Americans' detached view of northern states. But it is not just the weather. The border region within many of these states is less prosperous than are their respective states as a whole. The lack of porosity truncates their economic region, and limits on cabotage accentuate the economic differential. Transportation networks that serve the trans-border markets are less well-developed than the transportation services in either the Canadian or American domestic markets.

The effects of market truncation are visible and associated with the long stretches of geography that make the border permeable to trade. The four regions of the northern U.S. border are the Northwest, Northern Great Plains, the Upper Great Lakes, and the Northeast. A measure of economic distress of communities in the United States can serve as a proxy for the negative effects of market truncation. Figure 9.3 presents a map by the Economic Innovation Group.[5] Three of the four areas along the northern U.S. border and the southwestern U.S.-Mexico border stand out as economically truncated markets. They are circumscribed by ellipses superimposed on the map.

The individual counties inside these highlighted areas were taken from a map of all U.S. counties.[6]

There are thirty-seven of these "truncated counties" inside the ellipse demarking the Northwest region, 109 inside the Upper Great Lakes region's ellipse, 80 inside the Northeast region's, and 70 in the Southwest region's, for a total of 296. The level of economic well-being (or distress) of each was calculated from per-capita income tables published by the U.S. Census Bureau, and the numbers for the counties was aggregated in each of the four regions.[7]

The results are shown in table 9.1. The per capita income ranges from a low of $23,840/year (for the northwest) to a high of $26,351/year (for the northeast), with the mean for all 296 truncated counties in all four regions at $25,651/year.

By themselves these numbers signify relatively little. But what matters is the counties' standing relative to those in *non-truncated* regions. The per-capita income of *all* counties in the states is used as a basis for comparison to the truncated market area. In other words, the truncated counties are compared with those within the same states but farther from the border. If the hypothesis were true, the truncated regions' per-capita income would be lower than that of those states as a whole, which is less exposed to the border effect.

This is indeed the result. For example, the mean per-capita income for the three states in the northwest region (including all of their counties)

Figure 9.3. Economic Distress and Truncated Border Communities

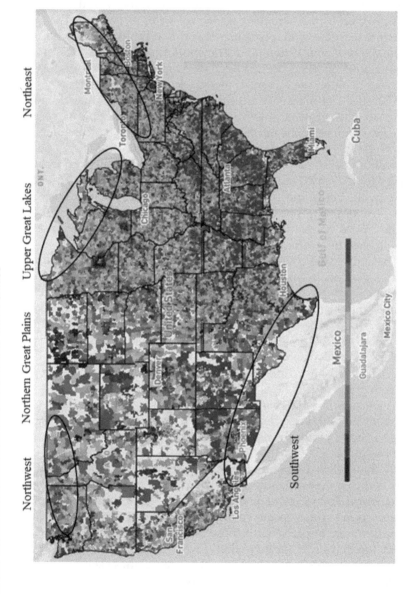

Source: Adapted from Richard Florida and Aria Bendix, "Mapping the Most Distressed Communities in the U.S.," February 26, 2016, www.citylab.com.

Table 9.1. Mean Income per Capita of Truncated and Non-Truncated Counties in Four Regions

Region	*All* counties in region			Truncated counties in region			
					Mean income/capita		
	Mean income/ capita				Absolute		Relative to all counties in region
	U.S. $/year	Rank on list of all U.S. counties (out of 3143)	No.	US $/year	Rank on list of all U.S. counties (out of 3,143)	US $/year	Rank on list of all U.S. counties (out of 3143)
Northwest Washington, Idaho, Montana	29,771	415	37	23,840	1318		-17.1% -903 places
Upper Great Lakes Minnesota, Wisconsin, Michigan	27,518	558	109	26,165	804		-4.9% -246 places
Northeast Pennsylvania, New York, Vermont, New Hamp., Maine	30,752	283	80	26,351	761		-14.3% -478 places
Southwest California, Arizona, New Mexico, Texas	27,731	532	70	25,930	962		-8.4% -430 places
All four regions	28,549	439	296	25,651	913		-10.2% -474 places

is $29,771/year, while the corresponding amount for the 37 truncated counties in that same region is almost $6,000 less – a shortfall of 17.1%. For the four regions as a whole, the per-capita income in the 296 truncated counties is almost $3,000 less than the mean for all counties in those same fifteen states – a shortfall of 10.2%. That is not massive, but it is not trivial either.

There is another way to look at things. The U.S. Census Bureau data can be arranged to rank the per-capita income of all 3,143 counties in descending order. If the northwest region were a county, its $29,771/year income would place it 415th on the national list, in other words in the 86th percentile nationally. But the truncated counties in the northwest, with a mean per-capita income of $23,840, would place 1318th on the list. That is equivalent to being in the 61st percentile nationally. The difference is 903 places, or 25 percentage points!

It is a similar story, albeit not quite as pronounced, for all four regions taken together. If they were a county, they would place 439th on the national list. That is in the 85th percentile. But their truncated counties, all 296 of them, would place 913rd on the list. That is in the 70th percentile. The difference is 474 places, or fully 15 percentage points.

The exception among the border regions is the Northern Great Plains where incomes along the boundary are similar to incomes in the remainder of the U.S. states. Certainly the absence of natural barriers, like the mountains and lakes makes north-south truck-based trade easier. Ciuriak observes that both consumers and producers living near the border benefit from the North American Free Trade Agreement (now renamed USMCA). The porous border gives U.S. producers the benefit of a circular market. This reduces their locational disadvantage relative to U.S. producers farther south. Similarly, consumers who are located at the periphery of a market are the most expensive to serve. Reduced profitability means that they are more likely to experience shortages, delays, and higher prices. "An open Canadian border thus puts the northernmost US communities in the middle of a regional economy rather than being on the periphery. These benefits are over and above the direct benefits of exports to Canada."[8]

Clearly this analytical method has many weaknesses and should be interpreted more as illustrative than definitive. For example, the work of identifying counties within the ellipses in each region is imprecise, because some counties straddle the perimeter of the ellipse. Probably more significant, this method does not take into account the differences between some groups of counties in terms of geography, population, presence or absence of transportation routes through them, and the nature of economic activity on the other side of the national frontier. Also the relationships on the northern and southern borders of the United States are quite different. Walther observes that the need to combine security and trade is nowhere more evident than at the U.S.-Mexico border.[9] This paradox has created a "gated world" where trade is focused through key gateways, like Laredo, Texas. All those things affect the rate of economic activity within the truncated regions surely as much as truncation itself does. It is beyond the scope of this chapter to separate out the effects of the influencing factors. But suffice to say the data strongly suggest that truncation is likely to be a real phenomenon and the hypothesis, at the very least, is not disproven.

The Nature and Role of Cabotage

Cabotage and cross-border transport are often confused. Although both involve regulations that affect the trade of transportation services, cross-border transportation is already free in most jurisdictions, yet cabotage is generally prohibited. The difference lies in the origin and destination of the shipments. If the origin and destination are in different countries, it constitutes cross-border transport and is permitted. (Cross-border air travel is a separate case, for reasons described later.) But if the origin and destination are in the same country, it amounts to cabotage and foreign carriers are prohibited. That usually applies to freight and passengers alike.

The economic impacts of cabotage restrictions are the reduction of competition, inefficient routing options or equipment relocation, and lost opportunities for regional integration. The reduction of competition is obvious, because denying cabotage preserves the entire domestic transportation market solely for the national carriers of each country. In general, more competition is viewed as positive because it forces the market participants to search for efficiency and to provide better customer service.[10] A priori it is difficult to estimate the benefits that cabotage could provide, but experience with other forms of economic deregulation provides evidence to suggest that shippers would have access to lower freight rates and to new (and better) service offerings.

In general, the impact of cabotage is inversely related to the network economies of the transport mode, and to that mode's ratio of fixed-to-variable costs. In other words, the lower the mode's network economies, the higher the impact will be from allowing cabotage; and the higher a mode's ratio of fixed-to-variable costs, other things being equal, the lower the effect of cabotage.

There are two general categories to consider. First, high-fixed-cost sectors with increasing network economies from scale, like the railways and pipelines, are not threatened much by foreign competition. These industries own their infrastructure and can permit or exclude domestic or foreign operators from using it. Railways and pipelines cooperate with their foreign counterparts to interchange traffic freely. In the case of the railways, foreign railcars move freely in both directions. Foreign train crews are restricted, but foreign railcars circulate freely. Technically speaking, equipment like railcars and intermodal containers are imported duty-free on a temporary basis, with a time limit for their re-exportation. We will say more about that later on.

The second category consists of transportation sectors that are subject to relatively flat economies of scale and that have a low ratio of

fixed-to-variable costs. They are more susceptible to foreign competition. Truckload (TL) trucking and bulk marine carriers are the two
modes most sensitive to cabotage competition because neither one
owns or provides any infrastructure, such as the roads or ports they
use. The fixed costs of that infrastructure are supported by the public.
The efficiency of TL trucking and bulk marine does not increase with
fleet size, because the individual vehicles have virtually no interaction
with each other. That means adding another vehicle to the fleet will
have minimal impact on average costs. And so, as constant-cost industries, they have minimal network economies to give them an advantage
over foreign-owned trucks and ships.

There is an in-between category as well. Air transport, marine intermodal (containers), and less-than-truckload (LTL) trucking are
only moderately sensitive to cabotage competition. Although the public provides their infrastructure, these modes have a higher ratio of
fixed-to-variable costs and they enjoy significant network economies.
For example, the addition of another airplane service to a new destination can feed traffic to the entire network of the airline. Of course,
the quid pro quo for a national government granting cabotage is the
opening of reciprocal access for its carriers to the markets of the participating countries. Consequently, all trucking, marine, and air carriers are exposed to more competition if cabotage is permitted, but they
too can extend their networks to bordering markets. In short, allowing
cabotage creates larger markets for every carrier, and simultaneously
allows more carriers to compete for the traffic.

Rail Transport

As mentioned above, railways own their own infrastructure and can
allow or prevent other carriers from using it. That makes rail cabotage
an entirely different phenomenon than it is with trucking or airlines,
where all lanes and routes are open to everyone, and where on-board
crews, usually nationals of the carrier's host country, stay with the vehicle or aircraft and therefore do gainful work while it travels in the
foreign country.

But with railways, crews usually change at or near the border, and so
the engineer and conductor from one country do not travel and work in
the foreign country. In other words, they do not stay with their vehicle
very far, if at all, when it crosses the border, so the crewing restrictions
in anti-cabotage rules are effectively moot.

Regarding the vehicles themselves, the a-territorial rules governing interchange of freight cars are very liberal. A load picked up by an

originating railway gets handed off to a connecting railway as a matter of routine, because the equipment is designed to be interchangeable and the North American network is designed to function that way. The handoff usually occurs at a classification yard. Railway X brings a train-load of cars to a yard, where it sorts them. If Car Y on the inbound train is going someplace not served by the originating railway, then a connecting railway will take Car Y out of the yard on one of its departing trains. If the connecting railway is in a different country, there is some cross-border paperwork to be done but not much else.

Few if any railway freight yards sit right on the border, so handoffs usually occur *close to* the border but not on it. For example, a southbound CP train originating in Winnipeg would take all the cars on the train to Minneapolis. The crew members might change at the border, or, perhaps run through and disembark in Minneapolis, as long as they returned north on a subsequent train.

Southbound cargo originating in northern Minnesota would be loaded onto freight cars at the shipper's site, then picked up by a U.S.-based CP train or by a short-line railway, all crewed by Americans, then taken to the classification yard in Minneapolis, and from there added to a southbound main-line train. Let's say it is a CP train. By then it would have a U.S. crew, so the crewing part of normal restrictions on cabotage would not apply. The locomotives assigned to the train in Winnipeg probably would stay on the train at least as far as the CP tracks went, i.e., beyond Minneapolis. So the Canadian locomotives would be pulling a train with freight cars picked up in the U.S. and destined for delivery somewhere in the U.S., but it violates no anti-cabotage provision. From a cabotage standpoint the nationality of the freight cars is a non-issue and things are not much different for locomotives.

Regarding territoriality and a-territoriality, there is not much to ponder. There are three main reasons. First, a primary characteristic of railway transport is that railcars need to be sorted and resorted in classification yards, so there is relatively little opportunity for main-line trains to pick up local traffic when they pass shippers' sites in a foreign country. Second, historical operating practices required crew changes every division point or two, usually about 120 miles apart, so crews are dissociated from their vehicles and employment issues for foreigners are practically moot. And finally, railway rolling stock is built to be interchangeable among all North American railways under the auspices of a tri-national standards-setting body, the Association of American Railroads. That removes any opportunity for countries to fiddle with equipment specs as a disguised way of aiding protectionism.

The upshot is that, to a first approximation, border regions are disadvantaged little, if at all, by anti-cabotage rules pertaining to railways.

Truck Transport

A small amount of cabotage is granted to U.S. and Canadian truckers, as long as the movement is done in conjunction with a cross-border truck trip. In Canada, this is called a "repositioning move"; in the United States it is known as a "return trip – outward move." There are several conditions a trucker must meet. A truck returning home can carry domestic goods between the point at which it dropped off its international load and another point in the other country, but that portion of the trip must be "incidental" to an international move. U.S. regulations state that this must be part of a "regularly scheduled" international move; Canada says that the domestic freight must not exceed 30 per cent of the weight and value of the total load.

Cabotage is permitted in Canada as part of a repositioning move, but only if the vehicle is en route to pick up a scheduled load for export and the drop-off point of the goods is in a direct line with the pickup point of the export load. The domestic pickup and delivery must allow for only "minor deviations" from the international route. Most important, only one incidental (domestic) move is permitted per international trip, and the move must follow a route consistent with the international route of the imported or exported goods.

In the United States, the "return trip – outward move" provision restricts Canadian tractor-trailers to hauling domestic (U.S.) goods on "regularly scheduled" international runs and forces them to proceed northward (i.e., "outward") while moving these goods. U.S. cabotage restrictions also apply to empty equipment, which causes significant extra expenses for Canadian trucking firms. Truckers like to drop off full trailers and leave them for unloading, then take another empty trailer to their next customer. But in the United States, the Canadian carrier has to pay a U.S. trucker to pull the empty trailer to the next job, while the Canadian tractor and driver tails along behind. From an efficiency standpoint that leaves a lot to be desired. Canada has no restrictions on the relocation of empty truck trailers.

Cabotage gets tangled up in tax and immigration issues as well as in competition acts, regulations on foreign ownership of domestic industries, and provincial/state motor vehicle legislation. Prokop outlines how the insistence of the Canadian Finance Department that the GST (goods and services tax) be applied on the value of U.S. trucks was enough to kill a proposal by the Canadian and American trucking associations in 1997.[11] Since that time nothing has emerged, nor was cabotage considered in the NAFTA talks of 2017–19. Immigration rules that apply to truck drivers further limit the potential for competition. Under the two countries' respective immigration acts, a foreign driver must have a work visa, or be a dual citizen or half-indigenous by blood in order to engage legally in cabotage.

Trade impacts are never one-sided. Current cabotage regulations mean that the services allowable in truck transport biases trade patterns of Canada and Mexico. Canada tends to trade with the northern U.S. states, and Mexico is more inclined to trade with the southern states. Shippers in the northern states are denied opportunities for lower-cost southbound domestic freight on trucks heading for Mexico. Similarly, U.S. shippers in the southern states cannot take advantage of north-bound Canadian trucks to move their domestic freight towards the Canadian border.

Truck shipments between Canada and Mexico would be more economic if their carriers could transport each other's freight into the United States on their return. But that is not allowed. Although such freight is technically "international," U.S. immigration and employment regulations block Canadian trucks from carrying Mexican freight to U.S. destinations. This case has never been tested for the Mexican trucks coming to Canada, because the United States has yet to allow Mexican trucks to cross their territory to the northern border. On the southern border, trucks clear customs on the U.S. side, which means the move is then treated as "domestic" rather than international. Canadian freight is cleared *at* the U.S. border.

Transportation companies try to avoid empty moves whenever possible, for obvious reasons. But balanced trade-flows (full loads in both directions) are the exception rather than the rule. Consequently, in one direction the vehicles are all full, and the rates are attractive to the trucker. In the transportation industry, this is referred to as the "fronthaul" leg. But in the reverse direction, some vehicles are forced to return empty. And even if a return load *can* be secured, the rates are always low and may not cover much more than the variable costs, so the trucker makes little or no contribution towards overhead and profit. The undesirable traffic lane is always called the "backhaul" leg on a round-trip circuit.

One option that carriers can use to minimize the economic penalty of the backhaul is to search for triangular routes that provide two fronthaul legs and only one backhaul leg.[12] A well-known example in trucking is illustrated in figure 9.4. In this case, Canadian truckers operating between Halifax and Toronto can take a low-value load (like seafood or

Figure 9.4. Triangularization of Cross-Border Routes to Minimize
Backhaul Loads

potatoes) on what would be considered a backhaul to New York, where
there is a huge demand for these commodities. At New York or the surrounding area they can obtain fronthaul loads to Toronto (or Montreal).
The triangle is completed with another fronthaul load back to Halifax
(or elsewhere in Atlantic Canada). This routing is permissible because
the base of the triangle is in Canada (a domestic trip), and the other two
loads are cross-border trips. Similarly, an American carrier could take a
load from New York to Chicago, then back to Toronto and home to New
York with the triangle's base located in the United States.

Cabotage regulations make triangular routing difficult because the
base of the "triangle" always must be in the domestic market. Opportunities for triangular routes across the Canada-U.S. border are limited,
forcing many freight trucks to travel farther empty. Almost nothing
good comes from this. It means more trucks on the road – wasting fuel,
exacerbating congestion (particularly at border crossings), generating
unnecessary emissions, and running up costs. If it were permitted,
cabotage would give Canadian and U.S. truckers more choices. This
would give shippers more choices and lower their costs.

In the United States, with a much larger geographic market, square-
shaped routing is also employed in addition to triangular patterns. If
cabotage were allowed, such square patterns could be used in border
regions as well as in central parts of the country, but without it the base

of the routing triangle has to be in the home country. This impediment to obtaining efficient routes through the border regions truncates the operations of the carriers and forces them to concentrate on serving their main domestic routes, leaving the border regions under-served and their industries less well-connected to potential markets.

If Canadian or U.S. carriers cannot find a backhaul load across the border to their home, they must return empty. The impact is asymmetric – it is harder on Canadian carriers – because most Canadian markets lie within 350 kilometres of the U.S. border. Even at worst, American carriers have only a short empty drive back to where loads are legally available. But for Canadian truckers the penalty of running empty kilometres is much greater because they could be as far as 2,500 to 3,500 kilometres from their home market. California is one of the few southern U.S. states served by Canadian trucks as much as the northern states are. This is made possible by triangularization. The Calgary–Los Angeles-Toronto route for refrigerated trucks is a good example. From Calgary, refrigerated trucks take beef to California, then go to Toronto with a load of fresh vegetables, and then westbound home to Calgary with freight that may not even need to be refrigerated.

A border region with many opportunities for better truck routing is the Saskatchewan-Manitoba/North Dakota–Minnesota border area. Some triangular routes already exist. For example, canola seed is carried south from Manitoba to crushing plants in the United States, with trucks carrying soymeal back to Saskatchewan, where they pick up lentils or fertilizer to haul back to Manitoba. Local (i.e., geographically constrained) triangular routes can be made to work, but many more options would be available to serve border regions if cabotage allowed the vehicles to run on the "square."

In terms of territoriality and a-territoriality, a lot boils down to individual countries and their sub-jurisdictions using their regulatory authority to circumscribe what is allowed intramurally. There is little or no suggestion the constraints they impose have anything to do with "strategic" sectors or whether the public or private sector owns the infrastructure, and everything to do with protectionism of employment opportunities for crews. Because drivers usually remain with their vehicles, the prohibition of cabotage targets them specifically. Standards-setting does not help truckers escape the cabotage restrictions, either, because many states and provinces set their own limits on truck weights and dimensions. These are limits that favour certain configurations over others. The "universal" truck – one that can cross any jurisdictional boundary – is almost always the lowest common denominator from an efficiency standpoint. Trucks that take full advantage of the regulations within their jurisdiction automatically have a commercial advantage over competitors from jurisdictions with different rules.

Short Sea Shipping and Tourist Cruises

Discussions of marine cabotage regulation almost inevitably involve the U.S. Merchant Marine Act of 1920 (commonly called the "Jones Act"). It is an egregious example of protectionism. It states that cargo may not be transported between two U.S. ports unless it is carried by vessels built and registered in the U.S., owned by citizens of the United States, and crewed by Americans. The Jones Act is only the latest in a history of U.S. restrictions on marine cabotage. In 1817, Congress barred foreign-flagged ships from engaging in American coastal trade. Even this was not a new form of protectionism. Britain's Navigation Act of 1651 restricted trade with its North American colonies to British ships.[13]

Canadian policy is no different in its main intention, and its origins also date back to British regulations, in this case as set out in the Treaty of Paris, 1763. British laws against cabotage were continued under the British North America Act, 1867, that formed the Canadian Confederation. The Coasting Trade Act, 1934, set out the modern legal framework that limits the carriage of passengers and freight in Canadian waters. However, waivers apply to the Coasting Trade Act that permit the importation of non-duty-paid foreign ships for up to twelve months, as long as Canadian ships are not available or suitable to provide adequate service. Another key difference is that Canada does not have a Jones Act–style, "built in Canada" regulation, but it does require an operator to convert the vessel to meet unique Canadian specifications.[14] This waiver does not, however, permit foreign crews to staff the ships. Nationality requirements apply to the officers and crew.[15]

Marine transportation on the Great Lakes provides an example of regional market truncation. The Canada-U.S. border divides the Great Lakes, with the exception of Lake Michigan, which lies entirely within U.S. territory. As illustrated in figure 9.5, Canadian and American ships can pick up or drop loads only at ports in which either the origin or destination is their country, or on a triangular route in which the base is located in the home country. This means that two separate national fleets must be maintained. It also means that efficient "milk runs" cannot be developed. The resulting higher-than-necessary cost of marine transport biases shippers towards rail or truck, and reduces the advantage of location for the cities on the Great Lakes. Liberalizing cabotage on the Great Lakes could increase traffic density, lower shipping costs, improve the economy of cities on the lakes, and remove the bias towards

Figure 9.5. Permissible Triangular Shipping Routes in the Great Lakes

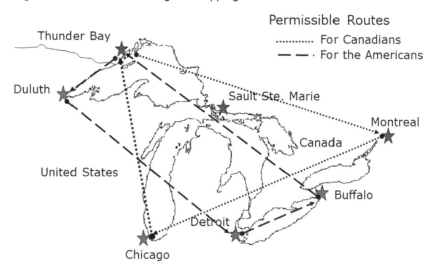

rail and trucking, with the result that congestion would decrease on busy highways and rail corridors.

Under the Coasting Trade Act, Canada had a punitive customs tariff of 25 per cent on non-NAFTA-built ships, but it was removed in the 2010 federal budget, although only for ships over 129 metres long. Cruise ships that could use the Great Lakes for passenger traffic are below this threshold and consequently still face this punitive barrier. The original restrictions were designed to protect Canadian-built and Canadian-flagged overnight passenger cruise ships, but are so out of date as to be pointless. No Canadian cruise ship has sailed on the Lakes since 1965,[16] so there is nothing to protect.

Lack of cabotage rights for Canada and the United States impedes the development of cruise ship tourist services on the Great Lakes. This is a significant obstacle to economic activity, particularly with respect to attracting foreign cruise-ship lines to develop such a tourist service. It is a missed opportunity of potentially substantial importance. Cruises in the Mediterranean, Caribbean, the British Columbia–Alaska coast, and on the river systems of Europe have grown rapidly. Cruises are likely to continue to be popular because the world's aging population has both the time and means to enjoy them. The Great Lakes–St. Lawrence Seaway offers an experience equal to or better than any foreign offerings, albeit the season may be restricted to the

May–September period. Three companies do offer cruise services on the U.S. side of the Great Lakes, but they are expensive and offer limited service to Canada.

The cruise market on the Great Lakes is even more restricted for non-Canadian and non-U.S. cruise lines than it is for ships of the two countries. A domestic carrier can go from port to port on its own side of the border. But with cabotage prevented, a foreign carrier would have to alternate the pick-up and drop-off of passengers between the United States and Canada. They could not originate and return passengers to different ports – or even the same port! – on either side of the lakes. The cabotage problem is aggravated by inadequate border security facilities. In part, this may be a chicken and egg argument, but customs agencies also resist the development of any new services unless they are provided on a user-pay basis. The combined impact of these restrictions makes Great Lakes cruises less attractive to tourists because they would entail longer sailing times between ports on the Great Lakes itinerary restrictions.

Local actions to raise awareness of cruising opportunities on the Great Lakes are desirable, but by themselves they are unlikely to deliver much in terms of results. The no-cabotage problem lies at the national level and involves trade agreements. The recent Comprehensive Economic Trade Agreement (CETA) between Canada and the European Union includes language that could permit some European shipping lines or their subsidiaries to operate feeder services, but cabotage opportunities available to them are limited to transport between Halifax and Montreal. Still, this offers some hope that cabotage could be expanded with European carriers.

Cabotage restrictions have a long history steeped in economic and military justifications, but if they were ever relevant it is doubtful they are today. The U.S. Jones Act, for example, restricts all U.S. coastal trade to American-built, -owned, and -operated ships. Canada's cabotage restrictions recognize more exceptions than American ones do, and allow some ships to operate between Canadian ports on waiver programs. Mercogliano explains that the U.S. views its merchant marine as a "Fourth Arm of Defense," but this basis for policy hardly stands up to scrutiny.[17] Even during the Vietnam War, the U.S. Defense Department chartered foreign ships to move oil and other strategic supplies because the domestic fleet lacked the necessary capacity. And more recently, the Jones Act has been under fire because it delayed the provision of relief needed by Puerto Rico after Hurricane Maria destroyed much of the island's infrastructure in 2017. Senator John McCain spoke out on the need to terminate this protectionist measure, but did not get very far.

Marine cabotage in the United States remains firmly entrenched and resistant to reform.

As with trucking, the territoriality vs. a-territoriality question largely boils down to protectionism by national and sub-national governments of employment within their borders. But U.S. limitations on the country in which a vessel must be built constitutes American determination to extend its legal authority beyond its territorial boundaries.

Passenger Air Transport

The air transport industry is particularly complex. Like land travel, it has two distinct elements, passengers and freight. But unlike trucks, buses, and private automobiles, whose carriage of passengers and freight work independently of each other, a substantial amount of cargo is carried in the bellies of commercial passenger aircraft and so the two elements are intermixed. There are of course dedicated air freight transport carriers – couriers (such as UPS and FedEx) and strictly cargo airplanes. The air cargo industry has grown rapidly globally and in North America in recent years, particularly as a result of the creation of global cross-border supply chains that operate on a lean-production basis, but traditional cabotage restrictions remain unhelpfully in place.

The United States moved more aggressively than Canada to liberalize its aviation policies. With the deregulation of its domestic aviation market in 1978, Washington launched an international policy that relied on market forces in areas such as pricing and charters. In the mid-1980s, this was followed by an "Open Skies" policy framework and the conclusion of bilateral agreements with many foreign governments. That framework allows unrestricted routes to and from all cities in the United States and the partner country. No limits are made on the number of airlines permitted to operate, or the number of flights they can offer. Fares are based on airlines' marketing decisions and, apart from possible cases of monopolistic pricing, the regulator will not intervene.

In 1985, Ottawa followed Washington's 1978 initiative towards deregulating air transport. But it was only ten years later, in 1995, that Canada signed a bilateral "Open Skies" agreement with the United States that allowed for more liberalized movement of passengers throughout North America. A separate agreement applied to purely cargo services.

The Canada-U.S. Open Skies Agreement led to a sharp increase in the capacity of scheduled airline services between the two countries – with an increase of 25 per cent in the first year alone! But the Open Skies agreement did not give the foreign air carriers any cabotage rights for

Figure 9.6. Transborder Air Connections in the Northern Plains of Canada and the United States

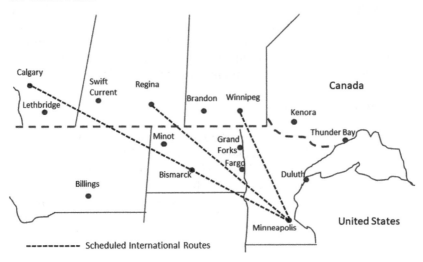

passengers or cargo. Nor did it permit airlines from the United States or Canada (or investors, either) to establish airlines in the other country or to own more than 25 per cent of an airline's voting shares. The latter was an indirect way of preventing even a vestige of cabotage by foreign interests. In amendments to the Canadian Transportation Act passed in 2018, the Government of Canada raised this ownership limit to 49 per cent.

The difficulty that cabotage restrictions pose to the construction of economic routes confronts regional air and bus markets. An example is the cross-border air transport market between the Rocky Mountains and Lake Superior. Figure 9.6 presents a map of the area that identifies the major urban centres and some of the main scheduled cross-border air routes between them. Anyone wishing to travel between two smaller communities must go through Calgary, Regina, or Winnipeg on the Canadian side of the border, and through Minneapolis on the U.S. side. This creates long, roundabout routes that extend the travel time much beyond the physical separation of these communities. Inevitably, routes involve two or more stops and always result in higher airfares. For example, anyone wishing to fly from Thunder Bay, ON, to Duluth, MN (a 278-kilometre trip) would have to fly Thunder Bay–Winnipeg-Minneapolis-Duluth (a 1,440-kilometre trip). The only alternative would be to drive a car, because cabotage restrictions make it impossible to provide regional cross-border bus routes.

Air cabotage is by no means a dodgy concept, suitable for practising only clandestinely. It has gained formal acceptance in principle (if not necessarily in practice) as ICAO's "8th Freedom of the Air." More cabotage would open up competition to additional second- and third-tier air carriers and help to create efficient networks that serve communities on both sides of the border. For example, Bearskin Airlines, which connects Winnipeg and Kenora to Thunder Bay, could extend that route to include a simple and efficient hop to Duluth, as long as it could take traffic from Duluth to another probably underserved U.S. city, like Fargo, before returning to Canada. But Duluth and Fargo will have to get used to being connected to each other by U.S. carriers only. That seems a shame. The current prohibitions on cabotage have the impact of isolating communities along the border from greater economic and social integration and reducing interconnectedness and economic prosperity for all of them.

ICAO is a counterweight to what otherwise would be a higher degree of territoriality in the governance of this transport market space. Bilateral negotiations are underway worldwide almost continuously under ICAO's framework for Freedoms of the Air. Each country seeks reciprocal (or better) benefits from trans-border flights and, as noted above, rarely grants cabotage rights to foreign airlines. In other words, were it not for the international framework established under ICAO, territoriality based restrictions almost certainly would now extend beyond cabotage and as far into cross-border traffic as the more muscular countries of the world could arrange to their own benefit.

Conclusion

Experience over many years has consistently shown trade theory's prediction that reducing trade barriers improves the economies of nation states on both sides of a border. There are two ways of looking at barriers. The first consists of costs of transporting goods across borders. This can be considered to be a "natural barrier" to trade, as opposed to the second type of barrier, political ones, like customs duties or non-tariff barriers (e.g., discriminatory regulations). But the two types sometimes are intermixed. Restrictions on cabotage augment the "natural barrier" to trade by increasing the cost of transportation. Customs and labour regulations are territorial but occur away from the border and are largely invisible. Arguably, the greatest impact of cabotage restrictions occurs in the border regions. Where cabotage is not allowed, efficient cross-border transportation networks are difficult to build or sustain.

The importance of cabotage is associated with the permeability or porosity of the border. Land crossings that serve as gateways on well-established corridors are not greatly affected by cabotage restrictions, but the regions adjacent to the crossings can be. If it were easier for carriers to obtain cargo or passengers on both sides of the border, fewer vehicles would be operating empty within the region, which would greatly increase transport efficiency and lower its costs.

The impetus to sustain trade barriers unfortunately is much greater than the motivation for change. The benefits of removing trade barriers are diffused and difficult to identify. But by contrast, a narrow group of vested interests that represent the main beneficiaries of limiting competition are vocal and well-organized. Cabotage also runs up against military strategic interests that consider a vibrant civilian transportation system as vital to defence. This is exemplified in the long-standing U.S. protection of its own marine industry, under the Jones Act.

The effects of the 2019 USMCA agreement remain uncertain. Progress on issues like cabotage have been overshadowed by efforts merely to sustain the degree of openness that has been achieved in the past. President Trump of the United States went beyond negative rhetoric to announce unilateral tariffs on steel and aluminium imports based on national security. The threat that Canadian steel and aluminium trade could possibly represent to the United States was never explained and almost certainly never could be. Long-needed improvement in transport efficiency will have to wait for future trade discussions. Meanwhile, truncated border regions will have to live with the economic disadvantages visited on them by protectionism designed to maintain transportation service jobs and, in effect, a beggar-thyself as much as beggar-thy-neighbour policy.

NOTES

1 Olivier Walther, "Border Markets: An Introduction," *Journal of Urban Research* (2014): 10.
2 R. Scott Hacker, Börje Johansson, and Charlie Karlsson, eds., *Emerging Market Economies and European Economic Integration* (Northampton, MA: Edward Elgar Publishing, 2014).
3 Walther, "Border Markets"; Hacker et al., *Emerging Market Economies*.
4 Max-Stephan Schulze and Nikolaus Wolf, "On the Origins of Border Effects: Insights from the Habsburg Empire," *Journal of Economic Geography* 9 (2009): 117–36.
5 Richard Florida and Aria Bendix, "Mapping the Most Distressed Communities in the U.S.," *CityLab*, 26 February 2016, https://www.citylab.com/equity/2016/02/mapping-distressed-communities-in-the-us/471150/.

6 Mapchart.net, "United States – Counties – Mapchart" (2018), https://
   mapchart.net/usa-counties.html.
7 United States, Census Bureau, "2009–2013 American Community Survey
   5-Year Estimates" (Washington, DC: Surveys of Economic Characteristics,
   Demographics and Housing, and Households and Families, 2015).
8 Dan Ciuriak, "The March into Trade Wars: US Policy Aims and the Implications for Reconciliation" (Toronto: C.D. Howe Institute, 2 August 2018), 9.
9 Walther, "Border Markets."
10 Joseph Monteiro and Benjamin Atkinson, "Cabotage: Are We Ready? How
   Is It Dealt With in Various Sectors of Transportation in Canada?" Canadian
   Transportation Research Forum, 44th Conference Proceedings, Winnipeg,
   2009. www.ctrf.ca.
11 Darren Prokop, "In 1988 We Freed Trade. Now Let's Free Transport," Policy Options 20, no. 5 (1999): 37–40.
12 Richard Beilock, Robert Dolyniuk, and Barry Prentice, "Encouraging
   Development through Better Integration of U.S. and Canadian Transportation: The Open Prairies Proposal," Regional Economic Development 2, no. 2
   (2006): 73–86.
13 Bahar Barami and Michael Dyer, "Assessment of Short-Sea Shipping;
   Options for Domestic Applications, Appendix G" (Washington, DC: Volpe
   National Transportation Systems Center, US Department of Transportation, 23 December 2009), 8.
14 J.R.F. Hodgson and Mary R. Brooks, "Towards a North American Cabotage
   Regime: A Canadian Perspective," Canadian Journal of Transportation 1, no.
   1 (2007): 19–35.
15 Monteiro and Atkinson, "Cabotage: Are We Ready?"
16 Mark P. Fisher, "Remarks to the House of Commons Finance Committee,
   Pre-Budget 2018 Consultations" (Windsor: 19 October 2017). Fisher is president and CEO, Council of the Great Lakes Region.
17 Salvatore R. Mercogliano, "Fourth Arm of Defense: Sealift and Maritime
   Logistics in the Vietnam War" (Washington, DC: Naval History and Heritage Command, Department of the U.S. Navy, 2017).

10 National Security and Economic Security: Distributed vs. Hierarchical Management of Domestic and Critical Infrastructure Security in Canada and North America

GEOFFREY HALE

National policies governing domestic security – euphemistically known as "homeland security" in the United States and "public safety" in Canada – have always faced tensions between the diffused nature of "threats" or "risks" and political imperatives for policy and resource coordination. Integrating these objectives becomes more challenging when patterns of vulnerability are embedded within broadly diffused networks of economic and social activity. Such challenges are multiplied when national governments invoke broader concepts of "national security," whether associated with traditional concepts of national defence, including great power conflicts, significant risks to public safety or social cohesion, or contemporary issues of securing telecommunications, other critical infrastructure, and intellectual property.

The impact of the "9/11" terrorist attacks led to pressures for greater inter-agency coordination in the United States, leading to the formation of the Department of Homeland Security (DHS) in 2003 from twenty-two predecessor agencies.[1] Partial emulation followed in Canada with the creation of Public Safety and Emergency Preparedness Canada and the subsequent publication of Ottawa's 2004 *National Security Strategy*.[2] However, significant asymmetries persist between the two countries' institutions and policies, including the distribution of responsibilities within and across particular executive departments and agencies, despite a narrowing of specific policy differences through the bilateral Beyond the Border (BTB) process. The management of these differences varies from sector to sector – whether in policy fields traditionally associated with "security" or new vulnerabilities arising from critical infrastructure networks. In some cases, they are mitigated by networks of public (and sometimes private) sector actors with shared professional outlooks and technical expertise. In others, constitutional and institutional structures, particularly those resulting from differences in federal systems, establish conditions and limits for cross-border cooperation.

This chapter examines the domestic and intermestic challenges of domestic, cross-border, and wider international cooperation within and beyond North America in the context of this volume's broader exploration of territorial and "aterritorial" determinants of policy. Its first section investigates broader conceptual challenges arising from overlapping, but often different approaches to "homeland security," "public safety," and national security in each country. Its second section addresses the overlapping, but conceptually more diffuse field of critical infrastructure security, with particular attention to electricity and pipeline networks.

It also discusses the evolution of institutional mandates and philosophies of central government agencies in responses to changing risks to domestic security. It assesses the impact of these policies in facilitating or complicating cross-border trade, travel, and interaction of citizens and businesses. Finally, it addresses the challenges of imposing centralized, command-and-control systems of coordination on decentralized networks of economic activity, particularly critical infrastructure, which face growing risks as a result of the ubiquity of the internet and "connected" information technology systems. As a result, conventional issues of data security, emergency management, and business continuity increasingly spill over into the domains of national security and law enforcement, and vice versa. Different approaches to assigning regulatory responsibilities across departments in each country creates selective opportunities for horizontal policy cooperation. However, the proliferation of regulatory actors and their clusters of competing interest groups largely precludes cross-border policy harmonization in many areas.

Domestic Security, Asymmetrical Interdependence, and Evolving Domestic Regimes

Canadian domestic security or public safety policies continue to evolve in the context of evolving risk and threat assessments. Canada retains considerable discretion in its choice of security policy instruments, subject to domestic constitutional disciplines and the workings of Canadian federalism, as demonstrated by responses to the 2020 COVID pandemic. However, interdependence with the United States ensures that its political leaders and security officials must consider risks of what Charles Doran calls "inter-vulnerability": the potential or actual "costs to one's neighbors of one's own policy choices" and vice-versa.[3]

Domestic or "Homeland" Security coordination moved from a largely ad hoc policy function to a central strategic preoccupation of the U.S. government following the 9/11 terrorist attacks of September 2001.[4] The White House issued successive Homeland Security Strategies in 2003, 2007, 2010, and 2014 under an evolving framework of quadrennial reviews. The 2018 review, while "in progress," had not been submitted to Congress by late 2020. The 2003 and 2007 strategies focused primarily on threats of terrorism, although high-profile failures of federal-state coordination after Hurricane Katrina (2005) underlined the limits of this approach. The Obama administration began its shift towards an all-hazards approach in the 2010 review. This approach continues to guide U.S. policies under the Trump administration, despite the latter's penchant to use "national security" levers, however inconsistently, to project U.S. power in international economic disputes.

Homeland Security policies are further embedded within an evolving national security strategy addressing a wide range of national, homeland, economic, and environmental security challenges.[5] This strategy explicitly addresses factors that affect American and global economic prosperity. However, its approach to balancing these objectives has evolved substantially under the Bush, Obama, and Trump administrations. Presidents may also convey policy guidance on particular national or homeland security issues to the Executive Branch through a variety of presidential directives. The legal and administrative forms of such directives are analogous to executive orders in other policy fields.[6]

The design and implementation of such strategies may cultivate shared interests with allied and other countries to enable their governments to build and sustain domestic political and interest group support. Alternately, they may be expressions of "grand strategy" in the tradition of state-centred great power politics: one that "tries to change the existing relationship a country has with another nation or group of nations to something more favourable. Other countries [may] help, hinder or distract the implementation of the ... strategy but [are] unimportant in themselves."[7] A key distinction between the two approaches – or the location of particular policies along a continuum – is the extent to which strategies involve attempts at unilateral imposition of U.S. policies on other countries, as currently under the Trump administration, rather than the negotiation of bilateral or plurilateral agreements and processes based on shared or overlapping interests. As a result, although DHS and other U.S. agencies engage in extensive reciprocal cooperation with counterparts in other countries, U.S. homeland security policies remain resolutely territorial.

Protecting the United States and its citizens from terrorism remains the principal objective of U.S. domestic homeland security policies. However, the 2014 *Quadrennial Homeland Security Review,* echoed in the 2017 *National Security Strategy,* identifies four other key missions: to "secure and manage" U.S. borders, "enforce and administer … immigration laws, "safeguard and secure cyberspace," and "strengthen national preparedness and resilience" – in response to emergencies and challenges to critical infrastructure, including the continuity of government operations.[8]

Implementing this mandate involves a challenge of bureaucratic politics and culture: reconciling the top-down, command-and-control cultures of national security, intelligence, law enforcement, related immigration agencies, and sometimes other regulatory agencies with the highly decentralized, more collaborative governance structures and cultures of emergency management professionals (a primarily provincial/municipal jurisdiction in Canada) and owner/operators of critical infrastructure.[9] This challenge grows substantially when national agencies attempt to coordinate and administer policies with counterparts in other countries – a process described by Slaughter as horizontal policy integration.[10] Such issues become particularly salient when the latter depend on the cooperation and expertise of other regulatory and economic actors operating within decentralized domestic (or intermestic) networks for their effectiveness. These transnational networks are both instigators and consequences of what have been described as "bottom-up integration."

As a result, the effectiveness of strategic policy or "high politics" is often heavily dependent on "low politics" – the interest-driven policies and processes of individual government agencies, private sector, and other societal interests. Informal security cooperation and low-level institutionalization of that cooperation typically respect the territorial integrity of national legal systems while recognizing the practical interdependence of government agencies in a world of extensive trade, travel, and virtual communication.

Inter-agency coordination of domestic security issues shifted to the long-established National Security Council in 2010, reflecting cross-partisan proposals from major Washington think tanks.[11] The political salience of homeland security after 9/11, the continuing interaction of domestic hot-button issues of terrorism, immigration, and border controls throughout the 2000s, and persistent inter-agency competition within the sprawling U.S. federal bureaucracy have all reinforced pressures for centralized oversight and direction of homeland security strategy and its implementation.[12]

At the same time, the realities of international interdependence arising from globalization have led several U.S. agencies to project their
security and other policy priorities through specialized international
agencies responsible for maritime commerce (the Container Security
Initiative of 2003), air travel (IATA electronic passport standards and
coordinated national systems for advanced screening of travellers), and
global financial infrastructure (money-laundering, state-owned or influenced firms based in U.S. strategic competitors) among others, along
with law enforcement cooperation and ongoing intelligence gathering
of major allies (e.g., "Five Eyes"). These security-driven initiatives have
subsequently been complemented by U.S. international regulatory cooperation intended to encourage the complementary domestic regulatory systems with major U.S. trading partners, as discussed in chapter
4, and more sector-specific initiatives of specific agencies. These policies
often provide a reciprocal context for policy cooperation by Canada
and allied states. However, they also involve extraterritorial projections
of U.S. power, inviting combinations of accommodation, balancing, or
countervailing policies among other countries. Farrell and Newman
have characterized the exercise of this leverage on foreign-based firms
with U.S. operations or exposure to the U.S. financial system as "weaponized interdependence."[13]

As a result, U.S. security and regulatory policies involve overlapping
and interlocking domestic and international strategies with widely varying degrees of coherence. For measures affecting Canada, the wide
range of related U.S. government departments and agencies has necessarily involved decentralization in policy design and implementation,
despite efforts at coordination within North America. A Congressional
Research Service report identified thirty federal "entities" with homeland security responsibilities in 2011, with DHS accounting for only 52
per cent of overall U.S. federal expenditures in this field, and the U.S.
Defense Department an additional 26 per cent.[14] The greater the diffusion of governmental responsibility for a particular policy field or
sub-field, the greater the difficulty of pursuing collaborative policy coordination with friendly governments. Departmental or agency leaders
sometimes respond to these challenges by pursuing shared objectives
through relatively decentralized, uncoordinated approaches sometimes described as horizontal "soft law" governance.[15] Homeland security and related border policies under the Trump administration remain
heavily focused on the Mexican border, although growing geopolitical
tensions with China have increased bordering pressures on Chinese
trade and investment, especially in technologically sensitive sectors,
as noted in chapter 5. However, changes to U.S. Temporary Protected

Status policies applying to non-naturalized residents of Haitian (2017) and Central American (2018) origin, along with weaknesses in U.S. visitor visa processes that facilitate transit of other migrants intending to settle in Canada, noted in chapter 6, prompted sharp increases in irregular migration to and asylum applications in Canada, overloading its refugee application and adjudication system.[16]

Canadian Domestic Security Policies

National (including domestic) security policies in Canada reflect a different dynamic than in the United States, balancing international and domestic policy goals within a different set of political constraints. Canada's National Security Policy of 2004 summarized the Martin government's official priorities as: "i) protecting Canada and Canadians at home and abroad; ii) ensuring Canada is not a base for threats to our allies; and iii) contributing to international security."[17] In practice, domestic realities have fostered an all-hazards approach to domestic security (or "public safety") policies under the Martin, Harper, and Trudeau governments, but with widely varying centralization or decentralization, depending on particular subsectoral realities.

Three dominant realities have shaped Canadian federal policy priorities and implementation since 2001 in ways that attempt to blend the accommodation of aterritorial forces and domestic (or "territorial") priorities. First, effective cooperation with the United States is seen as critical to maintaining trade and travel flows on which Canada's prosperity and economic security largely depend. Second, such cooperation is frequently subject to Canada's capacity to preserve sufficient policy discretion to accommodate differences in national legal systems and sector-specific domestic interests to sustain domestic political legitimacy and sovereignty. Third, Canada's highly decentralized federal system limits Ottawa's legal capacity and political will to negotiate international agreements that intrude into areas of sovereign provincial jurisdiction. These realities frame both the context and terms of horizontal policy cooperation described above.[18]

The 2004 National Security Policy identified eight major threats or risks impinging on Canada's national security: (1) terrorism, (2) proliferation of weapons of mass destruction, (3) failed and failing states, (4) foreign espionage, (5) natural disasters, (6) critical infrastructure (CI) vulnerability, (7) organized crime, and (8) pandemics. In practice, subsequent concerns over cyber-security have come to encompass espionage, vulnerability of CI and intellectual property (IP), and identity theft – paralleling similar U.S. concerns. Federal policies to reduce risks of pandemic illness, sparked by the 2003 SARS epidemic that threatened to overwhelm Toronto's public health system, have evolved into broader approaches to public health, particularly the strengthening of Canada's food safety system paralleling the U.S. Food Safety Modernization Act of 2011, as noted in chapter 18.

In practice, successive federal governments have pursued "all-hazards" approaches to policies within their direct jurisdiction – particularly on terrorism, espionage, and other national security issues. However, issues related to organized crime, public health, natural disasters, CI vulnerability, and other emergency management priorities have typically been subject to the vagaries of intergovernmental relations and federal budgetary priorities.[19] Emergency preparedness and management functions of governments typically reflect systems of network governance within and across functional areas of public and private sector activity, many of which are under direct provincial jurisdiction. The global coronavirus (COVID-19) pandemic of 2020 clearly demonstrated the pressures on and limitations of federal border and related emergency management policies in such circumstances, but also the adaptive capacities of governments in both Canada and the United States.

Approaches to domestic security-policy coordination within Canada have typically been more decentralized, although subject to variations in prime ministerial governing styles and what one public administration scholar has called "bolts of electricity" from the Prime Minister's Office (PMO).[20] Under the Harper and Trudeau governments, the Privy Council Office (PCO), or, on occasions, senior PMO officials have managed high-level domestic and bilateral policy coordination with the United States, with officials of line departments such as Public Safety, Natural Resources Canada, Agriculture and Agri-Food, and Health Canada or their respective agencies typically managing operational issues. On matters of high political salience, PMO or PCO officials have discretion to take direct operational control from line departments.

On cross-border issues, the Harper government invested considerable political capital in pursuing the BTB and Regulatory Cooperation Council (RCC) initiatives with the Obama administration after 2010–11. However, such efforts became decreasingly noticeable as cross-border differences over energy and environmental policies, particularly pipeline issues, became more obtrusive. The result was a reversion from central government coordination of bilateral relations to more decentralized cross-border relations between related subject-matter agencies.[21]

Similar patterns have persisted under the Trudeau government, especially since Donald Trump's 2016 election. Although Canada-U.S.

security cooperation continues, particularly at operational levels, it has been overshadowed by the two governments' diverging economic security agendas: different priorities in renegotiating NAFTA, Canadian ratification of CETA, and recalibration of TPP after U.S. withdrawal, and competing perspectives of numerous trade remedy actions. After initial inconsistencies in regulating foreign investment (especially from Chinese or Russian sources) in telecommunications and other high technology sectors with national security implications, the Trudeau government vetoed a Chinese state-owned SOE proposed takeover of construction giant Aecon in 2018 on national security grounds.[22] Thirteen of nineteen transactions subject to national security reviews between 2014–15 and 2018–19 fiscal years, and seven of twelve transactions withdrawn or blocked, involved Chinese-based firms.[23]

Growing U.S.-China geopolitical competition has spilled over into Canadian policies governing telecommunications security on least three levels: concerns over links between global telecom giant Huawei and Chinese security agencies, particularly in supplying equipment for advanced ("5G") wireless telecommunications systems in Canada and other industrial countries, Huawei's use of Canadian researchers in developing related intellectual property, and linkage between Chinese government retaliation against Canadian citizens and exporters arising from U.S. efforts to extradite Huawei executive Meng Wanzhou to face criminal changes in that country.[24] Major Canadian telecom providers have tacitly acknowledged growing security and geopolitical risks by shifting their 5G procurement from Huawei to E.U.-based Nokia and Ericsson.[25]

Cross-Border Security Cooperation

Cross-border security cooperation between the United States and Canada has tended towards horizontal bilateralism, emphasizing departmental level engagement on an issue-by-issue basis. These processes also reflect ongoing professional relationships between military, law enforcement, or intelligence professionals based on shared professional cultures and practices, rather than "binational" processes in the sense of joint policymaking and implementation on major policy issues. This distinction reflects Brister's observations on persistent distinctions between political and professional approaches to bilateral cooperation on security.[26] Issues of national security, border management, and immigration are sufficiently sensitive in both countries that neither government can be seen to "contract out" sovereignty to binational bodies, except in very limited circumstances in which each government employs other policy instruments to maintain its discretion to act in the national interest – as with distinctions between NORAD's mandate, and those of NORTHCOM and Canada Command.[27] In this sense, the legal forms of security cooperation remain formally territorial, while the frequently aterritorial character of security threats enforce the necessity of functional interdependence.

The failure of the Bush administration's trilateral Security and Prosperity Partnership (SPP) process (2005–8) informed the Obama administration's subsequent efforts to engage both Mexico and Canada in separate bilateral arrangements on border security and modernization after 2010 with somewhat greater transparency and public engagement – although still keeping their respective Congresses and Parliament at arm's length. The Trump administration's unpopularity in both neighbouring countries limits incentives for Canadian or Mexican governments to pursue new high-profile security cooperation for the foreseeable future, even while continuing substantial professional and operational cooperation under the political radar.

Evolving Risks, Threats, and Approaches to Governance

The passage of time since the creation of DHS and Public Safety Canada in 2003 has seen the emergence of significant new domestic security threats, and modest increases in cooperative approaches to risk and threat assessment. Three significant shifts address risks of "homegrown" terrorism resulting from internet-driven radicalization, evolving approaches to bilateral and international cooperation in monitoring international travellers and migrants, cyber-security threats arising from foreign (or transnational) state- and non-state actors, transnational and domestic criminal organizations, and domestic "hackers." The last category spills over into the widely distributed, sectorally varied world of critical infrastructure (CI) security.

Law Enforcement and Border Security

Post-9/11 security measures responded initially to risks of jihadist terrorism, along with persistent U.S. domestic concerns over the breakdown of border management and immigration enforcement along its southern border with Mexico.[28] Subsequent geopolitical developments led to tactical shifts by jihadi groups by the mid-2000s involving the largely internet-based recruitment of "homegrown" operatives and sympathizers to carry out terrorist acts in Western countries.[29]

These tactics prompted both strategic and tactical responses by Western governments involving the extensive use of "signals intelligence," including the collection and analysis of telecommunications "metadata" by national intelligence agencies, extensive intelligence-sharing, particularly among national agencies of the "Five Eyes" group, and their law-enforcement counterpart (FELEG).[30] Similar information sharing takes place among Canada and other members of the "Border Five" – paralleling the "Five Eyes" network, and between Canada and the European Union.[31] Cox describes the former as "a cooperative, complex network of linked autonomous intelligence agencies" whose "individual intelligence organizations follow their own nationally legislated mandates,"[32] with both formal and informal cooperation protocols. However, as with professional relationships between U.S. and Canadian militaries noted above, a shared professional ethos and outlook provides the "glue" to bilateral intelligence cooperation. Cox also notes that "the Five Eyes national assessment community is professionally tight, bound by gravities of trust and confidence. Heads of national assessments meet at least annually and joint working groups are formed when needed to address relevant issues of mutual concern. Inter-agency contact is routine at working levels, where the default inclination is to consult widely before assessments are finalized and provided to government."[33]

This author's own research suggests the same principle applies to Canada-U.S. law-enforcement cooperation.[34] In addition to FELEGs, high-level discussions of law enforcement priorities occurs through semi-annual meetings of the U.S.-Canada Cross-Border Crime Forum (CBCF), which is co-chaired by Cabinet officers responsible for law enforcement and border security in each country. Since 1997, CBCF has provided inter-agency coordination to bilateral working groups on operational issues relating to "organized crime, counter-terrorism, smuggling, economic crime and other emerging cross-border threats."[35] Ministers responsible for immigration have also participated in quintet ministerial meetings under the Trudeau government.[36]

Interviews with police and border security agencies in both countries suggest that law enforcement agencies have maintained relatively closed, hierarchical cultures with significant cultural and resource-related boundaries on inter-agency cooperation. However, the far-flung geography and varied law-enforcement challenges of Canada-U.S. border regions have contributed to significant regional innovations.

Differences in the legal systems of the two countries and sensitivities over national sovereignty and the "privileges and immunities" of law enforcement personnel impose constraints on the extent to which bilateral cooperation and/or coordination can be translated into binational (or jointly designed and managed) policy structures. Amendments to security legislation in each country expanding surveillance powers of security agencies have proven controversial, primarily as the result of concerns over civil liberties, debates about trade-offs inherent in external oversight over policy and security agencies, and periodic incidents that remind both Americans and Canadians of their vulnerability to politically motivated violence. The Trudeau government has rolled back some aspects of the Harper government's Anti-Terrorism Act of 2015, strengthening external review of security agencies.[37] These measures suggest a degree of territoriality in Canadian security policies, notwithstanding the extensive cross-border cooperation, including the Beyond the Border process discussed below.

Attempts at Coordination: Beyond the Border

The bilateral Beyond the Border initiative announced in 2011 has been one of the more prominent bilateral security initiatives between the United States and Canada. Its "perimeter security" approach has been intended to enhance security cooperation "within, at, and away from" national borders "while expediting lawful trade and travel" through initiatives coordinated between U.S. National Security Agency (NSA) and Canadian Privy Council Office (PCO) officials, in cooperation with respective national executive / line departments overseeing border security, transportation, immigration, and law enforcement in particular. As such, it has combined centralized oversight in each country with bilateral inter-departmental and inter-agency negotiations to implement rolling annual work plans. Major priorities identified in the initial strategy, such as the development of common risk and threat assessment tools for law enforcement and security officials in both countries have evolved into more targeted measures aimed at subsectors and approaches to the monitoring and facilitation of trade and travel:

- Developing policies and programs to identify and respond to potential offshore threats before they reach North America – notably the Integrated Cargo Security Strategy and harmonization of air cargo screening standards to reduce duplications in screening processes, and parallel food safety screening processes for third countries;
- Common approaches to passenger screening in each country to identify security risks and expedite in-transit travel of third-country nationals between the United States and Canada;
- Improving and extending coordination of trusted traveller and trader programs between the two countries;

- Coordinating investments in border infrastructure to facilitate border flows and, where feasible, facilities and technologies;
- Implementing an entry-exit program in stages using reciprocal data exchange to identify foreign travellers and visa holders leaving each country through the other, extending to include Canadian citizens and permanent residents following passage of Bill C-21 in 2018;
- Expanding the Shiprider program and exploring "next generation" cross-border law enforcement initiatives.
- Exploring options for pre-clearance of cross-border freight between the two countries;
- Conducting regional resilience assessments of border regions under a joint Action Plan for Critical Infrastructure agreed upon in 2012; and
- Expanding cyber-security cooperation and information sharing under a joint Cybersecurity Action Plan agreed upon in 2012.[38]

The range of initiatives under BTB is too broad to enable easy generalizations about its effectiveness. The process appears to have provided a useful framework to enable incremental extensions of existing initiatives – particularly trusted traveller and other traveller-screening initiatives, which offer relatively clear benefits to both countries and also allow for more efficient use of limited resources by national border and air travel security agencies without compromising security. A 2015 bilateral pre-clearance agreement to expand pre-clearance facilities and clarify reciprocal rights and obligations of security officials operating outside their respective jurisdictions was ratified by Congress in 2016 and Parliament in 2017. Canadian officials are exploring pre-clearance pilot projects at major U.S. tourism destinations. However, these processes have been less successful in limiting duplication of freight screening measures hoped for in business circles, or in overcoming national legal differences in the extraterritorial enforcement of border regulations.

The Harper government's decision to join in the administration of the U.S. "entry-exit" system, a long-resisted U.S. initiative seen by Canadian officials as detrimental to efficient border management, appears to have been closely related to the tightening of *Canadian* immigration enforcement measures introduced as part of a broader streamlining of Canadian immigration policies. In this sense, BTB has been a largely decentralized exercise driven by individual executive departments or agencies rather than by political leaders, or economic or societal interests. In general, the harmonization or mutual recognition of major border-related regulations appears to have been contingent on the willingness of domestic regulatory and administrative agencies to compromise or accept the standards proposed by the larger country. Similar patterns have been observed in the processes of the RCC. Cooperation and incremental extension of BTB measures continue under the Trump administration.

However, just as cooperation on national and border security reflects the interactions of two large, dispersed, national and subnational bureaucracies with responsibilities for other aspects of public safety and law enforcement, critical infrastructure security reflects the interaction of very different national – and subnational – regimes governing industries responsible for much of the central nervous system of the modern economy. As such, they tend to reflect the properties of those sectoral regimes – many of which are characterized by substantially more distributed (decentralized) forms of governance.

Critical Infrastructure Security: Policy Parallelism and Sectoral Segmentation

The nature of critical infrastructure (CI) security policies in each country differs significantly from most other security initiatives except for domestic aspects of cyber-security. Various elements of CI – major transportation facilities and networks, oil and gas and electricity transmission systems, financial and payments infrastructures, food inspection and safety systems, and key elements of environmental infrastructure such as water treatment systems for major urban areas – are subject to dispersed systems of national and, in many cases, subnational regulation. Ownership of these facilities is also widely dispersed, generally in private hands, although major utilities in most Canadian provinces are owned by provincial Crown corporations. The concept of CI security incorporates three major forms of risk management: those relating to normal operating risks, including potential technical failures, malicious human-induced risks, and natural disasters.

These risks are typically industry-specific and subject to technical mitigation, whether through the design of operating systems, specialized regulatory systems, or a combination of both. Security is (or is supposed to be) intrinsic to the operation of the particular sector. Questions of criminal law or state security generally apply only at the margins, in cases of overt negligence causing significant harm or malicious human action, whether by domestic or foreign actors, although major investments by foreign-based firms in information technology and telecommunications are often subject to national security reviews, as noted above. However, widespread civil disobedience disrupting major railway networks in 2020 has led to provincial legislation and calls for federal legislation criminalizing obstruction or disruption of critical infrastructure.[39] Nevertheless, most CI security questions are the purview of domestic regimes governing particular economic or industry sectors, which are widely dispersed in both countries, but in the contexts of very different federal distributions of power.

This section explores the challenge of critical infrastructure security and the major forces that have contributed to its evolution since the mid-2000s as applied to two major sectors: electricity transmission and pipelines. These realities are acknowledged in Canada's National Strategy for Critical Infrastructure[40] and similar initiatives in the United States. Regulatory cooperation on food processing is addressed in chapter 18.

CI security issues frequently overlap with those involving emergency management: the anticipation (and attempted prevention or mitigation of), preparation for, response to, and recovery from natural and human-induced events ranging from routine risks to potential disasters.[41] Emergency management networks tend to be driven by sectoral practitioners, backstopped to varying degrees by provincial governments, which exercise primary regulatory responsibility. Provincial governments exercise primary responsibility for emergency management, reflecting their jurisdiction over public safety, municipal governments, and the structuring of emergency services.[42]

The discipline of emergency management, which draws on a mix of risk management principles developed in the insurance industry, along with logistics and contingency planning from the armed forces. It seeks to develop a culture of continuous improvement through an "all-hazards" approach to contingency planning that seeks to identify and address routine risks appropriate to the development of standard-operating procedures and low-probability incidents with significant potential impact to reduce their probability, anticipate and prepare for potential response requirements.[43]

Canada's CI Action Plan identifies ten broad sectors – Energy and Utilities, Finance, Food, Government, Health, Information and Communications Technology, Manufacturing, Safety, Transportation, Water – through which it attempts to "build partnerships," promote "all-hazards" approaches to risk management, and facilitate "timely sharing and protection of information among key stakeholders."[44] However, the diversity of ownership, operating conditions, and jurisdictions reflected in these sectors is sufficiently diverse to require networked rather than centralized, hierarchical, command-and-control responses in identifying and responding to evolving vulnerabilities.

Provincial officials interviewed by the author cite the dynamic nature of security risks (especially relative to the proverbial "speed of government" and its regulatory processes), and the highly compartmentalized expertise of government officials relative to industry counterparts as disincentives to centralized regulatory control, particularly when sectoral regulators already have the legal and administrative resources to respond to significant system failures. Provincial responses to the global coronavirus pandemic of 2020, which have involved "whole of government" initiatives, especially in larger provinces, may ultimately prove an exception to this rule – while testing the effectiveness and adaptability of "business continuity" plans for critical infrastructure and many other areas of governmental and societal activity.

Electricity Networks

Critical infrastructure security in the electricity sector speaks to two principal sets of issues: the protection and reliability of generating facilities, with specific vulnerabilities varying according to the energy sources, processes, and related risks of particular types of power generation (e.g., hydroelectric, nuclear, natural gas, or coal combustion, etc.), and the securing of transmission systems against natural, technical, and human-induced disruptions. Regulatory issues related to the approval of new power plants or transmission corridors, which are closely related to highly contested environmental and land use issues, are beyond the scope of this chapter.

Electricity networks in North America differ from other forms of CI in that they are simultaneously regionalized, heavily interdependent, and subject to extensive state ownership, at least in Canada, although private ownership of power generation and transmission facilities is more frequent in the United States. Provincially owned generation and/or transmission utilities dominate markets in seven of ten Canadian provinces, excepting Alberta, Nova Scotia (Emera) and Prince Edward Island (Fortis). However, the two largest electric utilities in Alberta, Enmax and Epcor, which account for about 40 per cent of output, are municipally owned corporations with extensive operations outside their original market areas. The partial privatization in 2015 of Hydro One, Ontario's dominant transmission firm, has resulted in a provincially regulated mixed enterprise, if one subject to unpredictable policy shifts. The preponderant power of provincial governments in this field enforces some degree of territoriality, notwithstanding the substantial interdependence of provincial grids with broader regional grids.

Figure 10.1. Regional Electricity Coordinating Councils across North America

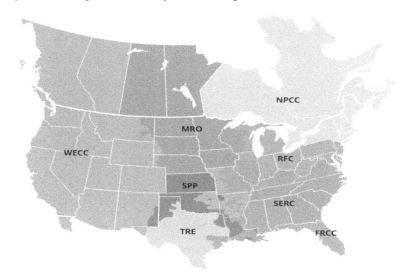

These realities contribute to a complex environment for multilevel governance. Preoccupations with electricity network security achieved continental scope with the "northeast blackout" of November 1965, which cut electricity to Ontario and all or part of eight New England and mid-Atlantic states, triggering voluntary electricity security arrangements among major utilities.[45] A more widespread incident in 2003, which spanned Ontario and seven U.S. states from Michigan to Massachusetts, triggered binational regulatory action to establish the North American Energy Regulatory Corporation (NERC), a non-profit regulatory and standards-setting organization overseen by U.S. and Canadian regulators, which in turn oversee eight regional entities, three of which involve cross-border oversight between the United States and bordering Canadian provinces (see figure 10.1). Major electric utilities on both sides of the border were deeply engaged in these processes, which, as a result of their highly technical and non-transparent character, are among the last politicized elements of electricity regulation in either country.

One reason for the delegated, nongovernmental character of "umbrella" electricity reliability standards is the presence of diverse, asymmetrical national and subnational regulatory systems in the United States, Canada, and Mexico. Until the 1990s, electricity generation and distribution in Canada was overwhelmingly the product of dominant

Table 10.1. Electricity Trade as Percentage of Generation, by Province, 2013–2017

	Interprovincial trade	Net Canada-U.S. trade	Net electricity trade
Canada	0.080*	0.093	0.093
Manitoba	0.020	0.263	0.283
New Brunswick	-0.049	0.147	0.097
Quebec**	-0.128	0.139	0.011
Ontario	-0.007	0.115	0.108
British Columbia	0.020	0.055	0.075
Nova Scotia	-0.053	0.003	-0.051
Saskatchewan	-0.004	-0.001	-0.004
Alberta	-0.026	-0.001	-0.027
Newfoundland and Labrador**	0.711	0.013	0.724
Prince Edward Island	-1.302	n.a.	-1.302

* Interprovincial exports as share of total domestic electricity generation
** Subject to 1969 Churchill Falls Agreement
Source: Statistics Canada, table 25-10-0021-01; author's calculations.

or monopolistic provincial Crown corporations. The Canadian Energy Regulator (formerly National Energy Board) has residual regulatory authority over interprovincial and international transmission lines. However, the dynamics of Canadian federalism, some of which are noted in chapters 13 and 17, have ensured continued provincial primacy in this field. Hale and Bartlett characterize electricity regulation in Canada as a series of concentric circles, with provincial control over electricity production and distribution at the core, embedded within wider (but somewhat thinner) layers of federal and North American regulation.[46]

Moreover, electricity trade is predominantly local and regional. Canadian net electricity exports averaged 9.3 per cent of electrical power generated in Canada (primarily in Quebec, Manitoba, British Columbia, and, more variably, Ontario, see table 10.1) in 2013–17, and 2 per cent of U.S. consumption in 2015.[47] Historically, Canadian imports of U.S. electricity account for a comparable proportion of Canadian consumption. Similarly, as noted in table 10.1, Canada's geographic vastness has limited dependence on interprovincial energy trade for most provinces, except for the highly varied circumstances of three of the four Atlantic provinces. The U.S. Federal Energy Regulatory Commission (FERC) imposed a measure of reciprocity in market access on Canadian utilities as part of its efforts to promote greater regionalization and increased market competition within the United States during the 1990s. However, the intensely politicized character of U.S. energy regulation, the activity of multiple competing interests that seek to influence periodic congressional energy legislation, and the highly protective approach of most Canadian provinces to their jurisdictions over electricity and other energy policies have made electricity reliability organizations an exception to the general rule of policy fragmentation.

The growing preoccupation of federal, provincial, and state governments with expanding renewable energy generation (especially wind and solar) increases tendencies towards continued regulatory fragmentation in the pursuit of electricity and related CI security. Definitions of renewable energy sources, formalized in U.S. federal and state renewable-energy portfolio standards, are highly politicized, varying significantly across jurisdictions. Intermittent sources of electricity generation, particularly wind and solar, impose significant pressures for increased transmission capacity – expanding technical pressures on independent authorities who frequently manage the balancing of loads on widely distributed transmission systems. Given the predominantly local, regionalized, and highly politicized character of electricity generation and transmission systems in both countries, these dynamics are unlikely to change for the foreseeable future.

Pipelines

Pipelines provide the principal form of CI for transmitting petroleum (or diluted bitumen), natural gas, and assorted refined products within North America. Trends towards economic deregulation of oil and natural gas production and distribution in the United States and Canada during the 1980s, combined with rapid changes in prices and other market conditions, contributed to the progressive integration of North American markets (partially excluding Mexico) during the 1990s and 2000s. CI security in the pipeline sector addresses three major sets of issues:

- The potential disruption of pipeline networks due to natural or human-induced hazards, ranging from the hurricanes that shut down about one-quarter of North American refining capacity in 2005 and 2006, to the Fort McMurray forest fire, which shut in about 30 per cent of Canada's oil production for several weeks in 2016, to occasional acts of sabotage and growing interest-group resistance to pipeline expansion or modernization;
- Physical and technical failures resulting in risks to public safety and the environment, particularly substantial leaks near major bodies of water;

- Failures in computerized control systems designed for remote monitoring and control of pipeline systems, whether due to human or technical error or potential cyber-sabotage.

Pipeline regulation in Canada is divided between the federal Canadian Energy Regulator, which oversees technical and environmental regulation of about 73,000 kilometres of interprovincial and international pipelines, and regulators within energy producing provinces, which set standards for more than 750,000 kilometres of intraprovincial pipelines.[48] Pipeline regulation in the United States is even more complex, with regulatory authority divided between FERC, the Pipeline and Hazardous Materials Safety Administration (PHMSA), which oversees railway and pipeline safety issues, the Environmental Protection Administration (EPA), with its wide-ranging responsibilities for environmental regulation, and diverse patterns of state and tribal government regulation, generally related to land use and local environmental risks. Most pipeline ownership is controlled by investor-owned firms, with both major Canadian-based firms such as TC Energy (formerly TransCanada), and Enbridge having extensive operations in both countries, along with several other midstream firms based in each country.

A key factor in the nature and extent of cross-border regulatory coordination is the requirement for cross-border pipelines to secure a presidential permit, whether for imports or exports hydrocarbons, on national security grounds. Such requirements also apply to transfers of ownership, thus potentially affecting the 2018 sale of the U.S. segment of Kinder Morgan's Trans-Mountain pipeline to the government of Canada.[49] During the 1990s and early 2000s, when the United States was heavily dependent on oil and gas imports and, latterly, attempting to reduce its dependence from politically unstable regions, State Department (or presidential) approval was a relative formality subject to more technical regulatory considerations. Subsequent politicization of such permits has reflected partisan and ideological polarization in the United States, as in Canada.

Regulatory coordination between the two countries is relatively informal, with regular meetings between CER and FERC officials, and parallel processes for ongoing improvements in pipeline standards, which are frequently the product of non-profit standard setting bodies such as CSA Group (formerly Canadian Standards Association). In Canada, standards may be established by regulatory reference or with modifications specified by regulatory agencies.[50] In the United States, these functions are performed by the Pipeline Advisory Committees to

PHMSA (2016), which are composed of equal numbers of government, industry, and public representatives under U.S. law.

However, the construction of new pipelines has become intensely controversial since the mid-2000s as the result of decisions by environmental groups to challenge the industry's legitimacy as part of a broader campaign to reduce fossil fuel use by constricting supply, and attempting to externalize the costs of adjustments to climate change: a major aterritorial pressure point. In Canada, despite greater efforts at federal policy coordination, these issues have been compounded by expanding judicial recognition of First Nations' land claims, and a duty of governments and industry to consult with First Nations on projects with the potential to affect treaty and traditional lands.[51] Provincial opposition in Quebec and British Columbia to pipeline expansion played a major role in the 2017 cancellation of the proposed Energy East pipeline and substantial delays in (and threatened cancellation of) the proposed expansion of the Trans-Mountain pipeline between Alberta and BC, despite federal regulatory primacy over interprovincial and international pipelines, culminating in Ottawa's purchase of the pipeline in 2018.

Cross-border pipelines in the United States still remain subject to extensive state-level and even local regulation, particularly on approvals for upgrading or rerouting pipelines, as with Enbridge's Line 3 linking Manitoba with Superior, WI, through the state of Minnesota. These processes frequently involve the management of environmental risks, as with Enbridge's Line 5 linking Wisconsin with Sarnia, Ontario's, petrochemical complex, across Lake Michigan's Mackinac Strait and the St. Clair River. They may also require negotiations with U.S. tribal governments, whose legal authority, administrative capacities, and willingness to enforce them have grown substantially since these pipelines were built in the 1950s and 1960s. Minnesota's acceptance of the rebuilding (and partial rerouting) of Line 3 in 2018, although further delayed by litigation, was directly linked to the deterioration of the sixty-four-year-old structure.[52] Enbridge's lacklustre response to a major 2010 oil spill on Michigan's Kalamazoo River made the company liable for US$1.2 billion in clean-up costs and more than US$250 million in federal and state legal penalties.[53] It has also led to calls for the replacement or removal of Enbridge Line 5, part of which runs across the bottom of Lake Michigan on its way from Superior, WI, to Sarnia, ON – demonstrating that while governance of CI security is resolutely territorial (if highly fragmented), threats to CI security may well transcend national borders.

As a result, while horizontal cooperation between national regulators continues on bilateral issues of pipeline security, state regulators are assuming growing importance due to controversies over land use and environmental issues. Broader domestic political conflicts have shifted opportunities for broader policy cooperation away from North American energy security towards the development of technologies that support new forms of renewable energy.

Conclusion: A Work in Progress

The decentralized, networked systems associated with critical infrastructure tend to lend themselves to the all-hazards approaches to security, which characterizes emergency management systems across a wide range of economic sectors and governmental actors. Such systems usually impose general internal requirements for the creation of internal responsibility systems and provide for periodic review by and interaction with regulatory authorities. By their very nature, they involve adaptation and integration with the structures and cultures of businesses and organizations of widely varying scale, operational challenges, and levels of organizational capacity. As such, from a policy perspective, they are more likely to lend themselves to horizontal cooperation across jurisdictions and within sectors, and to continuing "trial and error" approaches to policy and organization learning, reinforced by periodic focusing events, which provide incentives for policymakers and business leaders to identify weaknesses and opportunities for improvement.

However, the territorial imperatives of national security and the decentralized character of land use and environmental regulatory regimes in both countries have generally enforced territorial primacy in infrastructure regulation and development, notwithstanding the pressures of economic interdependence and market integration. However, the historically state-centred character of national security policies and growing societal polarization over energy and environmental issues militate against such outcomes, without having yet provided the means for effective domestic, let alone cross-border coordination on these issues.

NOTES

1 Peter Andreas and Thomas J. Biersteker, eds., *The Rebordering of North America* (New York: Routledge, 2003); Edward Alden, *The Closing of the American Border* (New York: Harper, 2008).
2 Canada, *Securing an Open Society: Canada's National Security Policy* (Ottawa: Privy Council Office, 27 April 2004).

3 Charles Doran, *Forgotten Partnership: U.S.-Canada Relations Today* (Baltimore, MD: Johns Hopkins University Press, 1984), 8, 53–4.

4 Alden, *Closing of the American Border*; Shawn Reece, *Defining Homeland Security: Analysis and Congressional Considerations*, CRS Report R42462 (Washington, DC: Congressional Research Service, Library of Congress, 8 January 2013), 4.

5 United States, White House, "Executive Order 13609: 'Promoting International Regulatory Cooperation,'" Washington, DC: Office of Management and Budget, 4 May 2012, https://www.federalregister.gov/documents/2012/05/04/2012-10968/promoting-international-regulatory-cooperation; United States, White House, *National Security Strategy of the United States*. Washington, DC: December 2017. https://www.whitehouse.gov/wp-content/uploads/2017/12/NSS-Final-12-18-2017-0905.pdf.

6 Harold C. Relyea, *Presidential Directives: Background and Overview*, CRS Report #98-611 (Washington, DC: Congressional Research Service, Library of Congress, 28 November 2008), https://www.fas.org/sgp/crs/misc/98-611.pdf.

7 Peter Layton, "Making a Canada National Security Strategy," *On Track* (Summer 2015): 37.

8 United States, Department of Homeland Security, *Quadrennial Homeland Security Review* (Washington DC: 18 June 2014), 6.

9 Christopher Bellavita. "Changing Homeland Security: What Is Homeland Security?" *Homeland Security Affairs* 4, no. 2 (2008), https://www.hsaj.org/articles/118; Jason Clemens and Brian Lee Crowley, eds., *Solutions to Critical Infrastructure Problems: Essays on Protecting Canada's Infrastructure* (Ottawa: Macdonald Laurier Institute, 2012), http://www.macdonaldlaurier.ca/files/pdf/Solutions-to-Critical-infrastructure-Problems-February-2012.pdf.

10 Anne-Marie Slaughter, *A New World Order* (Princeton, NJ: Princeton University Press, 2004).

11 David Heyman and James Jay Carafano, *Homeland Security 3.0: Building a National Enterprise to Keep America Safe, Free and Prosperous* (Washington, DC: Center for Strategic and International Studies and Heritage Foundation, 18 September 2008), http://www.heritage.org/research/reports/2008/09/homeland-security-30-building-a-national-enterprise-to-keep-america-safe-free-and-prosperous.

12 Alden, *Closing of the American Border*.

13 Henry Farrell and Abraham L. Newman, "Weaponized Interdependence: How Global Economic Networks Shape State Coercion," *International Security* 44, no. 1 (Summer 2019): 42–79.

14 Reese, "Defining Homeland Security," 1.

15 Slaughter, *New World Order*; Geoffrey Hale, *So Near Yet So Far: The Public and Hidden Worlds of Canada-US Relations* (Vancouver: UBC Press, 2012).

16 Canada, Citizenship and Immigration Canada, "2017 Asylum Claims" (Ottawa: 22 March 2018); Michelle Zilio, "Asylum-Seeker Surge at Quebec Border Chokes Refugee System," *Globe and Mail*, 12 September 2018, A1.

17 Canada, *Securing an Open Society*, vii.

18 Hale, *So Near Yet So Far.*

19 Andrew Graham, *Canada's Critical Infrastructure: When Is Safe Enough Safe Enough?* (Ottawa: Macdonald Laurier Institute, December 2011), http://www.macdonaldlaurier.ca/files/pdf/Canadas-Critical-Infrastructure-When-is-safe-enough-safe-enough-December-2011.pdf; confidential interviews, provincial and municipal emergency management officials, 2005–13; Johanu Botha, *Boots on the Ground* (Toronto: University of Toronto Press, forthcoming); Grant Robertson, "Going Dark: 'Without Early Warning You Can't Have Early Response': How Canada's World Class Pandemic Alert System Dailed." *Globe and Mail*, 25 July 2020, A12–14.

20 Donald A. Savoie, *Governing from the Centre: The Concentration of Power in Canadian Politics* (Toronto: University of Toronto Press, 2000).

21 Geoffrey Hale, "Regulatory Cooperation in North America: Diplomacy Navigating Asymmetries," *American Review of Canadian Studies* 49, no. 1 (2019): 123–49.

22 Robert Fife and Steven Chase, "Chinese Takeover Would Bar Aecon from Canada-U.S. Bridge Bidding," *Globe and Mail*, 5 April 2018, A1; Fife and Chase, "Aecon Deal Threatened Sovereignty, PM Says," *Globe and Mail*, 25 May 2018, A1.

23 Geoffrey Hale, "State-Owned and Influenced Enterprises and the Evolution of Canada's Foreign Direct Investment Regime," in *The Changing Paradigm of State-Controlled Entities Regulation*, ed. Julien Chaisse and Jędrzej Górski (Springer, forthcoming).

24 Robert Fife and Steven Chase, "Ottawa Won't Rush Huawei Decision Despite U.S. ban," *Globe and Mail*, 16 May 2019, A1; Fife and Chase, "China Is Changing the Geopolitical Climate. Canada Has to Mitigate and Adapt," *Globe and Mail*, 17 May 2019.

25 *National Post*, "As Ottawa Dithers, Telcos Shun Huawei" (editorial), 6 June 2020, A16.

26 Bernard J. Brister, "The Same Yet Different: The Evolution of the Post-9/11 Canada–United States Security Relationship," in *Borders and Bridges: Navigating Canada's Policy Relations in North America*, ed. Monica Gattinger and Geoffrey Hale, 82–99 (Toronto: Oxford University Press, 2010).

27 Joseph Jockel, *Canada in NORAD, 1957–2007: A History* (Montreal and Kingston: McGill-Queen's University Press, 2007).

28 Alden, *Closing of the American Border.*

29 Marc Sageman, *Leaderless Jihad: Terror Networks in the 21st Century* (Philadelphia: University of Pennsylvania Press, 2008).

30 The original "Five Eyes" group includes national agencies responsible for
    signals intelligence and, to varying degrees, cyber-defence and warfare,
    including the U.S. National Security Agency (NSA), Canada's Commu-
    nications Security Establishment (CSE), and counterparts in the United
    Kingdom, Australia, and New Zealand. See James Cox, "Canada and the
    Five Eyes Intelligence Community" (Calgary: Canadian Foreign Affairs
    and Defence Institute, 18 December 2012), https://www.opencanada.org
    /features/canada-and-the-five-eyes-intelligence-community/.

    The Five Eyes Law Enforcement Group is a law enforcement con-
    sortium chaired by the U.S. Federal Bureau of Investigation (FBI), in-
    volving the U.S. Drug Enforcement Agency, U.S. Homeland Security
    Immigration and Customs Enforcement (ICE), and the RCMP, national
    law enforcement agencies from the United Kingdom, Australia, and
    New Zealand. United Kingdom, National Crime Agency, "NCA DG at
    Five Eyes Law Enforcement Group Meetings in Washington" (London:
    17 June 2016).
31 National Security and Intelligence Committee of Parliamentarians, *Annual
    Report: 2019* (Ottawa: House of Commons, 12 March 2020), 133.
32 Cox, "Canada and the Five Eyes Intelligence Community."
33 Cox, "Canada and the Five Eyes Intelligence Community."
34 Confidential interviews, government of Canada 2006–15.
35 Canada, Public Safety Canada, "Integrated Border Enforcement Teams"
    (Ottawa: 1 December 2015).
36 Canada, Public Safety Canada, "Canada Hosts Five Country Ministe-
    rial Meeting and Quintet of Attorneys General" (Ottawa: 28 June 2017),
    https://www.canada.ca/en/public-safety-canada/news/2017/06/canada
    _hosts_fivecountryministerialmeetingandquintetofattorneysg.html; Can-
    ada, Public Safety Canada, "Five Country Ministerial Meeting and Quintet
    of Attorneys General Concludes" (Ottawa: 30 August 2018).
37 Kent Roach and Craig Forcese, "The Roses and the Thorns of Canada's
    New National Security Bill," *Maclean's*, 20 June 2017.
38 United States, White House, and Canada, Privy Council Office, "Beyond
    the Border Implementation Report (Washington and Ottawa: 19 December
    2013), https://www.dhs.gov/sites/default/files/publications/btb-canada
    -us-final_-_dec19_0.pdf; United States, White House, and Canada, Privy
    Council Office, *2014 Beyond the Border Implementation Report to Lead-
    ers*"(Washington and Ottawa: 13 May 2015), https://www.dhs.gov/sites
    /default/files/publications/15_0320_15-0745_BTB_Implementation
    _Report.pdf.
39 Alberta. *2020 – Bill 1 – Critical Infrastructure Defence Act*, Second Session,
    30th Legislature, 69 Elizabeth II; Erin O'Toole, "What Must Be Done," *Na-
    tional Post*, 26 February 2020, A9.

40 Canada, Public Safety Canada, "National Strategy for Critical Infrastruc-
    ture" (Ottawa: 15 December 2009), http://www.publicsafety.gc.ca/cnt
    /rsrcs/pblctns/srtg-crtcl-nfrstrctr/srtg-crtcl-nfrstrctr-eng.pdf.
41 George D. Haddow and Jane A. Bullock, *Introduction to Emergency Manage-
    ment*, 2nd ed. (Burlington, MA: Elsevier, 2006).
42 Daniel Henstra, ed., *Multilevel Governance and Canadian Emergency Man-
    agement Policy* (Montreal and Kingston: McGill-Queen's University Press,
    2013).
43 Geoffrey Hale, "Emergency Management in Alberta: A Study in Multi-Level
    Governance," in *Multilevel Governance and Canadian Emergency Management
    Policy*, ed. Daniel Henstra, 134–89 (Montreal and Kingston: McGill-Queen's
    University Press, 2013).
44 Canada, Public Safety Canada, "National Strategy for Critical
    Infrastructure."
45 G. Bruce Doern and Monica Gattinger, *Power Switch: Energy Regulatory
    Governance in the Twenty-First Century* (Toronto: University of Toronto
    Press, 2003).
46 Geoffrey Hale and Cailin Bartlett, "Managing the Regulatory Tangle: Criti-
    cal Infrastructure, Security and Distributed Governance in Alberta's Major
    Traded Sectors," *Journal of Borderland Studies* (2018) 34, no. 2 (2019): 257–79,
    https://doi.org/10.1080/08865655.2017.1367710.
47 Statistics Canada, "Electrical Power, Electrical Utilities and Industry:
    Annual Supply and Disposition," table 25-10-0021-01 (formerly CANSIM
    table 127-0008), (Ottawa: 19 July 2019); Doug Vine, "Interconnected: Cana-
    dian and U.S. Electricity" (Arlington, VA: Center for Energy and Climate
    Solutions, 2017).
48 Hale and Bartlett, "Managing the Regulatory Tangle."
49 Shawn McCarthy, "Federal Government's Proposed Trans-Mountain Deal
    Could Require Trump's OK," *Globe and Mail*, 17 July 2018, A4.
50 Greg Orloff, "CSA Group Safety Standards for Oil and Gas Pipelines: A
    Life-Cycle Approach (Toronto: CSA Group, January 2015); Hale and Bart-
    lett, "Managing the Regulatory Tangle."
51 Dwight G. Newman, *Revisiting the Duty to Consult* (Saskatoon, SK: Purich
    Publishing, 2014); Geoffrey Hale and Yale Belanger, "From Social Licence
    to Social Partnerships: Promoting Shared Interests for Resource and Infra-
    structure Development." Commentary #440 (Toronto: C.D. Howe Institute,
    December 2015).
52 Mike Hughlett, "State Regulators OK Certificate of Need for Controversial
    Enbridge Pipeline," *Star-Tribune*, Minneapolis, 29 June 2018.
53 Geoffrey Morgan, "Enbridge Settles in U.S. over 'Humbling' Spills," *Finan-
    cial Post*, 21 July 2016, FP1.

11 Environmental vs. Territorial Borders: Canada-U.S. Cooperation on Environmental Issues and the Resilience of Transboundary Governance

DEBORA VANNIJNATTEN AND CAROLYN JOHNS

Bilateral and regional cooperation on environmental issues remains among the most varied, complex, and challenging in Canada's international policy relations – particularly given recent political and policy developments in the United States. Yet, despite very different policy stances taken in the White House on environmental protection and climate change under the Trump administration, Canada had to continue to engage the Americans at all levels of governance in transboundary environmental cooperation. Not only are there decades-old, multi-scale policy and institutional legacies in which both countries have heavily invested, but Canada is also seeking to fulfil environmental protection commitments domestically and on the international stage that rely, in large part, on cooperation across the border. While President-elect Joseph Biden has environmental policy aspirations similar to those of the Trudeau government, Canada will still need to use all mechanisms available to pursue successful cooperation with its American neighbours.

This chapter addresses the "bordering" of environmental policy within North America, in the context of broader continental developments; it explores the architectures established to manage environmental problems that do not adhere to conventional borders, and it looks at the degree to which policymakers have sought to fashion this architecture to ecosystems and to the economic dynamics underlying environmental problems. We seek here to answer the central research question that the editors have posed for this volume: to what extent have developments in specific policy fields been shaped primarily by territorially defined forces, by territorial responses to U.S. or other transnational/aterritorial forces, or through more formalized international cooperation?

The first section of the chapter provides some insights into bordering processes from the border theory, new regionalism, and North American relations literatures. It highlights the distinction that is generally drawn between the conventional focus on territorial boundaries of state sovereignty (and political and diplomatic means for managing those boundaries) and aterritorial borders that evolve dynamically through the interaction of domestic and international economic, social, political, and environmental forces. These forces may result in networks or different kinds of institutions that can bridge or work alongside territorial institutions. We use this analysis to show how environmental policy is bordered in North America – through shared Canada-U.S. ecosystems and via integrated economic sectors and pathways.

We focus here on two different environmental policy spheres: the management of complex water systems, using the example of the Great Lakes basin, and climate change mitigation. These policy spheres engage policymakers at all scales of governance, across sectors and government and societal spheres. For both there are cross-border pressures for governance and management that are tied less to territory and legal and bureaucratic "turf," as suggested in chapter 1, and more to the dimensions of the problems being solved. The Canada-U.S. water management case provides the opportunity to observe cooperative infrastructure that has used a cross-border regional approach superimposed on the boundaries of a large-scale water basin; the regime balances territorial jurisdiction with formal aterritorial institutions and networking. We then compare it with Canada-U.S. climate change cooperation, which is a more recent phenomenon and has tended to operate more informally, through networks and pathways tied to energy production and use.

Finally, we attempt to provide initial insights into what makes transboundary architectures more resilient, especially as Canada attempts to continue its cooperative work with the United States, even when there are deeply rooted political differences. We would argue that the networks we see developing in both cases, which tend to be organized on an aterritorial basis and more focused on specific environmental problems, help formal institutions and governments adapt to the shared dimensions of problems and solutions. Networks allow for higher-level functioning through informal dimensions that integrate new knowledge more effectively and engage key actors beyond just state actors and government officials. However, our cases also provide some indication that, in order to achieve greater transboundary cooperation and even coordination, there needs to be a combination of mutually reinforcing aterritorial networks and more formal territorially based mandates based on legal authority. At higher scales (i.e., binational and global, as opposed to watershed), this combination is much harder to achieve, given the manner in which mandates and authority are distributed. It is also harder to accomplish when formal authority is not being exercised or when it is being subverted, as was the case with U.S. environmental policy under Trump. When formal authority is not exercised, there is more pressure on networks to carry the heavy load of cooperation. Much depends, then, on network attributes and strength.

The Bordering of Environmental Policy in North America 1: Bioregionalism, Ecosystem Management, and Cross-Border Regions

Border theory, as it has evolved over the past few decades, has pursued a re-imagining and redefining of "borders" as something other than merely the physical markers of state territory and sovereignty.[1] Moving away from the early study of borders as concerned with formal state frontiers, agreements, and intergovernmental institutions, border theory and studies of regionalism have sought manifestations of "bordering" at various geographical scales and in diverse economic, sociological, cultural, political, technological, and environmental dynamics. Often led by those who study European borders and regionalism, bordering processes are seen as reflecting flows of ideas, culture, and commerce, as well as strengthened discourses on shared institutions, consensus-building, and action-taking[2] within channels that span (or perhaps even disregard) national boundaries. Indeed, the European project can be understood only as a macro-scale attempt to overcome territorial boundaries through strategic application of policies and programs, symbols and resources, as well as the organization of economies, in ways that encourage the growth of extraterritorial networks and a sense of community untethered from national citizenship.[3] In North America, though in a much more modest way, it was thought (and hoped, by some) that the passage of NAFTA and the side agreements on environment and labour, along with a new recognition of increased flows of people, ideas, and culture across continental borders, would encourage shared purposes and dialogues that could challenge state-led exclusivism and asymmetric political and economic relations.[4]

Alongside these more fluid understandings of borders and bordering, however, is the consensus that the nation state has not become weaker but rather is becoming "redefined and relocated."[5] Predictions of a largely borderless world, where commerce, culture, and people follow their own impulses unencumbered by state constraints,[6] have not (and are very unlikely to) come to pass. At the same time, a *very* large literature tracks the many ways in which the state must now govern differently (i.e., with the help of others, both inside and outside government) and use different tools (which rely on collaboration domestically and across borders rather than conventional forms of authority) as a result of the intersectoral and intermestic nature of almost everything it does.

In this context, discussion continues about the role of networks in providing critical linking functions that are difficult for the formal apparatus of the state to undertake, particularly given its preoccupation with maintaining its "capacity for choice." In particular, those who study international relations argue that networks – collections of linkages across domestic agencies, scientific communities, and nongovernmental organizations that create relational channels of communication – can facilitate and manage growing transnational economic, political, and environmental interdependence in ways that support shared problem-solving.[7] While recent political developments might lead us to conclude that we are moving in the direction of stronger (in some cases, even authoritarian) state constraints that threaten to choke off transborder ties, the capillaries of connection – particularly via networks – are many and deep.[8]

A major objective for students of regionalism in North America has been to discover the presence and strength of transboundary "borderlands," and the dynamics and mechanisms that support them. The large research effort by the federal government's Policy Research Initiative in the mid-2000s sought to uncover the strength of economic, cultural, and political ties among subnational entities along the Canada-U.S. border, and contribute to "cross-border regional" borders that render national borders less important. These findings suggested that "too little attention has been paid to the regional and subnational dimensions of Canada-U.S. relations" and that "it is in cross-border regions that Canada-U.S. relations are the most dynamic and intense."[9] The final report also found that stronger and more diversified trade and economic linkages, shared socio-cultural values, and a proliferation of new institutions and networks for cooperation within cross-border regions (especially in the West, Great Lakes Heartland, and Atlantic Canada–New England) meant that shared problems were more likely to be solved within these regions by subnational officials and interests than in formal bilateral forums.[10]

Those who study Canada-U.S. environmental cooperation also advocate for unconventional bordering and regional cooperative architecture to match. Here the overriding concern is the development of management mechanisms that are matched to *human-ecosystem interactions*, rather than territorial boundaries. As Ronnie Lipschutz explains, "Some of the most important causes and consequences of global change ... are inevitably distributed in uneven fashion over a large number of bounded nation-states, cultures and societies, all of which complicates problem-solving. But the very real existence of complex social linkages underlines a fundamental problem in thinking about approaches to environmental protection. If existing borders are a problem, when and how are we to draw boundaries for managing resources so as to facilitate such protection and prevent damage.... For historical and economic reasons, the jurisdiction of all governments matches poorly to nature."[11] Perspectives on bioregionalism and ecosystem management provide insights into how we might imagine cooperation that is less constrained by sovereignty and territorialism and better suited to environmental realities.

As Taylor explains, "Bioregionalism is nothing less than the creation of sustainable human societies in harmony with the natural world," living in communities "within political boundaries redrawn to reflect the natural contours of differing ecosystem types."[12] Bioregionalism connects a concern for the intrinsic value of nature with socio-environmental considerations and specific resource management strategies.[13] The assumption here is that resource and environmental stewardship will be enhanced through regional and local efforts that are rooted in an attachment to place, regardless of formal boundaries. Higher-level (national) environmental management, it is believed, has failed to recognize these critical socio-ecological links. "Cascadia" is perhaps the most tangible manifestation of bioregionalism in North America. Linked to the independence movement "Cascadia Now,"[14] adherents argue that existing boundaries and borders do not fit the current linkages between people, place, and the environment; a jurisdiction that encompasses considerable sections of British Columbia, Washington State, and Oregon would be better able to represent the true interests of people in the region. Adherents argue for the need "to break down boundaries and borders that are arbitrary or negative; shift our actions and impacts locally."[15]

In a similar vein, for decades the ecosystem-based management (EBM) approach has advocated a reimagining of borders for environmental and resource policy, while recognizing that to actually do so "would require that the conventional [political] matrix be unraveled and rewoven in a new pattern."[16] EBM provides a framework for thinking about how we "move from the conventional approach" – resource management defined by pursuing maximum sustainable yields of a single resource within territorial borders – to focus on "managing essential ecological processes that sustain the delivery of harvestable resources and ecosystem services at multiple scales."[17] The ecosystems approach reflects social-ecological systems thinking, which focuses on linking human and ecological systems across scales and using the tools of multilevel governance.[18] It is, like bioregionalism, a place-based approach, though its emphasis on systemic interactions acknowledges and embraces a more complex reality. It defines management borders based on the need to preserve ecosystem integrity and advocates for working across scales, linking local managers and stakeholders that know the ecosystem best to national leadership but also across sectors through interagency cooperation.[19]

This approach, then, points to the need for networks that can link and integrate – across scales, agencies, sectors, and government and nongovernment organizations. Environmental protection in the twenty-first century requires an integration of efforts to aid in ecosystem management, which is most likely achieved by bringing together the diversity of management entities that are implicated in the management of an ecosystem.[20] Certainly this accords with bioregionalism, with its recognition that ecosystems, economic activities, and societal interactions are linked across borders, and that problems have shared roots and impacts. The research on cross-border regions also shows that policymakers and organized interests within the region are adept at working around or alongside formal institutions by creating new networks that can scale problem-solving architecture to the boundaries of problems.[21] However, these schools of thought also recognize the importance of linking more local efforts to leadership at higher scales, implying that authority and mandates – or the formal spine – do matter for supporting effective cooperation.

What might we look for, then, in the Canada-U.S. transboundary context, where environmental cooperation represents a significant and necessary undertaking by the two governments and myriad subnational governments along with nongovernmental organizations? To what extent has the cooperative architecture been able to operate with or alongside conventional territorial borders and create cross-border linkages that match the ecosystems and environmental problems that are shared and must be managed together? The analysis we provide below of two cases of North American environmental governance – shared water management and cooperation on climate change mitigation – illustrates a common concern with matching cooperative governance to the true dimensions of environmental problems. However, the manner in which this architecture has evolved to match the dimensions of each problem differs. This stems at least partly from their differing aterritorial dynamics. Whereas for water management the imperative is the need to manage a shared large-scale, regional ecosystem, in climate mitigation aterritorial forces reflect complex economic currents and linkages that run in all directions independently of national boundaries.

Case Illustration: Transboundary Water Management in the Great Lakes Basin

The Great Lakes is the largest freshwater basin on earth, containing roughly 20 per cent of the world's freshwater supply and 84 per cent of North America's supply. It is arguably the most significant of the many Canada-U.S. border water systems.[22] Like other large water systems, the Great Lakes are sensitive to the effects of a wide range of point- and non-point-source pollutants from their users, as well as human and economic activities in the region. For over forty years, it has been recognized that this large-scale ecosystem is experiencing pollution problems that require multi-scale, integrated management.

Yet this is a hugely complex task. The five lakes and their draining river systems collectively span more than 1,200 kilometres straddling the U.S.-Canadian border encompassing two provinces and eight U.S. states, more than three thousand municipalities, and hundreds of Indigenous communities.[23] Cooperative efforts in the Great Lakes have become highly institutionalized and formalized, beginning over a hundred years ago with the signing of the 1909 Boundary Waters Treaty. The treaty established the International Joint Commission (IJC) as a unique transboundary institution to resolve binational water disputes. Headed by six commissioners (three appointed by the U.S. president and three by the Canadian prime minister), the IJC acts on references from both governments to co-operatively address disputes over the use of water resources. However, the treaty also states that "the waters herein defined as boundary waters and waters flowing across the boundary shall not be polluted on either side to the injury of health or property on the other,"[24] and the IJC was given a mandate to address environmental concerns as well. The IJC thus became the guardian of a shared perspective on environmental management of the lakes.

Over the 1960s, growing public concern about the deterioration of water quality stimulated citizens to push for more concerted basin-wide action by Canadian and U.S. governments, and the binational architecture became more fully articulated. In 1972, the federal governments in Canada and the United States signed the Great Lakes Water Quality Agreement (GLWQA). In light of "the grave deterioration of water quality on each side of the boundary," the agreement aimed – quite ambitiously – to "restore and enhance water quality" as well as prevent future pollution.[25] In a manner similar to the 1909 treaty, it established the lakes as a shared "commons" and the two nations as jointly responsible stewards of this freshwater resource.

The agreement, which is an "executive agreement" between the two countries and does not bind them in the same way as the Boundary Waters Treaty, has been described as "unprecedented in scope."[26] It laid out basin-wide general objectives enjoining the signatory federal governments to keep the waters free of pollutants, and it included specific objectives to reduce these pollutants in the lakes. The parties also committed to joint implementation measures to meet these objectives, specifically to establish municipal and industrial pollution control programs and also to engage in binational cooperative programming. The IJC was to support achievement of the objectives by monitoring, collecting, analysing, and disseminating water quality data, and the operation and effectiveness of the agreement as a whole. The vision underlying the GLWQA, then, was aterritorial in that it focused on management and oversight of the shared basin, while implementation remained largely territorially based, with governments on both sides of the border having mandates and authority to implement and the responsibility to coordinate with other levels of government.

Revisions made to the GLWQA in 1978 and 1987 broadened the binational governance system in ways that created cooperative networks. The 1978 GLWQA introduced the more complex "ecosystem approach" into the water-quality management regime as well as a more concerted focus on point sources of pollution. Both meant that a wider range of authorities and interests would need to be involved. The 1987 revisions were even more substantial, introducing forty-three "areas of concern" (AOCs) to focus policy and clean-up on particularly polluted watersheds. The federal governments agreed to create multi-stakeholder groups that would implement Remedial Action Plans (RAPs) for each AOC in order to address "beneficial use impairments." The 1987 revisions also ushered in the development of Lakewide Management Plans (LaMPs) to address whole lake contamination by persistent toxic substances. These additions to the agreement were designed to fully embrace and link local, watershed-focused efforts with subnational, national, and international efforts in the region. The focus on AOCs, RAPs, and LaMPs created cross-border networks of government and nongovernment actors as well as local organizations. These networks were very productive in some areas, perhaps less so in others.[27]

Although there was progress during the 1980s in regulating and reducing point sources, leadership by the IJC and Canadian federal government began to wane by the mid-1990s.[28] This affected cross-border relations and thus what was being achieved.[29] While the Americans continued to follow through on statutory and budgetary commitments, as well as mandates in the GLWQA, the Great Lakes Legacy Act of 2002 and the Great Lakes Regional Collaboration in 2004, cross-border relations and activity waned as the Canadian federal government was largely absent from the region's environmental policy for over a decade.[30]

Particularly disruptive was the lack of Canadian support for agency personnel and nongovernmental organizations that had been working cooperatively with their counterparts across the border within networks to carry out the expanded list of GLWQA tasks.[31] These networks of Canadian and U.S. actors, which had focused on AOCs and other environmental work under the GLWQA, began to disintegrate. This meant that cleaning up AOCs – a challenging, long-term policy goal in any case – progressed very slowly; after twenty years, only three of the forty-three AOCs had been delisted. Some analysts argued that the increasingly local focus on AOCs actually provided an excuse for federal and even subnational governments to step back from leadership.[32]

With the advent of the Obama administration, however, this critical region became a higher priority on the U.S. agenda. In 2009 President Obama announced a plan to invest $475 million in the region, renegotiate the GLWQA, and launch the Great Lakes Restoration Initiative (GLRI). GLRI was established in 2010, and Congress has supported this initiative with major investments annually. These U.S.-based efforts were reinforced with the renegotiation of the GLWQA and re-signing by both federal governments in 2012. The 2012 GLWQA recommitted both federal governments to implementation using the same policy instruments and institutions from previous agreements, but with renewed focus on accountability and reporting to address continuing and new environmental problems in the basin.[33]

Significantly, cross-border efforts resumed after more than a decade of decline. The growth in subnational diplomacy and capacity since the 1987 agreement was signed[34] was recognized in the increasingly complex binational arrangements under the GLWQA, which, when combined with constitutional authorities and important operational responsibilities in water quantity management, have given states and provinces a central role in Great Lakes governance and subnational diplomacy.[35] These subnational roles are also increasingly important with respect to the work of AOCs, LaMPs, and the new "annexes" added to the GLWQA in 2012 to address groundwater, invasive species, and climate change.

Further, the new agreement clearly outlines that "the involvement and participation of State and Provincial Governments, Tribal Governments, First Nations, Métis, Municipal Governments, watershed management agencies, local public agencies, and the Public are essential to achieve the objectives of the Agreement."[36] The expanded list of issue-focused annexes under the 2012 agreement have become the focus of significant transboundary activity under the cross-border Great Lakes Executive Committee. A series of networks including federal, state, provincial, local, Indigenous, academic, and nongovernment actors have mobilized around annexes under co-leads from the U.S. EPA and ECCC. Public and nongovernment participation is being encouraged through each annex by having an Engagement Working Group and regular reporting. The result has been considerable cross-border network- and capacity-building to implement the current GLWQA. The web of networks at the basin level, working towards collective action and outcomes, has thus become more dense and more active.[37]

Over the 2012–18 period, there was progress on GLWQA commitments.[38] Some of this progress is specifically related to investments and efforts on the U.S. side under GLRI. Some relates to significant engagement by the federal bureaucracy on the Canadian side, the re-signing of the Canada-Ontario Agreement in 2014, and action by the Ontario government, including passage of the Great Lake Protection Act in 2015. Investments on the Canadian side, however, show a more mixed picture. Funding for water resources by Environment and Climate Change Canada actually declined from $92 million in 2014–15 (under the Harper government) to $70 million (2017–18) under the Trudeau government.[39] For its part, Ontario indicated it has spent $1 billion on Great Lakes programs and initiatives through the Ontario Ministry of Environment and Climate Change since 2007.[40]

But of course the political climate in the region changed significantly, again, under the Trump presidency. The U.S. EPA and the GLRI came under increased scrutiny, and the new administration specifically targeted the region for cuts. In its first budget the Trump administration tried to cut the federal expenditures under GLRI by 90 per cent from $300 million annually to $30 million, arguing that regional water initiatives are "primarily local efforts" and that state and local governments were capable of paying for them.[41] But significant pushback in Congress from both Democrats and Republicans held off cuts, illustrating the popularity of GLRI as economic and environmental policy with both parties and voters. Again, in February 2018, President Trump announced GLRI funding would be cut by 90 per cent, but the U.S. House Appropriations Committee supported legislation fully funding a Great Lakes restoration program for $300 million in 2018–19.[42] Then, in March 2019, Trump announced funding of the GLRI, which constitutes a reversal of the administration's repeated attempts to slash funding,[43] likely with an eye to the 2020 election. Indeed, in February 2020, Trump proposed a slight increase in funding to the Great Lakes Restoration

Initiative, at the same time that the U.S. House of Representatives approved bipartisan legislation to incrementally increase the federal appropriation over the coming five years.[44]

In another recent development, six new commissioners have been appointed to the International Joint Commission. Despite the trepidation of many around the basin, concerned that unsupportive commissioners might sideline binational environmental efforts, both the Canadian and U.S. appointees actually show a range of deep water management expertise and are more activist than might have been expected. The new commission has been characterized as a "fresh start" for the binational body and for shared efforts to address pollution in the basin.[45] In a surprising development, given the anti-climate policy stance of the Trump administration, the U.S. and Canadian IJC co-chairs will be focusing on how climate change is affecting shorelines in the Great Lakes basin.[46]

At the time of writing, significant ecological and governance challenges remain for Great Lakes shared water management, including fully engaging major water users such as agriculture and business, and coping with fluctuations in political will and leadership.[47] A major task confronting Great Lakes policymakers is how to achieve the promise of cross-scale integration of governance in the basin to link the cross-border regional imprint of subnational, national and local administrators to the basin-specific networks that are still being articulated – and in an era where the U.S. administration was not predisposed to support such networks by exercising its formal authority for environmental protection.

In fact, the Trump administration was not predisposed to support any environmental regulation and consistently proposed deep cuts to the Environmental Protection Agency, which did affect activities across all regions and programs, across all pollution media. Congress has thus far blocked these cuts each year. The administration then continued its efforts, with a 24 per cent reduction in its proposed 2021 budget to EPA programs, as well as cuts to research and science programs that are linked to water protection.[48] However, the Trump administration is not alone in failing to support environmental programming. The Ford government in Ontario – one of the Great Lakes' biggest jurisdictions – has also cut environmental and research budgets, restricted agency travel, and has been undertaking environmental regulatory changes to reduce restrictions on business. These changes place pressure on agency officials and researchers who are attempting to work across borders to respond to Great Lakes Water Quality mandates, while joint research and data-gathering efforts, and even in-person meetings to provide updates and discuss initiatives, have been curtailed. The COVID-19 pandemic has made the situation even more difficult, as the result of social isolation measures and the diversion of public monies to fight the outbreak.

Despite the current situation, the longer-term history of shared-water management demonstrates that territorially based authority and formal transboundary institutions can be successfully married to aterritorial networks. Balancing the two provides a way to govern this complex, transboundary governance system, and networks of agency officials and researchers, along with the nongovernmental groups and Indigenous peoples who are engaged in planning and policy, play a critical role in achieving this balance.

The Bordering of Environmental Policy in North America II: Dynamic Environment-Economy Pathways

Clearly, not all environmental problems facing Canada are place-based in quite the same way. Climate change, for example, is a global problem where contributions to GHG emission loadings are many and diffuse, and responsibility as well as accountability is much more difficult to sort out. However, in the same way that actions on both sides of the Canada-U.S. border can be linked to water pollution, thereby requiring joint action, climate change mitigation also implicates choices made on both the sides of the border, requiring joint action. More specifically, the integrated nature of the Canadian and American economies means that efforts to achieve a lower-carbon economy (which reduces greenhouse gas emissions) must take this integration into account in both policy design and implementation.

Programs to facilitate the transition towards a low-carbon economy, or LCE, will achieve higher energy efficiency, greater use of renewable energy, and increased support for and application of green technologies. An LCE must pursue these aims across all sectors – industry, agriculture, transportation, building and residential, and services – and through all stages of the production-to-consumer-to-waste cycle.[49] However, transitioning to a low-carbon economy is a highly complex undertaking, requiring widely variable yet co-evolutionary shifts across the many means of powering our economy and producing goods.[50] In other words, one mitigation policy tool will not fit all of the dynamics at play. More to the point, one jurisdiction cannot act alone or effectively. Transnational action and cooperation are required, particularly where economies are highly integrated.

Strategic discussions about the LCE thus often refer to the need to pursue and manage "decarbonization pathways" for decreasing the average carbon intensity of primary energy, tracking and transforming avenues of energy production and utilization across linked economies and societies, and undertaking policy interventions and technological applications. These pathways are not merely sectoral. They follow key drivers in how energy is produced and used. Here too, then, the nature of the environmental problem is defined aterritorially but not according to ecosystems or bioregions; climate change mitigation and GHG emissions reduction must of necessity influence the conduits of economic incentives that underlie the regional and global economies. In the discussion below, we address pathways to pursue GHG emission reductions across the Canada-U.S. border, each of which attempts to overcome the constraints of territoriality and bring policy into line with economic and consumptive dynamics.

Case Illustration: Canada-U.S. Climate Change Mitigation

Targeting sources that are major contributors to GHG emissions within the two countries through regulation, and having complementary standards across the North American market, can significantly reduce emissions and speed the transition to a low-carbon economy. Given that the lion's share of GHG emissions in North America come from the transportation and electricity/heat generation sectors,[51] there are a few obvious targets for mandated reductions in these pathways, including fuel-economy and tailpipe standards, low-carbon fuel standards (LCFS) to reduce the life-cycle emissions of the fuels sold, and limits on electric power generation. This would imply that GHG reductions should follow these pathways of energy use that have a significant aterritorial dimension – fuel production and refining, transportation and electricity generation, all of which are characterized by transnational dynamics – by encouraging policy and standards diffusion and alignment across the Canada-U.S. border.

The clearest example of cross-border coordination to reduce fuel use has been vehicle fuel-emission standards.[52] The December 2009 ruling by the U.S. EPA that GHGs endanger human health (pushed by California's ambitious controls) provided the Obama administration with the means to regulate emissions under the 1990 Clean Air Act, even in the face of congressional refusal to act on climate change, and the administration proceeded to regulate mobile sources. At every step of the way, the Harper Conservative government in Canada attempted to follow the American lead on vehicle standards. This is consistent with a longer-term trend in which Canada adopts U.S. standards, as a result of the integrated nature of automotive manufacturing. In some cases, California's standards are adopted by national governments in both countries – largely in response to the weight of the California economy. This has meant explicit harmonization of fuel economy standards and GHG tailpipe standards, accomplished largely through informal meetings of transportation agency officials and consultation with the automotive industry in the two countries. In 2010, at the same time that the U.S. federal government announced its new standards for model years 2011–2016, the Canadian government unveiled "harmonized" rules; in fact, in its notice of intent to regulate emissions from new cars and light-duty trucks, the government explicitly stated that "these regulatory standards will be equivalent to applicable U.S. fuel economy standards."[53] This pattern was repeated for an additional round of standards-tightening for cars and light-duty trucks for model years 2017–25, and also for heavy-duty vehicles for model years 2014–18 and 2018 and beyond.

In early 2017, however, the Trump administration began rolling back these ambitious regulations by proposing to freeze standards in 2020 at lower levels, and reduce the fines levied on car-makers for missing targets. This unleashed conflict and litigation with California and the thirteen other states that adopted the higher Obama-era standards. Interestingly, although some automakers' organizations initially lobbied the Trump administration to loosen standards, the industry more generally was opposed to the administration's plan to weaken them too much, arguing that differential standards across the country – which would have resulted from the Trump administration's actions – are the worst-case scenario for the industry.[54] They also argued that much weaker standards would get tied up for years in litigation. The environmental community also came out strongly against the administration's plans and sought to block some initiatives in court.

In mid-2019, four major automakers – Ford, Honda, Volkswagen, and BMW – concluded an agreement with California to continue to pursue more ambitious standards, with some flexibility provided for implementation. Canada, in June 2019, appeared to also be seriously considering meeting that state's higher fuel efficiency and GHG standards. The two jurisdictions signed a cooperation agreement committing them to "collaborate on the development of [their] respective greenhouse-gas regulations for light-duty vehicles that require meaningful improvements in vehicle efficiency every year" and to undertake additional measures to achieve zero-emission vehicle sales as well as collaborate on emissions testing and modelling research and programs related to alternative-fuel vehicles.[55] Members of the Canadian automotive industry, however, pushed for following Trump's approach, while environmental groups have been holding firm in favour of following a more stringent direction. The Democrats win in the 2020 election, however, strengthens the hand of emissions reduction forces.

On 31 March 2020 the Trump administration formally announced its final rule for a loosening of the Obama-era emissions standards, calling it "the largest deregulatory initiative of this administration."[56] The rule would weaken the changes to efficiency, from requiring that automakers' fleets average fifty-four miles per gallon by 2025, moving the target down to forty miles per gallon – which is even lower than the car industry is on track to achieve through current technological improvements.[57] The finalized rule drew rebukes from both former president Barack Obama and former EPA administrator Gina McCarthy. California immediately announced that it would take the federal government to court.[58] Interestingly, the group of states that have been supporting California's opposition to the loosening of air pollution rules has swelled to twenty-three. Industry response was muted, although the Ford vice president of sustainability stated, "We remain committed to meeting emission reductions consistent with the California framework and continue to believe this path is what's best for our customers, the environment, and the short- and long-term health of the auto industry."[59]

At the time of writing, the Canadian government had not yet decided on its response to the new U.S. standards; a statement from the Office of the Minister of Environment and Climate Change noted, "We will consider today's announcement by the U.S. EPA as well as California's standards, which together with 12 other states', represent 40% of the U.S. auto market.… Any decision that is made will be in consultation with provinces, territories, industry and more, to ensure that we enact the best regulations that work for Canadians."[60] With Joseph Biden in the White House, however, California will be given free rein to move forward with its stricter standards and there will be renewed pressure on Canada to follow suit. The about-face by General Motors, which had supported Trump's deregulatory policies, right after the election shows how the political landscape has changed.

Another example of an intent to pursue regulatory coordination in fuel use is the adoption of low-carbon fuel standards, or LCFS. California was first-adopter here, and there is some evidence of subnational policy emulation, particularly in California's effect on BC regulations and policy ruminations in Ontario (under the Wynne Liberals) as well as West Coast states.[61] Indeed, under the umbrella of the Pacific Coast Collaborative, also led by California, West Coast jurisdictions have discussed measures to support low-carbon and renewable-energy initiatives, including through transportation and fuel use. Under its 2016 Pan-Canadian Framework on Clean Growth and Climate Change, the Canadian federal government has moved to establish a national Clean Fuel Standard (CFS). In following through on its commitment under the framework to "develop a clean fuel standard to reduce emission from fuels used in transportation, buildings and industry," Environment and Climate Change Canada initiated stakeholder discussions via workshops and meetings in late 2016 and through 2017. Interestingly, the ECCC reached out to key decision-makers on California's Air Resources Board (CARB) to gain information on the California experience, but also with a view to ensuring the interoperability of a proposed Canadian LCFS with the California LCFS.[62] In this case, then, transboundary coordination takes on the cast of "diagonality," with California and the Trudeau government doing the heavy lifting. A regulatory design paper was released by ECCC in December 2018, and the proposed regulatory approach was released in June 2019.[63] Several measures proposed for the Clean Fuel Standard are modelled on California's approach, including calculation of average national fuel intensity, verification processes, and methods for calculating carbon intensity of biofuels.

Interestingly, the electricity generation sector has not shown much aterritorial activity; jurisdictions in Canada and the United States have *not* focused on regulatory harmonization, even though grids are linked across borders. Indeed, Canada and the United States have taken quite different approaches to address high emissions from coal-fired generating.[64] In 2015, the Obama administration introduced the Clean Power Plan, which set source-level and source-category-wide standards that individual units could meet through a variety of technologies and measures; states could choose their own methods for meeting the standards.[65] In Canada, rules brought in under the Canadian Environmental Protection Act (effective in 2015) require that new coal-fired power plants must bring GHG emissions down to the level of high-efficiency gas plants (thus requiring Carbon Capture and Storage [CCS]). Companies will also have to close plants built before 1975 by 2020, and any plant built after 1975 would have to close by 2030, unless equipped with CCS. The U.S. Clean Power Plan was being dismantled under the Trump administration, though this attempt to unravel the legal and regulatory framework for the coal sector will now be halted, and renewed attempts to reduce emissions will begin. The Trudeau administration, for its part, has doubled down on reducing the number of coal-fired power plants and developing cleaner generation capacity as well as instituting a new national carbon-pricing plan.

Indeed, one key tool to address emissions from electricity generation, as well as from heavy industry and oil and gas production, is carbon pricing, where significantly less progress has been made in mitigation and where global mitigation technology lags.[66] Here, the primary goal is to capture the "external costs" of emissions from the use of fossil fuels, namely the costs paid by the public for the negative impacts of emissions, such as extreme climate events, drought, and health-care costs.[67] Putting a price on carbon – through a carbon tax or a carbon trading system – highlights these negative externalities and sends out signals across the economy, encouraging polluters to shift to less carbon-intensive activities and incentivizing clean technology innovation.

Several subnational emissions trading systems (ETS) are operating in the United States and Canada,[68] including the Western Climate Initiative (WCI) and the Regional Greenhouse Gas Initiative (RGGI), both cap-and-trade systems. The WCI is cross-border, given that the California ETS has worked together with the Quebec ETS since 2014, when they had their first joint auction, and Ontario joined this system in 2017 (only to defect after the election of the Ford Conservatives in the summer of 2018). The other ETS, RGGI, integrates Connecticut, Delaware, Maine, Maryland, Massachusetts, New Hampshire, New York, Rhode Island, and Vermont, and focuses only on reducing CO_2 emissions in the power sector. It has been functioning since 2008, when it had its first auction.

Several other subnational governments have dabbled in carbon pricing. While British Columbia has had a carbon tax since 2008, rather than an ETS, it participates on the board of directors of the WCI, indicating a serious interest in ensuring that its actions complement those of its neighbours. Alberta, for its part, introduced its Specified Gas Emitters Regulations (SGER), which taxed emissions from industry-emitting GHGs over 100,000 tonnes at the level of $15 per tonne, as early as 2007. After it came to power in 2015, the Notley New Democratic government instituted a new program featuring a broad-based carbon levy on transportation and heating fuels (diesel, gasoline, and natural gas), an annual cap on oil-sands emissions, and a phase-out of coal-fired electricity. The province's carbon tax was intended to cover 78–90 per cent of emissions under this new plan. Large emitters would continue to be covered by the Specified Gas Emitters Regulations (SGER) system, though the levy would increase to $30 per tonne by 2018. However, since the election of the Kenney Conservatives in the spring 2019, the consumer carbon pricing system has been dismantled and a "pro-oil war-room" has been created by the new government. Yet the SGER remains in place; the Kenney government, after consulting with industries and declaring that it would look at ways to loosen the regulations,[69] merely rolled back the levy to $20 per tonne, with the proceeds to fund a Technological, Innovation, Emissions Reduction program to support emissions reduction in the oil and gas sector.

At the national level in Canada, the Trudeau government, under the Pan-Canadian Plan, has implemented a national carbon-pricing system across Canada, which was first rolled out in January 2019. The national plan covers the same emissions sources in each province. Initially, provinces could institute their own cap-and-trade or a carbon-tax approach, but the equivalent of a minimum federal price floor had to be met in carbon tax and cap-and-trade provinces, and this price escalates over time. For provinces that opted out of carbon pricing and had no system in place (Ontario, New Brunswick, Manitoba, Saskatchewan, and Alberta), the federal tax was to apply to them. It was thought that the presence of a national framework undergirding a rising carbon price will certainly affect the calculus of governments inside Canada and across the border. Re-election of the Trudeau government in the October 2019 election was due largely to its climate action plan (in the face of the Conservative Party's complete lack of a viable climate strategy) and certainly helped to cement the national carbon pricing plan, as well as other emission reduction initiatives, at least for the time being. However, the federal government has since accepted new Emissions Performance Standards in Ontario as an alternative to the federal pricing system. Further, New Brunswick has decided to institute its own carbon pricing system.

For carbon pricing, while legal frameworks undergirding such systems must be passed by territorial jurisdictions in North America, the impetus and the design of such systems is aterritorial. Carbon taxes or trading systems set a price for the entire economy or for specific sectors, and the application of carbon pricing can pose variable challenges for different sectors of the economy.[70] The needs of industries thus vary, and each must choose how best to respond to emission-reduction impulses according to sectoral and cross-border dynamics. Further, given the integrated nature of sectoral economies in North America, carbon price differentials across jurisdictions can place industries in higher price jurisdictions at a competitive disadvantage. The design of carbon pricing systems, then, requires aterritorial orientation in order to track sector-specific impacts and cross-border dynamics in the linked economy. This reasoning has inspired much of the opposition to the carbon pricing regimes in California, the U.S. Northeast, and British Columbia, as well as – before their dismantling – Ontario and Alberta.

Certainly, carbon pricing could be more aterritorial in its design and impacts than regulatory harmonization of general energy and fuel use.[71] Regulatory authority is, of course, firmly territorial (as recent electoral outcomes have shown), and the networks supporting harmonization are largely informal. Carbon pricing, however, assumes that economic dynamics must be given priority in mitigation efforts over regulatory authority. Moreover, each carbon pricing system requires a series of networks to support design and implementation; in the case of RGGI, these networks are interstate and now firmly articulated,[72] and in the case of WCI, associated networks are even cross-border. Governance thus entails linking jurisdictional efforts in a way that overcomes political constraints. This is one reason carbon pricing has faced such resistance in North America.

Environmental Policy, Aterritoriality, and Resilience

In the discussions above, we have examined shared cross-border governance in the Great Lakes basin as well as efforts at climate change mitigation to assess how well cooperation and architecture have overcome the constraints of territorial borders, and instead operate on the basis of ecological boundaries or along economic-environmental pathways. In this final section, we reflect on territorialism and aterritorialism in terms of transboundary cooperation and also consider how resilient these infrastructures may be even when Canada-U.S. relations are rocky.

In both cases, concerns about jurisdiction, legal mandates, and sovereignty loom large over transborder cooperation. In the Great Lakes case, the cross-border infrastructure is quite formalized, resting on a century-old treaty and binational commission, a renewed Canada-U.S. executive agreement, and institutions created to realize the aims of both the treaty and agreement. But this formal spine clearly cannot function without basin-level networks of agency officials working with their nongovernmental and scientific counterparts on issue-focused objectives aligned with AOC, lake-level, and annex institutions. By contrast, in the climate change mitigation case, there are few formal institutions – although international agreements and especially the Paris Treaty do provide a light framework for encouraging action. Instead, we see a set of cross-border networks towards the informal end of the spectrum – from policy emulation and diffusion, through to meetings of agency officials and more conscious coordination, and on to more articulated infrastructure supporting carbon trading initiatives such as RGGI and the WCI. Further, there are interactions between these cross-border interactions and networks, on the one hand, and Canadian federal officials and intergovernmental working committees that have been created under the Pan-Canadian Framework for Combating Climate Change, on the other. Interactions between California and the Canadian government on vehicle fuel emissions and clean fuel standards are examples of this diagonal dynamic.

So networks are critical in both cases. We would argue that networks, generally organized on an aterritorial basis and more focused on specific environmental problems and often transgovernmental in nature, help formal institutions adapt to the shared dimensions of problems and solutions. Networks allow for higher-level functioning through informal interactions that can integrate new knowledge more effectively into the transboundary system and engage critical actors beyond just state-centred actors and government officials. However, in one case – shared water management – these networks are bolstered by being linked to formal binational institutions; they draw their functional strength from mandates to achieve specific objectives and perform certain tasks as set out in formal agreements such as the GLWQA. The Great Lakes case is also characterized by a linking of national mandates and leadership to subnational and basin-level policy efforts through networks.

Certainly, the Great Lakes system is vulnerable to decreases in resources and political will. Although the asymmetry between the United States and Canada has not been a major issue during times of policy engagement, it has been an issue when one side of the border is not actively engaged at the national or subnational levels. Nevertheless, the density of the system of formal institutions and networks, working together across annexes, AOCs, and at the lake level, has considerable momentum, even with an unfriendly administration in the White House. The new U.S. commissioners appointed by the Trump administration, who show (somewhat surprisingly) both water-management experience and some environmental proclivities, highlight the force of this momentum.

The climate case tells us a different story, one where the lack of formal bilateral institutions to promote climate policy cooperation and move towards an LCE means that it can much less easily endure political attacks, as well as the ebbs and flows of political attention and will. One might argue that the formal spine for transboundary coordination and even harmonization may eventually be put in place, when harmonized standards become law across jurisdictions, or when the legislative basis for carbon pricing is passed by all neighbouring jurisdictions. Further, the build-out of carbon pricing systems requires that cooperative infrastructure for monitoring, tracking, and enforcing the system be put in place, and this would add to the density of networks in the climate policy system. However, formalization of climate policy, especially carbon pricing systems, is a high bar, and can be easily lowered; domestic legislation can be rapidly reversed by new governments – as in Ontario and Alberta. The example of the Western Climate Initiative shows us that the supporting infrastructure for carbon pricing can easily collapse when jurisdictions (e.g., Ontario) defect. Certainly, the entrance of the Canadian federal government onto the carbon pricing scene, which has thus far survived constitutional and electoral challenge (the Supreme Court has heard the constitutional challenge of the provinces and will provide its judgment in late 2020 or early 2021), will provide a foundation for carbon pricing infrastructure in North America. It is worth noting here that the scale at which cooperation is taking place figures into the equation, given that the formal architecture is harder to erect as the number of participating jurisdictions increases and as one moves from place-based to global environmental problems. Carolyn James's chapter in this volume expands on this lesson, using the example of Arctic resource management.

If we link these insights to the Continuum of Policy Relations introduced in chapter 1, one might argue – perhaps counter-intuitively – that strong bilateral institutions and informal networks are equally important in the pursuit of successful *aterritorial interactions*, given that they operate symbiotically. Reinforcing formal and informal elements in the system appears more likely to lead to effective cooperation, coordination, and perhaps even harmonization than a transboundary system that is heavily networked but where these networks are not rooted in formal bilateralism.

The recent discord between the federal governments in Canada and the United States and the rise of populist Conservative governments at the subnational level tested the resilience of all cross-border efforts, given the fundamentally different environmental policy orientations. The system is still fundamentally territorially based and oriented, but issue-focused networks are increasingly critical to pursuing water and climate policy objectives on the Canada-U.S. border. Aterritorial cross-border networks will only become more important in more fully engaging the diversity of authorities and interests in the future, and cementing relationships that are harder to break during fluctuations in political will. Providing continued, strong, and tangible support for these networks of people can do much to push policy action, even when one national government, or governments at the subnational level, are less supportive. It is thus imperative for Canadian governments at all levels to continue to invest in and support the "people-centred" aspects of Canada-U.S. environmental cooperation. This will likely be easier as a pro-environment Biden administration settles in. At the time of writing, the global coronavirus pandemic had shut down national economies and resulted in the transfer of massive public funds into emergency social provision and corporate bailouts. Environmental protection regimes may shortly be fighting for their very survival as heavily indebted governments transfer resources to provide immediate support for citizens and failing businesses, and expend their available political will on public health and economic measures. It is in times like these, however, that networks of people committed to collaborative environmental efforts become critically important in supporting the environmental protection system.

NOTES

1 James W. Scott, "Bordering, Border Politics and Cross-Border Cooperation in Europe," in *Neighbourhood Policy and the Construction of the European External Borders*, ed. F. Celata and R. Coletti, 27–47 (Basel, Switzerland: Springer International Publishing, 2015).

2 Henk Van Houtum and Ton Van Naserssen, "Bordering, Ordering and Othering," *Journal of Economic and Social Geography* 93, no. 2 (2002): 125–36.

3 For a discussion of the current demise of the European sense of community and the increasing emphasis on nation-state mechanisms, see Thierry Chopin and Jean-Francois Jamet, "The Future of the European Project," Fondation Robert Schuman Policy Paper: European Issues no. 393 (Paris: Fondation Robert Schuman, 24 May 2016).

4 Robert Pastor, "The Future of North America: Replacing a Bad Neighbor Policy," *Foreign Affairs*, July/August 2008, https://www.foreignaffairs .com/articles/north-america/2008-06-01/future-north-america; Pastor, *The North American Idea: A Vision of a Continental Future* (Oxford: Oxford University Press, 2011).

5 Dyah E. Kurniawati, "Intermestic Approach: A Methodological Alternative in Studying Policy Change," *PCD Journal* 5, no. 1 (2017): 147.

6 See, for example, Jan Ceglowski, "Has Globalization Created a Borderless World?" *Business Review* (March/April 1998): 17–27, http://people .tamu.edu/~aglass/econ452/Ceglowski_BordlessWorld.pdf; and Kenichi Omhae, *The Borderless World: Power and Strategy in an Interlinked Economy* (New York: Harper Business, 1999).

7 Ann Marie Slaughter, *A New World Order* (Princeton, NJ: Princeton University Press, 2004); Kal Raustiala, "The Architecture of International Cooperation: Transgovernmental Networks and the Future of International Law," *Virginia Journal of International Law* 43 (2002), https://papers.ssrn.com /sol3/papers.cfm?abstract_id=333381.

8 Neil Craik and Debora VanNijnatten, "'Bundled' Transgovernmental Networks, Agency, Autonomy and Regulatory Cooperation in North America," *North Carolina Journal of International Law* 41, no. 1 (2016): 1–40.

9 Canada, Policy Horizons Canada, *The Emergence of Cross-Border Regions between Canada and the United States* (Ottawa: Policy Research Initiative,

2009); Emmanuel Brunet-Jailly, Susan E. Clarke, and Debora L. VanNijnatten, "The Results in Perspective: An Emerging Model of Cross-Border Regional Co-operation in North America," *Leader Survey on Canada-US Cross-Border Regions: An Analysis*, North American Linkages Working Paper Series 012 (2006).

10 Canada, Policy Horizons Canada, *Emergence of Cross-Border Regions.*

11 Ronnie Lipschutz, "Bioregionalism, Civil Society and Global Environmental Governance," in *Bioregionalism*, ed. Michael Vincent McGinnis (London: Routledge, 1998), 102–3.

12 Bron Taylor, "Bioregionalism: An Ethics of Loyalty to Place," *Landscape Journal* 19, no. 1 and 2 (2000): 50.

13 Liette Gilbert, L. Anders Sandberg, and Gerde R. Wekerle, "Building Bioregional Citizenship: The Case of the Oak Ridges Moraine, Ontario, Canada," *Local Environment: The International Journal of Justice and Sustainability* 15, no. 5 (2009): 387–401. https://www.tandfonline.com/doi/abs/10.1080/13549830902903674?src=recsys&journalCode=cloe20.

14 Cascadia Now, "Welcome to Cascadia," https://www.cascadianow.org/.

15 Cascadia Now, "Welcome to Cascadia."

16 Linton Caldwell, "The Ecosystem as a Criterion for Public Land Policy," *Natural Resources Journal* 10, no. 2 (1970): 203–21.

17 Carl Folke, Thomas Hahn, Per Olsson, and Jon Norberg, "Adaptive Governance of Social-Ecological Systems," *Annual Review of Environment and Resources* 30 (2005): 443.

18 Norman L. Christensen, Ann M. Bartuska, James H. Brown, Stephen Carpenter, Carla D'Antonio, Rober Francis, Jerry F. Franklin, et al., "The Report of the Ecological Society of America Committee on the Scientific Basis for Ecosystem Management," *Ecological Applications* 6 (1996): 665–90; Fikret Berkes and Carl Folke, eds., *Linking Social and Ecological Systems: Management Practices and Social Mechanisms for Building Resilience* (Cambridge: Cambridge University Press, 1998); K.L. McLeod, J. Lubchenco, S.R. Palumbi, and A.A. Rosenberg, *Scientific Consensus Statement on Marine Ecosystem-Based Management* (Communication Partnership for Science and the Sea, 2005).

19 R. Edward Grumbine, "What Is Ecosystem Management?" *Conservation Biology* 8, no. 1 (1994): 27–38.

20 Roland Cormier, Christopher R. Kelble, M. Robin Anderson, J. Icarus Allen, Anthony Grehan, and Ólavur Gregersen, "Moving from Ecosystem-Based Policy Objectives to Operational Implementation of Ecosystem-Based Management Measures," *ICES Journal of Marine Science* 74, no. 1 (2017): 406–13.

21 Debora L. VanNijnatten, "Towards Cross-Border Environmental Policy Spaces in North America: Province-State Linkages on the Canada-US

Border," *AmeriQuests: The Journal of the Center for the Americas* 3, no. 1 (2006); VanNijnatten, "North American Environmental Regionalism: Multi-Level, Bottom-Heavy and Policy-Led," in *Comparative Environmental Regionalism*, ed. Lorraine Elliot and Shaun Breslin (New York: Routledge, 2011).

22 Lynne Heasley and Daniel MacFarlane, eds., *Border Flows: A Century of the Canadian-American Water Relationship*. Network in Canadian History & Environment, 2016, http://www.greatlakeslaw.org/files/Border_Flows.pdf.

23 Heasley and MacFarlane, *Border Flows*.

24 Treaty between the United States and Great Britain Relating to Boundary Waters, and Questions Arising between the United States and Canada, U.S.-Canada, 11 January 1909, Article IV.

25 Canada and United States, *Great Lakes Water Quality Agreement* (Ottawa, 15 April 1972).

26 United States and Canada, United States Environmental Protection Agency and Environment and Climate Change Canada, binational.net, "About the Great Lakes Water Quality Agreement," accessed 2018, https://binational.net/glwqa-aqegl/.

27 Mark Sproule-Jones, *Restoration of the Great Lakes: Promises, Practices, and Performances* (Vancouver: UBC Press, 2002).

28 Sproule-Jones, *Restoration of the Great Lakes*; Lee Botts and Paul Muldoon, *Evolution of the Great Lakes Water Quality Agreement* (East Lansing, MI: Michigan State University Press, 2005); Carolyn Johns, "Water Pollution in the Great Lakes Basin: The Global-Local Dynamic," in *Environmental Challenges and Opportunities: Local-Global Perspectives on Canadian Issues*, ed. Christopher Gore and Peter Stoett, 95–129 (Toronto: Emond Montgomery, 2009).

29 Canada, Office of the Auditor General, Commissioner of the Environment and Sustainable Development, *A Legacy Worth Protecting: Charting a Sustainable Course in the Great Lakes and St. Lawrence River Basin* (Ottawa: 2001); Canada, Office of the Auditor General, Commissioner of the Environment and Sustainable Development, Status Report, Chapter 7, *Ecosystems and Areas of Concern in the Great Lakes Basin*, 2008; State of the Lakes Ecosystem Conference (Erie, PA: 2011), https://archive.epa.gov/solec/web/html/.

30 Carolyn Johns and Mark Sproule-Jones, "Great Lakes Water Policy: The Cases of Water Levels and Water Pollution in Lake Erie," in *Canadian Environmental Policy and Politics: Prospects for Leadership and Innovation*, ed. D. VanNijnatten, 4th ed., 252–77 (Toronto: Oxford University Press, 2016).

31 Botts and Muldoon, *Evolution of the Great Lakes Water Quality Agreement*.

32 Sproule-Jones, *Restoration of the Great Lakes*; Thomas J. Greitens, J. Cherie Strachan, and Craig S. Welton, "The Importance of Multilevel Governance Participation in the Great Lakes Areas of Concern," in *Making Multilevel*

Public Management Work: Success and Failure from Europe and North America, ed. Denita Cepiku, David K. Jesuit, and Ian Roberge, 160–81 (Boca Raton, FL: CRC, 2012).

33 Debora VanNijnatten and Carolyn Johns, "The IJC and the Evolution of Environmental and Water Governance in the Great Lakes–St. Lawrence Basin: Accountability, Progress Reporting and Measuring Performance under the Great Lakes Water Quality Agreement," in *The First Century of the International Joint Commission*, ed. Murray Clamen and Daniel Macfarlane, 395–430 (Calgary: University of Calgary Press, *forthcoming*).

34 Carolyn Johns and Adam Thorn, "Subnational Diplomacy in the Great Lakes Region: Towards Explaining Variation between Water Quality and Quantity Regimes," *Canadian Foreign Policy Journal* 21, no. 3 (2015): 195–211.

35 Johns and Adam Thorn, "Subnational Diplomacy."

36 Binational.net, "About the Great Lakes Water Quality Agreement," https://binational.net/glwqa-aqegl/.

37 Carolyn Johns, "The Great Lakes, Water Quality and Water Policy in Canada," in *Water Policy and Governance in Canada*, ed. Steven Renzetti and Diane Dupont, 159–78 (Basel: Springer International Publishing, 2017).

38 Johns, "Great Lakes, Water Quality and Water Policy in Canada"; Carolyn Johns, "Transboundary Environmental Governance and Water Pollution in the Great Lakes Region: Recent Progress and Future Challenges," in *Transboundary Environmental Governance across the World's Longest Border*, ed. Stephen Brooks and Andrea Olive, 77–112 (East Lansing, MI: Michigan State University Press, 2018); United States and Canada, United States Environmental Protection Agency and Environment Climate Change Canada, *Progress Report of the Parties Pursuant to the Canada–United States Great Lakes Water Quality Agreement* (2017), https://binational.net/wp-content/uploads/2016/09/PRP-160927-EN.pdf; International Joint Commission, *First Triennial Assessment of Progress on Great Lakes Water Quality* (2017), http://ijc.org/files/tinymce/uploaded/GLWQA/TAP.pdf.

39 Canada, Environment and Climate Change Canada, *Departmental Plan 2017 to 2018 Report, Environment and Climate Change Canada* (2018), https://www.canada.ca/en/environment-climate-change/corporate/archive/archived-departmental-plans/2017-2018.html.

40 Ministry of Environment and Climate Change (2016), Ontario's Great Lakes Strategy. https://www.ontario.ca/page/ontarios-great-lakes-strategy

41 John Flesher, "Trump Budget Again Targets Regional Water Programs," *News Chief*, 13 February 2018, http://www.newschief.com/news/20180213/trump-budget-again-targets-regional-water-cleanup-programs.

42 Flesher, "Trump Budget Again Targets Regional Water Programs."

43 Jonathan Oostling, "How Lawmakers Convinced Trump to Reverse Great Lakes Funding Cut," *Detroit News,* 29 March 2019, https://www .detroitnews.com/story/news/local/michigan/2019/03/29/lawmakers -convinced-trump-reverse-great-lakes-funding-cut/3309275002/.

44 Steve Carmody, "President Trump Proposes Increase in Great Lakes Funding, Critics Point to Cuts to EPA," Michigan Radio, 10 February 2020. https://www.michiganradio.org/post/president-trump-proposes -increase-great-lakes-funding-critics-point-cuts-epa.

45 Robert Harding, "'Fresh Start' with New IJC Commissioners, Canada-U.S. Water Panel at Full Strength," Salmon Beyond Borders, 21 May 2019, https://www.salmonbeyondborders.org/news-articles/fresh-start-with -new-ijc-commissioners-canada-us-water-panel-at-full-strength.

46 Kevin Bunch, "New Commission Co-Chairs Focus on Climate Change and Resilient Shorelines," *Great Lakes Connection Newsletter: International Joint Commission,* 14 January 2020, https://www.ijc.org/en/new-commission -co-chairs-focus-climate-change-and-resilient-shorelines.

47 Johns, "Great Lakes, Water Quality and Water Policy in Canada."

48 Rebecca Beitsch and Rachel Frazin, "Trump Budget Slashes EPA Funding, Environmental Programs," *The Hill,* 10 February 2020, https://thehill .com/policy/energy-environment/482352-trump-budget-slashes -funding-for-epa-environmental-programs.

49 Regions for Sustainable Change, *Handbook: Tackling Climate Change by Shifting to a Low-Carbon Economy* (2016), http://rscproject.rec.org/indicators /index.php?page=tackling-climate-change-by-shifting-to-a-low-carbon -economy.

50 Timothy J. Foxon, *A Co-Evolutionary Framework for Analysing a Transition to a Sustainable Low Carbon Economy* (Leeds, UK: Sustainability Research Institute, University of Leeds, 2010), http://www.see.leeds.ac.uk/fileadmin /Documents/research/sri/workingpapers/SRIPs-22_01.pdf.

51 International Energy Agency, *CO2 Emissions from Fuel Combustion: Overview,* 2020, https://www.iea.org/reports/co2-emissions-from-fuel -combustion-overview.

52 Debora VanNijnatten, "Standards Diffusion: The Quieter Side of North American Climate Change Cooperation?" in *North American Climate Change Policy: Designing Integration in a Regional System,* ed. Neil Craik, Isabel Studer, and Debora VanNijnatten, 108–31 (Toronto: University of Toronto Press, 2013).

53 Canada, Department of the Environment, "Canadian Environmental Protection Act, 1999," *Canada Gazette* 143, no. 14 (2009): 842, http:// publications.gc.ca/collections/collection_2009/canadagazette/SP2-1-143- 14.pdf.

54 Coral Davenport, "Automakers Tell Trump His Pollution Rules Could Mean 'Untenable' Instability and Lower Profits," *New York Times*, 6 June 2019, https://www.nytimes.com/2019/06/06/climate/trump-auto-emissions-rollback-letter.html.

55 Keith Nuthall, "Canada Awaits Washington's Lead on Emissions Rules," Wards Auto, 13 September 2019, https://www.wardsauto.com/industry/canada-awaits-washington-s-lead-emissions-rules.

56 Devan Cole, "Obama Slams Rollback of Vehicle Emission Standards in Rare Rebuke of Trump," CNN News, 1 April 2020, https://www.cnn.com/2020/03/31/politics/barack-obama-fuel-standard-rollback-trump-administration/index.html.

57 Coral Davenport, "U.S. to Announce Rollback of Auto Pollution Rules, a Key Effort to Fight Climate Change," *New York Times,* 30 March 2020, https://www.nytimes.com/2020/03/30/climate/trump-fuel-economy.html.

58 Rachel Becker, "California to Sue after Trump Administration Eases Vehicle Emissions Standards" *Times of San Diego*, 31 March 2020, https://timesofsandiego.com/business/2020/03/31/california-to-sue-after-trump-aministration-eases-vehicle-emissions-standards/.

59 Becker, "California to Sue."

60 Kerry Banks, "Fuelling Controversy: What Do New U.S. Fuel Efficiency Regulations Mean for Canada's Climate Goals?" *Electric Autonomy*, 31 March 2020, https://electricautonomy.ca/2020/03/31/new-us-fuel-efficiency-regulations-impact-on-canada/.

61 Pacific Institute for Climate Solutions, "Briefing Note 2010 – 18: BC's Low Carbon Fuel Standard," 9 September 2010, http://www.pics.uvic.ca/sites/default/files/uploads/publications/BC%27s%20low%20carbon%20fuel%20standard%20.pdf.

62 Canada, Environment and Climate Change Canada, "Clean Fuel Standard: Discussion Paper" (Ottawa: February 2017), https://www.ec.gc.ca/lcpe-cepa/default.asp?lang=En&n=D7C913BB-1.

63 Environment and Climate Change Canada, "Clean Fuel Standard: Proposed Regulatory Approach" (Ottawa: June 2019), https://www.canada.ca/content/dam/eccc/documents/pdf/climate-change/pricing-pollution/Clean-fuel-standard-proposed-regulatory-approach.pdf.

64 VanNijnatten, "Standards Diffusion."

65 United States, Environmental Protection Agency, "Fact Sheet: Overview of the Clean Power Plan: Cutting Carbon Pollution from Power Plants" (2015), https://19january2017snapshot.epa.gov/cleanpowerplan/fact-sheet-overview-clean-power-plan_.html.

66 Marcela López-Vallejo, *Reconfiguring Climate Governance in North America: A Transregional Approach* (Surrey, UK: Ashgate Publishing, 2014).

67 World Bank, "Carbon Pricing," http://www.worldbank.org/en/programs/pricing-carbon.

68 López-Vallejo, *Reconfiguring Climate Governance*.

69 Robson Fletcher, "Alberta Invites Feedback on Plan to Relax Regulations and Lower Carbon Tax on Heavy Emitters," CBC News, 19 July 2019, https://www.cbc.ca/news/canada/calgary/alberta-ucp-tier-carbon-tax-large-emitters-regulation-1.5205646.

70 Elizabeth Beale, Dale Beguin, Bev Dahlby, Don Drummond, Nancy Olewiler, and Christopher Ragan, *Provincial Carbon Pricing and Competitiveness Pressures: Guidelines for Business and Policymakers* (Canada's Ecofiscal Commission, 2015), http://ecofiscal.ca/wp-content/uploads/2015/11/Ecofiscal-Commission-Carbon-Pricing-Competitiveness-Report-November-2015.pdf.

71 López-Vallejo, *Reconfiguring Climate Governance*..

72 Barry G. Rabe, "Building on Sub-Federal Climate Strategies: The Challenges of Regionalism," in *Climate Change Policy in North America: Designing Integration in a Regional System*, ed. Neil Craik, Isabel Studer, and Debora VanNijnatten, 69–107 (Toronto: University of Toronto Press, 2013).

PART THREE

Trans-Border and Cross-Border Regions

12 The Pacific NorthWest Economic Region: An Institutional Analysis of Effective Regional Governance

PATRICIA DEWEY LAMBERT

Each of the so-called *mega-regions* of North America[1] – the Northeastern United States/Eastern Canada region, the Great Lakes region, the Great Plains/Prairie region, the Pacific Northwest region, and the Arctic region – comprises a distinct and overlapping group of U.S. states and Canadian provinces/territories. Within each are transborder organizations within the public, private, and non-profit/community spheres to foster regional economic strength, enhance citizens' quality of life, and support cultural/social cohesion.

The lens of regionalism is used frequently in the study of U.S.-Canada cross-border linkages that are the result of transnational activity of subnational entities such as states, provinces, territories, and municipalities, and their citizens and businesses.[2] They engage in policy relations through provincial-state organizations, subnational agreements, and an array of other arrangements. As a result, patterns of multilevel governance manifest, leading to North American and sectoral models of regional integration through policy advocacy, parallelism, cooperation, coordination, and harmonization, as discussed in chapter 1.

This chapter synthesizes literature on theories of regionalism, multilevel governance, neo-institutionalism, and the public policy cycle to critically present current research on institutions established to address U.S.-Canada cross-border policy. Using the Pacific NorthWest Economic Region as a case study for analysis, the chapter then develops a conceptual framework to study the institutional structure and design of such cross-border entities. The chapter concludes with avenues for further research.

What Is the Pacific NorthWest Economic Region (PNWER)?

Among the cross-border networks, organizations, and other mechanisms that advance regional policy in the United States and Canada, the Pacific NorthWest Economic Region (PNWER) is recognized as the strongest and most successful.[3] A public-private partnership with wide-ranging elite participation drawn from industry, higher education, and government, PNWER comprises representation from five U.S. states and five Canadian provinces and territories that span three overlapping cross-border regions: the Cascadia–Pacific Northwest regions of coastal states and provinces, large elements of the Prairies/(Northern) Great Plains, and Arctic regions (see table 12.1). This international organization offers its members multiple pathways for engagement in regional public policy cooperation, agenda-setting, advocacy, and formulation through its semi-annual summits, program areas, working groups, capital visits, federal engagement, and legislative academies. Figure 12.1 summarizes PNWER's scope, mission, and goals.

Figure 12.1. Introducing the Pacific NorthWest Economic Region (PNWER)

Founded in 1991, PNWER is a public-private partnership chartered by the states of Alaska, Idaho, Montana, Oregon, and Washington; the western Canadian provinces of Alberta, British Columbia, and Saskatchewan; and Yukon and the Northwest Territories.

PNWER provides the public and private sectors with a cross-border forum for unfiltered dialogue, capitalizing on synergies between business leaders and elected officials to work to advance the region's global competitiveness.

PNWER is recognized by the U.S. and Canadian federal governments as the "model" for regional and binational cooperation because of its proven success.

Mission: To increase the economic well-being and quality of life for all citizens of the region, while maintaining and enhancing our world-class natural environment.

Goals:

- Coordinate provincial and state policies throughout the region
- Identify and promote "models of success"
- Serve as a conduit to exchange information
- Promote greater regional collaboration
- Enhance the competitiveness of the region in both domestic and international markets
- Leverage regional influence in Ottawa and Washington, DC
- Achieve continued economic growth while maintaining the region's natural environment

PNWER in the Context of Cross-Border Regionalism

PNWER is at the forefront of international institutions that have developed in the context of cross-border North American regionalism over the past several decades. Between 2004 and 2008, Canada's Policy Research Initiative published reports on their research on cross-border networks and organizations that encompassed distinct groups of states, provinces, and territories. This research demonstrates that "the density and nature of linkages vary widely in different border regions, ranging from primarily bottom-up and market driven in much of central Canada and the Great Plains region to deeper and more institutionalized on the west coast and, to a lesser degree, in the New England-Atlantic-Quebec region."[4]

The major cross-border regions of the United States and Canada are depicted in figure 12.2 and table 12.1, with regional groupings of states, provinces, and territories labelled as *West, Prairies–Great Plains, Great Lakes–Heartland*, and in the East two groupings labelled as *Quebec* and *Atlantic Canada–New England*. Alberta, Saskatchewan, and Montana are identified as belonging to both the *West* and the *Prairies–Great Plains*.

It is significant that the integrative agreements (NAFTA/USMCA, the Trans-Pacific Partnership, the Beyond the Borders initiative) among the three nations of North America have not been used to create shared political institutions. In particular, in contrast to the European Union, NAFTA has not led to the formation of the robust supranational institutions and continental governance mechanisms. As Stephen Clarkson points out, "North America was constituted with a set of deliberately weak continental institutions designed mainly to settle member governments' and their corporations' legal disputes arising out of NAFTA's provisions."[5] The United States–Mexico–Canada Agreement (USMCA), with its emphasis on managed trade and zero-sum trade relations in several sectors, may result in further weakening of supranational institutions of continental governance.

Although the future of trilateral North American governance appears to be questionable, alternative forms of governance can be found within a *dual bilateralism*[6] framework in which the United States works out separate policy arrangements with Canada and Mexico in various fields. Extensive cross-border regional networks between the United States and Canada provide fertile ground for investigating what might be viewed as a unique North American pathway towards integration. Instead of relying on the type of supranational and trans-national organizations that are found, for example, in the European Union, North American integration might be best understood through the development of

Figure 12.2. Canada–U.S. Cross-Border Regions

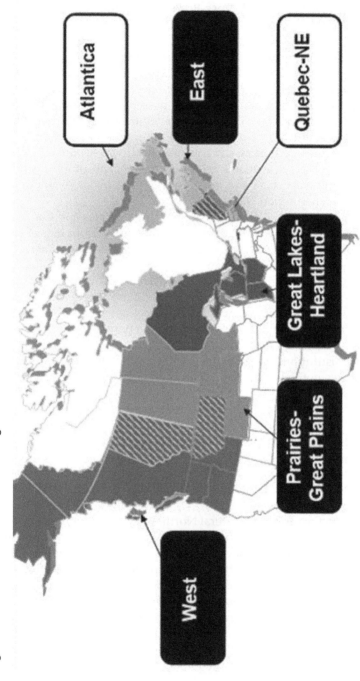

Source: Canada, Policy Research Institute, *The Emergence of Cross-Border Regions between Canada and the U.S.*, 5 (2016).

regional mechanisms for cross-border policy coordination – as noted by VanNijnatten and Johns in chapter 11. A key hypothesis of this chapter is that regional cross-border networks and organizations are moving the development of North America into a *beyond-NAFTA* framework, wherein NAFTA has been displaced as the central integrating framework of regionalism in the continent.[7]

A decade ago, it was found that "regional cross-border networks and organizations have proliferated since NAFTA, and provide a useful vehicle for binational business and community groups to work together on issues of mutual interest, often with the ultimate aim of problem-solving or creating local edges for success in the larger North American and global economies."[8] Formal and informal cross-border consultation and collaboration involve stakeholders from government, industry, and other nongovernmental organizations. Christopher Sands contends,

> We are seeing the first emergence of a new pattern of governance in North America. Cross-border regional ties suggest new networks of cooperation that can knit together the continent, filling a growing governance gap that has resulted from overcrowded national agendas and the pull of a globalized world that leads our federal governments to look outward. They suggest the need for new ways of thinking about policies and policy development, necessitating more than ever the use of a cross-border regional lens to recognize, understand, and better respond to the rising co-operative links and the increasing participation of regional players and local stakeholders in the practical problem-solving of common issues in the border areas of Canada and the United States.[9]

Indeed, a 2005 Canadian government report suggests that "cross-border regions are where North American integration is the most dynamic, and where the bridges of friendship, co-operation, and business are often first developed."[10] It can be argued that at least some regional actors and institutions are propelling North American integration by working towards policy cooperation and even coordination.[11] Regional cross-border organizations can facilitate economic integration and sometimes policy integration by promoting pathways to coordinate public policies. Scholars have identified distinct types of regional cross-border organizations and agreements found in the regions profiled in table 12.1. Mechanisms of regional governance can be as simple as the use of a memorandum of understanding or as complex as the establishment of an intergovernmental institution. Cross-border linkages operate at bilateral (state-province-territory) or multi-state/multi-province

Table 12.1. Clusters of Cross-Border Regionalism

Region	U.S. states involved	Canadian provinces and territories involved	Regional characteristics and representative cross-border policy issue areas	Representative regional institutions
West	Alaska, Washington, Oregon, Idaho, Montana	British Columbia, Alberta, Yukon, Northwest Territories, Saskatchewan	• Sense of remoteness from the federal governments of USA and Canada; • Strong cross-border institutions with engaged private sector; • Importance of environmental issues; • Innovative approaches to regional governance; • Strong sense of regional identity. Policy issues: energy, transboundary waters, border security, fisheries, forestry, tourism	Pacific NorthWest Economic Region Western Governors' Association and Western Premiers Conference Pacific Coast Collaborative
Prairies–Great Plains	Montana, Wyoming, North Dakota, South Dakota, Minnesota	Alberta, Saskatchewan, Manitoba	• A broad geographical area that is physically removed from major North American markets; • Stakeholders tend to practise forms of collaboration that require self-reliance; • Collaborative and cooperative activities are based on ongoing, informal, pragmatic, and often low-cost engagements involving high-level private and government officials; • Relationships sustained by a growing consciousness of shared interests; • Strong focus on trade corridors and transportation. Policy Issues: food safety, agriculture trade, water resources management, energy.	Western Governors' Association and Western Premiers Conference Council of State Governments – Midwest
Great Lakes–Heartland	Michigan, Indiana, Ohio	Ontario (Quebec)	• Maturity of economic linkages; • No overarching cross-border organization to provide leadership from regional stakeholders; • Strong emphasis on single-purpose organizations; • Interregional market competition; • Common challenges requiring closer cooperation;	Council of Great Lakes Governors Great Lakes Commission Council of the Great Lakes Region

Region	US states	Canadian provinces/territories	Characteristics	Organizations
East – Quebec	New York, Vermont, Maine, New Hampshire	Quebec	• Lack of regional identity, but strong sense of belonging to a North American community. Policy issues: conservation and ecosystem protection, water quality, pollution, tourism, transportation, forestry, border protection. • Mature and emerging economic linkages; • Importance of Quebec–New York corridor that is evolving into a binational region; • More emphasis on bilateral informal linkages; • Multilevel/multi-agency nature of networks; • Limited sense of regional identity. Policy issues: conservation and ecosystem protection, water quality, pollution, tourism, transportation, forestry, border protection.	Conference of New England Governors and Eastern Canadian Premiers
East – Atlantic Canada and New England	Maine, New Hampshire, Massachusetts, Connecticut, Rhode Island	New Brunswick, Nova Scotia, Prince Edward Island, Newfoundland and Labrador	• Economic prosperity is a key goal; • Importance of history and geography; • Strong stakeholder engagement, driven by different public and private networks; Need for enhanced infrastructure; • Importance of personal relationships; • Strong sense of regional identity. Policy issues: energy (oil), marine interests, border security, tourism.	Conference of Northeastern Governors – Eastern Canadian Premiers Atlantic Salmon Federation
Arctic	Alaska	Yukon, Northwest Territories, Nunavut	Policy issues: energy (oil), environment/climate change, mining, conservation, Aboriginal concerns.	Arctic Council Arctic Institute of North America

Sources: Modified from Dieudonné Mouafo, "Regional Dynamics in Canada–United States Relations" (paper presented at the Annual Meeting of the Canadian Political Science Association, Winnipeg, 3 June 2004) 10; Canada, Policy Research Institute, *Emergence of Cross-Border Regions*, 5; and Canada, Policy Research Institute, *The Emergence of Cross-Border Regions: Roundtables Synthesis Report* (Ottawa: Industry Canada, 2006), 5.

levels, and can be sector specific or multipurpose.[12] They can be characterized as *general purpose intergovernmental, single purpose intergovernmental* (with either an *environmental* or *infrastructure* focus), possessing a strong *civil society* orientation, or emphasizing a strong *city* orientation.[13] The most prominent of the cross-border networks are regional groupings of states and provinces that have created multi-jurisdictional intergovernmental institutions. These include important regional organizations that bring together U.S. governors and Canadian premiers: the Conference of New England ("Northeastern" since 2012) Governors and Eastern Canadian Premiers (est. 1973), the Council of Great Lakes Governors (est. 1983), and the Western Governors' Association and Western Premiers Conference (est. 1999).[14] Others, such as the Pacific Northwest Economic Region (est. 1991), the focus of this chapter, and the Council of the Great Lakes Region (est. 2001) are general purpose intergovernmental organizations with private sector participation and financing. Table 12.1 provides a snapshot of regional clusters in U.S.-Canada sub-national governance. While such organizations are significant in the sharing of information and coordinating intergovernmental policy advocacy, it is "difficult to evaluate the effectiveness of such links in enabling subnational governments to manage [cross-border] issues independently of Washington and Ottawa or to work together to influence them to reach mutually satisfactory alternatives."[15] Certainly, if PNWER is viewed as "the gold standard in advancing U.S.-Canada relations,"[16] it would be helpful to scholars, legislators, and industry leaders to better understand the institutional design and processes of this international organization that have contributed to its reputation. It is therefore surprising that a detailed institutional analysis of PNWER has never been conducted. This chapter seeks to begin to address this gap in scholarship by developing a conceptual framework (a way of seeing) that can be applied to continued empirical research on the institutional design of PNWER as well as to other models of regional governance in North America.

Key Theoretical Concepts in Regional Governance

Studying an international organization like PNWER requires development of an analytical model that provides a theoretical context for understanding how regional governance can be structured to succeed in North America. It follows that developing a conceptual framework for investigating the institutional structure and processes exemplified by PNWER would benefit from a review of relevant concepts and theories in regionalism, multilevel governance, and (neo-)institutionalism.

The prior section of this chapter introduced distinct regions of cross-border groupings of U.S. states and Canadian provinces and territories. Definitions of the terms *region* and *regionalism*, however, are contested in the literature. "Disputes over the definition of an economic region and regionalism hinge on the importance of geographic proximity and on the relationship between economic flows and policy choices."[17] Questions persist about whether regionalism pertains to the concentration of economic flows, or to inter-jurisdictional policy coordination, or to both. A pragmatic definition of *regionalism* is offered by J.H. Paterson as "a conscious subdivision, for whatever purpose, of a whole into parts; the identification of less-than-continental, or less-than-national, patterns that are clear enough, and significant enough, to be perceived."[18] The purposive dimension of this definition is key, since "regionalism, then, implies purpose – [a] purposeful community of interest."[19] This concept of regionalism suggests the need for a community of shared interests to be structured in some way to achieve members' goals for economic development and policy coordination.

Although structures and processes of regional governance seem to be faltering at the nation-state level, sub-national regional governance appears to be flourishing. Chapter 1 notes the use of *fragmegration* to explain the dynamics of governance in response to the effects of globalization in relocating the authority of nation states "in diverse directions, upward to supranational institutions, downward to sub-national entities, sideward to social movements, nongovernmental organizations, corporations, and a wide range of other types of collectives"[20] – although chapter 2 also notes the limits of such delegation. This concept can be useful in "analyzing the evolving political and administrative nature of policy relations and how that evolution is driving policy convergence and/or divergence in North America."[21] It follows that policy cooperation frequently takes place at the regional level, particularly when public policies formulated at the state or provincial level are supported by a critical mass of economic, social, environmental, and cultural interests that extend across state, provincial, and national borders.[22] American political scientist Earl Fry observes that "one should anticipate that transgovernmental linkages involving provincial, state, and municipal governments will continue to proliferate, a direct result of the thick network of cross-border interdependence which continues to expand."[23]

As states, provinces, territories, and, in some cases, cities expand their role in regional governance, literature on *para-diplomacy* is growing. This term refers to international political activity of sub-national governments and/or other entities, such as NGOs and transnational corporations.[24] The increase in stakeholders engaged in regional poli-cymaking underscores the local focus of most of the political sphere in North America. And, in the *intermestic* relationship between the United States and Canada, whereby traditional distinctions between interna-tional and domestic policies and politics are blurred,[25] it is imperative to recognize the "hidden wiring" of governance relationships at the province-state level.[26]

The growing "interaction of both integrating and fragmenting forces at multiple levels of social, economic, and political interaction"[27] may best be clarified through theories of multilevel governance. *Governance* refers to the full set of relationships that constitute systems for policy choice and action.[28] *Multilevel governance* refers to "a process of political decision making in which governments engage with a broad range of actors embedded in different territorial scales to pursue collaborative solutions to complex problems."[29] There is an extensive body of schol-arship engaging multilevel governance,[30] but when analysing regional governance in North America it may be most beneficial to focus on the strength of this theoretical construct in simplifying and clarifying the institutions and spatial configurations of policymaking. In this sense, multilevel governance can be applied to study a policy process that engages a variety of actors (governmental, nongovernmental, and/or quasi-governmental) located at different territorial scales, the outcomes of which are the product of decision-making processes.[31] Table 12.2 out-lines organizational forms used in multilevel governance across North American regions.

Multilevel governance and the similar theory of *network governance*[32] provide useful frameworks for understanding contemporary struc-tures of international and regional governance. Robert Keohane and Joseph Nye explain:

> The world system of the twenty-first century is not merely a system of unitary states interacting with one another through diplomacy, public international law, and international organizations. In that model, states as agents interact, constituting an international system. But this model's focus on the reified unitary state fails sufficiently to emphasize two other essential elements of the contemporary world system: *networks* among agents, and *norms* – standards of expected behavior – that are widely ac-cepted among agents.[33]

It is therefore crucial to consider the role of international institutions in influencing societal norms, values, and preferences in a geographic region.[34] *Institutions* are the "products of human design, the outcomes of purposive actions by instrumentally oriented individuals."[35] Focusing on the behaviour of actors and institutions in the policy process presents a tool for defining the rules of the game and understanding the actions of the players[36] in the public policy process. This theoretical lens involves assessing formal and informal choice frameworks for public preferences, policy goals, policy issue areas, and policy tools. In the *neo-institutionalist* perspective, institutions are viewed as "persistent and connected sets of rules (formal or informal) that prescribe behavioral rules, constrain activity, and shape expectations."[37] Institutions thus provide the framework within which international networks and norms influence the individual behaviour[38] that shapes public preferences and policy options at the local, state/province, regional, and nation-state level. In other words, actors and institutions are constituted by social environments, which give them form and legitimacy, which allows them to enable and constrain organizational structures and processes.[39]

There is nothing new about the idea that multilevel interactions in diverse geographic regions can lead to convergence in values and norms and can contribute to integration.[40] However, neo-institutionalist theories suggest that the formation of values and norms within regional networks can be mutually reinforcing for institutions and the actors engaged within them: "The argument is not that institutions cause an action. It is rather that they affect actions by shaping the interpretation of problems and possible solutions and by constraining the choice of solutions and the way and extent to which they can be implemented. While individuals, groups, classes and states have their specific interests, they pursue them in the context of existing formal organizations and rules and norms that shape expectations and affect the possibilities of their realization."[41]

The interplay of actors and institutions that leads to the formation of accepted values, norms, expectations, identity, and ideology has significant implications for North American institutions of regional governance. However, it is important to note the perennial cross-border disputes and growing contestation within distinct policy domains that illustrate constraints placed on the formation of values and norms within regional networks. For example, policy divergence on softwood lumber and oil pipeline expansion represents significant divergence in values and norms among diverse stakeholders in both countries. Regional convergence and cooperation in the public policies affecting these domains is not likely anytime soon.

Table 12.2. Multilevel Governance in North American Regions: Organizational Forms, Bi-lateral Processes, and Instruments Used to Facilitate Cooperation

Types of cross-border organizations	Bi-lateral processes to support collaboration	Instruments used to facilitate cooperation
General-purpose Intergovernmental Single-purpose intergovernmental (typically an *environmental* or *infrastructure* focus) Organization with a city orientation Organization with a civil society orientation Joint commission Joint task force Advisory council Collaborative research institute Think tank Coalition	Summits Conferences Symposiums Colloquiums Workshops Forums Institutes Joint operations Joint programs Joint research projects Joint training Joint boards or panels Exchange of personnel Minister-secretary meetings Meetings of legislators Working group meetings Routine working group communications	Convention/accord Joint declaration/resolution Framework agreement Mutual recognition agreement (MRA) Commission protocol Cooperation agreement Memorandum of understanding (MOU) Memorandum of cooperation Letter of intent Strategic plan (Mission, vision, goals) Research report Annual report Newsletters/websites White papers/grey literature Routine working group documents Educational materials Informational materials Advocacy materials

Sources: Modified from Mouafo, "Regional Dynamics"; as well as both Canada, Policy Research Institute, *Emergence of Cross-Border Regions: Interim Report* and *Emergence of Cross-Border Regions: Final Report*; and from Canada, School of Public Service, *Advancing Canadian Interests in the United States*, 22.

Models of Cross-Border Regional Governance in North America

A study on socio-cultural values found in the cross-border regions of the United States and Canada was published by Canada's Policy Research Initiative in 2005. This publication reported important regionalisms in values, identity, and ideology, with socio-value convergence concentrated along the Atlantic and Pacific coasts (also referred to as "blue America" for their consistent liberal political orientations). In fact, regional value linkages between Canada and the United States are usually stronger than national ones in these regions.[42] However, "it is in the West, which includes British Columbia, Alberta, Washington,

Idaho, Oregon, Montana and, in some cases, Yukon, as well as Alaska, that cross-border regional relations seem the most profound and diversified. Based on these economic, socio-cultural, and organizational links, a cross-border region has taken shape."[43]

Several major private and public sub-national regional institutions have been created, as noted in table 12.1. The strongest are in the West (PNWER and the Western Governors' Association and Western Premiers' Conference). In addition to international organizations, there are many additional linkages in North American regions through sector-specific organizations, informal networks, social groups, and communities of practice.

The West's extensive cross-border linkages tend to have a strong focus on trade and policy coordination in the industrial sectors of agriculture, energy, transportation, and tourism. The region has a dense network of cross-border organizations, with PNWER as the largest and most sophisticated. Additional binational organizations include the Western Governors Association, the Pacific Corridor Enterprise Council, and the Pacific Coast Collaborative. Sector-specific entities include Northwest Environment Watch, the International Mobility and Trade Corridor, and the Discovery Institute. A relatively high degree of collaboration takes place within and among these regional organizations, as well as in partnership with local think tanks and universities.[44]

Drawing on the Canadian government's Policy Research Initiative project of 2003–8 and his own research on regional governance and multilevel governance, Emmanuel Brunet-Jailly contends that the Pacific Northwest provides unique opportunities for investigating the public policy impact of regional governance institutions and mechanisms: the Pacific Northwest "is not the only case study but best illustrates these observations of rise of border regions spanning the U.S. Canadian border. It is in the Pacific North West that this emerging complex ideational construct spanning economic, social and cultural, and political elements and influencing policy making across the boundary line in increasingly numerous policy fields is most easily documented."[45]

Brunet-Jailly further hypothesizes that the prevalence, scope, and scale of four factors will lead to regional cross-border integration in North America: "(1) economic integration, (2) the convergence of socio-cultural values, (3) the formation of border-spanning institutions, and (4) the emergence of policy mechanisms that lead to policy parallelism."[46] This begs deeper analysis of the role of the Pacific NorthWest Economic Region in leading and facilitating regional governance involving all four factors.

Developing a Conceptual Framework for an Institutional Analysis of PNWER

If institutions of regional governance facilitate policy coordination, it is necessary to investigate the institutional structure, design, processes, and instruments that allow an international organization like the Pacific NorthWest Economic Region to perform this role. Neo-institutionalists posit that institutions are significant in the political realm because they "constitute and legitimize political actors and provide them with consistent behavioural rules, conceptions of reality, standards of assessment, effective ties, and endowments, and thereby with a capacity for purposeful action."[47] With PNWER, legislators, industry leaders, academics, and leaders from nongovernmental civil society groups are provided with an institutional framework within which they can effectively engage in public policymaking. This institutional framework provides for participation of elites in regional governance; for participation of those who are willing and able to "pay to play." It is noteworthy that many groups representing marginalized members of society (for example, First Nations and Native American tribal governments) – as well as many civil society groups – are not active participants in PNWER's organizational structure and routine gatherings, although PNWER officials may invite their participation on selected issues such as 2017 and 2018 discussions of the Columbia River Treaty revision.

PNWER's primary organizational goal is to provide an institutional structure and processes that facilitate elite stakeholder participation in regional policy cooperation. To understand how this organization works, it is beneficial to explore the relationship between the institutional constraints, the roles of actors, and the stages of the policy process.

To develop a neo-institutionalist model of the Pacific NorthWest Economic Region for better understanding the organization's impact on regional policy coordination, it is imperative to reference the robust scholarship on the *public policy process*, also referred to as the *policy cycle*. Where the general term *policy* refers to a purposive course of action by an individual or group to deal with a problem,[48] the term *public policy* is defined as a course of government action or inaction in response to public problems.[49] Because public policy is "made" by the decisions of a government or an equivalent authority, as emphasized by Anderson and Jones in chapter 2, the study of the process by which those decisions are made constitutes a long tradition in public policy research. Public policymaking can be disaggregated into discrete stages and sub-stages; the resulting sequence of stages is referred to as the *policy cycle* or the *policy process*. Public policy scholars make choices in researching one or more stages in the policy cycle, actors and institutions that constitute policy design and content, policy outcomes and impacts, or any number of other policy-relevant factors. There is a common sequence of five stages used in studying the policy cycle:

1 *Agenda-setting* – the process by which problems come to the attention of governments;
2 *Policy formulation* – the process by which policy options are formulated within government;
3 *Decision-making* – the process by which governments adopt a particular course of action or non-action;
4 *Policy implementation* – the process by which governments put policies into effect;
5 *Policy evaluation* – the processes by which the results of policies are monitored by state and societal actors.[50]

PNWER states in its communication materials that its first goal as an institution is to coordinate provincial and state policies throughout the region (see figure 12.1) – although as a nongovernmental body without delegated legal authority, it might be more accurate to say that it *encourages* such cooperation and coordination among participating governments, particularly in dealing with their respective federal governments. Most of the other goals listed can be interpreted as actions that might be taken to support this primary goal. This goal – to coordinate public policies across the region's states, provinces, and territories – is clearly viewed as directly supporting PNWER's mission to achieve continued economic growth while maintaining the region's natural environment. So how does PNWER as an institution lead in coordinating public policies?

PNWER prominently promotes itself as being "the preeminent binational advocate for regional state, provincial, and territorial issues."[51] The term *advocate* is important here, as it suggests an institutional focus on the first several steps of the policy process; namely, agenda-setting and policy formulation. PNWER actors identify and define problems, prioritize policy areas for engagement, and articulate proposed policy solutions that ultimately lead to governmental policy formulation. Indeed, the organizational structure, programmatic areas, and routine events established by PNWER emphasize opportunities for a wide array of stakeholders to identify shared problems and craft solutions. As a regional institution, PNWER offers an organizational structure that supports advocacy within each country's central government to manage certain aterritorial policy functions or, alternatively, mitigates territorially based policies with adverse effects on regional interdependence.

In each year, PNWER holds two meetings: a major *annual summit* in the summer and a combined *economic leadership forum* and *legislative leadership academy* in the fall. These major meetings rotate among the member states, provinces, and territories. The international organization coordinates timely major program areas and institutes throughout the year: the *Legislative Energy Horizon Institute*, the *Center for Regional Disaster Resilience*, a *NAFTA Modernization Task Force*, a *Preclearance Task Force*, and a major regional prevention initiative against *Invasive Species*. PNWER also coordinates advocacy visits each year for stakeholders to meet with legislative offices in Ottawa and Washington, DC.

The main body of work within PNWER takes place within *working groups*, of which there are nineteen at the time of this writing. Current topics of these working groups are Arctic Caucus, agriculture, border issues, cross-border livestock health, disaster resilience, energy and environment, energy, forestry, infrastructure finance, innovation, invasive species, mining, natural gas, ocean policy, trade and economic development, transportation, tourism, water policy, and workforce development. There is also an annual forum for university presidents within the PNWER organizational structure. "Working Groups focus on key regional issues throughout the year. They are led by a public sector and private sector co-chair, along with one lead PNWER staff member, and include public, private, academic and non-profit stakeholders.... Action items are developed by working groups at PNWER's two annual meetings. They represent concrete actions the working group will take to advance the priorities of the group. These Action Items are approved by the Executive Committee. Action Items constitute the working groups' and PNWER's priority work throughout the year."[52]

In short, by providing these well-organized forums within which leaders from the public and private sectors can meet regularly to share information, discuss challenges and opportunities, and articulate problem-solving strategies, PNWER is providing a framework for ground-up policy agenda-setting and policy formulation. Once agendas are set, PNWER working groups are intended to address action items and (with participation of additional stakeholders) formulate necessary policy changes. Competing interests are thereby managed by segmenting policy domains into working groups, which allow elite stakeholders to address policy differences and disputes. The working groups provide functional structures and processes to enable policy formulation and agenda-setting. The resulting new and improved policies are then advocated by policy subsystems throughout the region and, as necessary, among appropriate bodies in the two nations' federal governments.

Simply stated, public policies "are made by *policy subsystems* consisting of actors dealing with a public problem."[53] The nature and interaction of specific subsystems in participating jurisdictions vary widely across provinces, states, and policy subfields. The fundamental nature of policy processes also varies between Canadian provinces' Cabinet-parliamentary systems of government and U.S. states' divisions of power between executive and legislative systems. Diverse theories and models are used by scholars to analyse the structure and function of policy subsystems, ranging from *sub-governments* and *iron triangles* to *issue networks, policy networks, policy communities,* and *advocacy coalitions.*[54]

The *advocacy coalition framework* (ACF) developed by Paul Sabatier and his colleagues was created to study complex activities of policy actors in policy subsystems. "An advocacy coalition consists of actors from a variety of public and private institutions at all levels of government who share a set of basic beliefs (policy goals plus causal and other perceptions) and who seek to manipulate the rules, budgets, and personnel of government institutions in order to achieve these goals over time."[55] In other words, the ACF refers to "collections of actors sharing similar beliefs and coordinating their actions to achieve political goals."[56] An in-depth discussion of the ACF model and how it may be applied to analyse the policy subsystems exemplified by PNWER extends beyond the scope of this chapter. However, identification of the PNWER as an advocacy coalition is very useful in understanding its important role in the policy process. Michael Howlett and M. Ramesh clarify the concept in a way that can be directly applied to analysis of PNWER:

> An advocacy coalition includes both state and societal actors at the national, sub-national, and local levels of government. It also cleverly combines the role of knowledge and interest in the policy process. The actors come together for reasons of common beliefs, often based on their knowledge of the public problem they share and their common interests. The core of their belief system, consisting of views on the nature of human-kind and some desired state of affairs, is quite stable and holds the coalition together. All those in an advocacy coalition participate in the policy process in order to use the government machinery to pursue their (self-serving) goals.[57]

The general application of an ACF analytical scheme begins with describing how like-minded actors are organized to govern and engage in an institution, which provides the structure within which to participate in the policy process. In practice, PNWER may best be understood as a cluster of functional, policy-specific advocacy coalitions held together by an effective overarching institutional structure.

Governance of PNWER is provided by a delegate council comprising governors and premiers – or their designated representatives – as well as four legislators (state senators and representatives, and MLAs) from each member state, province, or territory. One delegate from each state, province, or territory serves as the jurisdiction lead on the PNWER Executive Committee. Service on the delegate council and/ or in other positions of PNWER leadership is a one-year commitment. The elected officers of PNWER include a president, two first vice presidents (one from the United States and one from Canada), and two second vice presidents (one from each country). Officers must be elected legislators, or an elected governor or premier, from participating jurisdictions at the time of their election and remain in office for their tenure as a PNWER officer. The PNWER president position rotates equitably among the member jurisdictions. Organizational leadership and governance matters are overseen by the Executive Committee, the Finance and Audit Committee, the PNWER Past Presidents' Committee, and the PNWER Foundation Board of Directors. The PNWER Foundation was created as a non-profit organization to act as the fiscal agent for PNWER.

PNWER also provides for private sector participation in governance through the Delegate Council's Private Sector Council (PSC), which includes membership from business, the non-elective public sector, NGOs, local governments, and other non-profit organizations. One PSC member is elected from each jurisdiction to serve as the chair of their PSC of their respective state, province, or territory, and this chair sits on the PNWER Executive Committee ex-officio. As explained in PNWER's *Officer's Handbook*, "The purpose of the PSC is to provide a forum for the private sector to meet and discuss issues of concern and communicate these through the working groups and the Private Sector Council to the PNWER Executive Committee. Furthermore, each state/province/territory delegation is encouraged to hold meetings as needed in their own jurisdiction and promote business leadership within all aspects of PNWER."[58]

With significant annual turnover in delegates and officers, evolving areas of programmatic focus, and working groups that change over time, the PNWER Secretariat located in Seattle is crucial in providing consistent executive leadership and a support infrastructure for the ongoing work of the international organization. The senior staff of PN-WER has been remarkably stable over the years; the executive director, Matt Morrison, has served as PNWER's CEO since 1994. The Secretariat "consists of the Chief Executive Officer, Chief Operating Officer, and Program Directors and Managers who oversee specific programs, working groups and projects, as well as other support staff."[59] It is noteworthy that the permanent administrative staff of this major regional organization consists of only the CEO, COO, two program director staff, three program managers/coordinators, an events manager, and an accountant. The arm's-length quality of staff leadership provides an opportunity for PNWER to "represent" member jurisdictions on selected issues on which there is consensus while still allowing considerable room for independent action by other stakeholders.

Managing PNWER for Public Value

It is evident from this discussion of PNWER's institutional goals, leadership structure, and main activities that the ultimate goal of the entire organization is to provide value to the public through regional governance. The major desired outcome of the entire transborder organization is to coordinate public policies that provide value to the economic development and environmental well-being of the region as a whole.

Extensive scholarship on strategic management in the public sector to create public value has been published by Mark Moore, and the concepts he discusses in his two major books on this topic[60] are very helpful in developing a conceptual framework for studying the institutional design and processes of regional governance organizations such as PNWER. In sum, Moore suggests that managing successfully for public value involves management of the authorizing environment that provides legitimacy and support to the organization, management of the operational capacity of the organization, and management of the desired outcome (the value to the public) to be created by the organization.

Drawing on schematics developed by Mark Moore and other scholars of public administration, figure 12.3 provides an overview of how these three key dimensions of managing for public value may be analysed in the institutional design and processes of PNWER. These three domains are each paired with the public policy cycle theoretical lens that might provide the best framework for analysing that domain. In considering legitimacy and support – that is, the authorizing environment necessary for PNWER leadership – the advocacy coalition framework (ACF) offers

Figure 12.3. Institutional Design and Processes of PNWER for Managing Public Value

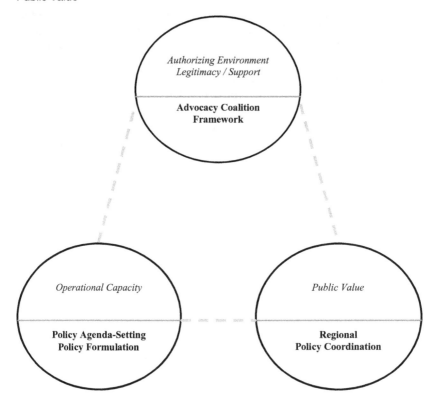

a lens for understanding how the exchange of values, norms, and learning may be facilitated among the elites who participate in the PNWER organization. Operational capacity has to do with resources available to the institution for management of agency programs and activities. As discussed previously, the main goal of PNWER's programs, activities, and communications instruments is to facilitate engagement of leaders from the public and private sectors in policy agenda-setting and policy formulation. Ultimately, it is a focus on the third domain – that of public value – where the desired outcomes of PNWER's impact on regional policy coordination can be assessed.

Conclusion and Avenues for Future Research

This chapter has presented a theoretical and conceptual framework for the institutions of regional governance across the continent, and it offers analysis of PNWER as a case study. The chapter depicts an idealized model of PNWER as an international institution that effectively functions as an advocacy coalition engaged in regional policy cooperation. In reality, however, PNWER's governance structure faces persistent challenges in mobilizing consent across an increasingly disparate group of stakeholders with different electoral cycles, political cultures, and partisan agendas.

The extensive review of literature conducted to develop this chapter led to the three findings below, upon which any conceptual lens used to study regional governance in North America should be based. In time, better understanding of the types of cross-border organizations, processes that are used to support regional collaboration, and instruments that are used to facilitate cooperation will help researchers to continue to develop theoretical models relevant to the unique North American system of regional integration.

1. The Strength and Political Orientation of Regional Identity Matters

The potential to develop strong institutions for regional governance with Canada appears to be concentrated in the Northwest United States, in a grouping of states, provinces, and territories in which there is a stronger sense of regional identity. Where there are strongly shared values, norms, identity, and ideology, it is more likely that shared institutions will be established to reflect and reinforce policy actors' belief systems. It follows that the relatively strong regional identity in the Northwest has enabled PNWER to grow and to be effective.

But cross-border regional identity in all areas of North America – including the Northwest – is not a unitary construct. Overlapping and competing interests and identities result from the presence of overlapping sub-regions (for example, in the Northwest this includes Cascadia, Rocky Mountains, the Great Plains, the Prairies, and the Arctic clusters of sub-regions). To accommodate multiple regional identities, PNWER necessarily functions in a multi-regional context that features economic and ideological differences. Competing regional interests manifest in policy on environmental interests, economic development interests, resource management, and land ownership. Even in a region that features a comparatively strong sense of regional identity, the competing and conflicting pressures of (sub-)regional identities must be taken into account in studying the institutions of regional governance. Analysing PNWER and similar institutions by using the advocacy coalition framework model will assist in better understanding the authorizing environment for regional governance.

2. Institutional Leadership, Structure, and Design Matter

It is imperative to assess the ways in which policy actors engage within institutions, and to analyse the institutional structure that provides for participation in the policy process. "Individuals, groups, classes, and states participating in the policy process no doubt have their own interests, but the manner in which they interpret and pursue their interests, and the outcomes of their efforts, are shaped by institutional factors."[61] Jurisdiction-specific, regional, and national institutions shape the interests of policy actors, but these interests are not always congruent. When interests conflict or diverge, jurisdictionally based interests will usually hold greater political power. A recent example of conflicting sub-regional interests in the Northwest led to a series of roundtable discussions (in 2016, 2017, and 2018) on proposed changes to the Columbia River Treaty. These roundtables allowed for helpful exchanges of deeply conflicting interests and priorities but did not result in regional policy resolution.

In North America, institutions of regional governance may be designed to provide for most effective participation by stakeholders in the policy agenda-setting and formulation stages of the policy process. This is certainly the case with PNWER, which provides an effective set of functional institutional structures and processes as well as considerable political will and leadership by regional stakeholders. Even if the focus of regional governance remains on policy agenda-setting and policy formulation, it would be beneficial for future research to better understand the relationship between operational capacity of regional institutions in supporting engagement in all five stages of the policy process.

3. Institutional (Policy) Goals, Processes, and Instruments Matter

Diverse bilateral processes and communications instruments can be used by regional institutions to support actors' engagement in the policy process. It is the purposeful and strategic use of institutional goals, processes, and instruments that allows an international organization like PNWER to facilitate the engagement of public and private sector leaders. Ultimately, the ground-up engagement through well-structured processes and communications lead to the articulation of specific policies that are advocated throughout the region and at the federal level. In an ideal model, these policies will be coordinated and harmonized, which is the specific desired outcome – that is, the *public value* – of the institution. Several of PNWER's initiatives illustrate this kind of success, such as improved transportation and traveller pre-clearance in the Cascadia corridor, the development of policy instruments to combat invasive species throughout the region, and enhanced cooperation among emergency management professionals.

In practice, however, competing interests of sub-regions and distinct jurisdictions are rarely fully reconciled within a regional institution of governance (for example, policy issues on forest management practices or oil pipeline expansion). When hindered by the deeply conflicting policy goals of its constituent jurisdictions, a regional institution will be constrained in its ability to formulate policy goals, instruments, and processes. As such, there are structural limitations in what a regional governance organization like PNWER can reasonably be expected to do.

As Randall Germain and Abdulghany Mohamed argue, regionalism "must have at its core a certain level of region-ness that is multidimensional and that embraces both ideational and material processes and arrangements…. These arrangements [are] composed of two key attributes: the policy horizon of decision makers and the depth and robustness of regional institutions."[62] The entire field of enquiry on North American models of regional governance is wide open for further research. All of the concepts, theories, and models discussed in this chapter beg further empirical study in their application to institutions like PNWER. Research on regional governance would also greatly benefit from more in-depth institutional analysis as well as the integration of public policy process research.

To conclude, findings from this research suggest that North American citizens' understanding of opportunities to engage regionally in advancing political, economic, social, and environmental initiatives has never been more important. Current isolationist inclinations of the U.S. federal government administration suggest that continued investment in cross-border policy coordination will necessarily continue to take place through regional governance, at least for the foreseeable future.

NOTES

The author thanks her two research assistants, Charissa Hurt and Jes Sokolowski, and is grateful for the valuable input provided from the book project editors and contributors, in supporting development and revision of this chapter.

1 See Joel Garreau, *The Nine Nations of North America* (Boston: Houghton Mifflin, 1981); and Colin Woodard, *American Nations: A History of the Eleven Rival Regional Cultures of North America* (New York: Penguin Books, 2011) for a historical background to North America's "mega-regions."

2 J.C. Day, James Loucky, and Donald K. Alper, "Policy Challenges in North America's Pacific Border Regions: An Overview," in *Transboundary Policy Challenges in the Pacific Border Regions of North America*, ed. Loucky, Alper, and Day, 1–37 (Calgary: University of Calgary Press, 2008).

3 For example, see Geoffrey Hale, *So Near Yet So Far* (Vancouver: UBC Press, 2012), 219; and Robert Pastor, *The North American Idea: A Vision of a Continental Future* (New York: Oxford University Press, 2011), 110.

4 Hale, *So Near Yet So Far*, 217.

5 Stephen Clarkson, "Continental Governance, Post-Crisis: Where Is North America Going?" in *North America in Question*, ed. Jeffrey Ayres and Laura Macdonald (Toronto: University of Toronto Press, 2012), 86.

6 Robert A. Pastor, "Beyond NAFTA: The Emergence and Future of North America," in *Politics in North America: Redefining Continental Relations*, ed. Yasmeen Abu-Laban, Radha Jhappan, and Francois Rocher, 461–76 (Peterborough, ON: Broadview, 2008).

7 Randall Germain and Abdulghany Mohamed, "Global Economic Crisis and Regionalism in North America: Regionness in Question?" in *North America in Question*, ed. Jeffrey Ayers and Laura Macdonald, 33–52 (Toronto: University of Toronto Press, 2012).

8 Canada, Policy Research Institute, *The Emergence of Cross-Border Regions between Canada and the United States: Reaping the Promise and Public Value of Cross-Border Regional Relationships – Final Report* (Ottawa: Industry Canada, November 2008), iv.

9 Christopher Sands, as quoted in Canada, Policy Research Institute, *Emergence of Cross-Border Regions*, 33.

10 Canada, Policy Research Institute, *The Emergence of Cross-Border Regions: Interim Report* (Ottawa: Industry Canada, 2005), 25.

11 Emmanuel Brunet-Jailly, "Cascadia in Comparative Perspectives: Canada-U.S. Relations and the Emergence of Cross-Border Regions." *Canadian Political Science Review* 2, no. 2 (2008): 105.

12 Mouafo, "Regional Dynamics."

13 See the compelling map of cross-border networks, which includes a clearly labe;led typology of these linkages, in Canada, Policy Research Institute, *Emergence of Cross-Border Regions ... Final Report*, 8.

14 See Canada, School of Public Service, *Advancing Canadian Interests in the United States: A Practical Guide for Canadian Public Officials* (Ottawa: 2004); as well as discussions by Hale, *So Near Yet So Far*, 198–223; and Greg Anderson, "Expanding the Partnership? States and Provinces in Canada-U.S. Relations 25 Years On," in *Forgotten Partnership Redux: Canada-U.S. Relations in the 21st Century*, ed. Greg Anderson and Christopher Sands, 535–74 (Amherst, NY: Cambria, 2011).

15 Hale, *So Near Yet So Far*, 219.

16 Former U.S. ambassador to Canada Bruce Heyman, quoted in Pacific Northwest Economic Region, *Annual Report 2017*, http://www.pnwer. org/uploads/2/3/2/9/23295822/2017_pnwer_annual_report_-_reduced _size.pdf, 4.

17 Edward D. Mansfield and Helen V. Milner, "The New Wave of Regionalism," *International Organization* 53, no. 3 (1999): 590.

18 J.H. Paterson, *North America: A Geography of the United States and Canada*, 8th ed. (New York: Oxford University Press, 1989), 197.

19 Paterson, *North America*, 197.

20 James Rosenau, as cited in Greg Anderson and Christopher Sands, "Fragmentation, Federalism, and Canada–United States Relations," in *Borders and Bridges: Canada's Policy Relations in North America*, ed. Monica Gattinger and Geoffrey Hale (New York: Oxford University Press), 43–4.

21 Anderson and Sands, "Fragmentation," 42.

22 Monica Gattinger and Geoffrey Hale, "Borders and Bridges: Canada's Policy Relations in North America," n *Borders and Bridges*, ed. M. Gattinger and G. Hale, 8–9 (Toronto: Oxford University Press, 2010).

23 Earl H. Fry, "Federalism and the Evolving Cross-Border Role of Provincial, State, and Municipal Governments," *International Journal* 60, no. 2 (2005): 482.

24 James P. Allan and Richard Vengroff, "Paradiplomacy: States and Provinces in the Emerging Governance Structure of North America," in *North America in Question: Regional Integration in an Era of Economic Turbulence*, ed. Jeffrey Ayres and Laura Macdonald, 277–82 (Toronto: University of Toronto Press, 2012).

25 Hale, *So Near Yet So Far*, 4.

26 Colin Robertson, "CDA_USA 2.0: Intermesticity, Hidden Wiring and Public Diplomacy," in *Canada among Nations 2007: What Room for Manoeuvre?*, ed. Jean Daudelin and Daniel Schwanen, 268–310 (Montreal and Kingston: McGill-Queen's University Press, 2008).

27 Hale, *So Near Yet So Far*, 198.
28 Laurence E. Lynn Jr., Carolyn J. Heinrich, and Carolyn J. Hill, *Improving Governance: A New Logic for Empirical Research* (Washington, DC: Georgetown University Press, 2001).
29 Christopher Alcantara and Jen Nelles, "Indigenous Peoples and the State in Settler Societies: Toward a More Robust Definition of Multilevel Governance," *Publius: The Journal of Federalism* 44, no. 1 (2013): 185.
30 In particular, see Ian Bache and Matthew Flinders, eds., *Multi-Level Governance* (New York: Oxford University Press, 2004); Liesbet Hooghe and Gary Marks, "Unraveling the Central State, but How? Types of Multi-Level Governance," *American Political Science Review* 97, no. 2 (2003): 233–43; Simona Piattoni, *The Theory of Multi-Level Governance: Conceptual, Empirical, and Normative Challenges* (New York: Oxford University Press, 2010); Robbie Waters Robichau and Laurence E. Lynn Jr., "The Implementation of Public Policy: Still the Missing Link," *Policy Studies Journal* 37, no. 1 (2009): 21–36; and Paul Stephenson, "Twenty Years of Multi-Level Governance: 'Where Does It Come From? What Is It? Where Is It Going?'" *Journal of European Public Policy* 20, no. 6 (2013): 817–37.
31 Alcantara and Nelles, "Indigenous Peoples," 189.
32 See Stephen Goldsmith and William D. Eggers, *Governing by Network: The New Shape of the Public Sector* (Washington, DC: Brookings Institution, 2004).
33 Robert O. Keohane and Joseph S. Nye Jr., "Introduction," in *Governance in a Globalizing World*, ed. Keohane and Nye Jr. (Washington, D.C.: Brookings Institution Press, 2000), 19.
34 Martha Finnemore, *National Interests in International Society* (Ithaca, NY: Cornell University Press, 1996), 3–6.
35 Walter W. Powell and Paul J. DiMaggio, eds., *The New Institutionalism in Organizational Analysis* (Chicago: University of Chicago Press, 1991).
36 Douglas C. North, *Institutions, Institutional Change and Economic Performance* (Cambridge: Cambridge University Press, 1990).
37 Robert O. Keohane, *International Institutions and State Powers: Essays in International Relations Theory* (Boulder, CO: Westview, 1989), 163.
38 In "new institutionalism" theory, *normative institutionalism* emerged as a specific sub-theory, in which an institution is not necessarily a formal structure, but rather is better understood as a collection of norms, rules, understandings, and routines. For further clarification, see James G. March and Johan P. Olsen, *Rediscovering Institutions* (New York: Free Press, 1989); as well as B. Guy Peters, *Institutional Theory in Political Science: The "New Institutionalism"* (New York: Continuum, 2005).
39 Ronald L. Jepperson and John W. Meyer, "The Public Order and the Construction of Formal Organizations," in *The New Institutionalism in Organizational Analysis*, ed. Walter W. Powell and Paul J. DiMaggio, 204–31 (Chicago: University of Chicago Press, 1991).

40 See, for example, the writings of Karl Deutsch in the 1950s and 1960s.

41 Michael Howlett and M. Ramesh, *Studying Public Policy: Policy Cycles and Policy Subsystems* (New York: Oxford University Press, 1995), 27.

42 Christian Boucher, *Toward North American or Regional Cross-Border Communities*, Government of Canada Policy Research Initiative Working Paper Series 002 (Ottawa: June 2005).

43 Canada, Policy Research Institute, *Emergence of Cross-Border Regions: Interim Report*, 20.

44 Canada, Policy Research Institute, *Emergence of Cross-Border Regions: Interim Report*, 20.

45 Brunet-Jailly, "Cascadia," 117.

46 Brunet-Jailly, "Cascadia," 108.

47 James G. March and Johan P. Olsen, "Institutional Perspectives on Political Institutions" (paper presented at the meeting of the International Political Science Association, Berlin, 1994), 5.

48 James E. Anderson, *Public Policymaking*, 7th ed. (Florence, KY: Wadsworth/Cengage, 2011).

49 For more detailed definitions of "public policy," see Thomas A. Birkland, *An Introduction to the Policy Process: Theories, Concepts and Models of Public Policy Making*,, 3rd ed. (Armonk, NY: M.E. Sharpe, 2011), chap. 1; Christopher M. Weible, "Introducing the Scope and Focus of Policy Process Research and Theory," in *Theories of the Policy Process*, ed. Paul A. Sabatier and Weible, 3rd ed. (Boulder, CO: Westview, 2014), 4–9; and Michael E. Kraft and Scott R. Furlong, *Public Policy: Politics, Analysis, and Alternatives*, 4th ed. (Los Angeles: Sage, 2012), 3–5.

50 Howlett and Ramesh, *Studying Public Policy*, 11.

51 Pacific Northwest Economic Region, *Annual Report 2017*, 4.

52 Pacific Northwest Economic Region, *Annual Report 2017*, 18.

53 Howlett and Ramesh, *Studying Public Policy*, 51.

54 Excellent introductory comparisons of these diverse models of policy subsystems can be found in Birkland, *Introduction to the Policy Process*; Howlett and Ramesh, *Studying Public Policy*; Paul A. Sabatier and Christopher M. Weible, eds., *Theories of the Policy Process*, 3rd ed. (Boulder, CO: Westview, 2014); and Kevin B. Smith and Christopher W. Larimer, *The Public Policy Theory Primer*, 2nd ed. (Boulder, CO: Westview, 2013).

55 Hank C. Jenkins-Smith and Paul A. Sabatier, "The Study of Public Policy Processes," in *Policy Change and Learning: An Advocacy Coalition Approach*, ed. Sabatier and Jenkins-Smith (Boulder, CO: Westview, 1993), 5.

56 Simon Matti and Annica Sanstrom, "The Rationale Determining Advocacy Coalitions: Examining Coordination Networks and Corresponding Beliefs," *Policy Studies Journal* 39, no. 3 (2011): 386.

57 Howlett and Ramesh, *Studying Public Policy*, 127.

58 Pacific NorthWest Economic Region. *Officer's Handbook: PNWER's Organizational Structure and Delegate Responsibilities* (Seattle: 2017), 4 .

59 Pacific Northwest Economic Region, *Officer's Handbook: PNWER's Organizational Structure and Delegate Responsibilities*. 2017, 8.

60 See Mark H. Moore, *Creating Public Value: Strategic Management in Government* (Cambridge, MA: Harvard University Press, 1995); and Moore, *Recognizing Public Value* (Cambridge, MA: Harvard University Press, 2013).

61 Howlett and Ramesh, *Studying Public Policy*, 51.

62 Germain and Mohamed, "Global Economic Crisis and Regionalism in North America," 35.

13 Cross-Border Constraints and Dynamics in the Northeast

STEPHEN TOMBLIN

Functionalist thinkers in the past[1] put much stock in the notion that territorial-jurisdictional boundaries and sources of power and sovereignty (whether nation states or federalism) would naturally decline and be replaced by new-modern, more policy- (evidence-) focused approaches to problem definition and resolution. However, pooling sovereignty and rebalancing functional-territorial interconnections thought critical to better policy practices and outcomes has not proven to be "inevitable" in an era of Trump's "America First" or Brexit. The failure to erect new functional regimes has raised new questions about the power of old territorial path dependencies in a new, more divided North America.

The focus of this chapter is understanding and explaining the recent further decline of cross-border coordination among Eastern provincial premiers and Northeastern governors. In the wake of two world wars, functionalism as a theory focused much attention on the obsolescence of territorial states and natural – even "inevitable" –logic of working collaboratively across borders and boundaries. The functionalist perspective was very much committed to contesting the idea of territorial state power and finding ways to accelerate functional collaboration. Much has changed since then. The objective of the chapter is to make the case that there has been a further tilt in the territorial/aterritorial balance in the Eastern Canadian–Northeast region of North America, and much of this can be explained by the actions of the premiers and governors coupled with the ongoing strength of old bilateral regimes they have inherited. This has especially been so for the regional energy file that has pitted hydro against the political and economic effects of the shale revolution and alternative forms of energy within and across national borders.

Several factors help to explain this decline in interprovincial, interstate, and cross-border collaboration through initiatives such as the

Coalition of Northeastern Governors and Eastern Canadian Premiers Conference. These factors are interdependent and consequently produce mutually reinforcing cause-and-effect relationships. They include:

1 Lack of political will and leadership by premiers and governors;
2 Persistence of self-contained, silo-based territorial institutions that have always worked against pressure to advance common identities, regional audiences, patterns of knowledge construction based on evidence (function), and have (by design) reinforced unilateral, territorial-first approaches to resolution of problem definitions;
3 Functional institutional structures and processes (interprovincial, interstate and cross-border activity generally) that lack autonomy, resources, political support, even visibility;
4 Challenge of super-partisanship in the United States and interprovincial competitive federalism in Canada that have heightened executive power and undermined "rule of law" powers critical for reining in bad political-territorial behaviour based on the aterritorial projection of domestic policies.

Function-versus-form debates, connected to different assumptions about where power lies, have deep historical roots but gained momentum during discussions over European integration. In the post-war era, in reaction against territorial conflicts over ideas that pointed to replacement of bilateral forms of territorial decision-making gained popularity.[2] But these changes and reforms have also been influenced by time and space, the "critical moments" and "accidents of history"[3] that produced "distinctive logics" and attitudes around zero-sum versus positive-sum relations across national and subnational borders.

Since the 1980s, attitudes towards the need for functional regions – and how bureaucracies work (or should work) – have changed dramatically with the spread of public choice and post-positivist ideas.[4] At the same time, globalization has created new political challenges. Border defensive actions have accelerated on both sides of the Canadian-American border. Competitive federalism has undermined efforts to work together across borders in Canada. Super-partisanship in the United States has had a similar impact. Regional collaboration in the Northeast continental region (particularly in energy-sector development and management) has declined in recent times. Much of this trend has occurred as a result of the deliberate weakening of the cross-border regional processes involved, and corresponding strengthening of inherited bilateral policy relations between jurisdictions.

To offer critical insights on bordering patterns, this chapter borrows from historical-institutional and path-dependency literatures to explore the history of struggle over regional collaboration in Eastern Canadian provinces and U.S. Northeast states. It argues that function has not proven to be "inevitable" and that territorial form has remained powerful and underestimated.

Historical-Institutionalism and Path Dependency

For generations, boosters of building functional regions talked a great deal about constraining territorial sources of political power that undermine wider political cooperation among jurisdictions. These calls for action were as much prescriptive as empirical. Whether in the case of modernization (whose advocates often thought federalism obsolete) or continental integration, these regional boosters often assumed that new, more functional realities and conditions would ultimately result in carving out more integrated universal systems of interest mobilization, and problem definition and resolution. But these assumptions about where power lies and how transformation occurs never proved out in practice.[5] For example, some past observers wondered why Canadian province-building survived in an era of modernization. In New England, for boosters of regionalization, there were other mysteries to be explored and explained – not least why certain "inevitable" transformations proved to be more difficult than originally assumed by their aspiring architects.[6]

Historical-institutionalism and path-dependency theorists paid closer attention to the power of past events and how these inherited structures and processes reinforced parallelism, and different expectations for politicians who continued to play to dissimilar audiences. It was always a question of form over function in where power lies and how this affects future behavioural patterns.

The explosion of the Canadian province-building literature originally focused on the role of the constitution, institutions, party systems, and other "state-centred" explanatory factors, generating much interest and debate.[7] Another literature of province-building and regionalism focused more on socio-economic factors. While both camps explored fundamental empirical differences, they were often as prescriptive as empirical in explaining why province-building persisted and what might be done to spawn new aterritorial forms and citizen identities. U.S. analysts also emphasized regionalism, but they focused more on finding ways to work across states, whether in New England, Appalachia, the South, or West, with the idea of building functional regions gaining momentum in the post-war era.[8] As a result, regionalism means different things on both sides of the border.

Integration Battles

Debates over interstate (bilateral) as well as intra-state (multilateral) patterns of regional functional coordination have been shaped and provoked by changing nation-centred policy regimes, but also U.S.-Canadian bordering strategies. Canada's abandonment of national developmental policies in the early 1980s, the weakening of pan-Canadianism and subsequent embracing of free trade had a huge impact. Ever since, provinces have played their role in a political dance that has become more and more bilateral and territorially competitive, especially in energy production, management, and transport.

More recently, south of the border, the combination of the rise of Maine's abrasive Tea Party governor (Paul LePage, 2011–19), followed by Donald Trump's 2016 election as president, has dramatically weakened the "regional club" and cause of functional integration in the Northeast.[9] Like President Trump, Governor LePage was sceptical of multilateral policy discussions and went out of his way to weaken partnerships. This has been reflected in the cutting of critical resources for regional agencies such as the Environmental Protection Agency (EPA), but also the rise of a more competitive, partisan brand of regional discourse. Multilateral, evidence-informed approaches to regionalization (popular in the past with strong leadership coming from reform-minded Democrats like President John Kennedy, Senator Ted Kennedy, and Governor Michael Dukakis) have fallen off the radar. Even so, there are clear signs that the region's other governors have worked in concert to oppose Trump and LePage's priorities, whether in resisting the opening of the federally controlled offshore to oil and gas development or the cutting of regional federal agencies, especially the EPA.[10] Some of this has changed with the recent election of a more moderate governor, Janet Mills, who took over the reins of power in January 2019.

Even so, it is clear that past national-continental strategies designed to tilt the territorial-functional integration balance and rein in defensive border behaviour, such as the launching of regional commissions in the United States (especially the 1963 Appalachian federal-state partnership sponsored by President Kennedy), or past encouragement from Ottawa for advancing regional integration in Western or Atlantic Canada, have not always worked out in practice.[11] In fact, such initiatives have often produced contradictory defensive reactions designed to preserve and defend territorial sovereignties from outside functional attack.

For example, during the 1960s and 1970s, New Brunswick's leaders showed an appetite for ideas of municipal restructuring, building stronger bureaucracies, and working across borders. But it was never enough to put the regional idea into action, since all provinces had to be on board. Successive Nova Scotia premiers flatly rejected recommendations of the 1970 Maritime Union report[12] calling for a European-style commission and regional parliament for the Maritimes and other "big bang" integration proposals over the decades.[13] Prince Edward Island, as a small island community, was never as enthusiastic as New Brunswick about opening up its borders. Hence, maritime union was a non-starter.

Similarly, New England governors have resisted periodic calls to erect powerful multilateral regional commissions or other critical instruments important to "big bang" policy changes. Instead, they have consistently preferred incremental changes to ensure any emerging regional vision was confederal by design, never threatening borders or local sovereignty.[14] For governors and premiers playing to different local audiences, controlling the "forms" of power have been central to directing their functions.

Early Border Connections

The New England governors have a long history of bordering together. New York has tended to be more of an outlier, even a competitor. These trends have often been reinforced by federal agencies' borders and programs, but also by New York's separate electrical grid system. These border actions are at times separate from the Canadian provinces, but the border mechanisms and practices that exist today in Atlantic Canada have been shaped by early U.S. regional regime ideas and practices. The Atlantic Provinces Economic Council, for example, was greatly influenced by the design of the New England Council.

Relations and interconnections among "eastern provinces" – the Maritimes, Quebec (which defines itself as a nation, not region), and sometimes Newfoundland and Labrador, have evolved in a complex series of rather informal, often well-camouflaged pathways that are difficult to map out, let alone navigate. Periodic attempts to coordinate bordering behaviour beneath the public radar have been affected by inherited bilateral forms of cross-border experimentation that have evolved incrementally.

For example, the New England Governors and Eastern Premiers Conferences that have been held since 1973 are interconnected in some ways while remaining separate entities. Provinces' participation in the latter has changed, further complicating matters. The province of New-foundland and Labrador left the Atlantic premiers' orbit in the 1960s only to return in the 1990s, while remaining part of the New England governors and Eastern premiers club. More recently, the New England governors and Eastern premiers function separately but are also housed in the larger Coalition of Northeastern Governors which left Boston in 2012 and is now located in Washington, DC.

Understanding bordering practices in the Northeast requires recognizing these separate but also interconnected axes, each with its own territorial-jurisdictional cocoon or silo. After decades of struggle, efforts to establish a New England regional commission have achieved little more than a confederal, bilateral game of Snakes and Ladders.[15] Similarly processes have played out in Canada, as efforts to integrate Maritime and Atlantic regionalism have never moved beyond confederal approaches to regionalism.

Cross-Border Regionalism: Northeast United States and Canadian Eastern Provinces

Regional integration has always played out differently across time and space. In New England, the ideas of reaching across state borders, advancing regional ideas, building regional interests, and strengthening regional institutions have an extensive historical background, as illustrated by Pierce.[16] Various organizations have promoted this movement at different times, including the New England Council (1925), New England Governors' Conference, the *Boston Globe*, New England Board of Higher Education, assorted federal-state partnerships and federal agencies designed by region (including the EPA), academics committed to the functional-regional cause, and even the New England Patriots football team.[17]

During the 1930s, the combined effects of the Depression, the New Deal, and the New England Council provided initial momentum, but regional integration never advanced very far. While the regional idea never went away, it lacked institutional-public support and the formal power necessary to transform political behaviours across jurisdictions or policy systems. Even during the 1960s, when President Kennedy's leadership on regional issues generated renewed momentum, regional initiatives faced entrenched structural-institutional constraints. Kennedy was deeply disturbed by the poor economic conditions in Appalachia and sought to promote development by investing in more rational, less territorial approaches to planning through regional commissions. But New England's more developed economy, while facing challenges, was not Appalachia. New England governors naturally wanted and encouraged the investment of new federal resources in the region, but had little interest in sacrificing their territorial sovereignty to a new regional experiment, making radical regionalization a non-starter. Eventual agreement on new regional commission was subject "to reflect[ing] and respect[ing] the divisions and traditions of the existing regime," enabling established "forms" to retain control over major "functions."[18]

The toothless New England Regional Commission that finally emerged lasted only sixteen years – despite providing resources for creation of the New England Governors' Conference and a number of regional networks. These resources facilitated the hiring of staff critical to advancing regional networks and more united approaches to defining and resolving interdependent problems, reflecting a strategic economic vision for New England during the early 1980s.[19] The location of their main office in Boston helped strengthen support for the regional mission at the centre, as did the emergence of the New England Congressional Caucus in 1972. In 1973, the launching of the New England Governors and Eastern Canadian Premiers Conference, with its annual meetings, formalized the connection between the Eastern premiers and New England governors. However, none of these regional experiments had a formal-legal base or source of independent power, resulting in limited policy capacity and legitimacy.

Similar outcomes resulted in the Eastern Canadian provinces, with their very different structural, ideological, socio-economic conditions, and models of federalism. Establishing new regional forms of cross-border governance and management did not prove to be automatic or "inevitable." Questions of regional integration emerged from much broader economic models of modernization, reflected in debates over Newfoundland and Labrador's 1949 entry into the Canadian federal orbit. In time, recognition of regional underdevelopment became a major driving force for regional ideas, whether for the Maritimes or the more contested notion of "Atlantic Canada." In the early 1950s, the latter was a voluntary affair aimed at Newfoundland and Labrador's integration within Canada but also strengthening an ailing regional economy. The Atlantic Provinces Economic Council (APEC) was formed in 1953, based on the model of the New England Council. APEC promoted the modernization-universal model and focused much attention on urbanization and unity, not diversity.

These ideas were an especially hard sell in Newfoundland, with its strong rural traditions, although New Brunswick was more open to regime change as it was in an extended state of economic-political crisis. Nova Scotia, enjoying a stronger economy, was more suspicious of radical changes. As a result, the regional integration debate created more diversity than unity as political leaders, playing to different audiences, squared off to defend and protect their interests.

Despite efforts to formalize structures and processes, and even for some to contemplate Maritime political union during the 1960s and 1970s, there was also much resistance to the concept of "Atlantic Canada," with Newfoundland pulling out of regional networks.[20]

Despite a maritime union study in the early 1970s,[21] discussions over the potential for a regional commission and even a common parliament, the only reform that emerged was a premiers' council that operated on the basis of unanimity. As a result, old path dependencies persisted. Despite efforts to coordinate, share information, even administer programs together, such initiatives have been limited over time. The number of regional agencies and programs have also declined as bordering priorities have lost institutional support and political resources. None of this bordering activity has posed much of a threat to traditions of provincial state development or maintenance.

Continental Border Regions

Cross-border relations between Canada's eastern provinces and U.S. northeast states have gone through various stages of change over time. The New England Governors and Canadian Eastern Premiers Conference (NEGC) that began in 1973 has been internationally recognized for its multi-jurisdictional regional policies and programs. While the running of annual conferences did help draw attention to cross-border ideas and interests, there was little significant change in strengthening cross-border institutional connections. In fact, if anything, these have been weakened as evidenced by significant staffing cuts in the 1990s and closure of the New England Governors office in Boston in 2012, which moved to Washington, DC, under the reconstituted Coalition of Northeastern Governors (CONEG). In the end, however, these annual meetings were never that significant. What mattered more was the networks and ideas that were generated and mobilized across policy fields as well as jurisdictions.

Despite this observation, the institutional foundations for bordering activity are generally not designed for easy flow. Regional pathways are without strong foundation and are roped into other more powerful institutional processes and pathways that operate with their own games, dynamics, and sets of rules.

A secretariat exists in Halifax to coordinate the activities, meetings, and programs of both the Maritime and Atlantic premiers. However, its responsibility is to respect and serve each jurisdiction independently, respectfully, but also bilaterally. For this reason, multilateral approaches, however limited, come about only after bilateral relations are clearly reinforced and protected. The Quebec government has its own machinery when it comes to any bordering activities involving other Eastern premiers. This adds further to the strengthening of the bilateral diversity model. Typically, agreement on common approaches to problem definition and resolution remain the responsibility of individual provinces, ensuring that bordering policy cycles remain in-house and under control. The spoke-and-wheel approach to problem definition, resolution, and governance has imposed strict limits on regional institutional or policy development.

The New England Governors' Conference first appeared in 1937 to ensure governors met regularly to share information, and define and attempt to resolve interdependent problems. It played an important role in a regional bordering game that worked in parallel with other regional interests and organizations, including the New England Council of States, New England Board of Higher Education, New England–Canada Business Council, and regionally organized federal agencies. But U.S. governors do not control their jurisdictions' budgets, legislatures, or own natural (including energy) resources in the same way Canadian provinces and premiers typically do. Governors have fewer resources to control agenda-setting. At the same time, traditions of local home rule, internal rural-urban regional divisions, different cultures, and policy legacies have all helped produce a pattern of decision-making that is more bilateral than multilateral.

Even when federal monies were flowing and the New England Governors' office in Boston employed a staff of over twenty, the notion of having each state operating independently with communication lines organized bilaterally was well entrenched. Multilateral ideas, interests, and institutions have never enjoyed much prominence. The more recent decline in New England regionalism began with a downward spike in political will and desire to facilitate good regional governance. Without strong leadership coming from the centre (Boston), it was only a matter of time before regionalization lost momentum.

When Michael Dukakis was governor of Massachusetts (1975–9, 1983–91) there was stronger support and political leadership at the centre. This reality was reflected in, for example, the hiring of competent, seasoned staff to have responsibility for regional policy files. The decision to hire good people sent a clear and powerful message to other governors that regionalization, working together on the basis of core values, common shared interests remained a top priority.

But this decline of regional collaboration began during Massachu-setts Governor Mitt Romney's term (2003–7). Initially, Romney ap-peared interested in working collaboratively across state boundaries on climate change and was less partisan, even gaining the nickname "Mr. Environment." However, his engagement dropped very quickly once he became interested in running for the White House.[22] Moreover, the practice of hiring junior staff only to manage regional files sent a clear message that regionalization was no longer a priority. Rising par-tisanship also did little to aid the cause of building a functional region in the Northeast. Regionalization was in a state of decline.

CONEG's reorganization into Coalition of Northeastern Governors in 2012 and the move of the New England Governors' organization to Washington, DC, from Boston did little to strengthen the regional cause. Geography matters and the fact most people never even knew this happened did little for regional governance. This new arrangement further isolated governors' regional activities from public view. It also removed local media oversight/analysis, while further reinforcing the power of partisan actors over brand control and identify formation. As a result, old path dependencies were strengthened, not weakened. It is worth noting that the governors never really informed the media that these significant internal changes were occurring, or that the regional club was moving from New England to the national capital region. None of this boded well for the regional functional cause.[23]

Adding New York to CONEG has created even more confusion on the meaning of region. New York, whose population is about 30 per cent larger than the combined population of New England, functions largely in its own orbit. Not surprisingly, it is often seen as a competitor to New England.

As for the Maritime-Atlantic premiers' cross-border regional exper-iments, these have also lost much momentum, reflecting the declining number of regional agencies and interest in pursuing new ones. For example, the idea of a regional energy corporation has never got off the ground. On the other hand, there are informal bilateral mechanisms in play with potential forms of communication due to the activities of the (interprovincial) Council of the Federation, National Roundtable on the Environment and Economy, Council of Energy Ministers, Canada–Nova Scotia Offshore Petroleum Board, Canadian–Newfoundland and Labrador Offshore Petroleum Board, Canadian Federation of Munici-palities, and Atlantic Provinces Economic Council. However, none of these institutions has overcome Canada's reputation for having weak peak intergovernmental institutions, with few venues for coordinating policy actions across jurisdictions.[24]

Province-building remains king, whether reflected in recent pipe-line disputes between British Columbia and Alberta, or recent efforts to decentralize decision-making and double oil and gas production off the coast of Newfoundland. All of this has strengthened an asymmetrical federal system without clear functional rules to rein in competitive territorial-partisan behaviour.

Recent changes on how energy projects (offshore or not) are to be assessed has added even further to the powers of energy-producing provinces.[25] Newfoundland's unilateral decision to increase, even double, offshore development led to clashes with the other members of the Northeast Governors' club who oppose offshore development.[26] But Newfoundland and Labrador's premier has little incentive to think about climate change and negative outcomes for adjacent subnational communities.

Energy Competition and Division

One area critical to good regional governance is energy production, and environmental restructuring. We will finish off our discussion on regional integration highlighting critical insights on some of the factors that have shaped patterns of inter-jurisdictional competition around energy.

Recently, competition over energy-environmental restructuring has intensified and become much more complicated and competitive as a result of upheavals in the U.S. political landscape and the fundamental rebalancing of energy needs and priorities in each country. Territorial-jurisdictional struggles in response to new technological developments (including shale oil and gas), pipeline development, alternative forms of energy, competing rules, regulations, and associated patterns of community definition and identity have boiled over in both countries. These shifts have further complicated regional bordering collaboration and efforts to pull together across both energy sectors but also jurisdictions organized on a competitive bilateral basis.

While territorial state competition over energy-environmental re-structuring has severely undermined cross-border collaboration in Canada, the United States continues to struggle with a brand of super-partisanship that has nearly brought the regional game to an end in the Northeast. As noted above, the rise of Paul LePage in 2010 contributed much to growing partisan bickering and division. LePage's propensity for attacking other governors has made it difficult to coordinate actions or maintain civility. It has been much more difficult to find common ground, build regional consensus, or inspire public confidence in such a setting.

In Canada also, zero-sum territorial games have become the norm, not the exception in interprovincial relations, as illustrated by New-foundland and Labrador's Muskrat Falls hydro development, contin-uing battles over pipeline development, and climate change policies. In both Eastern Canada and the U.S. Northeast states, regional energy issues have become more and more competitive – often within as well as between jurisdictions.

Recent changes in the U.S. electric grid management, reinforced by the shale gas revolution, have complicated cross-border relation-ships in significant ways. The shale gas revolution has exploded and significantly transformed the energy world. It has certainly been a game-changer for cross-border regionalism.[27] Such developments have had a dramatic impact on recent U.S. elections and patterns of partisan division, even the sudden rise of America as a more self-reliant and ag-gressive nation. All of this has produced much volatility and confusion for energy planners in North America and around the world.

It is clear that Canada's energy-producing provinces were not well prepared for the resulting energy revolution, whether the rapid growth of shale gas production, fracking-driven U.S. oil production, or other rapid technological developments. New forms of energy competition have made it more, not less difficult for the decision-makers to work together and produce common energy visions. Quite the reverse, the tendency has been to dig in and compete across both jurisdictions and energy sectors for survival rather than find ways to coordinate problem definition and resolution on energy matters.

North of the border, competitive interstate federalism in Canada has always been closely tied to provincial state control over energy management and production. The power of provincial states has been severely compromised by energy price deregulation trends south of the border, which have contributed to the shale revolution and al-ternative forms of energy production. The rapid transformation of the U.S. energy landscape has created huge adjustment problems for energy-exporting provinces. It has also created much confusion and social protest on the ground on both sides of the border. Revolutions are never smooth.

Shale gas was a game-changer with massive increases in production, whose contribution to U.S. electric power generation has increased from 10 to 35 per cent since 1990. Rising shale production has trans-formed both oil and gas markets, making the United States much more self-reliant and less interdependent. As a result, maintaining good re-lations with allies through multilateral institutions have become even a lower priority.

These developments have had major implications for hydro, oil and gas development, nuclear, and other sources in Canada's highly decentralized energy-policy community. Each province has produced its own energy reports, with premiers preferring to take advantage of their regulatory dominance within their jurisdictions to shape knowledge construction, which questions were asked, and what networks existed. They also continued to produce separate energy visions that focused little attention on competing external challenges, market conditions, rules or regulations, especially how green energy was defined south of the border. These patterns made it easier to defend the status quo and ignore new problems on the horizon. In some cases, these trends have brought about risky gambles, mistakes in budgets, and high costs for citizens largely excluded from discussions of policy priorities.

In the United States, energy-sector development and competition were not nested in common forums either. Each state approached energy sector development with different priorities. With each of these states operating bilaterally, there were intense battles over environmental restructuring, but also calls for building critical physical infrastructure essential to dissimilar forms of energy production. All of this created much volatility and challenges for those who played for dissimilar but also divided audiences.

The tendency for states and provinces to conduct their energy-planning exercises in isolation did little for the construction of common knowledge or systems of analysis required for better regional energy governance or management. While in the United States there were different partisan conflicts over shale, alternative forms of energy production, and offshore development, in Canada, these energy deliberations were very much focused on the competing territorial-jurisdictional priorities of the provincial states involved, but also the energy interests that helped anchor their hold on power.

To be sure, none of this made it easy to coordinate knowledge sharing or planning across either energy sectors or jurisdictions. Old path dependencies anchored by dependence on U.S markets, or province-first definitions for energy-environmental restructuring have produced limited prospects for functional integration. At a time when territorial and other forms of community identity are in constant collision, it has been difficult to ignore the political art of appealing to a limited audience while constantly building false narratives to justify political behaviour.[28]

The New England state trend towards price deregulation and movement away from long-term contracts (with the exception of Vermont) has produced a much more volatile and competitive energy game.

Policies have often been designed in ways that weaken the role of multilateral institutions. It is a different energy world, and that has created much volatility and greater risks for Canadian province-building traditions that have always been closely linked to energy management and development on the continent, whether oil and gas, or hydro. But rather than acknowledging these changing realities, the tendency has been for premiers to find ways to gain more control over public narratives and what the public is sold, pursuing expanded physical infrastructure in attempts to control energy futures.

Sometimes these knowledge gaps have created short-term popularity for premiers, but this has rarely contributed to good public policy or produced better outcomes. For example, the Muskrat Falls project in Newfoundland and Labrador was poorly designed and has resulted in high economic costs to citizens. Knowledge gaps on changing markets, rules, regulations, even community health that helped drive the project was very much connected to the dominance of various premiers who preferred operating independently behind closed doors rather than recognizing problems or facilitating contestation of evidence. Such political control over knowledge construction have seriously compromised efforts to focus on realities of interdependence or engage in evidence-informed debate. These limits on knowledge construction and cross-border insulation contributed significantly to policy failures and economic crisis currently facing Newfoundland, but the trend of producing different narratives is not new and appears to be accelerating with weakening of functional institutions. Historically, as demonstrated by the W.A.C. Bennett case study on physical infrastructural development, among many others, the tradition of hiding costs on provincial energy projects has occurred again and again in an interstate federal system with weak traditions of interprovincial coordination and oversight.[29]

Muskrat Falls is the tip of the iceberg in Canadian territorial conflicts and competition over energy, but a good illustration of institutional deficiencies associated with the combination of a Cabinet-parliamentary and interstate federal system. Speaking truth to power has never been easy in a highly fragmented, competitive federal system where premiers dominate most aspects of energy decision-making. Unlike governors, premiers not only control provincial legislatures and budgets, but they also enjoy much autonomy and capacity in the ownership and control over energy resources and their revenues. A major preoccupation of any government is to preserve control over territories and associational activities, within their respective jurisdictions, but these are shaped and influenced by inherited governance structures and processes, political resources, powers, and policy legacies.

The growth of independent, competitive, powerful provincial governments has resulted in the shaping, nurturing, and mobilizing of separate civil societies, patterns of communication, identity, and associational activities within provincial boundaries. Once these priorities and identities are implanted, it is often very difficult for those who inherit these legacies to reinvent and change direction. The sunk costs associated with the building of physical infrastructure, whether pipelines or those connected with hydro power, are difficult to reverse.

The Muskrat Falls hydro project launched in November 2010 was promoted aggressively by Premier Danny Williams as a means of sending a message to Quebec on interprovincial hydro transmission and gaining more control over hydro development. Many observers have suggested that from the beginning, Muskrat Falls has been more a reflection of political ambition than informed public policy, offering a critical case study on institutional conditions for building of physical infrastructure intended more to strengthen the provincial government's territorial control at the expense of informed policymaking. The Muskrat Falls initiative was a political response to bad memories associated with the inherited 1969 Churchill Falls contract.[30] It was designed to make the province less dependent on Quebec for energy transmission to other jurisdictions, especially the Maritimes and the United States. This goal took priority over effective cost control, environmental outcomes, technical capacity, and the effects of other jurisdictions' rules and regulations or energy needs. From the premier's perspective it was a gamble but a worthy one, given the history of the Churchill Falls contract, which had made the province money but also saw billions in profits flow to Quebec. Gaining access to energy markets through Quebec was considered a non-starter and, despite the rapid rise of shale production in the United States, legal questions about water flow in Labrador, and a variety of other salient issues, the premier persisted on defending and promoting the hydro initiative.

For generations, such zero-sum conflicts fought between these two provinces over power contracts has focused more on territorial-jurisdictional needs of the competing provincial states than the energy needs of citizens. These historical resentments have severely compromised energy governance between the two provinces and the country. In 2002, Liberal Premier Roger Grimes was close to a deal with Hydro-Québec that would have seen the development of Gull Island (a much bigger project) with transmission through Quebec. However, both provinces were too close to the end of their election cycles to close the deal. In Newfoundland, Opposition leader Danny Williams's strong criticism of making a deal with Quebec derailed the project, even though that would not have cost the province anything for building infrastructure. At the time, Grimes said there was little need for generating new sources of power for the island.[31] But such an integrated approach to defining and resolving interdependent energy issues proved to be controversial and a non-starter, despite potential benefits for both sides.

Williams won the subsequent 2003 election on a slogan of "No More Giveaways," with a 16.6 per cent swing from the Liberals. In 2007, he took an unprecedented 69.6 per cent of the popular vote in the 2007 election, winning forty-four seats, compared with three for the Liberals and one for the NDP."[32] His campaign to defend the province against Quebec on the hydro file, Big Oil on offshore resource development, and the Harper government on the federal fiscal front, which was all good politics, did add much to Williams', soaring popularity.

However, there were a variety of challenges standing in the way of successful implementation. But rather than dealing with them in public spaces, Williams adopted a strategy of disconnecting with civil society and restricting contestation, information flow, and public oversight. From the beginning of the Muskrat Falls launch, there was clear evidence of executive political action that was clearly designed to deflect outside contestation and policy-informed debates. For example, early reports by both the joint federal-provincial Environmental Assessment Panel[33] and Public Utilities Board[34] raised concerns about insufficient information to evaluate the merits of the project properly. Even so, the Harper government provided up to $6.4 billion in loan guarantees in 2011, a political move that Kathy Dunderdale relied upon in her successful re-election campaign. The Trudeau government provided an additional $2.9 billion loan guarantee to backstop rising cost overruns in 2016. Ottawa's decisions to offer this kind of financial assistance to both Newfoundland and Emera, the Nova Scotia utility responsible for building a transmission line across the Gulf of St. Lawrence, did reduce the province's share of project costs, but whether that was a good thing is open to debate.

The Muskrat Falls Inquiry Report of March 2020[35] carried out under the direction of a retired judge, Richard D. LeBlanc, was scathing in blaming former Nalcor CEO Ed Martin and his second-in-command, Gilbert Bennett for blocking information flow and following unprincipled decision-making behaviour. Martin was fired by Premier Ball in 2017 but received a $1.4 million severance package.[36] Meanwhile, Gilbert Bennett has remained in his old position as executive vice president.

The strategy of not properly informing citizens has continued with the rise and fall of various premiers, whether Kathy Dunderdale, Paul Davis, or Dwight Ball. Both Martin and former premier Danny Williams have continued to defend the Muskrat Falls project, suggesting that the Muskrat Falls Inquiry is biased and paints a false picture of what has transpired.[37]

These debates point to a growing democratic deficit, connected to abuse of executive power and rapidly rising public debt. These, in turn, have contributed to a state of fiscal crisis and public fears of much higher power rates in Newfoundland and Labrador. An ambiguous mitigation plan announced by Premier Ball and federal regional minister Seamus O'Regan in February 2020 and aimed at keeping electricity rates affordable did little to win back public trust that the twin problems of debt and high energy prices had been solved, particularly as "the finer points of accounting for a $30 billion funding gap over the coming decades have yet to be formalized."[38] "Under the existing agreement, over the next forty-five years it would cost $74 billion to pay for Muskrat, a scenario that would eventually force retail power rates up to 78 cents per kilowatt hour from the current 13 cents."[39] After a caucus revolt and much negative public reaction to his ambiguous and confusing mitigation deal with Ottawa, Ball decided it was time to step down,[40] initiating a leadership race for the minority Liberal government. The lack of a formal legal commitment to the mitigation plan by the federal government, also in a minority position, has generated much public anxiety.

Knowledge gaps matter and silence propagates ignorance. For example, in the new U.S. energy game, new regulatory structures often divide the production, transmission, and distribution of electricity. In New England, the ISO (Independent System Operator) New England operates a single grid, and emphasis is placed on spot markets to ensure that new cheaper forms of power generation serve the needs of customers, not governments. New rules on green energy and the removal of guaranteed contracts (with the exception of Vermont) have produced an energy decision-making structure that is by nature more volatile and competitive, and naturally poses more risks for energy producers, whether private utilities or public ones like Hydro-Québec and Nalcor. This approach to energy sector transformation was not designed to smooth out or stabilize cross-border energy collaborative relations and it has not. Complicating matters, the state of New York operates its own grid but for its customers, creating further limits for regional dialogue or coordination.

It is a new era that has spawned much competition between alternative forms of energy associated with competing interests and priorities. For instance, while Maine Governor LePage (2011–19) pushed for expanding energy production (except wind), Massachusetts, Rhode Island, and

Connecticut have been more focused on reducing energy use rather than generating more energy.[41] All of this has made energy development and investment risky, within and across jurisdictions that for the most part operate independently when conducting their energy plans and visions. Sadly, premiers (and provincial utilities) who control energy resources and tend to be on the hook for building energy corridors north-south find themselves trapped by legacies, isolated and in constant territorial-jurisdictional competition with other political leaders. These traps often generate big costs with few positive outcomes for citizens.

In the past, when regional governors and premiers met, it is telling that the map displayed offered a fuzzy boundary between Newfoundland and Labrador and Quebec.[42] The Labrador boundary dispute that goes back to a controversial 1927 British Court decision is very much connected to bad memories around the 1969 Churchill Falls contract, which was very much the impetus behind the launching of Muskrat Falls project in 2010.[43]

Decline of regional resources, staff, and bureaucratic autonomy for regional conferences, along with assorted regulatory changes, have done little to encourage policy coordination – as had limited media coverage of evolving regional organizations. Taken together, these changing conditions clearly demonstrate the limits of functionalism and why territoriality is increasing across subnational boundaries.

Quebec has shown little interest in policy coordination with Atlantic Canada on energy-environmental matters ever since its failed attempt of 2009 to negotiate a takeover of New Brunswick Power. This proposal would have changed the contest to expand hydro exports into New England energy markets. It provoked strong reactions from many New Brunswickers and Premier Williams. Clearly, none of these events have helped grow the functional cause aimed at closer regional integration.

California's partnership with Quebec and Ontario on climate change policies, including a shared trading system for carbon dioxide emissions, has attracted more attention than working with the Atlantic premiers. However, Ontario's populist premier Rob Ford scrapped the deal with Quebec and California shortly after his election in 2018.

The Muskrat Falls initiative was designed more for territorial expansionism than function and has created further divisions across the eastern provinces. New Brunswick is not operating within the Quebec energy orbit, and Nova Scotia – through its investor-owned, publicly regulated utility Emera – has become a critical actor in the Muskrat Falls energy project. In recent competition to enter the New England grid, Hydro-Québec competed directly against Nova Scotia's private utility, Emera, which partnered with New Brunswick Power to present its own bid. This zero-sum exercise, won by Quebec, was a very public and controversial exercise. Massachusetts opened its market to new power imports in 2015 after Entergy Corp. announced plans to close its Pilgrim nuclear plant, which had supplied about 18 per cent of the state's electricity consumption, by 2019.

However, the state of New Hampshire prevented the province from building required infrastructure to access the Massachusetts grid. Hydro-Québec has agreed to work with Central Maine Power Company to provide 9.45 terawatt hours to the Massachusetts grid.[44] The election of a new, more progressive governor, Janet Mills, in Maine and her endorsement (followed by that of the Maine Public Utilities Commission) of the proposed transmission project connecting Hydro-Québec with the Massachusetts grid, generated much relief for the initiative's boosters.[45] But the battle over infrastructural development is never over or easy. Opponents to the transmission corridor mobilized sufficient signatures to trigger a citizen-initiated referendum pending ongoing litigation. If ultimately successful following litigation, the vote would force Maine's Public Utilities Commission to reverse a certificate of public convenience and necessity required for the transmission project.[46]

Such competitive electric grid bidding across both jurisdictions and energy sectors has accelerated rather than reduced political rivalries and policy divisions, internally and externally. It is a new, more competitive energy world in the Northeast. Competing stakeholders' zero-sum tendencies,[47] combined with territorial traditions of provincial-state border defence, continually playing to different audiences with dissimilar needs, identities, and attitudes, have weakened efforts to give regional functional venues and processes much autonomy or legitimacy. Complicating matters has been the rise of social movements and protests against fracking and many other forms of energy production that have further limited efforts to bring different interests together and put ideas into action. Building energy infrastructure has proven to be very difficult, whether new regional pipeline connections for shale gas or hydro transmission lines. These developments have proven to be major constraints for energy-environmental bordering efforts.

Regional Status Quo: Prospects for Change?

To conclude, internal as well as external forces have contributed to a decline in interprovincial, interstate, and cross-border cooperation throughout the Northeastern region of the North American continent. For the most part, bordering has centred upon regular meetings of intergovernmental bodies, whether of the New England Governors' Conference (since 1937), the Council of Maritime Premiers (CMP, 1971), New England Governors and Eastern Canadian Premiers Conference (NEGECP, 1973), or the Council of Atlantic Premiers (CAP, 2000). By design, the bilateral spoke-and-wheel approach was never designed to advance good regional governance very far or motivate governors and premiers to shift power, resources, or levels of autonomy that would be necessary to give strong institutional support and power to regional ideas (including knowledge and identities), or to patterns of interest mobilization and communication.

From where they sit, governors and premiers' behaviours are influenced by powerful embedded structural forces (bilateral processes of decision-making that fuse or separate power, negative or positive citizen attitudes about public policy reform, old economic development dependencies, rural-urban divisions, and constantly changing national-international policies). These have shaped regional outcomes, or lack thereof. Recent conflicts over identities, and fundamental partisan and ideological struggles in the United States have created more division than unity, resulting in the abandonment of multilateral traditions. These trends have not made it easier for governors to work together. North of the border, territorial state struggles in Canada over energy sectors and infrastructure have accelerated in response to changing American energy circumstances. Still, this has been occurring at a time when both pan-Canadian and numerous cross-border mechanisms (including NEGECP) have lost much autonomy, independence, and power. It has not been a good time for improving regional governance.

Our journey has shown that while functional thinkers assume external forces and realities of increasing interdependencies should make border experimentation easier, in practice, unless or until the power of embedded processes and mechanisms are replaced or reformed, cross-border regional transformation and restructuring are bound to remain limited. It appears, again and again in the regional integration struggle in this part of North America, functionalism is not enough. There needs to be a focus on understanding where power lies and persists, and that is very much connected to old bilateral path dependencies, ideas, interests, and institutions that continue to operate in separate streams and have much in common with the old nation-to-nation two-row wampum belt.

NOTES

1 David Mitrany, *A Working Peace System: An Argument for the Functional Development of International Organization* (London: Royal Institute for International Affairs, 1943); Mitrany, *A Working Peace System* (Chicago:

Quadrangle Books, 1966); Mitrany, "The Prospects of Integration: Federal versus Functional?" in *International Regionalism: Readings*, ed. Joseph S. Nye Jr., 43–71 (Boston: Little, Brown, 1968).

2　Stephen G. Tomblin and Charles S. Colgan, eds., *Regionalism in a Global Society: Persistence and Change in Atlantic Canada and New England* (Toronto: Broadview, 2004); Ivo D. Duchacek, *The Territorial Dimension: Within, among, and across Nations* (Boulder, CO: Westview, 1986), 163–4; Mitrany, *Working Peace System*, 25; Martha A. Derthick, *Between State and Nation: Regional Organizations of the United States* (Washington, DC: Brookings Institute, 1974).

3　Carolyn J. Tuohy, *Accidental Logics: The Dynamics of Change in the Health Care Arena in the United States, Britain, and Canada* (New York: Oxford University Press, 1999).

4　Donald Savoie, *Breaking the Bargain: Public Servants, Ministers, and Parliament*, chap. 1 (Toronto: University of Toronto Press, 2003).

5　Alan Cairns, *Constitution, Government, and Society in Canada* (Toronto: McClelland and Stewart, 1988).

6　Paul Pierson, *Dismantling the Welfare State? Reagan Thatcher and the Politics of Retrenchment* (New York: Cambridge University Press, 1994); Alan Cairns, "The Embedded State: State-Society Relations in Canada," in *State and Society: Canada in Comparative Perspective*, ed. Keith G. Banting (Toronto: University of Toronto Press, 1986), 55.

7　Cairns, *Constitution, Government, and Society in Canada.*

8　Neal R. Peirce, *The New England States: People, Politics, and Power in the Six New England States* (New York: Norton, 1976); Kent A. Price, ed., *Regional Conflict and National Policy* (Baltimore, MD: Johns Hopkins University Press, 1982); Heinz G. Preusse, *The New American Regionalism* (Cheltenham, UK: Edward Elgar, 2004); Stephen J. Hornsby, Victor A. Konrad, and James J. Herlan, eds., *The Northeastern Borderlands: Four Centuries of Interaction* (Fredericton, NB: Canadian-American Center, University of Maine and Acadiensis, 1989); Edwin A. Gere, *Rivers and Regionalism in New England* (Amherst, MA: University of Massachusetts Press, 1968); Roger Gibbins, *Regionalism: Territorial Politics in Canada and the United States* (Toronto: Butterworths, 1982).

9　Based on interviews.

10　Based on interviews.

11　Stephen Tomblin, *Ottawa and the Outer Provinces: The Challenges of Regional Integration in Canada* (Toronto: Lorimer, 1995).

12　Nova Scotia, New Brunswick, and Prince Edward Island, *Report on Maritime Union.* Frediction: Maritime Union Study, 1970.

13　Tomblin, *Ottawa and the Outer Provinces*, 87–100.

14　Gere, *Rivers and Regionalism in New England.*

15 Gere, *Rivers and Regionalism in New England.*

16 Peirce, *New England States.*

17 Tomblin and Colgan, *Regionalism in a Global Society.*

18 Tomblin and Colgan, *Regionalism in a Global Society,* 89.

19 New England Commission, *The New England Regional Plan: An Economic Development Strategy* (Hanover, NH: University Press of New England for the New England Commission, 1981).

20 Tomblin, *Ottawa and the Outer Provinces,* 97–8.

21 Nova Scotia, New Brunswick, and Prince Edward Island, *Report on Maritime Union.*

22 Based on interviews.

23 Based on interviews.

24 Paul Pross, *Pressure Group Behaviour in Canadian Politics* (Toronto: McGraw-Hill Ryerson, 1975).

25 CBC, "NL Unveils Plan to Double Oil Production by 2030, Speed Up Development Process," CBC News, 19 February 2018, http://www.cbc.ca /news/canada/newfoundland-labrador/newfoundland-oil-plan-1.4541830.

26 CBC, "NL Unveils Plan to Double Oil Production"; interviews.

27 Jan H. Kalicki and David L. Goldwyn, eds., *Energy and Security: Strategies for a World in Transition,* 2nd ed. (Washington DC: Wilson Center Press with Johns Hopkins University Press, 2013).

28 Alex Marland, *Brand Command: Canadian Politics and Democracy in the Age of Message Control* (Vancouver: UBC Press, 2016).

29 Tomblin, *Ottawa and the Outer Provinces;* Stephen Tomblin, "Effecting Change and Transformation through Regionalization: Theory versus Practice," *Canadian Public Administration* 50, no. 1 (2007): 1–20.

30 James P. Feehan, "Challenge of the Lower Churchill," in *First among Unequals: The Premier, Politics, and Policy in Newfoundland and Labrador,* ed. Alex Marland and Matthew Kerby, 231–46 (Montreal and Kingston: McGill-Queen's University Press, 2014).

31 CBC, "'Muskrat Falls Danny's Biggest Mistake,' Says Roger Grimes," CBC News, 19 January 2017, http://www.cbc.ca/news/canada /newfoundland-labrador/roger-grimes-danny-williams-muskrat-falls -cost-1.3941494.

32 Jeffrey F. Collins and Scott Reid, "'No More Giveaways!' Resource Nationalism in Newfoundland: A Case Study of Offshore Oil in the Peckford and Williams Administrations," *International Journal of Canadian Studies* 41 (2015): 155–81; Rand Dyck, "Political Developments in the Provinces, 2005–2015," in *Provinces: Canadian Provincial Politics,* 3rd ed., Christopher Dunn, ed. (Toronto: University of Toronto Press, 2015), 47.

33 Canada and Newfoundland and Labrador, Canadian Environmental Assessment Agency and Department of Environment and Conservation, Joint

Review Panel, *Report of the Joint Review Panel: Lower Churchill Hydroelectric Generation Project* (Ottawa: August 2011).

34 Newfoundland and Labrador, Board of Commissioners of Public Utilities, *Reference to the Board: Review of Two Generation Expansion Options for the Least-Cost Supply of Power to Island Interconnected Customers for the Period 2011–2067* (St. John's, NL: March 30, 2012).

35 Newfoundland and Labrador, Commission of Inquiry Respecting the Muskrat Falls Project, "Muskrat Falls: A Misguided Project," The Honourable Richard D. LeBlanc Commissioner (St. Johns: March 2020).

36 Andrew Sampson and Peter Cowan, "Ed Martin Fired after Quitting Nalcor, Triggering $1.4 Severance, Says Dwight Ball," CBC News, 24 May 2016.

37 CBC, "Muskrat Falls Critics Fire Back after Danny Williams Bashes Inquiry Report," 12 March 2020; CBC, "An Unrepentant Ed Martin Goes on the Attack in Response to Muskrat Falls Inquiry Report," CBC News, 11 March 2020.

38 ICBC, "Ottawa, NL Agree to Cut Financing of Muskrat Falls, but How Isn't Clear," CBC News, 10 February 2020.

39 CBC, "Ottawa, NL Agree to Cut Financing."

40 Malone Mullin, "Dwight Ball Stepping Down as Newfoundland and Labrador Premier," CBC News, posted 17 February 2020.

41 Based on interviews.

42 Based on interviews.

43 James Feehan, "Electricity Market Integration: Newfoundland Chooses Monopoly and Protectionism," Atlantic Institute for Market Studies (November 2013): 1–9; Feehan, "Challenge of the Lower Churchill."

44 Canadian Press. "Hydro-Quebec and U.S. Utility Sign Agreement to Get Power to Massachusetts." *Globe and Mail*, 14 June 2018. https://www .theglobeandmail.com/amp/business/industry-news/energy-and -resources/article-hydro-quebec-and-us-utility-sign-agreement-to-get -power-to/; Julien Arsenault, "Hydro-Quebec Looks to Maine as Massachusetts Ends Northern Pass Hydro Project," City News Toronto, 28 March 2018. http://toronto.citynews.ca/2018/03/28/hydro-quebec-looks-to -maine-as-massachusetts-ends-northern-pass-hydro-project/.

45 Lori Valigra, "Opponents Take the First Formal Step to Bring CMP's Transmission Project to a Statewide Vote," *Bangor Daily News*, 30 August 2019.

46 Maine Legislature, "Public Utilities, *Maine Revised Statutes*, Title 35-A3132"; Peter McGuire, "CMP Corridor Opponents Submit Signatures for Referendum Vote," *Press-Herald*, Portland, ME, 3 February 2020.

47 Lester C. Thurow, *Fortune Favours the Bold: What We Must Do to Build a New and Lasting Prosperity*. New York: Harper Collins, 2003.

14 Attributes of Cross-Border Economic Policymaking in the Great Lakes Economic Region: Insights into Complex Bordering Processes at the Subnational Scale

KATHRYN BRYK FRIEDMAN

In the first decades of the twenty-first century, globalization continues to produce two territorially based trends. On the one hand, economic power is shifting out of and across jurisdictionally defined borders (both legal and geographic) through enhanced trade and foreign investment. Countries and companies are engaging in more commerce with counterparts across than globe than ever before. On the other hand, globalization also is deepening economic forces to subnational or regional spaces.[1] This means that territorially delimited subnational regions are acting as critical loci of economic governance and innovation. Economic growth and prosperity increasingly depend on subnational actors, whether at the state/provincial or substate/sub-provincial scales.

Within this subnational context, cross-border regions are becoming important players. These regions are increasingly competitive for firms and talent[2] and increasingly are viewed as hubs for facilitating innovation, creativity, and prosperity. The idea of "thinking and acting across borders" to enable innovative, competitive, and prosperous communities with a highly skilled workforce and well-paying jobs is long-standing. Actors at the state/provincial and local scales have always maintained a strong interest in cross-border relationships because they most directly experience both the costs and benefits.

That being said, economic policymaking in cross-border regions is challenging. With territorially defined borders, these regions have differences in legislative and judicial approaches, exchange rates, labour markets, wages, and administrative systems. Deliberating and effectively managing issues requires navigating and finding common purpose within two or more political, legal, social, cultural, and fiscal regimes; every process, policy, decision, and action must plot a course through at least two federal, state/provincial, and local systems, all with different priorities, resources, laws, policies, and institutions.[3]

Given this complex reality, how do cross-border regions realize their potential as economic innovation corridors?[4] What are the key attributes of cross-border economic policymaking? How do territorial and aterritorial forces play into cross-border economic policymaking? Gaining insight into these questions is important for both scholars and practitioners in the trenches – whether economic development experts, government officials, private sector representatives, or non-profits. For scholars, clarifying the nature of cross-border economic policymaking yields better understanding of bordering processes, which in turn helps frame solutions. Practitioners can then use this understanding to diagnose and tackle the challenges of cross-border economic policymaking with innovative solutions.

As a first step in attempting to answer these questions, this chapter uses a case study approach with the Buffalo-Niagara-Hamilton cross-border region as the unit of analysis. This region is the largest cross-border region along the Canada-U.S. border and is unique in North America for the depth and breadth of its cross-border assets. Whether part of a broader "Tor-Buff-Chester" mega-region[5] or as part of a Great Lakes cross-border region (Council of the Great Lakes Region), Buffalo-Niagara-Hamilton experiences tremendous flows of technology, people, and trade across the international border by sheer virtue of its geography.[6] With four international bridges and two airports, the region reigns as a major port of entry along the Canada-U.S. border, facilitating more than 15 per cent of commerce between two of the world's largest trading partners. Augmented by strong cross-border shopping, heritage, and tourism economies, an advanced logistics industry and sophisticated "soft" infrastructure – customs brokers, third-party logistics, warehousing, attorneys, insurance brokers, bankers, and the like – the region has unmatched potential for strengthening innovation by strategically leveraging cross-border economic ties. The challenge, however, is that much of the region's formal planning and public policy is territorially based, ignoring the cross-border nature of the region and its potential assets.

Section 2 proposes the key attributes for cross-border economic policymaking drawn from multidisciplinary literatures. Scholarly analysis of cross-border regions tends to occur in silos with very little attention paid to concepts or frameworks outside of a particular discipline. Given that a cross-border region itself represents a unit of analysis that is inherently transnational, subnational, regional, and local simultaneously, the author draws on literatures grounded in international relations and law, urban and regional planning, and public policy to draw out relevant attributes. Section 3 examines Buffalo-Niagara-Hamilton against these attributes, with section 4 briefly discussing the case study within the broader context of the Great Lakes Economic Region. Section 5 concludes by offering insights into the territorial and aterritorial "bordering" processes of cross-border economic policymaking and implications for cross-border regions in an increasingly globalized world.

A Review of the Literature: Key Attributes for Cross-Border Economic Policymaking

There is a rich literature with varying frameworks and concepts that touch on cross-border economic regions in disciplines cutting across the social sciences, law, management, and urban planning, among others. This chapter reviews the key frameworks and concepts from four disciplines: international law and international relations, urban planning, and public policy. Each of these literatures contends with different kinds of cross-border or cross-boundary collaboration and hence implicitly relates to the enquiry at hand.

International Relations and International Law

International relations scholars have long considered why nation states cooperate in an anarchical international system. Classical theorists focus on power and power politics, with nation states as the primary actor in the international arena. Other researchers focus on alternative collaborative arrangements as the defining feature of global governance. These scholars reject a reified concept of the nation state, suggesting instead that international relations are far more complex. In their seminal works, Robert O. Keohane and Joseph S. Nye grappled with the increasing complexity of international relations. In an attempt to distinguish between "interstate" and "transnational" relations, they first defined interstate relations as those global interactions that "are initiated and sustained entirely, or almost entirely, by governments of nation-states."[7] For them, the conceptualization of transnational relations was much broader; these relations "may involve governments, but [they] may not involve only governments."[8] These interactions could include nongovernmental organizations as well as other actors in the global system. In their later work, they further refined these concepts to reflect activity conducted by governmental subunits versus activity conducted by other actors. Thus, transgovernmental interactions were defined as those between or among governmental subunits across national boundaries. These were defined as "direct interactions among sub-units of different governments that are not controlled or closely guided ... by ... top leaders," which in turn are synonymous with "official government policy."[9] Transnational interactions were restricted to nongovernmental actors.[10]

Slaughter[11] built directly on this body of scholarship. She uses the concept of "transgovernmental networks" to explain international relations. According to her, transgovernmental activity includes networking among "separate, functionally distinct parts" of the state, including "courts, regulatory agencies, executives, and even legislatures."[12] For her, these networks existed at the nation state or supranational level and offered many benefits for international governance outside of formal, legalized institutions.

With focus on networks that exist either among national government officials or between these officials and supranational entities, Slaughter's concept of transgovernmental networks fails to capture subnational networks at the state, provincial, regional, or local scale – all of which can exert influence on foreign policymaking. Friedman[13] fills in this gap, documenting the myriad state, provincial, and local transgovernmental networks at play at the subnational scale. Raustalia[14] also notes the importance of other actors. For him, although the nation state and international organizations remain important forces in international relations, government-to-government contact occurs more and more in a less centralized manner. Acting through "adaptable and decentralized" networks, substate actors have the ability to harmonize standards, share information and cooperate on enforcement, creating flexible solutions to global governance problems.[15]

Specifically with respect to cross-border environmental collaboration, Friedman and Foster[16] identified the necessary conditions, capacities, organizational models, and experienced that drive successful cross-border initiatives. These scholars examined not only what it takes to start such a venture, but also how to sustain an initiative. Necessary elements include a clear purpose, staff and resource capacity, political will, and a codified and purposeful mission.[17] They also conclude that there is no optimal institutional or organizational structure for successful cross-boundary environmental collaboration. Rather, the structure of governance, rules, and operations may take a variety of forms, ranging from formal institutional structures to others that are more ad hoc and fluid.[18]

Although most of the aforementioned scholarship focuses on formalistic, territorial approaches, others find a more complex global landscape. According to these latter scholars, state actors and non-state actors often operate informally in transgovernmental networks to achieve joint policy goals. For example, in discussing the rising role of non-state actors in international law, Picciotto[19] notes the tension between traditional approaches that view the state as the only legitimate actor driving international relations (even if through non-traditional governmental channels) and approaches that see non-state actors as policy drivers. These non-state actors are evidenced in a dense network of international institutions and other governance arrangements that involve social activists, scientists, businesspeople, and others.[20] The informal nature of these non-state actors, coupled with the fact that conventional legal instruments generally do not address these actors, contributes to issues of legitimacy in international relations.[21] With respect to legitimacy and accountability, Dobner[22] argues that such global public policy networks are "powerful" and must be "constitutionalized" because of their more prominent role in global policy-making. Although he notes the increased presence of these actors on the international stage, he also is sceptical about their legitimacy and the ability to formalize them through constitutionalization. According to Dobner, two requirements found at the state level are not found in transnational networks: the concentration of power in the state and the division between public and private.[23]

Urban and Regional Planning

The concept of networks also emerges in urban and regional planning literature. These scholars conceptualize networks as interdependent structures, involving a number of "nodes," i.e., actors such as agencies and organizations, with multiple linkages or "ties," i.e., the interactions between nodes.[24] These networks, in turn, create opportunities for governmental and nongovernmental actors to address shared problems.[25] For these scholars, leadership within a network becomes important to work across political and organizational boundaries to achieve shared goals.[26] However, not just any kind of leadership will suffice. To the contrary, because networks are non-hierarchical in nature, traditional, structured leadership focused on advancing siloed interests is insufficient and inappropriate.[27] Drawing on relational leadership theories, these scholars explore three ways of thinking about leadership in networks: (1) collaborative leadership, in which network members share leadership functions at different times; (2) distributive leadership, in which network processes provide local opportunities for members to act proactively for the benefit of the network; and (3) architectural leadership, in which the structure of the network is designed to allow network processes to occur.[28] All three types of leadership are different from traditional forms of hierarchical leadership; all three also advance the shared interests of network members but in different ways.

For these scholars, a collaborative leadership style involves both leading and following. Like traditional leaders who need to inspire people to act, collaborative leaders must inspire as well, but they also must follow. These leaders don't consolidate power, they disperse power throughout the network and build participation. These leaders also emphasize dialogue and build relationships, respecting different viewpoints all with an aim towards building a common agenda.[29] Distributive network leadership, in contrast, is akin to a "flock" in that people work in unison to achieve the collective interest of the network as opposed to individual interests. Multiple opportunities are presented for different individuals within a network to lead.[30] Last, architectural network leadership is quite intentional in designing a system that allows a network to thrive.[31] These leaders establish rules that influence and guide behaviour but remain invisible to participants. These rules are important, according to these scholars, as they provide for who will form an agenda, who will deploy resources, and who will speak on behalf of the network.[32] Regardless of the type of leadership, the relational nature of networks and the need for a certain kind of leadership sometimes brings boundary organizations into play. "Bridging" or "boundary" organizations provide a forum for the interaction of different kinds of knowledge and the coordination of other tasks that enable co-operation such as accessing resources, bringing together different actors, building trust, resolving conflict, and networking.[33]

Another relevant strain of urban planning literature that focuses on cross-boundary relations is placemaking. Placemaking is the creation, renovation, and maintenance of a shared physical world through the marriage of expert professional knowledge and knowledge of place.[34] Implicit in this definition is the notion that collaborators outside of government are just as important as those within government channels. The practice of placemaking relies heavily on building relationships among public, private, and not-for-profit entities to overcome the fragmentation of official agency or territorial claims.[35] Placemaking involves not only government, but also delegating decision-making authority to non-elected collaborators, typically extra-governmental participants. Like the international relations context, accountability challenges can be seen as a function of the process through which stakeholders are involved in decision-making and are legitimate representatives of the community.[36]

Placemaking "cherishes" dialogue through a strong process to arrive at a common understanding of place.[37] It involves indispensable "imaginal" work – the idea of a place held in the imaginations of the people who live there. Through process, stakeholders can arrive at a place where they share discourse and have mutual understanding – both of which are central to a collaborative placemaking enterprise.[38]

Public Policy

The concept of a cross-border economic region also has roots in public policy literature. One understanding of cross-border economic regions or "industrial corridors" in this literature corresponds to geographic spaces functionally linked together via a shared export base, the flows of inter-firm relations, or the flows of labour force activities within a particular sector. For these scholars, a cross-border economic region is a geographic space comprising a cluster of surrounding communities sharing similar economic assets in a particular sector, such as manufacturing, tourism, life sciences, and agriculture.[39] Clusters, in turn, are defined as "geographic concentrations of interconnected companies, specialized suppliers, service providers, firms in related industries, associated institutions ... in particular fields that compete but also cooperate."[40] Thus, a core element of this conceptualization is that of geographical proximity: clusters are spatially localized concentrations of interlinked firms. Co-location is a central determinant of value creation that arises from networks of direct and indirect interactions among private businesses, and between firms, customers, local public agents (such as economic development officials, for example), post-secondary institutions, and related entities with vested interests in the economic well-being of their community.

Cross-border economic clusters can consist of a combination of geographically proximate cities and surrounding peripheries of smaller communities from two or more countries bound together by interlocking economic flows of products and skills that create a self-reinforcing interdependency and synergy.[41] Implicit in this definition is the centrality of clusters of shared assets in a particular sector, which provides the basis for surrounding communities to plan economic development investment priorities in leveraging natural resources, human capital, investment capital, and market access to sustain and enhance a cross-border region's economic well-being. Some suggest that cross-border collaboration creates opportunities to leverage a more diverse range of economic assets that can strengthen the strategic position of a cross-border region and its competitiveness in the global economy.[42]

A central feature of economic clusters in general is the notion of industrial agglomerations. This concept points to the vital role of external economies of scale.[43] There are two types of agglomeration. The first derives from urbanization economies, the second from localization economies.[44] Urbanization economies are benefits that accrue to cities by virtue of their population and market density, which make them

economically resilient and often self-sustaining. Localization economies, on the other hand, can be seen in agglomerations that typically
manifest specialization in a key industrial sector.[45]

The main implication of understanding agglomeration in terms of
urbanization and localization economies is that it directs the construct
of cross-border economic clusters to focus on specific assets within a
given geographic location. Such a construct allows for an economic
cluster made up of communities with demonstrable strengths in certain sectors – for example, life sciences, advanced manufacturing, or
tourism – to leverage their local resources, mobilize non-local public
resources, and attract private investment in scaling up their sectoral
strengths and potential for economic development and reinvention.
The concept of cross-border economic clusters thus draws our attention
to the reality of economic development as a highly varied and complex
territorial process.[46] The significance of this observation is that different types of clusters will manifest different economic and demographic
characteristics, and, therefore, possess different capacities to respond to
and cope with both external and internal shocks and changes.

The aforementioned literatures use concepts associated with both
territorial and aterritorial approaches to policymaking and hence
demonstrate the complexities of "bordering" in our contemporary
global environment. These literatures also offer insight into identifying the necessary attributes or characteristics for cross-border economic
policymaking and suggest four attributes of particular importance.
First, the presence of vertical and horizontal networks with multiple
stakeholders from across scale and sector is important. Both international relations and urban planning literatures suggest that networks
that comprise public sector and civil society (private sector, NGOs)
actors at the transnational, federal, state/provincial, and locale scales
are significant. Second, non-traditional leadership – whether collaborative, distributive, or architectural – that is non-hierarchical also is an
important attribute. Boundary organizations also seem to play a crucial leadership role in these cross-border networks. Third, a process
that leads to shared discourse and mutual understanding is central to
cross-border economic policymaking. Fourth, the public policy literature on cross-border regions reminds us that the presence (or absence)
of cluster assets is an important attribute for policy.

The Case Study: The Buffalo-Niagara-Hamilton Region

The nature of the problem – cross-border economic policymaking –
is inherently territorial, as set forth in section 1. Nonetheless, the
underlying hypothesis of this chapter is that an aterritorial approach

to cross-border economic policymaking is occurring alongside the territorial nature of the challenge, thus resulting in a complex interplay of bordering at the subnational, cross-border scale.

The focus of the case study is the Buffalo-Niagara-Hamilton cross-border region. This is a region in transition. Leaders on both sides of the border are proactively embracing change and leveraging assets to strengthen human capital and to foster creativity and innovation. On the U.S. side, Buffalo Niagara is creating purposeful transformation to support sustainable, liveable communities with a high quality of life for all residents through initiatives such as the build-out of the Downtown Medical Corridor; investment in cultural/heritage tourism and health sciences innovation (e.g., the Western New York Regional Economic Development Council Strategic Plan and Buffalo Billion); and stewardship of its natural assets (One Region Forward).

At the same time, parallel efforts are taking place on the Canadian side of the border. Just a stone's throw away, leadership in the Niagara Region and Hamilton are similarly committed to increasing economic opportunity and creating well-paying jobs by strengthening innovation (e.g., health sciences), natural heritage (Niagara Falls and Great Lakes), and cultural asset (tourism) strategies – all with an eye towards creating healthy, vibrant, prosperous communities.

If cross-border regions are becoming more important actors in the global economy, as Snyder et al.[47] suggest, the research questions become: How can leadership on both sides of the border approach innovation, creativity, and prosperity collectively? What are the opportunities? What is the "economic case" for a cross-border approach to economic policymaking? What are the benefits of coordinating approaches to create a cross-border innovation and community prosperity hub, as opposed to each side "going it alone?" What would a process for cross-border planning look like?

The idea of working with counterparts on each side of the Niagara River is not new to stakeholders in this region. Attempts to forge strong relationships and collaborate on economic policy date more than back twenty years. Yet three primary challenges exist. First, continuing momentum with no single entity to "own" the effort was difficult. There was no cross-border actor or entity from either the public sector or civil society with a mandate for crafting collaborative cross-border policy. Stakeholders liked the idea but each had their "day jobs" to attend to. Second, although stakeholders intuitively "knew" that cross-border collaboration on economic policy would yield greater gain, no one had "made the case" or demonstrated the real economic gains to be had by engaging in a cross-border approach. Third, policy changes at the federal level in each country compounded efforts to collaborate on cross-border economic policy. Although stakeholders viewed passage of the North American Free Trade Agreement positively and supportive of their efforts, they viewed border policies initiated in the post-9/11 environment as challenging, to say the least. These three factors resulted in intermittent efforts over the years.

In 2016, a committed team of seven Canadian and U.S. researchers from Brock University and the University at Buffalo initiated a multidisciplinary approach to understanding cross-border economic policymaking in the Buffalo-Niagara-Hamilton region.[48] Drawn from disciplines including international relations and international law, urban and regional planning, management, and public policy, the research team adopted qualitative and quantitative methods to answer some of the questions outlined above and gain mutual understanding of the problem.

First, the research team used quantitative methods to characterize the problem. Using the life sciences sector as a "proof of concept" for making the economic case for cross-border economic policymaking, the research team first assessed whether there was a cluster so that stakeholders had common understanding of the economic nature of this aspect of the cross-border regional economy. The team discovered that there are 1,025 life sciences companies in the region: 625 in Buffalo Niagara, 168 in Hamilton, and 242 in the Niagara region.

The research team also developed a number of indicators for measuring the cross-border life sciences cluster. The team gathered data and conducted analysis on demographics as an important component of assessing whether there was a cluster. In particular, the team examined population characteristics such as size and age distribution because these data point to the merits of agglomeration advantages and labour market size. For the case study at hand, the combined 2015 population total of the cross-border region was 2,393,077, with a median age of forty-four years.

The research team looked at the rate of attraction of new residents to the region. The data demonstrated that communities on the U.S. side of the border are losing people (Erie County: -0.95; Niagara County: -.3.11; Cattaraugus County: -9.22) while communities on the Canadian side of the border are gaining people (Hamilton CMA: 2.15).[49]

Another critical question often raised about economic clusters is how much pressure is on working-age (fifteen to sixty-four) residents to support non-working (zero to fourteen, sixty-five and over) residents. This is measured through the dependency ratio, i.e., the ratio of non-working to working residents. The data demonstrated that on both sides of the border in all counties (in the case of the United States) or census metropolitan areas (CMAs – in Canada), this ratio increased over the last five years, with the St. Catharines–Niagara CMA and Cattaraugus County experiencing the greatest increase in the ratio.

The prosperity and potential competitiveness of a cross-border region can be measured in terms of GDP per capita and the composition of key sectors. In the Buffalo-Niagara-Hamilton cross-border region, core industry strengths are health care, manufacturing, education, and tourism. Methodological issues precluded deriving a cross-border figure of economic performance; however, a proxy of average weekly wages was used and indicated that weekly wages on both sides of the border grew in 2010–15 by 11.37 per cent in Ontario and 7.65 per cent in New York. On the U.S. side of the border, this is greater than the U.S. wage growth between 2010 and 2015 (7.51 per cent) but on the Canadian side, it is less than the Canadian average of 12.94 per cent. Finally, on levels of education, which can serve as a proxy for a region's innovation environment, 90.5 per cent of the population has a high school diploma or more, and 26.75 per cent of the population has completed a bachelor's degree or more.

The research team next attempted to answer whether a cross-border approach to economic policymaking made economic sense. Researchers designed a multi-regional input-output model (MRIO) for the cross-border region based on Miller and Blair.[50] The team then conducted four simulation scenarios consisting of an investment of $100 million to life science–related industries. Two of the scenarios were a "business as usual" case, meaning that each side of the border used status quo, "go-it-alone" economic development policies and processes; the remaining two scenarios assumed some level of cross-border alignment of policy. The results demonstrated that one of the scenarios that assumed a cross-border approach to economic development would generate $135 million more in the life sciences sector and almost 100 more jobs than the "go-it-alone" approach. While the rate of job creation within the life sciences sector appears modest, the research team noted that these are typically higher-paying jobs. The research team also concluded that the spillover generated throughout the whole economy (other sectors outside life sciences) from indirect trade effects of the investment into the cross-border region would be large.

On qualitative methods, the research team designed a process for engaging cross-border stakeholders and strengthening the cross-border network in the region. Three workshops were held over eighteen months, with participation of eighty cross-border stakeholders from the public, private, academic, and non-profit sectors. Regarding the public sector, representatives from each level of government attended, including the U.S. Department of Commerce and Global Affairs Canada (federal); New York State Empire State Development and Ontario Ministry of Economic Development, Trade and Employment (state/provincial); and the Regional Municipality of Niagara and Niagara Falls, NY (regional-local). Also stakeholders from the two international bridge owners and operators – the Peace Bridge Authority and the Niagara Falls Bridge Commission – attended. Private sector stakeholders included everyone from those who ran individual companies to trade organizations that supported the idea of cross-border economic policy, such as the Buffalo Niagara Partnership, the Niagara Region Chamber of Commerce, the Hamilton Chamber of Commerce, the World Trade Center Buffalo Niagara, and the International Trade Gateway Organization. In terms of the non-profit sector, representatives from the Ontario Trillium Foundation and the John R. Oishei Foundation attended.

The workshops were designed, organized, and facilitated by Canadian and U.S. researchers. Understanding that no single entity "owned" the effort and using their legitimacy to convene, the researchers undertook a collaborative style of facilitation. During these workshops, stakeholder participants dialogued on the opportunities indicated by the quantitative results – that there are the makings of an economic cluster in the life sciences sector, and the economic case for a cross-border approach to economic policymaking is strong. Stakeholders indicated that these findings confirmed what they suspected – a cross-border approach to economic prosperity made sense. Yet they also dialogued on the challenges on how to get there. Three challenges were primary. First, stakeholders discussed "fatigue" – the fact that this effort had been ongoing, with fits and starts, over two decades with little progress and tremendous "interference" by federal policies in both countries. Second, they discussed the very real challenges associated with traditional bordering processes – no jurisdiction had the authority or legitimacy to tackle the problem. Third, stakeholders acknowledged that economic development was a competitive and not a collaborative enterprise, yet collaboration was required to craft cross-border economic policy.

The workshops were designed to help stakeholders frame a collective definition of the problem as well as a collective vision. Yet stakeholders were free to disagree with each other – in fact, the research team facilitator encouraged critique and confirmation. Perhaps the best example of this occurred during Workshop 3 when participants were discussing how to move forward. Many had just assumed that harmonizing policy on each side of the border was the way to move ahead. Nonetheless, a robust discussion ensured on whether or not this was realistic at the subnational cross-border scale. Instead of harmonizing policy, stakeholders agreed that "leveraging the border" – that is, strategically aligning assets and differences to achieve economic growth – was more realistic and doable. Stakeholders also acknowledged that more research was required to assess assets in this manner but strongly believed this was the correct approach.

The Great Lakes Context

In terms of the attributes of cross-border economic policymaking set forth in section 2, vertical and horizontal network of stakeholders from across sectors participated in the workshops. In terms of the different styles of network leadership,[51] leadership in the case study was collaborative, with emphasis on dialogue, and building relationships, value, and respect. The leadership also focused on building strong collaborative processes throughout (three workshops) and used data and analysis to find common ground and shared discourse. It could be argued, additionally, that the research team itself actually served as a boundary organization, mediating conflict and negotiating interests. Finally, the research team preliminarily identified the basis for an economic cluster in the life sciences sector – an important attribute for cross-economic policymaking.

It seems that the cross-border Buffalo-Niagara-Hamilton region is a microcosm of broader cross-border economic policymaking efforts within the Great Lakes St. Lawrence River Economic Region. Defining the geographic extent of the Great Lakes region in economic terms is a challenge, as the region's economic activity is highly integrated, relying on complex networks of supply chains that criss-cross the region, often irrespective of state, national, or watershed boundaries.[52] Scholars and commentators suggest that the region's economy is among the largest in the world with a gross regional product (GRP) of US$4.1 trillion, which represents nearly one-third of the entire U.S. economy, over twice the size of the entire Canadian economy, and roughly US$1 trillion larger than the German economy.[53]

In terms of networks, researchers suggest that the region has relatively strong civil and single-purpose intergovernmental networks, yet this region is highly decentralized, with some estimating more than fifty nongovernmental organizations working on similar issues with resulting "turf" battles. In terms of leadership in the economic space, one entity – the Council of the Great Lakes Region (CGLR) – has emerged as a strong voice for the region. Established in 2011 when over 250 regional leaders from government, business, labour, non-profit, and academic communities within the Great Lakes met in Detroit and Windsor for the Great Lakes Summit, CGLR is a member-based organization whose aim is to enhance regional collaboration and cross-border integration by bringing together stakeholders from the private, public, and not-for-profit sectors to advance effective, coordinated, and broadly shared responses to the region's common challenges. It could be argued that the CGLR serves as a boundary organization for the region, much like the research team in the Buffalo-Niagara-Hamilton case study. In terms of cross-border asset clusters, some researchers suggest that the region is generator of talent, knowledge, and innovation, with cross-border economic clusters including agri-food processing, advanced manufacturing, and the life sciences.[54]

Conclusions

More comparative case study research is required in order to conclude that the multidisciplinary attributes of cross-border economic policymaking highlighted in this chapter are valid. Nonetheless, this chapter offers several insights into the interplay between territorial and aterritorial forces in the context of cross-border regional economic policymaking in southern Ontario and Western New York. First, the attributes themselves are territorial and aterritorial. The cross-border network in the Buffalo-Niagara-Hamilton region is both vertical and horizontal, inclusive of actors from myriad sectors. A collaborative leadership style is breaking down silos, with the research team serving as a boundary organization. The process designed for each workshop not only encouraged critique and confirmation of various points of view, but also included an emphasis on data to create shared understanding and mutual discourse. Economic clusters represent a particular kind of economic space – which is defined both by territory and beyond territory. This focus illumines the imperative of creating the institutions, networks, and/or processes that will facilitate interaction, trust, and cooperation among constellations of actors within this shared geographic space. It is this mix of territorial and aterritorial attributes (and perhaps others) that will continue to unfold in the Buffalo-Niagara-Hamilton cross-border region and shape its economic future in the years to come.

Second, a paradox emerges when examining cross-border economic development at the subnational scale. On the one hand, as a cross-border region, Buffalo-Niagara-Hamilton has been shaped by territorially defined globalization forces that are making subnational regions more prominent players in the global marketplace. At the same time, aterritorial processes and policies are being employed in this cross-border region to strengthen its position in a globalized world, as evidenced by the networked approach taken by the research team. Thus, although territoriality remains constant in the globalized world of the twenty-first century – i.e., subnational jurisdictions are becoming more relevant – aterritorial process and policy may be necessary to ensuring that these regions actually achieve economic growth and prosperity.

Third, this weaving and comingling of bordering processes has produced another anomalous result. Some scholars suggest that spanning an international boundary consists of a framework or process of cooperation and coordination between two national states, between states or provinces of two national states, or between substate/subregional entities in two different national states for an identified purpose.[55] This proceeds "through a process of systematic harmonization of policy ... that leads to policy parallelism and is driven by the convergence of socio-cultural values, increased economic integration and the formation of border-spanning network institutions."[56] In line with this thinking, there is often a continuum of policy convergence, ranging from conflict on the one hand to harmonization on the other.[57] Nonetheless, stakeholders in the workshops rejected harmonization as the end goal. Rather than strive for harmonization of policy across the international border, stakeholders are acknowledging the territorial context within which they live and are determined nonetheless to "leverage the border" to their advantage. Clearly the international border separates two nations with significant differences, yet these differences actually present an opportunity – just as two people with complementary skill sets make good business partners, two communities with complementary assets can be stronger and better able to compete in the global economy.[58] This is a markedly different end game when compared to territorial directives for harmonization of cross-border economic policy like those that exist within the European Union (e.g., Interreg) and suggests perhaps that harmonization of policy makes sense only when driven by territorial forces from the "top down" rather than from the "bottom up." Each of these insights suggests that, when cross-border economic policymaking at the subnational scale is examined from any perspective, complex bordering forces and processes will continue to influence regions in the years to come.

NOTES

1 Organisation for Economic Co-operation and Development, *Globalisation and Regional Economies: Can OECD Regions Compete in Global Industries?* (Paris: Organisation for Economic Co-operation and Development, 2007),

https://www.oecd.org/gov/globalisationandregionaleconomiescano ecdregionscompeteinglobalindustries.htm; Michael Storper, *Keys to the City: How Economies, Institutions, Social Interactions, and Politics Shape Development* (Princeton, NJ: Princeton University Press, 2013); Kathryn Bryk Friedman, "Through the Looking Glass: Implication of Canada–United States Transgovernmental Networks for Democratic Theory, International Law, and the Future of North American Governance," *Alberta Law Review* 46, no. 4 (2009): 1081–97.

2 J.D. Snyder, Kelly Christopherson, Leslie Grimm, Anthony Orlando, and Aaron Galer, *Global Models of Binational Regional Collaboration: The Potential for Great Lakes Regional Innovation* (Lansing: Michigan State University, May 2014).

3 Kathryn Bryk Friedman and Kathryn Foster, *Environmental Collaboration: Lessons Learned about Cross-Boundary Collaborations* (Washington, DC: IBM Center for the Business of Government, 2011).

4 Snyder et al., *Global Models of Binational Regional Collaboration.*

5 Richard Florida, "Megaregions: The Importance of Place," *Harvard Business Review,* March 2008, https://hbr.org/2008/03/megaregions-the-importance-of-place.

6 Snyder et al., *Global Models of Binational Regional Collaboration.*

7 Joseph S. Nye and Robert O. Keohane, "Transnational Relations and World Politics: An Introduction," *International Organization* 25, no. 3 (1971): 332.

8 Nye and Keohane, "Transnational Relations and World Politics," 332.

9 Robert O. Keohane and Joseph S. Nye, "Transgovernmental Relations and International Organizations," *World Politics* 27, no. 1 (1974): 43.

10 Keohane and Nye, "Transgovernmental Relations and International Organizations," 41.

11 Anne-Marie Slaughter, "The Real New World Order," *Foreign Affairs* 76, no. 5 (Sept.–Oct. 1997): 183–97; Slaughter, *A New World Order* (Princeton, NJ: Princeton University Press, 2009).

12 Slaughter, "Real New World Order," 184.

13 Friedman, "Through the Looking Glass"; Kathryn Bryk Friedman, "New York State's Foreign Policy," in *The Oxford Handbook of New York State Government and Politics,* ed. Gerald Benjamin, 502–34 (Oxford: Oxford University Press, 2012).

14 Kal Raustiala, "The Architecture of International Cooperation: Transgovernmental Networks and the Future of International Law," *Virginia Journal of International Law* 43, no. 1 (2002): 1–92.

15 Raustiala, "Architecture of International Cooperation."

16 Friedman and Foster, *Environmental Collaboration.*

17 Friedman and Foster, *Environmental Collaboration.*

18 Friedman and Foster, *Environmental Collaboration.*

19 Sol Picciotto, "Fragmented States and International Rules of Law," *Social and Legal Studies* 6, no. 2 (1997): 259–79.

20 Picciotto, "Fragmented States and International Rules of Law."

21 Picciotto, "Fragmented States and International Rules of Law."

22 Petra Dobner, "On the Constitutionality of Global Public Policy Networks," *Symposium: Global Constitutionalism – Process and Substance – Kandersteg, Switzerland, January 17–20, 2008* (2009): 605–20.

23 Dobner, "On the Constitutionality of Global Public Policy Networks."

24 Mark T. Imperial, Sonia Ospina, Erik Johnston, Rosemary O'Leary, Jennifer Thomsen, Peter Williams, and Shawn Johnson, "Understanding Leadership in a World of Shared Problems: Advancing Network Governance in Large Landscape Conservation," *Frontiers in Ecology and the Environment* 14, no. 3 (2016): 126–34.

25 Imperial et al., "Understanding Leadership."

26 Jeffrey A. Alexander, Maureen E. Comfort, Bryan J. Weiner, and Richard Bogue, "Leadership in Community Collaborative Health Partnerships," *Nonprofit Management and Leadership* 12, no. 2 (2001): 159–75; Ronda C. Zakocs and Erika M. Edwards, "What Explains Community Coalition Effectiveness? A Review of Literature," *American Journal of Preventative Medicine* 30, no. 4 (2006): 351–61; Imperial et al., "Understanding Leadership."

27 Imperial et al., "Understanding Leadership."

28 Imperial et al., "Understanding Leadership."

29 Imperial et al., "Understanding Leadership."

30 Imperial et al., "Understanding Leadership."

31 Imperial et al., "Understanding Leadership."

32 Imperial et al., "Understanding Leadership."

33 Fikret Berkes, "Evolution of Co-management: Role of Knowledge Generation, Bridging Organizations and Social Learning," *Journal of Environmental Management* 90, no. 5 (2009): 1692–1702; Beatrice I. Crona and John N. Parker, "Learning in Support of Governance: Theories, Methods, and a Framework to Assess How Bridging Organizations Contribute to Adaptive Resource Governance," *Ecology and Society* 17, no. 1 (2012): 32–49.

34 Lynda H. Schneekloth and Robert G. Shibley, *Placemaking: The Art and Practice of Building Communities* (New York: Wiley, 1995).

35 Robert G. Shibley, Lynda H. Schneekloth, and Bradshaw Hovey, "Constituting the Public Realm of a Region: Placemaking in the Bi-National Niagaras," *Journal of Architectural Education* 57, no. 1 (2003): 28–42.

36 Shibley, Schneekloth, and Hovey, "Constituting the Public Realm of a Region."

37 Shibley, Schneekloth, and Hovey, "Constituting the Public Realm of a Region."

38 Schneekloth and Shibley, *Placemaking.*

39 Charles Conteh, Kathryn Friedman, and Barry Wright, "Strengthening Prosperity in Binational Corridors: Public Policy Lessons on Generating Innovation and Entrepreneurship" (prepared for the ICPP 3 2017 Conference, Singapore, 2017).

40 Michael E. Porter, *On Competition* (Boston: Harvard Business Press, 1998), 197.

41 Conteh, Friedman, and Wright, "Strengthening Prosperity."

42 Snyder et al., *Global Models of Binational Regional Cooperation.*

43 Yuko Aoyama, James T. Murphy and Susan Hanson, *Key Concepts in Economic Geography* (Thousand Oaks, CA: Sage, 2011).

44 Thomas L. Friedman, *The World Is Flat: A Brief History of the Twenty-First Century* (New York: Farrar, Straus and Giroux, 2005).

45 Sebastiano Brusco, "The Emilian Model: Productive Decentralisation and Social Integration," *Cambridge Journal of Economics* 6, no. 2 (1982): 167–84; Margherita Russo, "Technical Change and the Industrial District: The Role of Interfirm Relations in the Growth and Transformation of Ceramic Tile Production in Italy," *Research Policy* 14, no. 6 (1985): 329–43; Ash Amin and Nigel Thrift, "Neo-Marshallian Nodes in Global Networks," *International Journal of Urban and Regional Research* 16, no. 4 (1992): 571–87.

46 Bjørn T. Asheim, Philip Cooke and R.L. Martin, "The Rise of the Cluster Concept in Regional Analysis and Policy: A Critical Assessment," in *Clusters and Regional Development: Critical Reflections and Explorations,* ed. Asheim, Cooke and Martin, 1–29 (New York: Routledge, 2006); Ron Martin and Peter Sunley, *Economic Geography* (London: Routledge, 2008).

47 Snyder et al., *Global Models of Binational Regional Cooperation.*

48 Kathryn Friedman, Barry Wright, Charles Conteh, Robert Shibley, Jiyoung Park, Ha Hwang, and Carol Phillips, "Creating Prosperity in Binational Buffalo-Niagara" (unpublished presentation, 2017). See Cathy Majtenyi, "Brock, University at Buffalo Launch Cross-Border Economic Research Project," Brock, 2 November 2016, https://brocku.ca/brock-news/2016 /11/brock-university-at-buffalo-launch-cross-border-economic-research -project/.

49 Friedman et al., "Creating Prosperity in Binational Buffalo-Niagara"; Conteh, Friedman, and Wright, "Strengthening Prosperity."

50 $x = (I – C*A) ^ (-1)*f$ with x = total output; f = final demand (investment scenarios); I = identity matrix; and C = the magnitude of industry relationships between and within regions. In this model, C is defined by average travel time between businesses in different industry sectors between and within regions. A = IO/total output. Ronald E. Miller and Peter D. Blair, *Input-Output Analysis: Foundations and Extensions* (Cambridge; Cambridge University Press), 2009.

51 Imperial et al., "Understanding Leadership."

52 Maureen Campbell, Matthew J. Cooper, Kathryn Friedman, and William P. Anderson, "The Economy as a Driver of Change in the Great Lakes–St. Lawrence River Basin," *Journal of Great Lakes Research* 41, no. S1 (2015): S69–S83.

53 Campbell et al., "Economy as a Driver of Change."

54 John Austin and Brittany Affolter-Caine, *The Vital Center: A Federal-State Compact to Renew the Great Lakes Region* (Washington, DC: Brookings, 2006).

55 Emmanuel Brunet-Jailly, "Cascadia in Comparative Perspectives: Canada-U.S. Relations and the Emergence of Cross-Border Regions," *Canadian Political Science Review* 2, no. 2 (2008): 104–24.

56 Brunet-Jailly, "Cascadia in Comparative Perspectives."

57 Monica Gattinger and Geoffrey Hale, eds., *Borders and Bridges: Canada's Policy Relation in North America* (Oxford: Oxford University Press, 2010), 13; Geoffrey Hale, "Borders Near and Far: The Economic, Geographic and Regulatory Contexts for Trade and Border-Related Issues in Landlocked Alberta," *Journal of Borderland Studies* 34, no. 2 (2019), 157–80, https://doi.org/10.1080/08865655.2017.1315609.

58 William Anderson, "The Binational Advantage," *How It Works and (And Why It Matters)* (Windsor, ON: Cross Border Institute, 2016).

15 Canada's Arctic Boundaries and the United States: Binational vs. Bilateral Policymaking in North America

CAROLYN C. JAMES

Sovereignty is a term widely used by policymakers, academics, and the general public. While its importance is widely recognized, its exact meaning is rarely considered. Successful U.S. policymaking with Canada, however, often rests on meeting or engaging in purposeful ambiguity on sovereignty concerns. Therefore careful consideration of sovereignty is necessary to understand the dynamics of this relationship, particularly as it affects Arctic issues. Key sources of interstate disagreement include climate change, possible year-round navigation through the Northwest Passage, energy and related environmental issues, such as drilling in the U.S. Arctic National Wildlife Refuge, and rules governing offshore drilling. When combined, these issues can be "destabilizing."[1] The importance of sovereignty is reinforced by estimates that the Arctic may contain at least 90 billion barrels of currently recoverable yet undiscovered oil, estimated at 13 per cent of global totals.[2]

The goal of this chapter is a policy-relevant assessment of sovereignty and its role in bilateral Arctic relations between Canada and the United States. Specifically, it argues that an institutionalized, binational mechanism is essential to handle bilateral Arctic issues. It offers a solution to meet the primary concerns of Canada and the United States, protecting Canada's sovereign authority while assuring U.S. security concerns. It seeks to identify mechanisms to produce effective joint policy recommendations through processes acceptable to both states, reinforcing traditions of cooperation and reciprocity. Specifically, it argues that an "Arctic Joint Commission" (AJC) similar to the International Joint Commission (IJC), would complement existing bilateral and multilateral Arctic arrangements.

This analysis starts with the assumption that territorially based nation states are the primary actors in the international system, although the presence of subnational or aterritorial networks of interaction require them to pursue their goals with greater flexibility and creativity, as suggested in chapter 11.[3]

Arctic cooperation in North America has been relatively uneventful. By 1946 the United States had no interest in disputing Canadian sovereignty over the islands of the archipelago, although not making that recognition public immediately, including to Canadians.[4] The only land in the Canadian Arctic under dispute is Hans Island, a half-mile-square rock in Nares Strait claimed by Canada and Denmark. However, sovereignty over Arctic waters has been disputed, even politically disruptive, as each state has different priorities with implications for sovereignty claims. For Canada, Arctic sovereignty is a key issue and a fundamental component of national identity. Canada asserts that the waters of the Arctic Archipelago are domestic waters. For the United States, top priorities include security, freedom of the seas, and energy. Not surprisingly, it asserts that these waters are international straits. These differing priorities and claims combine to create a complex mixture of cooperation and conflict in the North American Arctic.

Often domestic concerns, such as state/provincial sovereignty, have impeded foreign policymaking. Trends in Canadian constitutional litigation suggest that the interaction of Indigenous rights and Canada's international policy relations will become increasingly important in coming years.

The chapter ultimately asserts that it is a geopolitical reality that fully separate and independent policy approaches to continental Arctic issues, in a variety of areas, are impossible for both countries.[5] North American geography ensures that U.S. national security and Canadian sovereignty cannot be treated by either government as distinct policy matters.[6] With the world's longest peaceful border, the two states are contiguous in the Arctic. Recognizing this joint necessity, this chapter's proposals seek to enable foreign policymakers to maintain a successful and respectful partnership in the North. The specific proposition guiding the research, the institutional structure hypothesis (ISH), asserts that if two countries institutionalize binational structures of cooperation, then mutually satisfying bilateral policy recommendations are more likely than otherwise.

The chapter unfolds in multiple sections. It begins with a brief discussion of bilateralism, binationalism, and different views of sovereignty, contending that a form of "interdependence sovereignty" best informs Arctic relations between the United States and Canada. It then examines the International Joint Commission (IJC) as a pre-existing, institutionalized binational entity that allows management of even unforeseen issues to avoid sensitive issues of sovereignty. The final sections present a summary of Arctic authority and control and concluding remarks.

The Need for Canada-U.S. Binationalism in the Arctic

The terms "binational" and "bilateral" are often used interchangeably in Canada-U.S. relations without necessarily clarifying the distinction. Following Nossal[7] and Lagassé,[8] pursuing bilateral cooperation typically involves "normal diplomacy," sometimes described as give-and-take negotiations.[9] Such interactions are often parochial, exemplified by diplomatic meetings in which representatives of each state seek to meet their respective country's needs or wishes, often according to predetermined instructions.

Binationalism, which involves a joint decision-making or administrative process, is non-parochial, facilitating coordination and even harmonization. Examples of Canadian-U.S. binational entities are NORAD, the North American Aerospace Defense Command, and the IJC. NORAD monitors North American airspace and maritime approaches for defence. NORAD's binational character enables Canada to be a "formally equal, sovereign continental defence partner" while ensuring the United States' security concerns.[10] Charron argues that NORAD "is of enormous benefit to Canada," with considerable potential in the Arctic that Canada cannot afford to replicate alone.[11] With the North American continent flanked by two oceans and a non-threatening Mexico, the Arctic provides the only land- or sea-based approach to Canada and the United States in which sovereignty could be challenged or security threatened.

Canada and the United States are two of five littoral Arctic states, joining Russia, Norway, and Denmark (Greenland). Three additional states, Sweden, Finland and Iceland, have territory, including Continental Shelf, close to or above the Arctic Circle.[12] These "Arctic Eight" form the core of the Arctic Council, the leading plurilateral body engaged with Arctic affairs.

Such entities can be problematic. Consider the May 2008 Illulissat Declaration[13] from the five littoral Arctic states. In it, these states assert that they are best placed to meet challenges, such as shrinking ice, due to climate change, "by virtue of their sovereignty, sovereign rights and jurisdiction." It further asserts that no additional "comprehensive international legal regime" needs to exist alongside the Arctic Council and International Maritime Organization.[14] The five littoral Arctic states also appear cool to additional actors participating in Arctic policymaking. At the same time, not all five may be necessary to address certain issues, such as search-and-rescue in the North American Arctic.

Created in 1996 with a focus on environmental issues, the Arctic Council serves many purposes. Relatively new, it can be viewed as a work in progress.[15] It has incorporated several Indigenous bodies as permanent participants enjoying "active participation and full consultation": the Aleut International Association, the Arctic Athabaskan Council, the Gwich'in Council International, the Inuit Circumpolar Council, the Russian Association of Indigenous Peoples of the North, and the Saami Council.[16] Thirteen other nation states have observer status, even without having territorial or maritime claims in the Arctic.[17]

The Arctic Council's jurisdiction does not extend to security. Indeed, Charron warns against expanding NATO's role in the Arctic, a possibility raised in a 2017 Canadian defence policy paper, as a probable provocation to Russia.[18] Climate change has increased shipping activity in the Arctic with a concomitant increase in security concerns, more than could be handled even if the council's mandate was broadened.[19] Consequently, an Arctic Joint Commission between the Canada and the United States would not replace the Arctic Council and other multilateral institutions. Rather, it would augment their important work.

An AJC also would not negate the role of other subnational actors, such as the state of Alaska, in other pre-existing entities, although some would argue that Alaska's influence in Washington is insufficient.[20] From a U.S. military perspective, the United States is "an Arctic nation, not simply a nation with an Arctic state."[21] Input from sub-federal governments and Indigenous communities is critical to any Arctic programs, especially under constitutional "duty to consult" principles related to Indigenous rights affirmed by Canadian courts in recent years.

A joint decision-making process similar to the IJC could also address unwelcome forms of unilateralism. As vice-president, Al Gore brought schemes before the Arctic Council that "angered" members of the Inuit Circumpolar Council and the Department of Indian Affairs and Northern Development. For example, policies amenable to the Antarctic were inappropriately applied to the Arctic, putting environmental issues ahead of human and community development.[22] Laws and policies for Antarctica, a continent without a permanent population, held few lessons for Alaska and the Canadian Arctic, home to some 800,000 people[23] in an area heavily affected by laws of the sea. Released on 19 November 2019, Canada's Arctic Policy Framework rests on the principle of "Nothing about us, without us," ensuring greater participation by Canadian Indigenous groups and institutions, as well as provincial and territorial governments.[24] These examples provide reminders that many Canadians, including the media, are intensely sensitive to perceptions of U.S. infringement on Canada's Arctic sovereignty.[25] Issues that might otherwise be settled easily can become politicized to the point that diplomacy and foreign policymaking become unnecessarily difficult. Clearly, for many issues, both Canada and the United States would benefit from a strong binational institution that can function without triggering sovereignty concerns.

Sovereignty in the Canadian/U.S. Context

The widely varying concepts of sovereignty used in different contexts make it essential to pin down a definition that can be applied effectively to the North American experience. As noted above, Canadians are extremely sensitive about threats to sovereignty (both real and perceived).

Sovereignty in the study of international relations usually concentrates on independence and authority, particularly domestic authority and self-governance without interference from another state, even if scholars such as Stephen Krasner refer to it as "organized hypocrisy."[26] In practice, absolute sovereignty simply does not exist, given variant levels of relative state power, competing motivations among national decision-makers, and limited international institutions for its enforcement.[27] Such realities lead Krasner to identify four types of sovereignty, each with its own characteristics.

The first is international legal sovereignty. This category concentrates on the de jure status of a state, often indicated by official recognition and centring on the existence of authority rather than control. It is related to the "act of state" doctrine or "juridical independence" in international law, meaning acts within one country cannot be judged in the courts of another country. States that have international legal sovereignty can join international governmental organizations, sign treaties, and enjoy diplomatic immunity for official government representatives. Confusion arises when states act as though they have these characteristics, or have been treated as though they do, yet lack full recognition. Canada exercised, but did not have, full international legal sovereignty when, as a dominion within the British Commonwealth, it signed the Versailles Treaty after the First World War and joined the League of Nations.[28]

A second category, Westphalian sovereignty, reflects basic information from the locale. A form of de facto authority, it acknowledges the rule of a given state on its own territory without external interference into its decision-making institutions and authority structures. Thus, in addition to authority, control is a consideration. This is the kind of sovereignty originally intended in the United Nations Charter and is the basis of arguments against non-intervention, despite its selective application in challenges to South African apartheid, smaller states' abuse of human rights, and interventions based on the failure of member states to protect their own citizens from genocide, war crimes and crimes against humanity, known as "responsibility to protect" (R2P).[29] Canada's Westphalian sovereignty is limited mainly through voluntary agreements with other countries. An example is the International Joint Commission (IJC), established by the 1909 Boundary Waters Treaty between Canada and the United States, to manage cross-border water issues. The IJC has functioned very effectively while remaining non-controversial for over a century by providing "mutual advantage of both nations."[30] Whether or not external actors exert influence upon domestic policies, and to what extent, relates to Krasner's third definition of sovereignty.

Domestic sovereignty relates to autonomy and self-government. Unlike international legal sovereignty and Westphalian sovereignty, the concept of control is more central than the notion of authority, either de jure or de facto: how authority actually is exercised within the state, by the state. Domestic sovereignty can become diffused when extensive policymaking power is loosely delegated to subunits, such as provinces or states, or within a confederation. Indeed, longstanding constitutional precedents preclude Canada's federal government from enforcing provisions of international treaties in areas of independent provincial jurisdiction.

Canada historically has had great sensitivity to this kind of sovereignty. As the decline of the British Empire overlapped with the rise of American global power, sensitivity to domestic sovereignty essentially shifted focus from the United Kingdom to the United States.

A contemporary focal point of Canadian domestic sovereignty sensitivities is in the Arctic. This is evidenced by policies to solidify control through the increased presence and activities of the federal government, such as the Canadian Rangers, part of the Canadian Forces (CF) Reserve.[31] Their missions include national security, safety, and environmental roles. Members primarily are local, often from Indigenous communities. The Canadian Rangers are well-suited to fit into an institutionalized binational structure between Canada and United States for meeting challenges in the Arctic, in both decision-making and execution, confirming Canada's sovereign authority and also contributing to U.S. security.

Krasner's final category is interdependence sovereignty. Again, the emphasis is on control, but across national borders rather than within a state. Globalization and modern technologies can challenge interdependence sovereignty, although this kind of transnational phenomenon is not new.[32] While authority is not inherent in the determination of interdependence sovereignty, an inability to control border traffic (human, material, or monetary) can eventually result in a loss of authority as well. Alternatively, control can be exercised without formal authority.[33] It is this area of sovereignty that may best explain Canada-U.S. relations in the Arctic, as well as Canada's sovereignty sensitivity in this region. However, rather than Krasner's description of an interdependence as a result of globalization or improvements in transportation, *geopolitics* has made security policy cooperation an immutable reality in North America. The geopolitical situation between Canada and the United States in North America leans towards support for ISH.

To bolster operational security in the Arctic, Canada and U.S. policymakers are more effective when approaching security policy challenges interdependently. Further, to function effectively, security interdependence requires a reduced sovereignty sensitivity, which can be achieved more readily through institutionalized, preferably binational structures of cooperation that satisfy domestic sovereignty concerns for each state. This kind of voluntary relationship would not affect the right to self-government, but would require mutual respect for the principle of supreme authority and territorial independence, further supporting ISH. Sovereignty interdependence necessitated by continental geopolitics best describes the Canadian-U.S. relationship, rather than technology or globalization, thus broadening Krasner's definition. As noted above, the IJC is a fully institutionalized binational structure of cooperation that consistently functions effectively yet does not challenge the domestic sovereignty concerns of either state, thus supporting the assertion that institutionalized binational structures of cooperation are more likely to produce bilateral policy recommendations that provide satisfying results.

Specific applications of Krasner's work on sovereignty often lean towards a concentration on the Canadian constitution[34] or do not concentrate on the Canadian-U.S. relationship itself.[35] This study suggests that Canada and the United States, by virtue of immutable geopolitics, are in a policy relationship of interdependence sovereignty in the Arctic. This kind of interdependence also applies in areas beyond security, such as the environment, energy, and rights and protection of Indigenous peoples, who form majorities of the populations of two of Canada's three northern territories.

McRae writes, "The notion that Canadian sovereignty in the Arctic is fragile and that there is a threat that it will be 'lost,' particularly as a result of action by the U.S., appears to be hardwired into any public discussion of the issue in Canada. Any new Arctic strategy will thus require effective advocacy and diplomacy at both the domestic and international levels."[36] The IJC has functioned for over 100 years between Canada and the United States as a fully institutionalized binational mechanism for preventing and resolving transboundary disputes, particularly those related to common water resources. As such, its binational institutional characteristics are an effective foreign policy instrument without triggering sovereignty sensitivity, at least partly because the convention that the IJC investigates issues only on which *both* central governments have agreed to joint terms of reference. With the assumption that if two states institutionalize binational structures of cooperation, then mutually satisfying bilateral policy recommendations are more likely than otherwise, the IJC would seem to be the model for which to assess other policymaking entities.

International Joint Commission

The IJC comprises six members: three selected by the Canadian prime minister and three by the U.S. president. Its purpose is to provide the "machinery"[37] for meeting its "two main responsibilities: regulating shared water uses and investigating transboundary issues and recommending solutions."[38] The success of the IJC is ironic in two ways. First, it exists to handle border issues. Elsewhere in the world, border issues often are extremely dangerous. They can touch the core of a state's sovereignty concerns and can be the spark for armed conflict, as for many years between India and Pakistan. Second, the IJC was created on the heels of the 1903 Alaska Boundary Dispute with the 1909 Boundary Waters Treaty. In fact, Burpee noted in 1940, "It is certain that the Treaty and the Commission mean even more to the people of the Dominion. They mean, in fact, just this much more, that they represent a voluntary arrangement between a small country and a powerful neighbor that puts both on a footing of absolute equality."[39] While keeping a "low profile,"[40] it was able to "dispos[e] of controversies at the beginning before they have ceased to be personal and nations have become excited and resentful about them."[41] It also meant going around Great Britain, as "once the Americans come to deal directly with us, they will play the game fairly. It is only because we have got John Bull along that they bully us. Once get him out of the game [sic] and there will be no prestige in tackling the little fellow who will kick their shins."[42]

Consider an attempt to set up a commission to deal with water among the United States, Mexico, and Canada in 1895. Rather than start direct communication between Ottawa and Washington, DC, on a resolution, Canada was forced to engage in a cumbersome and time-consuming process. It began 8 January 1896 with an Order in Council, to the Committee of the Dominion Privy Council, to advise the governor general, to instruct the British ambassador in DC "regarding the wishes of the Canadian government," to be forwarded to the U.S. secretary of state. Secretary of State Olney responded to the governor general that the United States "does not lack interest" but didn't have any information on border water issues, so would not comment on the U.S. position. Not only did it take until 17 March 1896 to get a poor answer to a relatively simple question, somewhere in the process the word "Mexico" was dropped and never would be reinstated in the development of the IJC.[43]

The IJC thrives and continues today as an exemplary instrument to achieve binational policymaking. As with a potential Arctic Joint Commission, neither state would be bound to accept any of the recommendations, ensuring sovereignty.[44] In addition, how policies are administered within each state is entirely up to that state. Existing domestic structures, such as regulatory regimes and federal and subfederal laws, need not be altered or substituted.[45]

There are other institutionalized relationships between Canada and the United States, such as the Permanent Joint Board on Defense (PJBD) created out of the 1940 bilateral Ogdensburg Agreement. The PJBD is noteworthy for its original and truly "permanent" aspect, "indicating a shift in Canada from Great Britain to the United States that would continue beyond the immediate emergency."[46] As noted above, the North American Aerospace Defense Command (NORAD) is another example of successful binational cooperation between Canada and the United States, continuing to fulfil their stated purposes while (mostly) staying out of the public and political limelight.

Two aspects of former Canadian ambassador (1953–7, 1959–62) and IJC member Arnold Heeney's career worth noting here was his friendship and professional collaboration with an American counterpart, Livingston T. Merchant, and his tenure on the IJC.[47] Heeney wrote extensively about the anti-Americanism in Canada in his day,[48] both in government and among the public, noting that it often was promoted by (or originated in) the media. Towards the end of his career, Heeney served on the IJC, recognizing that "each country is affected by the other's actions in lake and river systems along the border. The two countries cooperate to manage these waters wisely and to protect them for the benefit of today's citizens and future generations." Heeney's comments are telling: "I was not long in the commission before it became clear to me that the IJC method of finding solutions to problems between the two countries was different from that to which I had been accustomed. Whereas in normal diplomacy the representatives of each side, under instruction from their respective governments, strove for their national advantage as a buyer and seller in a commercial transaction, the six commissioners of the IJC approached the problems put before them with the object of reaching solutions which would be to the mutual advantage of both nations."[49]

In the 1960s, Heeney and Merchant were asked to put together a recommendation for bilateral Canadian-U.S. relations. Published in 1965 after an eighteen-month effort, the "Principles for Partnership" document received blistering criticism from the Canadian media.[50] The controversy revolved upon a single sentence in paragraph 81: "It is the abiding interest of both countries that, wherever possible, divergent views between the two governments should be expressed and if possible resolved in private, through diplomatic channels." As Heeney recalled, charges in the press included that the report "was a proposal to gag the Canadian government and to prevent Canadian public criticism of American external policies. Approval of such a principle would be taken as acceptance by Canada of satellite status. Headlines in the Canadian press and comment by Canadian columnists employed such terms as 'lap dog.' I was charged with having been 'conned' by Merchant into recommending a shackling of the legitimate expression of Canadian views. 'Quiet diplomacy' became a term of abuse."[51] Obviously the IJC has not been deemed an unequal relationship that engenders the kind of sovereignty sensitivity Heeney had perceived in his critics.

While not all of the discussion was negative, the point is that the prospect of muzzling Canadian views in order to promote U.S. interests indicated a very high sovereignty sensitivity. Why has the IJC worked so well for over a century and yet other issues receive such an emotional response? Doran wrote in 1984 that the IJC was a vehicle for research, dealt with disagreements that were amenable to "technical solution[s]," and avoided hot topics in the political and economic spheres.[52] For most states in the international system, border disputes are the basis for enormous attention and often instigate crises and even conflict. Yet Canada and the United States, a prototypical "security community,"[53] share a continuous border of more than 5,500 miles without a single military installation with the specific purpose to protect one state from the other. The IJC deals with border issues regularly. It could be argued that almost any bilateral relationship could be reduced to apparently "technological" fixes, if there was low sovereignty sensitivity. This study proposes that the long, peaceful relationship between Canada and the United States is not based on just an extended experience of cooperation that has nurtured an atmosphere of reciprocity and interdependence. Binational institutions have facilitated a reduction in sovereignty sensitivity. Can institutionalized, binational structures for policymaking recommendations between Canada and the United States move Arctic policies into this realm?

Does the situation today mandate a more urgent need to resolve Canadian and U.S. disagreements about Canadian sovereignty in the Arctic? A report to Department of Foreign Affairs and International Trade (now Global Affairs Canada) questions the rhetoric of the Arctic states, including members of the Arctic Council.[54] The report asserts that the actions of Arctic states do not match their peaceful, cooperative rhetoric.

Norway has a growing and upgraded Arctic-capable fleet that includes submarines, a research ship equipped to conduct electronic and signal intelligence, and a combat ship.[55] Norway's main security concern in the Arctic is Russia. In the effort to make much of the Arctic "a Russian lake," Doran asserts that "North America cannot be defended from the Arctic except in terms of close cooperation with Canada."[56] In the European North, Canada has sent troops to Latvia in response to Russia's "A2/AD bubble" in the Scandinavian and Baltic region, serving as what one scholar calls a "NATO tripwire."[57] A2/AD refers to anti-access/area-denial, a strategy to block an adversary from operating in a given region, or at a minimum shift a balance of power away from that adversary. In addition to the Arctic, Russia's interest in the Baltics certainly includes its naval installation at Kaliningrad. Partly in response to Russia's 2014 annexation of Crimea as well as other aggressive actions, a 300-strong U.S. Marine Corps presence was begun in Norway in January 2017. In June 2018, Norway agreed to an additional 400 troops. From 25 October to 7 November 2018, NATO conducted its largest joint military exercise since 2002.[58] "Trident Juncture" practised defending Norway from Russia and involved over 45,000 troops, including 18,000 Americans and 2,000 Canadians. The neighbourhood is becoming potentially more dangerous. This kind of multinational military exercise is a start, but Canada and the United States need a binational instrument to provide bilateral policy recommendations that concentrate on North American needs, and has the ability to provide this advice swiftly.

The IJC can serve as a model for North American relations in the Arctic. In the realm of Canadian-U.S. relations, sovereignty often is a bewildering factor, particularly in instances that promote sovereignty sensitivity among Canadian policymakers and the general public, rather than the relatively obscure characteristics of IJC activities. Given geopolitical realities and conditions of interdependence sovereignty, applying the IJC model to the Arctic could reduce sovereignty concerns.

Authority and Control in the Arctic Archipelago Waters

Krasner's four categories of sovereignty – international legal, West-phalian, domestic, and interdependent – form a useful tool in making the case for an institutionalized, binational entity for Canada and the United States to begin handling Arctic issues in their respective spheres of authority and control. In short, Canada has the upper hand when it comes to authority by virtue of its acknowledged sovereignty over islands of the Arctic Archipelago, while the United States has the greater capability (and potential) to assert control. For example, in a 2018 U.S., U.K., and Canadian Arctic military exercise, the Canadians did not possess a submarine capable of manoeuvring under the ice. While the U.S. military suffered a decline under the Obama administration, it still overwhelms the capabilities of most other states, even before the budgetary increases secured by the Trump administration. In the Arctic, however, the United States maintains a higher potential simply because it has greater resources and funding that can be directed there.

Canada has lagged between the stated intentions of its governments and actual provision of bases, equipment, and personnel. However, Canada's standing on legal authority is much more solid. Concerted efforts to assert a government presence in the Arctic began in earnest under the Laurier government, a direct result of the 1903 Alaska Boundary Dispute.[59] While Canada established an official presence in the Arctic, primarily through Royal Canadian Mounted Police posts, it was not until the Second World War that more extensive activity took place.[60] During the war, the United States had de facto control in the Yukon and Northwest Territories, as its personnel outnumbered Canadians and functioned under U.S. laws.[61] Extensive U.S. presence throughout the Arctic continued into the Cold War period. Not just Americans were involved. Into the 1960s, more non-Canadian companies were exploring and leasing Arctic oil and gas fields than Canadian companies.[62]

As a result of the work of explorers and other forms of Canadian presence, activity, and official acts of government, Canada's claims over the islands of the Arctic Archipelago as sovereign territory are overwhelmingly accepted internationally. Sovereignty over the Arctic waters, however, remains in dispute. It remains "one of the most significant bilateral legal issues in the Canadian–United States relationship."[63] This issue is relevant in many policy areas, including the environment, Indigenous rights, and access to resources. Primary concerns for the United States include unimpeded navigation, both military and merchant.[64]

The status of a waterway directly affects the kind of transit that is considered permissible. In international law, an international strait is defined as "water corridors meeting the geographical requirements of being less than 24 miles wide (or 12 miles from each coast) and of linking two parts of the high seas."[65] In this situation, a foreign ship would not need to ask permission for transit passage. In a state's internal or territorial waters, however, the principle of innocent passage applies and can be suspended. Under innocent passage there can be no threat to "the peace, good order or security of the coastal State"[66] and should be, outside unexpected needs, "continuous and expeditious."[67] Therefore, the legal status of the Northwest Passage has direct relevance to rights of passage.

A state enjoys the same full sovereignty on internal waters as it does on land. Foreign vessels have no inherent right of transit passage. There is no agreement on whether the Northwest Passage is a truly internal or an international strait, particularly between Canada and the United States. The two sides on this argument differ on whether a waterway can be designated according to potential versus actual use. Few ships have transited the Northwest Passage, yet the United States views it as an international strait.

In the effort to delineate Canada's sovereign authority in Arctic waters, as early as 1907 Senator Pascal Poirot suggested using the sector principle in the Arctic Archipelago, but this approach never received official support.[68] The matter remained "theoretical" until the 1969 transit of the Northwest Passage by the SS *Manhattan*, a Humble Oil (Exxon) tanker.[69] After months of extensive Canadian and American cooperation, the *Manhattan* became the first commercial vessel ever to make it through.[70] In addition, the *Manhattan* had an official Canadian observer aboard and was escorted by a Canadian icebreaker, the CCGS *John A. Macdonald*. In fact, "On an operational level, the voyage was a dramatic display of the kind of Canada-U.S. cooperation that had long typified joint Arctic operations." Lajeunesse suggests that "political questions of sovereignty were simply not a serious concern and were easily dismissed when they surfaced."[71]

What went wrong? The *Manhattan* was also escorted by the USCGC *Northwind*, a military vessel. The U.S. refused to make an official request to send the U.S. Coast Guard ship through Canada's "internal waters," as it might appear to be recognition that the passage was *not* an international strait.[72] The *Manhattan*'s passage was very significant and more than newsworthy. The permission aspect was grossly underestimated by the Department of External Affairs and erupted into a public uproar, forcing Ottawa to move forward in asserting its authority in Arctic waters.[73]

Canada responded with the passage of the 1970 Arctic Waters Pollution Prevention Act, which "amounted to the unilateral extension of maritime jurisdiction 100 miles into the high seas, for the purposes of environmental protection."[74] The United States objected to the act, arguing that a state could not unilaterally alter international customary law. The argument was technically valid, but the United States was the only state to make a formal objection.[75]

Canada needed the strength of a treaty, or "codified law," to underscore its claims. This was accomplished via UNCLOS III (United Nations Conference on the Law of the Sea), whose negotiations lasted from 1973 to 1982. At these "tricky" negotiations "the subtlety and sophistication with which the Canadian delegation played its Arctic hand at UNCLOS III [was] one of the most interesting stories of the Conference."[76] Canada's delegation often outnumbered the Soviet Union and many Western European states and at times included members of the Cabinet.[77]

The result was Article 234, the "Arctic Exception," written by Canada and achieved without directly mentioning the Arctic itself.[78]

Ice-Covered Areas

Coastal States have the right to adopt and enforce non-discriminatory laws and regulations for the prevention, reduction and control of marine pollution from vessels in ice-covered areas within the limits of the exclusive economic zone, where particularly severe climatic conditions and the presence of ice covering such areas for most of the year create obstructions or exceptional hazards to navigation, and pollution of the marine environment could cause major harm to or irreversible disturbance of the ecological balance. Such laws and regulations shall have due regard to navigation and the protection and preservation of the marine environment based on the best available scientific evidence.[79]

Article 234 gives Canada the right to pass laws and regulations in the Arctic waters that technically deal with environmental issues, but allows a broad application to other issue areas, and successfully "linked Canada's national interests to broader international interests in pollution control and prevention."[80] The United States has never ratified UNCLOS, but according the James Kraska of the U.S. Navy War College, it "scrupulously follows the rules in the convention."[81]

A U.S. ship again caused an uproar over Canada's Arctic sovereignty in 1985 with the passage of the USCGS *Polar Sea*, a heavy icebreaker, through the Northwest Passage, again without formal approval from Ottawa. On 1 January 1986, an Order in Council went into effect that introduced the straight baseline principle for determining Canada's claim over Arctic waters. Unlike the sector principle, straight baselines are drawn from headland to headland along coastlines, including islands. The waters on the landward of the baselines are claimed by Canada as internal waters. This and other claims rest heavily on the concept of historic title claims.[82] In 1988 the United States and Canada signed the Arctic Cooperation Agreement, a one-page document that essentially allows both states to continue their respective assertions vis-à-vis Arctic sovereignty while agreeing to disagree. U.S. ships going through the Northwest Passage would request the consent of Canada, which would be granted, but these requests would not constitute any recognition of Canadian claims to internal waters.

This policy stalemate has little chance of producing effective cooperation between Canada and the United States as the challenges produced by climate change increase in the Arctic. U.S. Secretary of State Mike Pompeo's May 2019 speech to the Arctic Council raised the ire of many Canadians, although criticism was based as much on delivery as it was on content. Pompeo described Canada's claim to the Northwest Passage is "illegitimate," violating what some scholars see as the core of the 1988 agreement. Fen Hampson felt it was a "stunning rebuke" of the agreement to disagree.[83] David Bercuson, while critical of the speech, asserts that the U.S. claim is valid, but more importantly Canada and the United States need to move forward, together, to delineate how the Arctic waters are managed.[84] Alastair Allan specifically has recommended a Northwest Passage treaty between Canada and the United States, as "its disputed status opens the northern Arctic to catastrophic risks."[85]

Conclusion

"Talleyrand's observation that geography determines diplomacy has special relevance to Arctic questions. Canada's engagement in the Arctic is a consequence of its geography, the same that has us sharing the North American continent with the world's greatest power. Our Arctic policy, like our broader foreign and security policy, is shaped by this basic reality."[86]

Interdependence sovereignty, defined by Krasner[87] as based upon technology and globalization, has been expanded in this chapter to include North American geopolitical realities. In Canadian-U.S. relations, issues that involve Arctic sovereignty easily can become a noteworthy and often antagonistic policy debate at the national level, thus impeding bilateral relations. This chapter argues that a binational institution to coordinate policymaking recommendations for the Arctic can result in policies that can be effective while avoiding an appearance of provocation on either side. Can an institutionalized binational approach to disagreements between Washington and Ottawa over the Arctic be de-politicized to the point that managing them is reduced to apparently "technical" fixes? There are myriad examples of Canadian-U.S. cooperation and coordination in recent years. In the Arctic, however, differences based on Canada's sensitivity to sovereignty and U.S. national security priorities have impeded further institutionalization of the relationship. The "unfinished business"[88] in the Arctic has yet to become the kind of permanent relationship, exemplified by Ambassadors Heeney and Merchant, of consistent diplomatic "patience" and "mutual understanding"[89] akin to the IJC.

The suggestion has been made that Canada's military cuts that began under Prime Minister Pierre Trudeau resulted in greater U.S. unilateralism in the Arctic as confidence in Canada's ability to exert control in the region waned.[90] Reversal of that trend began under the Martin government and continued under Prime Minister Harper, although action never met their stated ambitions. Meeting the ultimate objective of identifying effective and acceptable bilateral policies related to continental security in the Arctic, based on a tradition of cooperation and reciprocity, requires understanding the precise nature of Canadian sovereignty in terms of geopolitical interdependence.

Wartime events in the Arctic showed that Canada and the United States can work together effectively under the most extreme circumstances. The formation of joint forces in the Second World War was a result of geopolitical facts. Both states have a Pacific coastline and borders along that coastline. There was no way to avoid interdependent action against the Japanese presence in North America in 1943. Arguably, the Canada-U.S. border has become even more surreal in its character, even with the post-9/11 thickening, in terms of cross-border activity. Outside a national emergency, Canada and the United States should be able to achieve a binational institution for making policy recommendations in the Arctic, similar to the IJC on Canada's southern and western borders, which produces effective interdependence on what might become hot topics for politicians, the press, and social media.

Another geopolitical reality is that most of the North America's Arctic lands and islands are Canadian soil. However, the United States maintains, overall, the greater ability and potential to operate under the Arctic ice. The status of Arctic waterways is under dispute, yet the need to protect the peoples and resources of this region will not wait. Combined Canadian-U.S. Arctic policymaking, including foreign policies, needs the support of a binational institution in order to be fully operational in North America. This could produce a real and effective administrative policy making viability based upon mutual respect and a reciprocal relationship of equals. Applying Krasner's concept of interdependence sovereignty, broadened to include geopolitical factors, it becomes evident that the most effective way to approach Canadian-U.S. Arctic relations is via an institutionalized binational structure such as the International Joint Commission. An IJC-type commission for the Arctic, an Arctic Joint Commission, can avoid politicization of issues while mollifying Canada's sovereignty and U.S. security concerns.

NOTES

1 Malcolm N. Shaw, *International Law*, 7th ed. (Cambridge: Cambridge University Press, 2014), 387.

2 United States, Energy Information Administration, "Arctic Oil and Natural Gas Resources," 20 January 2012, http://www.eia.gov/todayinenergy/detail.cfm?id=4650.

3 See VanNijnatten and Johns, chapter 11 in this volume.

4 Patrick James, *Canada and Conflict* (Oxford: Oxford University Press, 2012), 104; Adam Lajeunesse, *Lock, Stock and Icebergs: A History of Canada's Arctic Maritime Sovereignty* (Vancouver: UBC Press, 2016), 39.

5 Important issues in the Arctic include environmental protection, Indigenous rights, energy, resource extraction (fishing, oil and gas, seabed mining), scientific research, and illegal interdiction (human and material).

6 Charles Emmerson, *The Future History of the Arctic* (New York: Public Affairs, 2010), 109.

7 Kim Richard Nossal, "The IJC in Retrospect," in *The International Joint Commission Seventy Years On*, ed. Robert Spencer, John Kirton, and Kim Richard Nossal, 124–30 (Toronto: University of Toronto, 1981).

8 Philippe Lagassé, "A Common 'Bilateral' Vision: North American Defence Cooperation, 2001–2012," in *Game Changer: The Impact of 9/11 on North American Security*, ed. Jonathan Paquin and Patrick James, 193–211 (Vancouver: UBC Press, 2014).

9 Arnold Heeney, *The Things That Are Caesar's: Memoirs of a Canadian Public Servant* (Toronto: University of Toronto Press, 1972), 187.

10 Lagassé, "Common 'Bilateral' Vision," 195–6.

11 Andrea Charron, "Contesting the Northwest Passage: Four Far-North Narratives," in *Border Flows: A Century of the Canadian American Water Relationship*, ed. Lynne Heasley and Daniel Macfarlane (Calgary: University of Calgary Press, 2016), 86, 87; see also Rob Huebert, "NATO, NORAD and

the Arctic: A Renewed Concern," in *North of 60: Toward a Renewed Canadian Arctic Agenda*, ed. John Higginbotham and Jennifer Spence, 91–7 (Waterloo, ON: Centre for International Governance Innovation, 2016).

12 Michael Byers, "Arctic Region," *Max Planck Encyclopaedia of Public International Law*, 2010, as cited in Shaw, *International Law*, 386.

13 "Ilulissat Declaration," Ilulissat, Greenland, 27–9 May 2008, https://cil .nus.edu.sg/wp-content/uploads/2017/07/2008-Ilulissat-Declaration.pdf.

14 Alan H. Kessel, "Canadian Arctic Sovereignty: Myths and Realities," in *Governing the North American Arctic: Sovereignty, Security, and Institutions*, ed. Dawn Alexandrea Berry, Nigel Bowles and Halbert Jones, 242–6 (New York: Palgrave Macmillan, 2016).

15 Heather Exner-Pirot, "Canada's Arctic Council Chairmanship (2013–2015): A Post-Mortem," *Canadian Foreign Policy Journal* 22, no. 1 (2016): 84–96; Svein Vigeland Rottem, "A Note on the Arctic Council Agreements," *Ocean Development & International Law* 46 (2015): 50–9.

16 Arctic Council, "Permanent Participants," https://arctic-council.org/en /about/permanent-participants/; Andrew Stuhl, *Unfreezing the Arctic: Science, Colonialism, and the Transformation of the Inuit Lands* (Chicago: University of Chicago Press, 2016), 48.

17 Arctic Council, "Observers" (Tromso, Norway: 31 July 2019). https:// arctic-council.org/en/about/observers/non-arctic-states/.

18 Andrea Charron, "NATO, Canada and the Arctic," Canadian Global Affairs Institute (September 2017), https://d3n8a8pro7vhmx.cloudfront.net /cdfai/pages/1793/attachments/original/1505784594/NATO_Canada _and_the_Arctic.pdf?1505784594; Canada, Department of National Defence, *Strong, Secure Engaged: Canada's Defence Policy* (Ottawa: 2017), http://dgpaapp.forces.gc.ca/en/canada-defence-policy/docs/canada -defence-policy-report.pdf.

19 Heather Exner-Pirot, "Defence Diplomacy in the Arctic: The Search and Rescue Agreement as a Confidence Builder," *Canadian Foreign Policy Journal* 18, no. 2 (2012): 196.

20 Chanda L. Meek and Emily Russell. "The Challenges of American Federalism in a Rapidly Changing Arctic," in Berry, Bowles, and Jones, *Governing the North American Arctic*, 165–79.

21 U.S. Congress, House, Subcommittee on Coast Guard and Maritime Transportation of the Committee on Transportation and Infrastructure, *Implementing U.S. Policy in the Arctic*, 109th Cong., 2nd sess., 2014, 9 (Peter V. Neffinger, Vice Commandant, U.S. Coast Guard).

22 Stuhl, *Unfreezing the Arctic*, 149–51.

23 Charron, "Contesting the Northwest Passage," 101.

24 Government of Canada. "Canada's Arctic and Northern Policy Framework," 2019, https://www.international.gc.ca/world-monde

/international_relations-relations_internationales/arctic-arctique/index
.aspx?lang=eng.

25 For example, see Franklyn Griffiths, Rob Huebert, and P. Whitney Lacken-
bauer, eds., *Canada and the Changing Arctic: Sovereignty, Security and Stew-
ardship* (Waterloo, ON: Wilfrid Laurier University Press, 2012); P. Whitney
Lackenbauer, ed., "Canadian Arctic Sovereignty and Security: Historical
Perspectives," Calgary Papers in Military and Strategic Studies: Occasional
Paper Number 4 (Calgary: Centre for Military and Strategic Studies, 2011);
Lajeunesse, *Lock, Stock and Icebergs*; Berry, Bowles, and Jones, *Governing
the North American Arctic*; and Elizabeth Riddell-Dixon, *Breaking the Ice:
Canada, Sovereignty, and the Arctic Extended Continental Shelf* (Toronto:
Dundurn, 2017).

26 Stephen D. Krasner, *Sovereignty: Organized Hypocrisy* (Princeton, NJ:
Princeton University Press, 1999).

27 Krasner, *Sovereignty*, 3.

28 For an argument against the existence of international legal sovereignty,
see John Agnew, "Sovereignty Regimes: Territorial and State Authority in
Contemporary World Politics," *Annals of the Association of American Geography* 95, no. 2 (2005): 437–61.

29 For instance, see United Nations, "United Nations Office of the Special
Advisor on Prevention of Genocide," http://www.un.org/en
/genocideprevention/; Thomas G. Weiss, *What's Wrong with the United
Nations and How to Fix It* (Malden, MA: Policy, 2009), 19–38.

30 Heeney, *Things That Are Caesar's*, 187.

31 P. Whitney Lackenbauer, "From Polar Race to Polar Saga: An Integrated
Strategy for Canada and the Circumpolar World," in Griffiths, Huebert,
and Lackenbauer, *Canada and the Changing Arctic*, 99–101.

32 Krasner, *Sovereignty*, 13.

33 Krasner, *Sovereignty*, 10.

34 Stephen McBride, "Quiet Constitutionalism in Canada: The International
Political Economy of Domestic Institutional Change," *Canadian Journal of
Political Science* 36, no. 2 (2003): 251–73.

35 Philpott's definition of authority – "supreme legitimate authority within a
territory" – is used in Elliot-Meisel's *Arctic Diplomacy*; Daniel Philpott, "Sovereignty: An Introduction and Brief History," *Journal of International Affairs* 48,
no. 2 (1995): 357; Daniel Philpott, "Usurping the Sovereignty of Sovereignty?"
World Politics 53 (2001) : 306; Elizabeth B. Elliot-Meisel, *Arctic Diplomacy: Canada
and the United States in the Northwest Passage* (New York: Peter Lang, 1998), 4.

36 Donald McRae, "Rethinking the Arctic: A New Agenda for Canada and
the United States," in *Canada Among Nations 2009–2012: As Others See Us*,
ed. Fen Osler Hampson and Paul Heinbeckers (Montreal and Kingston:
McGill-Queen's University Press, 2010), 254.

37 L.M. Bloomfield and Gerald F. Fitzgerald, *Boundary Waters Problems of Canada and the United States: The International Joint Commission (1912–1958)* (Toronto: Carswell, 1958), 15.

38 International Joint Commission, https://www.ijc.org/en/who/role.

39 Lawrence J. Burpee, "Good Neighbours," in *Contemporary Neighbours* (Toronto: Ryerson, 1940), 21.

40 John E. Carrol, "Patterns Old and New," in *The International Joint Commission Seventy Years On*, ed. Robert Spencer, John Kirton, and Kim Richard Nossal (Toronto: University of Toronto, 1981), 50.

41 Elihu Root as quoted in Lawrence J. Burpee, "The International Joint Commission," in *Canada in World Affairs*, Canadian Institute of International Affairs (Toronto: University of Oxford Press, 1941), 229.

42 George Gibbons, as quoted in A.D.P. Heeney, "Along the Common Frontier: The International Joint Commission," *Behind the Headlines* 26, no. 5 (1967): 5.

43 Chirakaikaran Joseph Chacko, *The International Joint Commission between the United States of America and the Dominion of Canada* (New York: Columbia University Press, 1932), 70–3.

44 Nossal, "IJC in Retrospect," 125–8.

45 Wendy Dobson, *Shaping the Future of the North American Economic Space: A Framework for Action*, Commentary, no. 162 (Ottawa: C.D. Howe Institute, 2002), 15; Daniel Schwanen, "Deeper, Broader: A Roadmap for a Treaty of North America," in *The Art of the State: Thinking North America* 2, no. 4, ed. Thomas J. Courchene and Donald J. Savoie (Montreal: Institute for Research on Public Policy, 2004), 12, https://irpp.org/wp-content/uploads/2014/08/schwanen_roadmap.pdf.

46 R.D. Cuff and J.L. Granatstein. *Ties That Bind: Canadian-American Relations in Wartime from the Great War to the Cold War* (Toronto: Samuel Stephens Hakkert, 1997), 101.

47 Merchant served as US ambassador to Canada (1961–2, 1956–8) and was under-secretary for political affairs (1959–61).

48 Heeney, *Things That Are Caesar's*.

49 Heeney, *Things That Are Caesar's*, 187.

50 A.D.P. Heeney and Livingston T. Merchant, *Canada and the United States: Principles for Partnership* (Ottawa: Queen's Printer, 1965).

51 Heeney, *Things That Are Caesar's*, 194.

52 Charles F. Doran, *Forgotten Partnership: US-Canada Relations Today* (Baltimore, MD: Johns Hopkins University Press, 1984), 13.

53 Karl Wolfgang Deutsch, *Political Community and the North Atlantic Area: International Organization in the Light of Historical Experience* (Princeton, NJ: Princeton University Press, 1957).

54 Rob Huebert, *The Newly Emerging Arctic Security Environment* (Ottawa: Canadian Defence and Foreign Affairs Institute, 2010). See also Alun Anderson, *After the Ice: Life, Death, and Geopolitics in the New Arctic* (New York: HarperCollins, 2009); Griffiths, Huebert, and Lackenbauer, *Canada and the Changing Arctic*.

55 Siemon T. Wezeman, "Military Capabilities in the Arctic: A New Cold War in the High North," SIPRI Background Paper (October 2016), 12–13, https://www.sipri.org/sites/default/files/Military-capabilities-in-the -Arctic.pdf.

56 Charles F. Doran, "Additional and Dissenting Views," in *Arctic Imperatives: Reinforcing U.S. Strategy on America's Fourth Coast*, chaired by Thad W. Allen and Christine Todd Whitman (Washington, DC: Council on Foreign Relations, 2017), 52.

57 Alexandra Lanoszka, "From Ottawa to Riga: Three Tensions in Canadian Defence Policy," *International Journal* 72, no. 4 (2017): 529.

58 John Vandiver, "Largest NATO Drill in 16 Years Brings Carrier, US Forces to Norway," *Stars and Stripes*, 9 October 2018, https://www.stripes.com /news/largest-nato-drill-in-16-years-brings-carrier-us-forces-to-norway -1.550983#.

59 Janice Cavell, "'A Little More Latitude': Explorers, Politicians, and Canadian Arctic Policy during the Laurier Era," *Polar Record* 47, no. 4 (2011): 290.

60 William R. Morrison, *Showing the Flag: The Mounted Police and Canadian Sovereignty in the North, 1894–1925* (Vancouver: UBC Press, 1985).

61 Shelagh D. Grant, *Polar Imperative: A History of Arctic Sovereignty in North America* (Vancouver: Douglas & McIntyre, 2010), 13.

62 Grant, *Polar Imperative*, 13–14.

63 Donald Rothwell, "The Canadian-US Northwest Passage Dispute: A Reassessment," *Cornell International Law Journal* 26 (1993): 332.

64 Canada and the United States also disagree about the boundary in the Beaufort Sea. As this dispute is primarily about access to resources and not navigational freedom or related U.S. security concerns, it has not been included in this study.

65 Robert Dufresne, *Controversial Canadian Claims over Arctic Waters and Maritime Zones*, PRB 07-47E (Ottawa: Parliamentary Information and Research Service, 2008), 7.

66 United Nations, "United Nations Convention on the Law of the Sea," Montego Bay, Jamaica, 10 December 1985, http://www.un.org/depts/los /convention_agreements/texts/unclos/unclos_e.pdf, article 19.

67 United Nations, "United Nations Convention on the Law of the Sea," article 18.

68 Under the sector principle, lines are drawn from Canada's most eastern and western points of land in a wedge shape up to the North Pole.

69 McRae, "Rethinking the Arctic," 245.

70 The *Manhattan* carried a single, symbolic barrel of oil. Interest in the Northwest Passage to transport oil became obsolete with the eventual use of pipelines.

71 Lajeunesse, *Lock, Stock and Icebergs*, 144-5.

72 Ross Coen, *Breaking Ice: The Epic Voyage of the SS Manhattan through the Northwest Passage* (Fairbanks: University of Alaska Press, 2012), 66–7.

73 Lajeunesse, *Lock, Stock and Icebergs*, 146–7.

74 Stephen Toope, "Power but Unpersuasive? The Role of the United States in the Evolution of Customary International Law," in *Untied States Hegemony and the Foundations of International Law*, ed. Michael Byers and Georg Nolte (Cambridge: Cambridge University Press, 2003), 309.

75 Richard B. Bidler, "The Canadian Arctic Waters Pollution Prevention Act: New Stresses on the Law of the Sea," *Michigan Law Review* 69, no. 1 (1970): 25–6; Louis Henkin, "Arctic Anti-Pollution: Does Canada Make or Break International Law?" *American Journal of International Law* 65, no. 1 (1971): 134–5.

76 Douglas M. Johnston, *Canada and the New International Law of the Sea* (Toronto: University of Toronto Press, 1985), 17.

77 Riddell-Dixon, *Breaking the Ice*, 180–1.

78 Douglas M. Johnston, ed., *Arctic Ocean Issues in the 1980s: Proceedings, Law of the Sea Institute, University of Hawaii and Dalhousie Ocean Studies Programme, Dalhousie University, Halifax, Nova Scotia Workshop, June 10–12, 1981, Mackinac Island, Michigan* (Honolulu: Law of the Sea Institute, University of Hawaii, 1982), 12.

79 United Nations, "United Nations Convention on the Law of the Sea," 115.

80 Toope, "Power but Unpersuasive?," 310.

81 William Gallo, "Why Hasn't the US Signed the Law of the Sea Treaty?" Voice of America News, 6 June 2016, https://www.voanews.com/a/united-states-sign-law-sea-treaty/3364342.html.

82 Donat Pharand, "The Arctic Waters and the Northwest Passage: A Final Revisit," *Ocean Development and International Law* 38, no. 1–2 (2007): 3–69.

83 Canadian Press, "Pompeo Says Canadian Claim to Northwest Passage Is 'Illegitimate,'" *Ottawa Citizen*, 7 May 2019, https://ottawacitizen.com/news/canada/pompeo-says-canadian-claim-to-northwest-passage-is-illegitimate/wcm/162bffe0-b328-4160-ab82-0903ba315d33.

84 David J. Bercuson, "Pompeo Is Blunt, But Right, about Canada's North," *National Post*, 21 May 2019, https://nationalpost.com/opinion/david-j-bercuson-pompeo-is-blunt-but-right-about-canadas-north#comments-area.

85 Alistair Allan, "Canada Needs a Deal with the U.S. to Control the Northwest Passage," *National Post*, 8 August 2019, https://nationalpost.com/opinion/canada-needs-to-cut-a-deal-with-the-u-s-to-control-the-northwest-passage.

86 William C. Graham, "The Arctic, North America, and the World: A Political Perspective," in Berry, Bowles and Jones, *Governing the North American Arctic*, 18.

87 Krasner, *Sovereignty*.

88 Lajeunesse, *Lock, Stock and Icebergs*.

89 Asa McKercher, "Principles and Partnership: Merchant, Heeney, and the Craft of Canada-US Relations," *American Review of Canadian Studies* 42, no. 1 (2012): 79.

90 James, *Canada and Conflict*.

PART FOUR

Sectoral and Subsectoral Issues

16 Whither Canada's Automotive Industry? Policy, Trade, and Regional Competitiveness

BRENDAN A. SWEENEY

Despite the absence of a homegrown automaker, Canada consistently ranks among the world's leading producers of vehicles. Suffice to say, the automotive industry plays a critical role in Canada's political economy. The economic importance of the automotive industry to Canada – and to other automotive-producing nations – is well recognized by policymakers and industry stakeholders alike. This is primary due to the high-wage employment generated directly by vehicle assembly plants and the high economic multiplier benefits that the automotive industry provides as a result of related upstream and downstream activities.[1] Following nearly half a century of growth between the 1950s and 1990s, Canada's automotive industry underwent significant restructuring. This restructuring was the result of a confluence of territorial (e.g., public policy) and aterritorial (e.g., new systems of production organization) forces emanating from a variety of scales. This chapter examines how these factors influenced the size and organization of Canada's automotive industry in the twenty-first century and Canada's position within broader international automotive industry production networks.

To do so, the chapter draws upon the economic geographic concept of global production networks (GPNs), which recognizes the influence of the complexity of a number of actors, including capital, the state, labour, and consumers in shaping the international political economy. The chapter demonstrates how the combination of territorial and aterritorial forces shape nationally based competitive advantages, and how these advantages increase and diminish vis-à-vis the competitive advantages of other nations integrated into the same production networks. It is also instructive and timely in that it demonstrates the importance of international trade agreements in developing national and supranational competitive advantages within international production networks, and it engages in debates on the future of manufacturing industries in high-wage nations and the efficacy of public policy tools used to support those industries.

The chapter is organized into three subsequent sections. The first introduces the economic geographic concept of global production networks (GPNs) and discusses how it has been applied to the automotive industry. The second examines the development of Canada's automotive industry since the 1960s. It focuses on the combination of forces that helped Canada develop competitive advantages to grow the industry in the latter half of the twentieth century, the decline of the industry between the early 2000s and the recession of 2008–9, and the modest recovery of the industry since the recession and contemporary challenges facing Canada's automotive industry. The third discusses the combined impacts of territorial and aterritorial forces in the context of competitive advantages, public policy, and the future prospects for Canada's automotive industry. A conclusion follows.

Global Production Networks and the Automotive Industry

The economic geographic theory of global production networks (GPNs) is an important conceptual development in the comparative political economy and economic development literatures. GPNs are "organizational arrangement[s] comprising interconnected economic and non-economic actors, coordinated by a global lead firm, and producing goods or services across multiple geographical locations for worldwide markets"[2] whose "nodes and linkages extend spatially across national boundaries."[3] GPN theory emerged partly in response to the limitations of national-institutional approaches to comparative political economy (e.g., varieties of capitalism) and of linear and firm-centric GVC and GVC approaches that underemphasize the role of the state and other place-specific actors, including labour.[4] The recent work of Coe and Yeung[5] – stylized as "GPN 2.0" – plays an important role in the development of GPN theory. Of particular importance is their recognition of the dynamic configurations of GPNs, the intersections between GPNs through suppliers, the state, and other strategic partners, and the aggregation of GPNs through these intersections to form distinct regionally bounded industries or sectors.

Several researchers engage with GPN theory as a means to observe patterns and hierarchies in the global automotive industry.[6] These studies follow the work of Sturgeon and Florida,[7] who, as part of a study of the internationalization of the automotive industry, developed typologies of automotive-producing nations based on several criteria, including production costs and market size. This work provided a useful starting point for future studies of the economic geography of the automotive industry that eventually relied on GVC or GPN theory. However, the typologies developed by Sturgeon and Florida fail to capture the complexities and fluidities that accompanied the rapid evolution of the automotive industry in the twenty-first century. For example, their "integrated periphery" typology includes both Mexico and Canada, both of which were conceived of as low-cost suppliers integrated primarily with the U.S. automotive industry. Yet such typologies became tenuous at best in the years following their study, as Canada's per capita market for vehicles and unit labour costs began to resemble those of the United States much more than Mexico (whether they ever truly did resemble Mexico's is another question – more on this below).

A subsequent study by Sturgeon et al.[8] engages with GPN theory, although its explicit conceptual focus is on GVCs. Yet this study is notable for several reasons. First and foremost, it recognizes the usefulness of GPNs as a tool to "systemize observed patterns in global industries."[9] While the authors ultimately settle on GVCs as their preferred conceptual tool for their study, largely because of its simplicity and focus on the buyer-driven and lead firm-centric nature of the automotive industry, they do not dismiss the value of GPNs. Second, they recognize the strength and resilience of the highly regionalized production that is common throughout the global automotive industry, and far less common in many other producer-driven industries. This, as Klier and Rubenstein[10] note, became even more evident following the recession of 2008–9, and is testament to the strength of both territorial and aterritorial factors in shaping automotive production networks and to the need of a framework that can properly conceptualize these complex factors. Third, they recognize that the eventual form of the industry where it touches down in specific national contexts is at once similar in some ways to those of other national contexts and varies significantly as a result of territorial forces. This is in many ways similar to Peck and Theodore's[11] notion of variegation.

The work of Sturgeon et al.[12] is extremely useful. However, its emphasis on GVCs over GPNs suffers from one significant flaw for which the authors cannot be faulted: timing. This piece was published on the eve of the recession of 2008–9, which significantly affected the organization and governance of the global automotive industry. Since the recession, the power of national and subnational governments has increased vis-à-vis that of the global automakers, through both their direct involvement in providing assistance as several automakers emerged from bankruptcy in 2009. to more sustained public policies designed to support the automotive industry across major automotive-producing nations. It is largely for this reason that this chapter draws upon GPNs as the primary conceptual tool to situate and explain the evolution of a nationally bounded but internationally integrated industry. This is similar to the work of Rutherford and Holmes,[13] who argue for a more explicit recognition of the manner in which nationally bounded industries are integrated into GPNs in studies of industrial organization and workplace governance. Rutherford and Holmes's work is consistent with the conceptual emphasis of this chapter: that GPN theory is particularly useful in providing a framework to reconcile that territorial and aterritorial forces that shaped the organization, politics, and geographies of Canada's automotive industry.

Canada's Automotive Industry

This section examines the evolution of Canada's automotive industry from the 1950s to the present. In so doing, and in addition to broader developments in the organization and geography of the global automotive industry, it emphasizes the importance of trade agreements and public policies in shaping the automotive industry in Canada. It is organized into three subsections. The first examines the growth of Canada's automotive industry between the 1960s and the late 1990s. The second examines the contraction of the industry as a result of diminished competitive advantages between the early 2000s and the recession of 2008–9. The third examines the limited recovery of the industry following the recession and the attempts of policymakers and industry stakeholders to sustain production in Canada in an increasingly challenging competitive environment.

Growth and Competitive Advantages: 1960s–1990s

Following the Second World War, Canada's automotive industry was dominated by the branch plants of U.S.-based automakers. As a result of high tariffs on imported vehicles and parts, these automakers produced vehicles in small batches primarily for the Canadian market.

Most vehicles sold in Canada were built in Canada, and opportunities to realize scale economies were limited. As a result, and following the recession of the late 1950s, Canada's automotive industry entered a period of crisis marked by falling employment and a decrease in the Canadian production-to-sales ratio.[14] In response, Canada and the United States implemented the Canada–United States Automotive Products Agreements (better known as the Auto Pact) in 1965. The Auto Pact provided duty-free trade in vehicles and vehicle parts so long as U.S.-based automakers assembled one vehicle in Canada for every one they sold and met domestic value-added requirements. It had the effect of integrating the automotive industries of Canada and the United States (most of which was located in Ontario and the Great Lakes states) and allowed Canadian assembly plants to achieve economies of scale. Through the combination of the threat of punishing tariffs on non-compliant companies and lower Canadian labour costs – themselves related to the implementation of publicly funded health care in Canada in the 1960s[15] – GM, Ford, Chrysler, and AMC made significant investments in Canadian manufacturing facilities. This resulted in increased vehicle production (figure 16.1) for over a decade and an improved production-to-sales ratio in the years following the implementation of the Auto Pact.

Canada's automotive industry faced challenges in the late 1970s and early 1980s. By the late 1970s, U.S.-owned automakers were building vehicles in sufficient volume to satisfy their Auto Pact obligations, but had made few significant investments in new capital since the mid-1960s. Meanwhile, the oil crises caused gasoline prices to rise, prompting a market shift to smaller vehicles, a segment in which the U.S.-owned automakers were weak. Automakers were further challenged by the severe recession of the early 1980s. Indeed, North American governments, including Canada and Ontario, provided financial support to Chrysler as it verged upon bankruptcy in 1979.[16] Through all of this, automakers without production facilities in North America, particularly Japanese firms specializing in more fuel-efficient vehicles, increased market penetration. The combination of these factors led to widespread job losses among U.S.-based automakers with operations in Canada in the late 1970s and early 1980s.

Canadian policymakers responded to these shocks by developing tools to encourage investments by non-U.S. automakers, particularly those headquartered in Japan. These included voluntary restraints on the export of vehicles from Japan and tariff waivers and duty drawbacks for non-U.S. automakers that built assembly plants in Canada.

Figure 16.1. Canadian Vehicle Production, 1960–2017

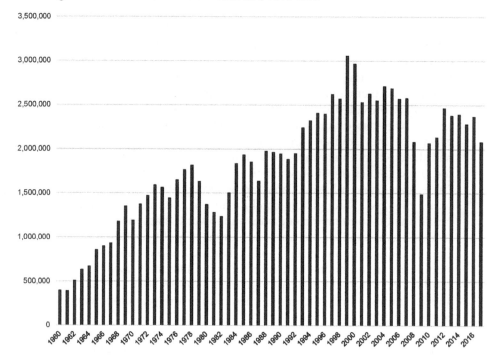

They also began to offer incentives for greenfield investments (notably to Ford, who built a new engine plant in Essex County in 1978), and, in one instance, to support capital expenditures in existing facilities (in the form of a low-interest loan to GM to support capital expenditures at the firm's Ste. Thérèse, Quebec, assembly plant). Beyond that, policymakers and industry actors deployed a concentrated campaign of aggressive and persistent overtures to automakers headquartered outside of the United States.[17]

The response to these policy overtures was swift and considerable. Between 1984 and 1986, five automakers (Toyota, Honda, Suzuki, Hyundai, and AMC-Renault) announced intentions to build full-scale vehicle assembly plants in Canada. The entrance of non-U.S. automakers in the 1980s dramatically changed the profile of Canada's automotive industry. The result was that automotive-related production, employment, and GDP increased considerably during the latter half of the 1980s. This trajectory of growth continued during the 1990s as new entrants increased their production capacity (notably as a result of Toyota and Honda building new assembly plants in the late 1990s).

In 1999, Canadian vehicle production exceeded three million units annually. Only the United States, Japan, Germany, and France – all of which have multiple "homegrown" automakers – built more cars than Canada in the last year of the twentieth century.[18] While Canada's proximity to the United States and favourable exchange rates aided growth and renewal, the growth was due in larger part to the ability of policymakers to develop innovative responses to shocks and disruptions at two critical junctures in the development of Ontario's automotive industry.

Contraction and Recession: 2001–2011

Canadian vehicle production peaked in 1999, when over 3 million units were built at fifteen assembly plants (all but one of which were located in Ontario). Vehicle production fell shortly thereafter and averaged just over 2.6 million between 2000 and 2007 (figure 16.1), by which time Canada had become the world's eighth-largest vehicle-producing nation. This was the result of several factors. Notably, the WTO struck down the Auto Pact in 2001, ruling that it favoured U.S.-based automakers at the expense of Toyota, Honda, and Suzuki. During the years leading up to this decision, Canadian policymakers, who relied on other competitive advantages and had not used Auto Pact safeguards to wrest investments from U.S.-based automakers for over two decades, dismissed the potential long-term effects of such a ruling. However, the impacts of the WTO's ruling were almost immediate. GM, Ford, and DaimlerChrysler each closed a Canadian assembly plant between 2002 and 2004 (including GM's Ste. Thérèse, Quebec, assembly plant, the last Canadian automotive assembly plant outside of Ontario).

In the post-2000 period, Canada's competitive advantages in healthcare costs and currency exchange rates diminished vis-à-vis the United States. This occurred in a relatively short time. Sturgeon and Florida's[19] study ruminated on the "diversion" of automotive investment away from the United States, Germany, France, and United Kingdom and towards low-wage nations like Canada, Mexico, and Spain throughout the 1980s and 1990s. That Canada was considered a low-wage competitor for investment akin to Mexico less than twenty years ago is not quite accurate. Wages of (at that point, mostly unionized) automotive assembly workers in Canada have always been higher than those in Mexico, and have long provided the means for a relatively high standard of living. This is perhaps not surprising, considering Canada is a prosperous nation. Yet there is truth to the notion that labour costs in Canadian automotive assembly plants were highly competitive relative to those in assembly plants in bordering states such as Michigan, Ohio, and Indiana. However, and fuelled largely by inflated commodity prices,[20] the value of the Canadian dollar increased steadily throughout the 2000s, achieving or exceeding parity with the U.S. dollar on several occasions between 2007 and 2010. Canada's advantage in health-care costs also eroded in the latter part of the 2000s, as GM, Ford, Chrysler, and the union representing autoworkers in the United States (the UAW) negotiated collective agreements that restructured and largely removed automakers' responsibilities for retiree health-care costs.[21]

Two other phenomena further diminished Canada's competitive advantages. The first was a result of what Mordue and Sweeney[22] refer to as the "commoditization" of automotive assembly. This commoditization refers to the increasing ubiquity of automotive manufacturing technologies and systems of production organization once thought to be the preserve of advanced industrial economies. These technologies and systems have diffused across high- and low-wage jurisdictions over the past two decades, narrowing cross-national differences in manufacturing capabilities and production quality. As a result, investment into low-wage regions increased substantially since 2000. In North America, this has meant more capital investment in the southern United States and in Mexico. In the context of the EU, this has meant more capital investment in East-Central Europe, Turkey, and North Africa. In Asia, this has meant more capital investment in Vietnam, Thailand, and Malaysia. This has had a profound impact on where vehicles are produced throughout the world, as evidenced by a dramatic shift in the location of vehicle production.

As a result, Canada's automotive industry shifted from a trajectory of growth to one of decline between 2001 and 2007. Its competitive advantages eroded and its then-largest automakers closed several assembly plants. Annual vehicle production fell by over half a million units. However, there was some optimism. With financial support from Canadian governments, Toyota announced the construction of a new assembly plant in 2005 (the plant came online in 2008, and is the last greenfield assembly plant in Canada to do so).[23] GM, Ford, and DaimlerChrysler also announced large investments in their remaining Canadian assembly plants. Each company received substantial financial contributions from Canadian governments for these investments as a means to secure an automotive manufacturing base. While Canadian governments had provided incentives for new production facilities since the late 1970s,[24] this marked only the second time that they provided automakers with incentives to keep plants open, and the first time that the Ontario government participated in such a program.[25] The practice of providing financial support for necessary periodic investments in existing production facilities remains a pillar of Canada's automotive policy.

This optimism was short-lived, as the recession of 2008–9 severely affected Canada's automotive industry. GM closed its Oshawa pickup truck assembly plant, its Windsor transmission plant, and a parts manufacturing plant in St. Catharines. Ford announced the closure of its St. Thomas assembly plant, although that plant would continue to produce vehicles until 2012. Over the same time, more than 200 independent supplier plants closed, while many others reduced production and employment. This led to over 50,000 jobs lost in Canada's automotive industry.[26] Moreover, union density and average nominal earnings decreased, leading to diminishing job quality.[27] In addition, Canada's automotive trade balance shifted from a surplus to a deficit beginning in 2007.[28] This was the result of fewer vehicle shipments to the United States (although Canada maintains an overall trade surplus with the United States) and an increase in vehicle and automotive parts imports from the United States, Mexico, the EU, and Asia. In short, Canada's position within international automotive industry production networks changed considerably during the 2000s as a result of diminished competitive advantages, large investments by automakers in lower-wage jurisdictions, and widespread restructuring brought on by the recession of 2008–9.

Recovery and Mounting Challenges: 2012–Present

Canada's automotive industry underwent modest recovery between 2012 and 2016. While production and employment would never completely rebound to the levels of the late 1990s, annual vehicle production averaged over 2.3 million units and employment increased from approximately 121,000 to 139,000.[29] Although automakers made no investments in new production facilities, most invested in their existing facilities with the support of government incentives. Over this time, FCA and Toyota emerged as Canada's largest automakers, overtaking

Figure 16.2. Canadian Manufacturing Employment by Automaker, 2000–2017

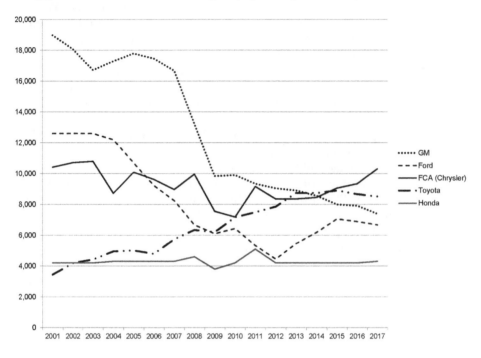

GM and Ford (figure 16.2). The proportion of vehicles produced by Japanese automakers also increased significantly over this time; Toyota and Honda currently account for nearly half of all vehicle production in Canada (figure 16.3).[30]

The modest recovery of Canada's automotive industry between 2012 and 2016 was certainly welcome news for those whose livelihoods depend on vehicle assembly and automotive parts production. However, Canada continues to face challenges maintaining automotive industry production and employment. While Canada's automotive industry has always been concentrated in southern Ontario, the economic benefits were once more dispersed. Today, employment and production are increasingly concentrated in specific communities *within* southern Ontario (e.g., Windsor-Essex, Waterloo Region, Oxford County, Wellington County).[31] This concentration is increasingly evident following the scheduled closure of GM's Oshawa, Ontario, assembly complex in the fall of 2019. The automotive industry, by extension, is therefore less relevant

Figure 16.3. U.S.- and Japanese-Based Vehicle Production in Canada,
2000–2017

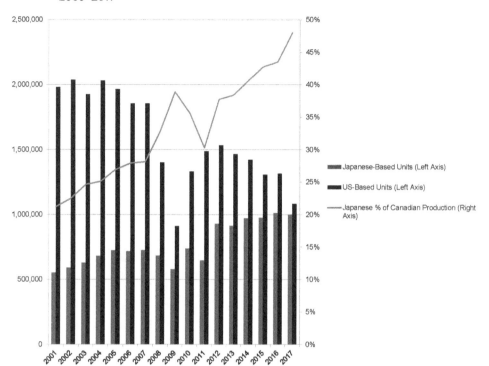

economically and politically to those living outside of southwestern
Ontario. Many indicators of job quality (e.g., hourly wages, union
density, job permanence) continue to diminish in relative and nominal terms.[32] Capital expenditure is far lower than it was prior to the
recession (figure 16.4). This raises concerns regarding productivity
rates, which must remain high if Canada is to compete with lower-cost jurisdictions.

Moreover, Canadian governments have focused on incentivizing
investments in automotive R&D as a means to increase employment
and investment. While there are some examples of success (e.g., GM
invested in a software development facility in Markham, Ontario),
total Canadian automotive R&D spending and employment remains
relatively low (figure 16.5), there are no proven linkages between R&D
spending and manufacturing investment or employment, and there
have been no major changes in public policies designed to encourage
investments in automotive R&D.

Figure 16.4. Automotive Capital Expenditures (Millions), 2002–2017

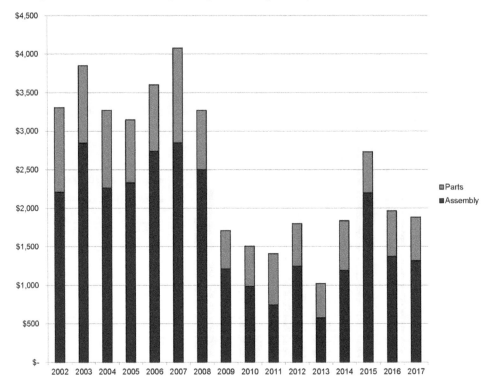

Concerns over the quantity and quality of employment, capital, and R&D expenditures, and the fate of individual manufacturing facilities are important. However, these issues are lower on the current priority list than what will likely determine the fate of Canada's automotive industry: international trade. Canada's automotive industry is export-oriented and relies heavily on the duty-free access to the U.S. market provided by the North American Free Trade Agreement (NAFTA). While the mounting trade deficit[33] and the potential impacts of the CPTPP[34] are of concern to Canadian policymakers and industry stakeholders, they pale in comparison to the challenges that Canada faced when renegotiating a trade agreement with the United States and Mexico, which included the threat of a potentially devastating 25 per cent tariff on vehicle imports. Industry observers are generally optimistic that USMCA will provide a return to normalcy in automotive production and trade within North America, if with higher border-related transaction costs.

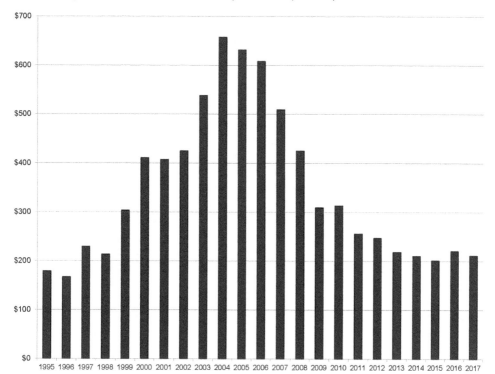

Figure 16.5. Automotive R&D Expenditures (Millions), 1995–2017

Discussion and Future Prospects for Canada's Automotive Industry

By the end of the twentieth century, Canada's automotive industry had emerged as the world's fifth-largest, despite the absence of a domestic automaker. This was the result of a combination of territorial and aterritorial factors, including proximity to the United States, automakers' hesitance to locate high-value export-oriented vehicle assembly plants outside of countries with advanced industrial economies and lucrative markets, favourable trade agreements, and innovative public policies that encouraged investments and helped develop competitive advantages in labour costs and quality. Yet in the first decade of the twenty-first century, these factors changed considerably, which led to contraction in Canada's automotive industry and growing concern over its future prospects. This is the result of automotive investments in lower-cost jurisdictions and the diffusion of production technologies and systems that were once the purview of wealthier nations to those jurisdictions, changes to international trade agreements (partly in response to the ascendancy of the automotive industry in lower-cost jurisdictions), and the inability of Canadian governments to develop new public policy tools to help them overcome diminished competitive advantages or develop new competitive advantages necessary to attract and retain automotive investment.

One reason that the Canadian government has struggled to develop and implement new public policies to encourage investments and sustain or grow production and employment is because of uncertainty regarding the political and economic value of the automotive industry. As noted, investments in automotive manufacturing are valuable to governments as the result of the economic benefits (e.g., jobs, tax revenues) associated with direct employment and with upstream and downstream activities related to vehicle production. Moreover, and as Helper et al.[35] note, manufacturing industries like automotive are especially valuable in that they provide high-wage jobs for workers who would otherwise earn the lowest wages and are a significant source of commercial innovation and R&D.

For decades, Canada's automotive industry certainly provided the former. Automotive manufacturing employed a large proportion of the labour force that had relatively low levels of education. For Canadian governments, this was doubly beneficial. Not only was there an industry willing to employ a significant proportion of the population, that industry did not require Canadian governments to make significant investments in those workers' education. Autoworkers simply went to work, earned a decent wage, and began paying employment (and presumably consumption) taxes at a relatively young age. Upon retirement they received a competitive pension and continued to pay taxes. Everybody won. As for the former, Canada's automotive industry depended on foreign automakers who invested relatively little in commercial R&D. Even as policy priorities shifted, any investments in automotive R&D were hard-won.

This raises important questions: What if automotive industry jobs continue to diminish in quality? And what if the automotive industry continues to invest relatively little in commercial innovation and R&D? Will Canadian governments continue to provide significant investment incentives in order to secure investment and maintain lower-quality employment? Or will they use their limited resources to incentivize investments from other industries or simply direct them to other portfolios? While the economic benefits that the automotive industry provides continue to justify the cost of government-provided investment incentives, industry stakeholders and policymakers should at the very least be aware that they may reach a point where they can no longer justify these costs if job quality continues to diminish and if investment in commercial innovation and R&D remains low. What is unclear is how much longer this rent-seeking behaviour will make economic sense if Canada's automotive industry remains on its current trajectory. It also remains unclear what will happen to the over 100,000 people employed in the automotive industry if we reach this point.

Recently, concern over international trade agreements emerged as the largest issue facing Canada's automotive industry. Had the United States replaced NAFTA with a tariff wall, the effects on Canada's automotive industry would be devastating. The homogeneity of the industry in Canada has done little to assuage the challenges created by new (or old) international trade agreements. In fact, it was Japanese-based automakers, frustrated with the preferential treatment that U.S.-based automakers received under the Auto Pact, who led to the WTO ruling that struck it down in 2001. The three U.S.- and two Japanese-based automakers with operations in Canada have since taken different positions on other international trade agreements with South Korea and the Pacific Rim (i.e., the CPTPP). U.S.-based automakers tend not to favour free trade with Asian nations and argue that it will hurt Canada's automotive industry, while Japanese-based automakers favour the agreement and argue that any negative effects will be minimal. They also note that one good turn may beget another, and that fostering good trade relations with Japan is more likely to encourage investments in Canada by Japanese-based firms than the opposite. Unifor (formerly the CAW), the union representing the majority of unionized Canadian autoworkers, adds another legitimate voice to the mix, while at the same time diminishing the ability to develop a unified voice for Canada's automotive industry. Moreover, and where once Unifor's predecessor, the CAW, could purport to speak on behalf of at least the majority of Canadian autoworkers, they can no longer do so as automotive industry union density continues to decrease.[36] That said, all three stakeholder groups likely agree that amicable trading relationships within North America will remain critical to the success of Canada's automotive industry.

Conclusion

This chapter has examined the confluence of territorial and aterritorial factors that have shaped Canada's automotive industry since the 1960s. It concludes by noting that where Canada's role in international automotive production networks was well-defined for most of the latter half of the twentieth century, it has become less well-defined since the early 2000s. While Canada continues to build a substantial number of vehicles primarily for export to the United States, it has lost its position as North America's lower-cost automotive-producing jurisdiction. Since this time, Canadian policymakers and industry stakeholders have struggled to identify the competitive advantages that remain and leverage them to encourage or sustain further investments. Such investments are critical to the future of the industry. Moreover, uncertainty regarding the impact and form of international trade agreements upon which Canada's automotive industry depends looms large. At the time of writing, USMCA has been ratified but implementation processes and capacities remain works in progress, with continued potential for disruption. We can only assume that time will tell regarding the single most important factor in the economic sustainability of Canada's automotive industry.

NOTES

1 Tomasz Rachwal and Krzysztof Wiedermann, "Multiplier Effects in Regional Development: The Case of the Motor Vehicle Industry in Silesian Voivodeship (Poland)," *Quaestiones Geographicae* 27 (2008): 67–80; Lei Zhang, Henry Kinnucan, and Jing Gao, "The Economic Contribution of Alabama's Automotive Industry to Its Regional Economy: Evidence from a Computable General Equilibrium Analysis," *Economic Development Quarterly* 30, no. 4 (2016): 295–315.

2 Neil M. Coe and Henry Wai-chung Yeung, *Global Production Networks: Theorizing Economic Development in an Interconnected World* (London: Oxford University Press, 2015), 1–2.

3 Neil M. Coe, Peter Dicken, and Martin Hess, "Global Production Networks: Realizing the Potential," *Journal of Economic Geography* 8, no. 3 (2008): 274.

4 Neil M. Coe, Martin Hess, Henry Wai-chung Yeung, Peter Dicken, and Jeffrey Henderson, "Globalizing Regional Development: A Global Production Networks Perspective," *Transactions of the Institute of British Geographers* 29, no. 4 (2004): 271–95; Neil M. Coe and Martin Hess, "Global Production Networks, Labour, and Development," *Geoforum* 44, no. 1 (2013): 4–9; Jeffrey Henderson, Peter Dicken, Martin Hess, Neil M. Coe, and Henry Wai-chung Yeung, "Global Production Networks and the Analysis of Economic Development," *Review of International Political Economy* 9, no. 3 (2002): 436–64.

5 Coe and Yeung, *Global Production Networks*.

6 See, for example, Timothy Sturgeon, Johannes Van Biesebrock, and Gary Gereffi, "Value Chains, Networks, and Clusters: Reframing the Global Automotive Industry," *Journal of Economic Geography* 8, no. 3 (2008): 297–321; Tod D. Rutherford and John Holmes, "Manufacturing Resiliency: Economic Restructuring and Automotive Manufacturing in the Great Lakes Region," *Cambridge Journal of Regions, Economy and Society* 7, no. 3 (2014): 359–78.

7 Timothy J. Sturgeon and Richard Florida, *Globalization and Jobs in the Automotive Industry: Final Report to the Alfred P. Sloan Foundation* (Cambridge, MA: International Motor Vehicle Program, Center for Technology, Policy, and Industrial Development, Massachusetts Institute of Technology, 2000).

8 Sturgeon, Van Biesebrock, and Gereffi, "Value Chains, Networks and Clusters."

9 Sturgeon, Van Biesebrock, and Gereffi, "Value Chains, Networks and Clusters," 6.

10 Thomas Klier and James M. Rubinstein, "Restructuring of the U.S. Auto Industry in the 2008–2009 Recession," *Economic Development Quarterly* 27, no. 2 (2013): 144–59.

11 Jamie Peck and Nik Theodore, "Variegated Capitalism," *Progress in Human Geography* 31, no. 6 (2007): 731–72.

12 Sturgeon, Van Biesebrock, and Gereffi, "Value Chains, Networks and Clusters."

13 Rutherford and Holmes, "Manufacturing Resiliency."

14 Dimitry Anastakis, *Autonomous State: The Struggle for a Canadian Car Industry from OPEC to Free Trade* (Toronto: University of Toronto Press, 2013); Greigory D. Mordue, "Doors Closed and Opportunities Missed: Lessons from Failed Automotive Investment Attraction in Canada in the 1980s," *Canadian Public Policy* 43, no. S1 (2017): S43–S56.

15 See Pradeep Kumar, *From Uniformity to Divergence: Industrial Relations in Canada and the United States* (Kingston, ON: Queen's IRC, 1993).

16 Dimitry Anastakis, "Industrial Sunrise? The Chrysler Bailout, the State, and the Re-industrialization of the Canadian Automotive Sector, 1975–1986," *Urban History Review/Revue d'histoire urbaine* 35, no. 2 (2007): 37–50.

17 Greigory Mordue, "Unanticipated Outcomes: Lessons from Canadian Automotive FDI Attraction in the 1980s," *Canadian Public Policy* 36 (2010): S1–S29; Mordue, "Doors Closed and Opportunities Missed."

18 Organisation Internationale des Constructeurs d'Automobiles.

19 Sturgeon and Florida, *Globalization and Jobs in the Automotive Industry*.

20 Jim Stanford, "The Geography of Auto Globalization and the Politics of Auto Bailouts," *Cambridge Journal of Regions, Economy and Society* 3, no. 3 (2010): 383–405.

21 Harry C. Katz, John Paul MacDuffie, and Frits K. Pil, "Crisis and Recovery in the U.S. Auto Industry: Tumultuous Times for a Collective Bargaining

Pacesetter," in *Collective Bargaining under Duress: Case Studies of Major North American Industries*, ed. Howard R. Stanger, Paul F. Clark, and Ann C. Frost, 45–80 (Ithaca, NY: Cornell University Press, 2013).

22 Gregory Mordue and Brendan Sweeney, "The Commoditisation of Automotive Assembly: Canada as a Cautionary Tale," *International Journal of Automotive Technology and Management* 17, no. 2 (2017): 169–89.

23 See Charlotte Yates and Wayne Lewchuk, "What Shapes Automotive Investment Decisions in a Contemporary Global Economy?" *Canadian Public Policy* 43, no. S1 (2017): S16–S29.

24 See Anastakis, "Industrial Sunrise?"

25 See Mordue and Sweeney, "Commoditisation of the Automotive Assembly."

26 Brendan Sweeney and Gregory Mordue, "The Restructuring of Canada's Automotive Industry, 2005–2014," *Canadian Public Policy* 43, no. S1 (2017): S16–S29.

27 Shannon Miller, "Job Quality and Labour Market Demographics in Canada's Automotive Industry, 2001–2017" (MA thesis, McMaster University, 2018).

28 John Holmes, "Whatever Happened to Canada's Automotive Trade Surplus? A Preliminary Note." Automotive Policy Research Centre Research Briefs (Hamilton, ON: March 2015).

29 Brendan Sweeney, "A Profile of the Automotive Manufacturing Industry in Canada, 2012–2016," *Automotive Policy Research Centre Research Briefs* (Hamilton, ON: April 2017).

30 See Gregory Mordue and Brendan Sweeney, "The Economic Contributions of the Japanese-Brand Automotive Industry to the Canadian Economy, 2001–2016" (report prepared for the Japan Automobile Manufacturer's Association of Canada, 2017).

31 Sweeney, "Profile of the Automotive Manufacturing Industry."

32 Miller, "Job Quality and Labour Market Demographics."

33 See Holmes, "Whatever Happened to Canada's Automotive Trade Surplus?"

34 See Jeffrey Carey and John Holmes, "What Does the Trans-Pacific Partnership Agreement Portend for the Canadian Automotive Industry?" *Canadian Public Policy* 43, no. S1 (2017): S30–S42; Brendan Sweeney and John Holmes, "Renegotiating NAFTA: What's at Stake for Canada's Automotive Industry?" Centre for International Governance Innovation Research Briefs (Waterloo, ON: 2017).

35 Susan Helper, Timothy Krueger, and Howard Wial, "Why Does Manufacturing Matter? Which Manufacturing Matters?" *Metropolitan Policy Program Paper* (2011): 1–53.

36 Sweeney and Mordue, "Restructuring of Canada's Automotive Industry."

17 Canadian Energy in North America and Beyond: Between an Economic Rock and a Progressive Hard Place

MONICA GATTINGER

"In order to get [the Trans Mountain Pipeline Expansion project] done, we need to deal with the political uncertainty, and the only way, in our estimation, that that can be done is through exerting our jurisdiction by purchasing the project." So said Canadian Finance Minister Bill Morneau on 29 May 2018, just two days before the 31 May deadline laid down by proponent Kinder Morgan for the government to provide reassurance that the pipeline could be built. That the Government of Canada would become the owner of a pipeline was virtually unthinkable a few short years ago. Canada has a market-based system for the transportation of oil and gas after all – regulated yes, but ultimately private-sector investment driven.

The Trudeau Liberal government's purchase of the Trans Mountain pipeline – a project to secure market access for Canadian energy resources to international markets beyond North America – underscores vividly that all has not been well in the market, political, and policy climates for Canadian energy. This has been so for quite some time. There has been much uncertainty and unpredictability in energy markets, politics, and policy. Companies are taking notice and rethinking their investment choices and strategies – including pulling out of energy investment in Canada, as Kinder Morgan has done with the Trans Mountain pipeline. All of this is having a profound effect on Canadian energy relations in North America.

Canadian governments are increasingly between an economic rock and a progressive hard place in the field of energy. Economically, the transformation of North American oil and gas markets in response to rapid production increases in light tight oil and shale gas in the United States means Canada needs to access new international markets for its oil and gas resources. In this book's assessment of shifting borders as factors in Canada's international policy relations, energy "borders" for Canada are no longer just those along the forty-ninth parallel, but those on the country's East and West Coasts for export to international markets beyond North America as well. This is a fundamental change for a country that has historically sold virtually all of its energy exports to the United States. This is the "economic rock" facing Canadian energy.

But Canadian energy also faces a progressive "hard place." The rise of environmental nongovernmental organization (ENGO), local community, and Indigenous opposition to energy projects (particularly but not exclusively fossil fuels) is transforming the investment climate for energy in North America and exerting tremendous pressure on policymakers to redraw policy boundaries between competing energy imperatives (market, environment, and security). In this context, the investment climate is becoming increasingly uncertain, unpredictable, and unstable. Having a regulatory permit in hand to construct a project is far less likely to mean that a project will be built. Political, legal, and social opposition to projects is proving increasingly effective at delaying construction and driving up costs – even shutting down projects entirely.

As a result, policy borders are also on the move in Canada's energy relations in North America: boundaries between energy policy imperatives are in flux. Governments are struggling with where, exactly, to draw the lines between economic, security, and environmental/social imperatives and what decision-making processes to use for energy projects. The number and range of concerns triggered by energy development seem ever on the rise – everything from climate change to Indigenous culture and reconciliation, to gender impacts. Community and citizen expectations for engagement and consultation in project decision-making also seem on the upswing.

These twin market (economic) and environment/social acceptance (progressive) transformations have generated uncertainty, unpredictability, and energy project delays for investors. They have also confronted policymakers with extraordinarily complex and challenging decisions: how to develop energy resources in ways that find workable balance points between market, environmental, security, and social acceptance imperatives, and how to get energy projects built in a socially acceptable and timely fashion in the twenty-first century. The impact of the COVID-19 global health pandemic on these market and environmental/social issues is still uncertain. Economically, oil prices plummeted sharply in early 2020 in response to demand declines and a price war between Saudi Arabia and Russia. Politically, it is not yet clear whether the far-reaching macro- and microeconomic impacts of the virus will reshuffle policy boundaries towards economic imperatives and away from progressive concerns.

For Canada, addressing all of these challenges effectively is far more than an academic exercise. The economic potential of the country's energy resources is rich and could contribute substantially to domestic, North American, and international economic development, competitiveness, security, and reduction of greenhouse gas emissions. But without public and investor confidence in the country's decision-making systems for energy development, i.e., finding a way between the economic rock and progressive hard place, this potential is unlikely to be realized.

This chapter addresses this issue, analysing its impact on borders for Canadian energy in North America and beyond. It begins with a brief overview of contemporary energy policymaking, highlighting the increasingly complex and multifaceted nature of energy policy. Governments must find workable balance points between market, environmental, security, and social acceptance imperatives. It then expands on the market and environment/social acceptance transformations of recent years. It explores the shifts in North American energy markets, the "economic rock" that has propelled the first transition in energy borders: from bilateral Canada-U.S. to multilateral Canada–international borders in energy trade, and the corresponding need for Canadian energy to secure access to international markets. The following section expands on the environmental and broader social acceptance issues, the "progressive hard place" propelling the second transition in energy borders: shifting policy borders between market, environmental, and social acceptance imperatives, and growing demands for consultation and engagement in energy project decision-making.

The final section assesses Canada's performance on the energy file, examining the Harper Conservative (2006–15) and Trudeau Liberal (2015–present) governments' efforts to diversify markets for Canadian energy resources beyond North America and to address social and environmental concerns about energy. The chapter argues that neither government has found a workable durable balance between competing energy imperatives. The challenges of doing so are likely to grow substantially following the election of a minority government in the fall of 2019 and the far-reaching economic, social, environmental and political impacts of COVID-19. If Canada cannot find its way between the economic rock and progressive hard place, it will fail to capitalize on the energy opportunities before it.

Contemporary Energy Policymaking: Governments Are Making a MESS of Energy Policy[1]

Energy policymaking is becoming ever-more challenging, multifaceted, and complex. Governments must address four increasingly demanding energy imperatives that have layered on one another: market, environment, security, and social acceptance. Addressing all four effectively at the same time is extraordinarily difficult.

The first imperative is energy markets. In the 1970s, Western industrialized countries began to focus on liberalizing energy markets to strengthen their efficiency and competitiveness. In Canada, oil and gas liberalization began in the 1980s. It included price deregulation, increasing competition in the upstream and downstream energy systems, trade liberalization with the Canada–U.S. Free Trade Agreement, and unbundling functions within energy firms to create open, non-discriminatory access for other companies to their services and facilities.[2] Liberalization of the electricity sector began in the 1990s. Many provinces introduced competition into electricity generation and wholesale/retail sales.[3] Similar moves were undertaken in the United States,[4] while in Mexico, liberalization of the energy sector has been much slower, with far more gradual "opening" of the oil, gas, and electricity sectors to competition, private sector involvement, and foreign investment.[5]

The second imperative, environment, emerged in the 1980s, and refers to the environmental impacts of developing energy resources (exploration, production, transportation/transmission, and consumption). The environmental imperative is broad. While recent years have seen global climate change ascend to the top of many environmental policy agendas, this imperative also comprises local impacts on land (e.g., biodiversity and ecosystem health), human health (e.g., toxicology), and water (e.g., water quality and diversion). Given that environmental impacts often extend beyond political borders, governments have addressed many of these issues through international agreements like the United Nations Framework Convention on Climate Change, or the Canada–United States Air Quality Agreement. In Canada, all levels of government make policy to address the environmental impacts of energy, including, for example, the federal carbon pricing plan (the Pan-Canadian Framework on Clean Growth and Climate Change, discussed later in the chapter) and provincial carbon pricing instruments in BC, Alberta, Quebec, and Ontario (Alberta and Ontario repealed their economy-wide carbon taxes after the election of Conservative governments); federal species-at-risk legislation, and provincial water quality and land use and remediation regulations for oil, gas, and electricity production/generation.

Energy security is the third imperative. As a net energy exporter, for Canada, security is understood less in terms of energy supply disruptions (although eastern regions depend on foreign oil imports), and more in terms of consumer, commercial, or industrial vulnerability to price spikes. In the aftermath of the September 2001 terrorist attacks, energy security also encompasses the physical and cyber-security of critical energy infrastructure like pipelines, nuclear power stations, and refineries, as discussed by Geoffrey Hale in chapter 8. It includes electricity reliability as well – ensuring that the lights come on when people flip the switch. North America has the largest interconnected grid in the world, one increasingly reliant on information and communications technologies (themselves vulnerable to cyber-attack). Electrification of residential, commercial, and industrial energy needs may help to reduce greenhouse gas emissions, but it places new demands on the energy system.

The fourth imperative energy policymakers face is social acceptance – or sometimes lack of acceptance – of energy development. Since the mid-2000s, public opposition to energy projects has become much more frequent, contentious, and protracted. It has also grown in scope and complexity, from opposition based mainly on local concerns (often described as NIMBY, "not in my backyard") to broader concerns of regional, national, or global scope surrounding issues like climate change or Indigenous rights. This latter form of opposition to energy development, discussed further, below, is far more challenging for policymakers, regulators, and industry to address. In some cases, opponents are against energy development "in principle," and conventional ways of addressing concerns (changes in project siting or routing, compensating affected parties, etc.) may be ineffective – even counterproductive. Regardless of the reasons underpinning social acceptance challenges, central to this new imperative are growing demands from civil society to provide meaningful opportunities for engagement and consultation – even real decision-making power – in project decision-making. For Indigenous communities in Canada, many of these demands are enshrined in the constitution and a growing body of case law, although much uncertainty remains over the nature and scope of Indigenous peoples' rights to be consulted and accommodated.

Taken together, these four imperatives – market, environment, security, and social acceptance – are the energy "MESS" that policymakers must attend to in the twenty-first century. Governments must find workable balance points between market, environmental, and security imperatives, in ways that garner social acceptance. This can include explicit consideration of trade-offs and working to align market, environmental, and security imperatives in ways that create virtuous – not vicious – circles.

Analysing Canada's energy relations in North America through the lens of the energy MESS reveals two key transformations underway: one dealing with markets (the economic rock) and the other with environmental and social acceptance (the progressive hard place). Borders for energy trade are on the move: they are shifting from the forty-ninth parallel to Canada's coastlines because of changes in energy markets. Policy borders between market, environmental, and security imperatives are also in flux as a result of changes in energy politics, environmental concerns, and social acceptance challenges. The next two sections address each of these transformations in turn.

The Economic Rock

Market Imperatives: From Canada-U.S. to International Borders for Canadian Exports[6]

For all of Canada's energy-commodity exporting history, the focus of energy trade has been on the Canada-U.S. border: energy resources have been exported to a single market, the United States. Recent years have seen that change dramatically, with increasing interest and efforts to export Canadian energy resources to international markets beyond North America – chiefly Asia. This transition in Canada's energy borders has been propelled by rapid and transformative changes in the North American energy marketplace.

Beginning in the mid- to late 2000s, the "shale revolution" profoundly reshaped North America's energy landscape. The capacity to profitably develop the continent's massive reserves of unconventional oil and gas (shale/tight oil and shale gas) with the technologies of hydraulic fracturing ("fracking") and horizontal drilling have forever altered the oil and gas reserve and production pictures in North America – especially in the United States. As of 2018, the United States possessed 43.8 billion barrels of proved oil reserves, an all-time high and an increase of more than 100 per cent in just ten years (proved reserves were 19.1 billion barrels in 2008).[7] In natural gas, the picture is even more striking: the increase in proved reserves of shale gas skyrocketed almost fifteenfold in just over a decade, from 23.3 trillion cubic feet in 2007 to 342.1 trillion cubic feet in 2018.[8] Proved reserves of all forms of natural gas (conventional and unconventional) in the United States in 2018 were a historic

504.5 trillion cubic feet,[9] more than doubling in just over a decade, thanks to shale gas. Most of the new oil and gas producing regions are in the South (Texas, Louisiana, New Mexico, and Oklahoma), Midwest (principally North Dakota and Montana), and Northeast (Pennsylvania, West Virginia, and Ohio).

For Canada, the changes are less about oil, given the country's long-standing proved reserves of 170 billion barrels, largely in Alberta's oil sands, but light/tight oil resources in Saskatchewan and BC are now also being developed. In gas, meanwhile, the country's potential is substantial: in 2017, the country's proved reserves stood at 69 trillion cubic feet, but marketable resources (gas accessible with more drilling) were estimated at 1,220 trillion cubic feet, and more than two-thirds of that is unconventional.[10] While unconventional oil and gas reserves can be found across the country, the major producing regions are BC, Alberta, and Saskatchewan. Other jurisdictions – Quebec and New Brunswick – have placed moratoria on shale gas development. Canada also has rich reserves of natural gas liquids (propane, butane, ethane, pentanes plus), with production chiefly in BC and Alberta.

The shale revolution is reshaping Canada-U.S. energy relations fundamentally. The United States has gone from being hydrocarbon scarce to hydrocarbon rich, with oil and gas production climbing steeply. Crude oil production increased from 5.2 million barrels per day in 2005 to 12.2 million barrels per day in 2019 – a level well beyond the American oil production peak of 9.5 million barrels per day in the early 1970s.[11] As a result of the shale revolution, the United States has become the largest oil producer in the world, with daily production estimated at 12.3 million barrels per day in the first five months of 2020.[12] The increase in production prompted the U.S. Congress to do something in late 2015 that would have been unthinkable a mere decade ago: lift the ban on exporting crude oil beyond North America. The first exports of American crude oil to international markets beyond North America took place in 2016. While the United States will continue to be a net importer of petroleum products – its domestic requirements exceed 20 million barrels per day[13] – import volumes are predictably on the decline. The country reduced petroleum imports by more than 4.5 million barrels per day between 2005 and 2019 (from 13.7 to 9.1 million barrels per day),[14] with the majority of the decline accounted for by reduced imports from OPEC countries, from 5.6 million barrels per day in 2005 to just 1.6 in 2019.[15]

U.S. natural gas production has climbed from 18.1 trillion cubic feet of dry (consumer grade) natural gas in 2005 to 33.7 trillion cubic feet in 2019.[16] This has had a predictable impact on natural gas trade: the

United States has been a net importer of natural gas for decades, with most imports coming from Canada. Now, the United States exports natural gas in increasing volumes (including to Eastern Canada from shale deposits in the Northeast): gas exports grew sixfold over the last decade, rising from 729 billion cubic feet in 2005 to 4.7 trillion cubic feet in 2019.[17] By 2017, the United States was a net exporter of natural gas – the first time it exported more gas than it imported since the 1950s.[18]

All told, the "shale revolution" calls into question the size and viability of the U.S. appetite for Canadian energy. Indeed, the United States is now consumer *and* competitor for Canada in oil and gas. But Canadian energy requires the infrastructure needed to bring oil and gas resources to markets beyond North America, as discussed below. To date, the main impact of the shale revolution for Canadian energy exports has been on the natural gas sector, where U.S. imports of Canadian natural gas have declined from 3.7 trillion cubic feet in 2005 to 2.7 trillion cubic feet in 2019.[19] Canada has fared better than other gas exporters in this environment though – Canadian imports edged out imports from virtually all other countries, with the proportion of natural gas imports from Canada rising from 85 to 98 per cent between 2005 and 2019.[20]

In petroleum, exports have grown from 2.2 million barrels per day in 2005 to 4.4 in 2019.[21] Canada has again fared relatively well, compared to other countries: Canadian petroleum imports have come to represent a much greater proportion of overall petroleum imports to the United States, rising from 16 per cent of imports in 2005 to almost half (49 per cent) in 2019,[22] edging out OPEC suppliers. But, as noted below, oil prices have dipped (precipitously so in the first part of 2020), and U.S. oil production is slated to continue to grow (assuming prices rebound and remain close to pre-2020 levels), posing real challenges for the future of Canadian oil.

Electricity imports, for their part, have declined in recent years. They grew from 41.5 terawatt hours in 2006 to a peak of 68.5 terawatt hours in 2015, but declined year-over-year to 51.5 terawatt hours in 2018 (the proportion of U.S. electricity imports from Canada declined from 97 to 88 per cent in this same period as Mexican exports to the United States grew).[23] As noted below, future prospects for Canadian electricity exports to the United States may continue to weaken.

The development of unconventional resources has had a predictable impact on oil and gas prices: it has increased volatility and placed strong downward pressure on prices in North America relative to international markets. The price of a barrel of West Texas Intermediate (WTI), the main marker for oil in North America, peaked at about $100 a barrel in 2008, dropped downward to just over $60 in 2009, rebounded to close to $100 for the following four years, and then dropped to about $50 a barrel in 2015, where it hovered until 2018.[24] Prices in 2018 and 2019 fluctuated between $50 and $70, but in early 2020, they dropped to a staggering $20 per barrel in response to the twin impact of COVID-19 on global demand and a price war between Saudi Arabia and Russia on global supply.

With rapidly expanding U.S. production, average annual WTI prices trailed those for a barrel of Brent, the main European marker, with average annual differentials approaching $20 between 2010 and 2014 (daily differences in some periods were far higher; differentials have narrowed since 2014). The discount between WTI and Western Canadian Select (WCS), the main marker for oil sands oil, has grown, reaching almost $40 in December 2013, after which time it levelled off to approximately $10, thereafter increasing sharply in 2018 to $25 or more before Alberta mandated production cuts in early 2019.[25] Price differentials and volatility can be very challenging for oil sands producers, who face a "double discount" when there are differences between WCS and WTI, and between WTI and Brent. Oil sands producers also tend to have a higher break-even point than lighter oil from unconventional formations. In early 2020, the twin impact of the coronavirus and the Saudi-Russia price war was catastrophic for oil sands producers, with WCS crashing to a jaw-dropping $5 or less per barrel. Prices strengthened in the summer as Saudi Arabia, Russia, and other producers instituted production cuts and as countries gradually reopened their economies, but they remained lower than they were in 2018 and 2019 (in August 2020, a barrel of oil sold for just over $40 for WTI and a little more than $30 for WCS; Brent sold for roughly $45).

Price differentials between natural gas in North America and gas in international markets have also grown, with prices diverging first in about 2008, reaching peak differentials between 2011 and 2014. In 2012, the average price of natural gas at the U.S. Henry Hub was under $3 per million BTUs, compared to over $9 in the United Kingdom and more than $16 for Japan LNG.[26] By 2019, differentials had narrowed, but North America still had by far the lowest average annual gas prices: about $2.50 per million BTUs in the United States, compared to roughly $4.50 in the United Kingdom and almost $10 for Japan LNG.[27] Prices plunged to $2 in early 2020, given record production in the United States, and remained deeply depressed worldwide through the summer as the result of demand collapse brought on by COVID-19. The ultimate impact on natural gas production and prices due to changes in the global oil market and the economic impacts of COVID-19 have yet to be understood, but they are likely to translate into reduced production, consumption, and prices in the short term, potentially followed by tightening markets and increased prices in the medium term as production trails a return of demand.[28]

All told, while Canada has come to represent a larger share of U.S. oil and gas imports, import volumes in natural gas are on the decline, Canadian petroleum production potential may outpace U.S. demand, and producers are facing price differentials and grim prices in the short term at least. Even before the price shocks of 2020, the situation appeared poised to become even more challenging. The U.S. Energy Information Administration (EIA) projected the United States would become a net exporter of energy (all forms combined) in 2020 and would remain that way until 2050.[29] The EIA also projected a further decline in natural gas imports from Canada in the decades ahead, along with an increase in U.S. gas exports to Eastern Canada from shale deposits in the Northeast.[30] It also forecasted that electricity generation prices would decrease as a result of fuel switching to cheap natural gas and lower generation costs for renewable power (wind, solar).[31] This could reduce the competitiveness of Canadian electricity in the U.S. market, leading to further export declines. Time will tell how the economic impacts of the coronavirus and the Saudi-Russia price war will affect these projections. In the short term, they have substantially increased pressure on energy companies.

This, in a nutshell, is the economic rock facing energy producers in Canada: they need to look to markets beyond the United States to secure their current and future prosperity. It is no accident, then, that recent years have seen multiple major project proposals (pipelines, liquefied natural gas facilities) to carry Canadian energy to international markets beyond North America to secure new markets and higher prices. The main proposals tabled for oil have sought to bring oil sands oil east to eastern Canadian and export markets (TC Energy's[32] Energy East pipeline and Enbridge's Line 3 replacement and Line 9 reversal), and west to BC's coast for export to Asia (Kinder Morgan's Trans Mountain Pipeline Expansion Project and Enbridge's Northern Gateway pipeline). For gas, multiple LNG export proposals have been tabled in BC. TC Energy's Keystone XL pipeline, while destined for the United States, aims to facilitate Canadian oil producers' access to U.S. refineries on the Gulf Coast. Remarkably, only one of these projects – Enbridge's Line 9 reversal – is in service (the Canadian portion of Enbridge's Line 3 replacement is in service, but the U.S. portion still faces delays). Two have been cancelled (Northern Gateway because the government rejected the proposal, and Energy East because TC Energy pulled the plug). All the others – even those that have received regulatory approval like

Kinder Morgan's Trans Mountain Pipeline Expansion project – have experienced substantial delays. Both Keystone XL and Trans Mountain are moving into construction, but further protests and legal challenges are likely. In all instances, a fundamental reason projects have not gone forward or have been delayed is lack of social acceptance and support – not lack of a business case. Projects have been subject to multiple delays, whether from court cases launched by opponents, lack of clarity over policy and regulatory rules, or intergovernmental disputes over jurisdiction. The result has been companies cancelling projects, ballooning project costs, and, for Trans Mountain, a Canadian government purchasing the project to avoid project cancellation. Keystone XL has followed a similar path: in March 2020, the Alberta government purchased a US$1 billion stake in the project and provided US$4 billion in loan guarantees to enable TC Energy to move forward with construction in the beleaguered market context.

Why have these projects not proven to be "no brainers," as former prime minister Stephen Harper quipped early in his first mandate when referring to the Keystone XL pipeline? The next section turns to this question.

The Progressive Hard Place: Environment and Social Acceptance of Energy Development

Policy Borders between Core Energy Imperatives Are on the Move[33]

In addition to the transition in energy borders brought about by the market transformations described above (the *M* of the energy MESS), Canadian energy is undergoing transitions in energy borders propelled by environmental concerns and social acceptance (or lack thereof) of energy development (the *E* and second *S* of the energy MESS). These two forces are placing tremendous pressure on governments to reshape the scope and relative weighting of the four components of the energy MESS. Environment and social acceptance imperatives are becoming broader (encompassing more considerations, e.g., gender-based or cultural impacts) and more heavily weighted in the balance policymakers strike between the four imperatives of the energy MESS. They are also having an impact on the ways in which government decisions are made on energy projects, propelling much greater emphasis on expanding opportunities for consultation and engagement. The ultimate impact of these changes (if they continue – the U.S. federal government under President Trump and some provincial jurisdictions in Canada are bucking the trend) on energy borders remains to be seen, but early signs are that the effect may be to stall or even reduce energy development (whether renewable or non-renewable) in North America.

How Did This Come About?

The rise of social acceptance (or lack thereof) of energy development is intertwined with the ascendance of environmental imperatives in energy policy and politics. In recent years, numerous high-profile and protracted conflicts over energy development have erupted in Canada. While oil pipelines are the flashpoint for these controversies – think Keystone XL, Northern Gateway, Trans Mountain, Energy East – energy projects of all descriptions (oil and gas, renewable, linear, non-linear) have faced widespread and extensive opposition. Whether wind farms in Ontario, large-scale hydro in British Columbia, or shale gas exploration in New Brunswick or Quebec, conflicts over energy development are flaring up at a seemingly increasing pace. Controversies play out mainly in the regulatory process for individual projects, even though many of the concerns raised by opponents – including climate, reconciliation with Indigenous peoples, or other environmental effects – extend well beyond the remit of individual projects, and, frequently, of energy regulators.

There is no *single* reason why social acceptance of energy development has become so salient, but rather, a *multiplicity of factors*. Many of the factors interact in ways that exacerbate the overall challenge; several extend well beyond the energy decision-making system. What follows lays out the key factors, presenting them as two kinds of issues: horses (that have left the barn) and elephants (in the room). The third animal in this metaphor is "sitting ducks": energy decision-making processes, which have come under increasing stress and strain as a result of policy gaps (the elephants) and social and value changes (the horses). Canada is not alone in facing the public confidence challenge – all Western industrialized democracies, particularly those with large energy resource bases, are confronting this issue. The United States is a case in point. The Keystone XL pipeline continues to face social acceptance challenges, as have projects like Yucca Mountain for the long-term storage of spent nuclear fuel, Quebec Hydro projects to bring hydropower from Canada to the U.S. Northeast, and the Dakota Access pipeline.

Canada's energy decision-making apparatus was built largely in the early post-war period, a time when Canadian energy resources were being developed in large quantities for domestic and international (American) markets. But much has changed. Today's context for energy decision-making would hardly be recognizable to decision-makers from the 1950s or 1960s. In addition to the deregulation and liberalization of energy markets and growing concerns over the environmental impacts of energy discussed above, are extensive, widespread, and permanent social and value changes since the 1950s, along with changes in technology. There is no turning back the clock on these changes – the horse has left the barn.

Social and value change, in conjunction with new information and communications technologies, have had significant impacts on political, economic, and societal governance, including the energy decision-making system. Five key changes stand out. First, public trust in government, industry, and experts has declined across Western industrialized democracies since the 1950s.[34] Successive results of the Edelman Trust Barometer document this change, with the 2017 annual study noting that "trust is in crisis around the world," including trust in government, industry, NGOs, and the media.[35] Second, citizens' deference to authority (elite, government, industry, medical, etc.) has also declined.[36] People are less likely to accept and believe what "the experts" have to say on everything from their health, to the environment, governance, and the economy.

Third, people are becoming increasingly preoccupied with risk, especially human-induced risk, and risk tolerance levels are on the decline.[37] Moreover, different people can have utterly different views on what constitutes meaningful risk. In some instances, this is because understandings of the nature of risk have not risen proportionate to the preoccupation with it; in others, differences in risk perception and risk tolerance are rooted in fundamental value differences.[38]

Fourth, and of tremendous importance to the energy sector, citizens have a greater desire to be involved in public decision-making processes that affect them.[39] They want to have a say in all manner of government decisions from the international, to national, provincial, and local. Finally, social values have become more individualistic than group/community-oriented over the years, with individual or small group interests able to trump community/national interests. There is much greater fragmentation and more visible lack of consensus over what, precisely, constitutes "the national interest" and how best to determine it.

Accompanying these changes are transformational developments in information and communications technologies, particularly the rise of social media. These changes have created unprecedented opportunities for unmediated and instantaneous communication between citizens and citizen groups, enabling rapid widespread mobilization and instantaneous sharing of information (along with misinformation/disinformation).

The impact of these changes can be far-reaching. First, citizens may be less likely to trust that governments make fair, unbiased, balanced decisions – governments can be seen as co-opted by special interest groups, notably industry. Second, people may lack confidence in expert opinion and scientific evidence, giving more weight to evidence from sources they trust (close friends, social media campaigns, celebrities, or NGOs), regardless of their knowledge or expertise than to the "experts." All evidence – from scientific to individual opinion and belief systems – may be perceived as equal and deserving equal weighting in decisions. All told, citizen trust in the source of evidence may be more important than its rigour.[40]

Third, governments are trying to "open up" decision-making processes to respond to demands for citizen involvement, but this can generate real and perceived tensions between participatory democracy (citizen involvement) and representative democracy (elected or appointed officials making decisions). It can also generate tensions between the imperatives of a system based largely on markets and private capital and those of a democratic political system. And in a democracy, there are multiple avenues to try to overturn or influence public decisions (lobbying, campaigns, the courts, etc.).

Fourth, when citizens' preoccupations are centred more on individual/local interests than on national/group interests, appeals to the "national interest" or broader regional/group interests may get less traction or even fall on deaf ears. Finally, perceptions of risk can trump realities of risk and risk mitigation. Citizens and leaders may not consider the impact of rejecting an individual energy project on the larger physical or market energy systems (e.g., rejecting an oil pipeline could result in more crude oil flowing through less economic and environmentally sustainable means of transportation; rejecting a hydro or other renewable power project might stymie efforts to reduce the carbon intensity of the electricity system).

The transformation in the information and communications environment – notably the rise of social media – magnifies and intensifies these tendencies. The internet, Twitter, Facebook, and other social media platforms have created tremendous opportunities for unmediated and instantaneous communication between citizens and citizen groups, with all the promise, prospects, and perils this entails. Unprecedented vistas of easy, widespread communication are constantly unfolding, enabling rapid mobilization of social movements and groups. They also enable the rapid dissemination of partial or poor information – or worse, misinformation or disinformation – and can quickly generate highly polarized and polarizing conflicts, along with wholesale opposition to energy development.

All told, what confronts decision-makers is a new world of energy decision-making, one that is far more complex, interconnected, volatile, and prone to polarization, fragmentation, distrust, and misinformation, and far less controllable. And the possibility of opposition to energy projects of all descriptions – neatly captured in the term "Blockadia"[41] – seems set to become the new normal.

These social, value, and technological changes are driving change in the level and nature of social acceptance of energy development (the second S in the energy MESS), and are interacting with concerns over the environmental impacts of energy. Energy projects are often opposed for reasons stemming from broader questions of public policy (notably around the environment) well beyond the merits or demerits of an individual energy project.[42] Such conflicts frequently play out in the regulatory process for an individual project, which is ill-equipped to address the issues if they lie outside the scope of a regulator's mandate or of the project assessment.

In Canada, policy gaps arise in three key areas: climate change, Indigenous issues, and cumulative effects.[43] On climate change, the absence of adequate forums for and perceptions of meaningful government action on climate over the last couple of decades has resulted in concerns over climate being played out in the regulatory system through opposition to individual projects. Advocacy in this space can be highly polarized and polarizing. It includes sharp targeting of the oil and gas industry itself, notably the oil sands. Exacerbating this challenge is the tendency for governments over the years to continue to make commitments on climate change that cannot practically be met in physical, economic, or political terms. This generates scepticism and a lack of confidence that governments take the issue seriously.

On Indigenous issues, inadequate government movement on reconciliation with Indigenous peoples in Canada can result in energy projects being opposed by Indigenous authorities or community members on the basis of concerns that extend well beyond energy policy, regulation, and development (e.g., rights and title; clean drinking water; social, health, and education issues; the viability of traditional subsistence economies). This policy gap is exacerbated by a lack of clarity and shared understandings of the legal context for Indigenous involvement in energy projects, notably, what court decisions mean for rights, title, and the duty to consult and accommodate and the scope and nature of Indigenous governments' authority. This situation has also generated a polarized and polarizing context, with the terms of debate often framed in the language of "vetoes."

On cumulative effects, the lack of adequate regional planning forums and mechanisms like strategic environmental assessments to address the effects of successive projects in geographic, environmental, social, and temporal terms can likewise generate opposition to individual projects for reasons that extend well beyond the project per se. Public concern can centre on the combined effects of projects or development in regions where regional planning mechanisms are inadequate or entirely absent.

The impacts of these policy gaps have been felt most keenly by the regulatory system and individual project proposals, and, ultimately, have lessened public confidence in energy decision-making. Because regulators cannot address issues beyond their mandates, and individual project proponents are limited by the extent to which they can address these broader issues on their own, public frustration mounts, and confidence in public authorities (policymakers, regulators) and industry (individual companies, entire industry sectors) can weaken.

All told, energy decision-making processes have been sitting ducks in the context of these changes. Policymakers and regulators have been critiqued along all the lines just noted: lack of confidence in the impartiality of evidence and of decision-makers, in the degree of inclusiveness and transparency (or not) of decision-making processes, and in the extent to which larger policy issues are (or are not) considered in decision-making. Opposition and lack of confidence in regulatory processes have raised individual project decision-making to the political level, where leaders have been called upon to override the regulatory process, often generating the worst of all worlds: a riskier, more uncertain, less fact-based process in which both the public and investors have even less confidence.

So how is Canada faring in this context?

Are Canadian Governments Finding Their Way between the Economic Rock and the Progressive Hard Place?

The short answer to this question is no. In the years since these market and environmental/social transformations have been underway (roughly since the mid-2000s), neither the Harper Conservative (2006–15) nor the Trudeau Liberal (2015–present) governments have found a way between the economic rock and progressive hard place. Neither have they been able to strike a workable or durable balance between market, environmental, security, and social acceptance imperatives.

The Harper government over-weighted market imperatives in ways that amplified social opposition to energy development, while the Trudeau government has weighted environmental and social imperatives in ways that make the investment environment even less clear, predictable, and stable. This section analyses each government's actions on Canadian energy, focusing on their approach to the energy MESS in North American context. The conclusion explores what the election of a minority government in Fall 2019 and the macro and microeconomic impacts of COVID-19 and oil price wars are likely to hold for the future.

THE HARPER CONSERVATIVE GOVERNMENT (2006–2015)[44]

The Harper government's policy interventions on energy focused first and foremost on the market imperative of the energy MESS. Securing and maintaining market access for Canadian energy exports was the government's primary objective. Early in its first mandate, Prime Minister Harper touted Canada as an "emerging energy superpower," stating, "We are a stable, reliable producer in a volatile, unpredictable world," and that "industry analysts are recommending Canada as 'possessing the most attractive combination of circumstances for energy investment of any place in the world.'"[45]

Environmental imperatives – particularly climate – took a back seat to market imperatives both at home and abroad. This was made clear early in the government's first mandate when Rona Ambrose, minister of the environment and chair of the United Nations Framework Convention on Climate Change (UNFCCC) meetings in Bonn, surprised delegates and many Canadians when she stated that Canada would not meet its Kyoto GHG reduction targets.[46] She called for the second post-2012 phase of the Kyoto Protocol to use voluntary targets, establish lengthier deadlines, and include exceptions for Canada's resources.[47] This was followed in 2007 by revised Canadian targets for GHG emissions reductions: a 20 per cent reduction in Canada's 2006 levels by 2020. The new target was critiqued for halving the country's original Kyoto commitment of reducing emissions to 6 per cent below 1990 levels over the 2008–12 period.[48] This was followed in turn by the government's 2015 GHG emissions reduction commitment: a 30 per cent cut in emissions from 2005 levels by 2030. This time, the government was critiqued by environmental groups because the objective was less ambitious than that of the United States, because regulations for the oil sands – Canada's fastest growing source of emissions – had yet to be announced and because the government was not on track to meet its prior climate commitments.[49]

Indeed, no comprehensive climate plan was ever announced by the Harper government. Instead, it increasingly distanced itself from putting a price on carbon that applied across the country, whether through a cap-and-trade system (to which the government was committed early in its first mandate) or a carbon tax. On the latter, when Liberal Party leader Stéphane Dion campaigned on a proposal to put in place a carbon tax in the 2008 federal election, Prime Minister Harper came out strongly against it, referring to the proposal as an "insane" idea that would "screw everybody" and comparing it to the National Energy Program of 1980.[50] The Conservatives committed to establishing a cap-and-trade system between 2012 and 2015, but this commitment was reversed in the wake of the global financial crisis and recession, which knocked cap-and-trade off the policy agendas of both the United States and Canada. By 2014, the prime minister said it would be "crazy" to establish climate regulations on the oil and gas sector, and made it clear that his government put the economy over the environment, saying, "We are just a little more frank about [doing it], but that is the approach that every country is seeking."[51]

The Harper government encountered the most difficulty on the social acceptance imperative. The Conservatives' third term in office (2011–15) saw energy rise substantially on the policy and political agendas. Opposition to the Keystone XL pipeline was increasingly joined by opposition to other major pipeline projects to carry Canadian oil to international markets: Enbridge's Northern Gateway pipeline, Kinder Morgan's Trans Mountain Pipeline Expansion proposal, Enbridge's Line 9 pipeline reversal, and TC Energy's Energy East pipeline.

The government responded to opposition by doubling down on its market access objectives. Beginning in 2012, it rolled out a suite of policy measures under the mantle of Responsible Resource Development, some aspects of which addressed opponents' concerns, others of which intensified opposition. The government committed to streamlining regulatory review for major projects by reducing duplication between federal and provincial environmental assessment processes, decreasing the time to process applications (including the establishment of timelines), strengthening protection of the environment and enhancing consultation with Indigenous peoples in environmental assessments.[52] It also committed to provide additional funding to the Major Projects Management Office, which it had established in 2007 to coordinate federal reviews with the aim of shortening timelines; established measures to strengthen tanker safety (inspections, frameworks for oil spills and emergency response, research to better understand spill risks) and increase inspections of pipeline safety; altered the regulatory framework governing the (then) National Energy Board, by requiring that applications rejected by the NEB come to Cabinet for review and possible reversal, leaving the ultimate fate of pipeline proposals directly in the hands of the prime minister and Cabinet, and, in the wake of the train derailment in Lac Mégantic, which saw train cars loaded with oil from the Bakken shale formation explode on the town's main street, killing forty-seven people and devastating the community, the government worked to strengthen railcar safety in tandem with the United States.

Responsible Resource Development was roundly critiqued by opponents for placing the economy over the environment. What's more, the government's policy style rubbed many of its opponents (and possible allies) the wrong way. Instead of working with industry, other governments, civil society, and Indigenous groups to identify balance points between market, environment, and security imperatives that would garner social acceptance, the government pre-determined where balance-points points lie, and pursued its market access objectives in ways that generated conflict.

Under-weighting environmental and social concerns, ironically, stymied realization of the government's market access objectives. In the absence of a forum to make meaningful progress on climate change and other social issues, environmental NGOs turned their efforts to a forum where they could get traction: blocking oil sands pipeline projects in the regulatory system.[53]

Systematic energy collaboration in North America was not high on the Harper government's agendas – nor those of its U.S. or Mexican counterparts. In 2008, Ottawa tried to engage the Obama administration in an energy security/climate change deal at the PM's first meeting with the president,[54] but to no avail. Instead, the government emerged with the far less ambitious Clean Energy Dialogue (CED), which focused on science and technologies to reduce GHG emissions. As for putting a price on carbon, the Obama administration moved ahead on this in 2009 with seemingly little reference to Canada. However, as noted above, neither government carried through on cap-and-trade, which was overtaken by the global financial crisis and recession.

Instead, the Harper government engaged bilaterally on key files on an ad hoc basis, notably on American opposition to development of Canadian oil sands and the Keystone XL pipeline. Ottawa contended with mass protests, high-profile advertising campaigns, and policy and legislative measures against the oilsands south of the border. Although the federal and Alberta governments emphasized that GHG emissions from the oil sands are comparable to those of conventional oil producers exporting to the United States, opposition did not dissipate. The Harper government's largest challenge was opposition to TC Energy's Keystone XL pipeline to carry oil sands crude to refineries on the U.S.

Gulf Coast. The government staunchly supported the project, making regular trips to Washington and other U.S. locations to lobby for its approval by the Obama administration. But domestic pressure to reject the pipeline was potent, relentless, and ultimately effective: the Obama administration turned down the project in the fall of 2015, although President Trump later reversed this decision.

THE TRUDEAU LIBERAL GOVERNMENT (2015–PRESENT)

The Trudeau Liberals came to power in 2015 on a platform intended to offer an approach to governing different from that of the Harper Conservatives. On energy, they made it clear they would strike a more progressive balance between market, environmental, and social acceptance imperatives, with Justin Trudeau regularly stating that environmental and market imperatives could be reconciled. "The choice between pipelines and wind turbines is a false one. We need both to reach our goal, and as we continue to ensure there is a market for our natural resources, our deepening commitment to a cleaner future will be a valuable advantage."[55] The Liberals also committed to make meaningful progress on climate change policy, prioritize reconciliation with Indigenous peoples, and give communities more say in energy project decision-making. On the last, Trudeau went so far as stating, "Governments grant permits, communities grant permission."[56]

The Liberal government's approach was like a three-legged stool: action on climate change (notably carbon pricing), support for market access (especially pipelines), and strengthening public confidence in energy decisions (reforming project decision-making).

On climate, in December 2015, Environment Minister Catherine McKenna (re)committed Canada to a reduction in greenhouse gas emissions of 30 per cent below 2005 levels by 2030 at the Paris meeting of the UNFCCC Conference of Parties. McKenna also committed to maintaining the rise in global temperatures "well below" 2 degrees Celsius compared to pre-industrial times, with the ambition to limit the rise to 1.5 degrees. This was rapidly followed by intensive engagement with the provinces and territories to create the Pan-Canadian Framework on Clean Growth and Climate Change in March 2016. The framework committed all governments to measures to address climate change, including establishing a price on carbon equivalent to $20 per tonne in 2019, ramping up to $50 per tonne by 2022. Under the framework, governments can develop their own plan, or, failing that, have the tax imposed by Ottawa through a federal "backstop." The government also committed to phasing out coal-fired electricity generating stations by 2030.

On market access, in the fall of 2016, the government announced its final decision on three major pipeline proposals that had received regulatory approval from the National Energy Board (now the Canada Energy Regulator): Enbridge's Northern Gateway pipeline, Kinder Morgan's Trans Mountain Expansion Project, and Enbridge's Line 3. The government rejected Northern Gateway, stating that it did not have social support and that the risks to land and water were too great; in the months that followed, it tabled and passed legislation establishing a moratorium on crude oil tanker traffic along the northern BC coastline (Bill C-48 Oil Tanker Moratorium Act). Ottawa approved the other two pipelines. In the case of Trans Mountain, it committed to enhance tanker safety substantially – a major concern about the project – through the $1.5 billion Oceans Protection Plan.

The government also moved quickly on strengthening public confidence in energy decision-making by establishing expert panels to consult Canadians and make recommendations to the government about how to reform environmental assessment and the National Energy Board. For projects in the regulatory process, the government put in place "interim principles," including the principle that existing applications would not need to recommence the regulatory process. The government also included the principle that the upstream climate impacts of a project would be a factor in its decision on whether or not to approve it (decisions on the three pipelines above were made under this interim regime).

Both expert panel reports recommended sweeping reforms to energy decision-making, including much deeper consultation and engagement, the consideration of a much broader range of impacts (environmental, social, cultural), and the development of more comprehensive pre-planning for projects. In response, the government tabled Bill C-69 (An Act to enact the Impact Assessment Act and the Canadian Energy Regulator Act, to amend the Navigation Protection Act and to make consequential amendments to other Acts). Passed into law in August 2019, it established the Impact Assessment Agency (a reformed Canadian Environmental Assessment Agency) to assess a much broader range of impacts, created the Canadian Energy Regulator (a reformed National Energy Board), created requirements for much earlier, deeper, and broader consultation and engagement, and established shorter timelines for project decision-making.

On the one hand, the Trudeau government implemented this three-pronged approach to address climate change, market access, and decision-making. It made solid progress on all three fronts with, respectively, the Pan-Canadian Framework, the approval of two major pipelines, and legislated reforms to energy decision-making. On the other hand, contentions over energy and environmental issues grew substantially over the course of its first mandate (2015–19) and have continued into its second minority mandate.

When the Liberals were elected in the fall of 2015, the provinces and Ottawa were largely singing from the same song sheet, and the three-pronged approach had strong support from most provinces. Government support for climate action was also at a high globally, notably with President Obama in the White House. All of this was to change in the intervening years. The election of President Trump marked a turn in global support for climate action, with the president committing to pull the United States out of the Paris Accord. Trump's keen focus on "America First" placed competitiveness in sharp focus for Canada's relations with the United States.

The election of a minority NDP government in British Columbia in 2017 also marked a turning point. The new government, dependent on the Green Party for support, committed to using "every tool in the toolbox" to oppose the Trans Mountain pipeline, including attempting to regulate increased oil shipments through the province. The escalating stand-off between the Alberta and BC governments led to a "summit" between Prime Minister Trudeau and the two warring premiers. Kinder Morgan decided that enough was enough and sold the pipeline to the federal government. Thereafter, a Federal Court of Appeal decision in 2018 quashed Ottawa's approval of the project over, among other things, inadequate consultation of First Nations communities by the government. These mounting delays stoked growing frustration in Alberta, a province that felt it was doing its fair share to address climate change with a carbon levy on large emitters, a provincial carbon tax, and a 100 MT cap on emissions from the oil sands.

The election of Conservative premiers in several jurisdictions, notably Ontario (2018) and Alberta (2019), also changed the political landscape significantly. When the Pan-Canadian Framework came into being, Saskatchewan was the lone province opposing the carbon tax. Over time, more provinces joined the fight against Ottawa, including with court cases involving Ontario, Saskatchewan, Alberta, and Manitoba challenging the constitutionality of the federal tax. Relative peace in federal-provincial relations was shattering. To date, the Saskatchewan and Ontario Courts of Appeal have found in favour of the federal government, while the Alberta Court of Appeal found the federal carbon tax ultra vires in its current form. The issue will be heard by the Supreme Court in the fall of 2020.

Conflict also grew over Bills C-48 and C-69. Both were mainly welcomed by the environmental community (some felt they didn't go far enough), but industry sounded multiple alarm bells, particularly over C-69, noting the legislation would drive investment out of Canada by increasing uncertainty and unpredictability in decision-making.[57] Subnational governments increasingly added their voices to industry opposition. The premiers of Alberta, Saskatchewan, Manitoba, Ontario, New Brunswick, and the Northwest Territories penned an open letter to the prime minister in June 2019 calling on the government to make major amendments to Bill C-69 and to scrap Bill C-48. They said the bills intervened in areas of provincial/territorial jurisdiction and would bring resource development to a halt in their borders. Alberta Premier Jason Kenney, elected in 2019, famously called Bill C-69 the "no more pipelines bill" on the campaign trail and once in office. Despite mounting opposition from industry and provincial-territorial governments, however, Ottawa forged ahead with both bills, passing them into law before the October 2019 federal election. Not surprisingly, this stoked further frustration in the West, affecting the results of the 2019 federal election, discussed below.

In sum, in their first mandate, the Liberal government began to address the elephants in the room: policy gaps on climate, relations with Indigenous peoples, and cumulative effects. They also moved on reforming energy decision-making processes in keeping with social and value changes, notably by democratizing energy project decision-making through greater public engagement and consultation. But, like the Harper government, they did not find a way between the economic rock and progressive hard place facing Canadian energy. Where the Harper government prioritized market objectives to the detriment of environmental and social acceptance imperatives, the Liberal government seems to have taken the opposite approach: prioritizing environmental and social acceptance objectives to the detriment of market imperatives. Despite the passage of Bill C-69, the establishment of a nationwide carbon tax, and the approval of two major pipelines, the investment climate for Canadian energy remains uncertain, and intergovernmental conflict over energy and environment have intensified.

Investors have taken notice and changed their strategies accordingly – as evidenced by Kinder Morgan's 2018 de-acquisition of the Trans Mountain Pipeline. Shell, Exxon, Equinor (formerly Statoil), and ConocoPhillips have all pulled out of the oil sands since 2015. TC Energy pulled the plug on its Energy East project in 2017, and Encana (now Ovintiv) moved its head office to the United States in late 2019. In early 2020, Teck Resources retracted its regulatory application to develop the Frontier oil sands mine, citing uncertainty in environmental and Indigenous policies as a major contributing factor in an open letter to the government. Meanwhile, Canadian energy companies, including AltaGas, TC Energy, Enbridge, Ontario Power Generation, and Fortis, expanded their North American investment footprint, particularly in the U.S. pipeline and utilities sectors. While these decisions are also influenced by changing market dynamics and business strategy, political, policy, regulatory, and investment uncertainties in Canada have also played a part.

All told, the Trudeau government has fared little better than its Conservative predecessor in bringing Canadian energy resources to market. The movement towards construction of the Keystone XL and Trans Mountain Expansion pipeline projects may change this, but further legal and political challenges are likely. Protracted controversy over energy infrastructure projects seems to be the new norm.

Fall 2019 saw one of the most striking examples of opposition yet. Protests against the Coastal Gaslink project to carry gas to LNG Canada's $40 billion export terminal in British Columbia brought supply chains to a virtual halt across the country. Protestors erected rail blockades in multiple choke points in the national rail network in solidarity with a small group of hereditary chiefs opposed to the project – despite support from the community's chief and council, and all of the communities along the Coastal Gaslink pipeline route.

As for Canada's energy relations in North America, beyond the seeming "win" for Ottawa when the Trump administration approved the Keystone XL pipeline, little else of consequence seems to have transpired on the North American energy collaboration front. Some modest collaboration took place in the Liberals' first year in power through meetings of the North American energy ministries in early 2016 (focused mainly on electricity) and the North American Leaders Summit in June 2016. The Leaders' Statement discussed increasing clean power, reducing methane emissions, strengthening vehicle efficiency, and the like. But the Clean Energy Dialogue lost steam, and once President Trump was inaugurated, little on energy beyond the Keystone XL pipeline decision has transpired. Instead, bilateral conflict and contention over everything from steel tariffs, to supply management, to the G7, to climate, to NAFTA has occupied the bilateral and trilateral agendas.

Many subnational governments remain active on the energy front. Quebec's cap-and-trade agreement with California continues, although Ontario's Ford government abandoned it in 2018 and the Trump administration challenged it. Hydropower-exporting provinces like BC, Manitoba, Quebec, New Brunswick, and Newfoundland and Labrador have worked to gain greater access to U.S. markets, but fuel switching from coal to cheap natural (shale) gas in the United States may tighten market opportunities, as noted above. And Alberta engaged with U.S. governors as Keystone worked its way through state regulatory processes. However, the fate of Canadian energy development and exports destined for markets beyond North American remains very uncertain.

Conclusion: What Might the Future Hold for Canadian Energy?

The economic, environmental, and social worlds of Canadian energy have transformed fundamentally – probably permanently – in recent years. Borders for Canadian energy are on the move, both in trade as oil and gas producers look to access markets beyond North America, and in policy, as decision-makers redraw boundaries between market, environmental, and social imperatives and rethink how they make energy decisions. Canadian governments have been unable to strike a durable balance between market, environmental, security, and social acceptance imperatives of the energy MESS. They have over-weighted either market imperatives or environmental/social acceptance imperatives – and remain stuck between an economic rock and a progressive hard place.

To be fair, the challenges are substantial and complex. Building new energy infrastructure can be daunting at the best of times. Doing so in the context of concerns over climate change, developing infrastructure with which Canada is less accustomed (e.g., LNG facilities, expanding tanker traffic), and building in areas that have not previously hosted energy facilities raises myriad concerns. Add to this a context in which people increasingly want a say in decisions that affect them and are less trusting of institutions, and the stage is set for controversy, uncertainty, and political wrangling of all types.

And the challenges are on the upswing. The 2019 federal election was a divisive campaign, with energy and environment handled in ways that divided Canadians. Pipelines, carbon pricing, oil and gas production, and climate action became "either-or" wedge issues, with support or opposition split along partisan lines. The Liberal Party was returned with a minority government that doesn't hold a single seat in Alberta and Saskatchewan, and very few in Manitoba. The Green, NDP, and Bloc Québécois Parties seek ambitious climate action, while the Conservative Party platform committed to repealing the carbon tax and Bill C-69. Moving into 2020, things became even more challenging as the global health pandemic and the Saudi-Russia price war drove oil prices down to levels few producers in Canada – if any – could withstand. In the short term, share prices in the oil and gas sector cratered, companies pulled back on investment, and production was shut in. Where all of this will go in the medium to long term is a very open question. And what it will mean for Canada's energy relations in North America is likewise unclear.

Governments must recognize that trade and policy borders are shifting. They need to "re-border" policy in ways that strike a workable, durable balance between energy MESS imperatives. The single-minded pursuit of economic imperatives won't succeed in the twenty-first century: failure to adequately address environmental and social concerns fuels opposition to energy development. Likewise, pursuing social or environmental objectives to the exclusion of economic considerations is destined to fail. While the COVID-19 crisis could see Canadians prioritize economic over progressive imperatives, social and environmental concerns won't disappear. Balance is key.

Energy infrastructure projects crossing national and sub-national borders require creative approaches to social acceptance. For local and especially Indigenous communities, this includes robust impact and benefits agreements; meaningful involvement in project design, construction, and monitoring; and, increasingly, equity ownership. For federal and provincial governments, creativity includes overcoming "mental" borders to collaboration. Canada needs consensus-building to capitalize on energy-resource opportunities and to address climate change.

Governments also need to develop a clear vision for Canada's energy future in an age of climate change. Canadians want to see climate change action alongside environmentally responsible oil and gas development, but partisan polarized debates have stymied productive discussion of the country's domestic and export energy economies, particularly for oil and gas. Looking forward, the pandemic and volatility in global energy markets will influence debates over Canada's energy future. Summer 2020 saw forceful calls from environmental advocates to use federal stimulus spending to aggressively transition the country away from oil and gas. If Ottawa heeds these calls without sufficient attention to market imperatives, recent history suggests the move will generate bitter conflict with the provinces, particularly those that produce oil and gas. Instead, political leaders should work together to articulate approaches that unite – not divide – the country.

This is a tall order, but it is needed to build and maintain public and investor confidence in Canada's energy decision-making. Without it, Canada will remain stuck between an economic rock and a progressive hard place and will fail to successfully navigate the changing energy world. For the market, environmental and social performance of Canadian energy, the stakes couldn't be higher.

NOTES

This chapter draws on scholarly literature, government and nongovernment documents, media coverage, and my close engagement with the energy community (policymakers, regulators, industry, environmental NGOs, Indigenous leaders, and municipalities) as chair of Positive Energy (www.uottawa.ca/positive-energy), an action research initiative focused on strengthening public confidence in energy decision-making in Canada. Any errors of fact or interpretation are mine alone.

1 For a more fulsome discussion of the energy policy MESS framework, see Monica Gattinger, "Canada–United States Energy Relations: Making a MESS of Energy Policy," *American Review of Canadian Studies* 42, no. 4 (2012): 460–73.

2 André Plourde, "The Changing Nature of National and Continental Energy Markets," in *Canadian Energy Policy and the Struggle for Sustainable Development*, ed. G. Bruce Doern, 51–82 (Toronto: University of Toronto Press, 2005).

3 Plourde, "Changing Nature of National and Continental Energy Markets."

4 See, for example, Paul Joskow, "Restructuring, Competition and Regulatory Reform in the U.S. Electricity Sector," *Journal of Economic Perspectives* 11, no. 3 (1997): 119–38.

5 See, for example, Isidro Morales, "The Twilight of Mexico's State Oil Monopolism: Policy, Economic, and Political Trends in Mexico's Natural Gas Industry," Harvard Kennedy School Belfer Center for Science and International Affairs and Center for Energy Studies, Rice University's Baker Institute, December 2013.

6 For a comprehensive treatment of the "shale revolution," see Monica Gattinger and Rafael Aguirre, "The Shale Revolution and Canada-US Energy Relations: Game Changer or Déjà-Vu All Over Again?," in *International Political Economy*, ed. Greg Anderson and Christopher J. Kukucha, 409–35 (Toronto: Oxford University Press, 2015).

7 United States, Energy Information Administration, "US Crude Oil and Natural Gas Proved Reserves, Year-End 2018" (Washington, DC: December 2019), table 5, 27. https://www.eia.gov/naturalgas/crudeoilreserves /pdf/usreserves.pdf.

8 United States, Energy Information Administration, "US Shale Proved Reserves" Release date, 12 December 2019 (Washington, DC: 2019), https:// www.eia.gov/dnav/ng/hist/res_epg0_r5301_nus_bcfa.htm.

9 United States, Energy Information Administration, "US Crude Oil and Natural Gas Proved Reserves," table 9, 35.

10 Canada, Natural Resources Canada, "Natural Gas Facts," Energy Facts, last modified 2 July 2020, https://www.nrcan.gc.ca/energy/facts/natural -gas/20067.

11 United States, Energy Information Administration, "Monthly Energy Review, September 2020" (Washington, DC: June 2020), table 3.1, 59. https://www.eia.gov/totalenergy/data/monthly/pdf/mer.pdf.

12 United States, Energy Information Administration, "Monthly Energy Review, June 2020" (Washington, DC: June 2020), table 3.1, 59.

13 United States, Energy Information Administration, "Monthly Energy Review, June 2020," table 3.1, 59.

14 United States, Energy Information Administration, "Monthly Energy Review, June 2020," table 3.1, 59.

15 United States, Energy Information Administration, "Monthly Energy Review, June 2020," table 3.3c, 66.

16 United States, Energy Information Administration, "Monthly Energy Review, June 2020," table 4.1, 101.

17 United States, Energy Information Administration, "Monthly Energy Review, June 2020," table 4.1, 101.

18 United States, Energy Information Administration, "Monthly Energy Review, June 2020," table 4.1, 101.

19 United States, Energy Information Administration, "Monthly Energy Review, June 2020," table 4.2, 102.

20 United States, Energy Information Administration, "Monthly Energy Review, June 2020," table 4.2, 102.

21 United States, Energy Information Administration, "Monthly Energy Review, June 2020," table 3.3d, 67.

22 United States, Energy Information Administration, "Monthly Energy Review, June 2020," table 3.3d, 67.

23 United States, Energy Information Administration, "Table 2.13 Electric Power Industry: US Electricity Imports from and Electricity Exports to Canada and Mexico, 2006–2016," Electricity, Data Tables, https://www.eia.gov/electricity/annual/html/epa_02_13.html; and United States, Energy Information Administration, "Table 2.14 Electric Power Industry: US Electricity Imports from and Electricity Exports to Canada and Mexico, 2008–2018." https://www.eia.gov/electricity/annual/html/epa_02_14.html.

24 United States, Energy Information Administration, "Spot Prices (Crude Oil in Dollars per Barrel)" (Washington, DC: 2018). https://www.eia.gov/dnav/pet/pet_pri_spt_s1_m.htm.

25 Government of Alberta, "Oil Prices" (Edmonton: Alberta Energy, 2020), http://economicdashboard.alberta.ca/OilPrice.

26 BP, *Statistical Review of World Energy*, 2020, https://www.bp.com/content/dam/bp/business-sites/en/global/corporate/pdfs/energy-economics/statistical-review/bp-stats-review-2020-full-report.pdf.

27 BP, *Statistical Review of World Energy*, 2020.

28 Nick Cunningham, "Natural Gas Prices Could Double Next Year," Oil-price.com, 29 March 2020, https://oilprice.com/Energy/Energy-General/Natural-Gas-Prices-Could-Double-Next-Year.html.

29 United States, Energy Information Administration, *Annual Energy Outlook 2020 with Projections to 2050* (Washington, DC: 2020), 9–11. https://www.eia.gov/outlooks/aeo/pdf/AEO2020%20Full%20Report.pdf.

30 United States, Energy Information Administration, *Annual Energy Outlook 2020 with Projections to 2050*, 58.

31 United States, Energy Information Administration, *Annual Energy Outlook 2020 with Projections to 2050*, 68–74.

32 TransCanada changed its name to TC Energy in 2019.

33 This section draws on the following sources: Monica Gattinger, "Public Confidence in Energy and Mining Development: Context, Opportunities/Challenges and Issues for Discussion" (paper presented at the National Workshop on Public Confidence in Energy and Mining Development, 8–9 June 2016, in preparation for the Energy and Mines Ministers Conference, Winnipeg, 2016); Michael Cleland and Monica Gattinger, *System under Stress: Energy Decision-Making in Canada and the Need for Informed Reform* (Ottawa: University of Ottawa, Positive Energy, 2017); and Michael Cleland and Monica Gattinger, with Rafael Aguirre and Marisa Beck, *Durable Balance: Informed Reform of Energy Decision-Making in Canada* (Ottawa: University of Ottawa, Positive Energy, 2018). Interested readers are encouraged to consult the original sources for a comprehensive analysis of challenges to Canada's energy decision-making system from environmental and social acceptance issues.

34 See Anthony Giddens, *Consequences of Modernity* (Cambridge: Polity, 1990).

35 See Edelman, *2017 Edelman Trust Barometer: Global Results* (2017), https://www.edelman.com/research/2017-edelman-trust-barometer.

36 See Neil Nevitte, *The Decline of Deference: Canadian Value Change in Comparative Perspective 1981–1990* (Toronto: Broadview, 1996); Nevitte, "The Decline of Deference Revisited: Evidence after 25 Years" (paper presented at Mapping and Tracking Global Value Change: A Festschrift Conference for Ronald Inglehart, University of California, Irvine, 11 March 2011).

37 See Anthony Giddens and Christopher Pierson, *Making Sense of Modernity: Conversations with Anthony Giddens* (Redwood, CA: Stanford University Press, 1998).

38 See Mary Douglas and Aaron Wildavsky, *Risk and Culture: An Essay on the Selection of Technological and Environmental Dangers* (Berkeley: University of California Press, 1982).

39 See Frank Fischer, *Reframing Public Policy: Discursive Politics and Deliberative Practices* (Oxford: Oxford University Press, 2003); and Loïc Blondiaux and Yves Sintomer, "L'impératif délibératif," *Politix* 15, no. 57 (2002): 17–35.

40 What's more, recent research in social psychology and political science underpins the fact that people use "motivated reasoning" when forming their opinions on controversial public issues, selecting evidence that aligns with their world views and values and dismissing what doesn't. See Dan Kahan, Ellen Peters, Maggie Wittlin, Paul Slovic, Lisa Larrimore Ouellette, Donald Braman, and Gregory Mandel, "The Polarizing Impact of Science Literacy and Numeracy on Perceived Climate Change Risks," *Nature Climate Change* 2, no. 10 (2012): 732–5. Paradoxically, this tendency rises with the level of education, and efforts to "educate" people about issues using "the facts" can backfire, entrenching them more firmly in their positions and further polarizing debates. Kahan et al., "Polarizing Impact."

41 See Naomi Klein, *This Changes Everything: Capitalism vs. the Climate* (Toronto: Knopf, 2014). For the progressive application of Blockadia tactics to clean energy projects like largescale hydropower, see George Hoberg, "Pipeline Resistance as Political Strategy: 'Blockadia' and the Future of Climate Politics (paper presented at the Annual Conference of the Canadian Political Science Association, 2–4 June 2015, University of Ottawa).

42 For a comprehensive study of community levels of satisfaction with energy project decision-making processes, see Michael Cleland, with Laura Nourallah and Stewart Fast, *Fair Enough: Assessing Community Confidence in Energy Authorities* (Calgary and Ottawa: Canada West Foundation and University of Ottawa [Positive Energy], 2016).

43 Cleland, with Nourallah and Fast, *Fair Enough.*

44 This section draws on Monica Gattinger, "The Harper Government's Approach to Energy: Shooting Itself in the Foot," in *The Harper Era in Canadian Foreign Policy: Parliament, Politics, and Canada's Global Posture,* ed. Adam Chapnick and Christopher J. Kukucha, 151–66 (Vancouver: UBC Press, 2016).

45 Stephen Harper, "Address by the Prime Minister at the Canada-UK Chamber of Commerce," London, 14 July 2006, http://www.dominionpaper.ca/articles/1491. The analyst in question is Henry Groppe. See David J. Deslauriers, "Oil Forecasting Legend Paints Dire Energy Picture," *Hard Assets,* 6 June 2005, http://www.resilience.org/stories/2005-06-06/energy-headlines-june-6-2005-part-two.

46 Bill Curry, "Opposition Parties to Force Tories to Meet Kyoto Targets," *Globe and Mail,* 16 May 2006, http://www.theglobeandmail.com/news/national/opposition-parties-to-force-tories-to-meet-kyoto-targets/article18162531/.

47 Curry, "Opposition Parties to Force Tories."

48 Matthew Bramley, "Far from Turning the Corner: Canada's Conservative Government Has Substantially Shifted Its Position on Climate Change, but Is Its Policy Response Too Timid, Too Complex and Likely to Be

Superseded?" *Carbon Finance*, 20 June 2008, http://www.pembina.org/op
-ed/1661.

49 Shawn McCarthy, "Ottawa Commits to 30-Per-Cent Cut in GHGs but No
Regulations for Oil Sands," *Globe and Mail*, 15 May 2015, http://www
.theglobeandmail.com/news/national/ottawa-commits-to-30-per-cent-cut
-in-emissions-but-not-for-oil-sands/article24453757/.

50 CBC News, "PM: Dion's Carbon Tax Would 'Screw Everybody," 20 June
2008, http://www.cbc.ca/news/canada/pm-dion-s-carbon-tax-would
-screw-everybody-1.696762.

51 Les Whittington, "Stephen Harper Says Economy Trumps Climate Action,"
Toronto Star, 9 June 2014, http://www.thestar.com/news/canada/2014/06/09
/stephen_harper_says_economy_trumps_climate_action.html.

52 Canada, Department of Finance, *Jobs, Growth and Long-Term Prosperity:
Economic Action Plan 2012* (Ottawa: 2012), 92.

53 Cleland and Gattinger, *System under Stress.*

54 Shawn McCarthy, "Ottawa Swoops in with Climate-Change Offer," *Globe
and Mail*, 6 November 2008.

55 Joanna Smith, "Trudeau Wants Canada to Play Key Role in Fighting Climate
Change," *Star*, 2 March 2016. https://www.thestar.com/news/canada
/2016/03/02/canada-will-play-leading-role-in-new-economy-trudeau-says
.html.

56 CBC News, "Trudeau: 'Governments Grant Permits, Communities Grant
Permission' Prime Minister Justin Trudeau Reacts to Quebec Seeking an
Injunction against the Energy East Pipeline." n.d. https://www.cbc.ca
/player/play/2684686536.

57 See Shawn McCarthy, "Canadian Energy Industry Slams Liberals' Environmental Assessment Rules," *Globe and Mail*, 2 April 2018.

18 Is NAFTA's Northern Border Thickening for Agri-Food Products?

WILLIAM A. KERR AND JILL E. HOBBS

As trade barriers such as tariffs have been progressively removed in multilateral forums and preferential trade agreements, further trade liberalization requires changes that reach deeply into domestic regulatory competency and territoriality. Anti-globalization activists often decry the erosion of sovereignty that such liberalization entails. This mingling of domestic and trade policy is termed *intermestic* and suggests that policy initiatives need to be cognizant of the concerns of both spheres. The intermestic nature of trade liberalization can complicate the policymaking environment. As a result, despite the good intentions of policymakers when trade agreements are signed, progress may be slower than was envisioned prior to a trade agreement coming into force. In some areas of policy, despite the cross-border nature of the issues, the development of policy is only notionally intermestic, while the reality is that policy development remains de facto domestic in response to territorially defined forces, or alternatively can be dominated by international initiatives where significant aterritorial forces stimulate cooperation.

Provisions for preferential trade agreements such as the United States–Mexico–Canada Agreement (USMCA) and the European Union (EU) are made within the multilateral trade architecture of the World Trade Organization (WTO). This is because preferential trade agreements can go farther down the path of liberalization than can be achieved multilaterally and they can extend the range of areas subject to liberalization. Of course, in the process of going *faster and farther* than the multilateral system, preferential trade agreements cannot contravene what has been agreed at the WTO. The North American Free Trade Agreement (NAFTA), for example, removed almost all tariffs between the United States, Canada, and Mexico. Further liberalization, however, requires cooperation over what are largely regulatory issues. NAFTA had less success in removing regulatory barriers to trade in agri-food products compared to, for example, the EU, where the creation of the *single market* has removed a large number of regulatory barriers, largely through harmonization.

This chapter examines the question of whether the border thickened for trade in agri-food products under NAFTA, particularly in food safety. Following the introduction, the chapter explains the nature of NAFTA's aterritorial architecture in the context of agriculture and food, and then examines how border issues pertaining to food safety and agri-food markets have evolved in the years following NAFTA inception.

Food safety is a particularly sensitive area in regulatory cooperation and one where governments guard their sovereignty closely. Unlike many barriers to trade whose motivation is economic protection, barriers to trade arising from divergence in food safety standards have a deep political rationale. Consumers negatively affected by a failure in a food safety regime may not care if the source of the incident is in a foreign portion of the supply chain; they expect their domestic politicians to ensure the safety of their food, no matter its origin; hence, it is important to have control over the food safety regulatory regime. Giving up sovereignty over the regulatory system for food safety is very difficult for policymakers, and policy is strongly shaped by these territorially defined forces.

While the retention of sovereignty is central to regulatory cooperation on food safety standards, there are other issues that also inhibit regulatory cooperation. As a result, within the structure of NAFTA, little progress was made on removing barriers to trade arising from diverging regulatory standards on food safety. As long as there are such regulatory non-tariff barriers, borders will remain thick. Further, if consumer concerns about food safety are heightened, governments may react with non-cooperative policies that increase the barriers to trade and thickness of a border. Since the inception of NAFTA in 199 consumer concerns about food safety have increased.[1] On the other hand, the architects of NAFTA tried to institutionalize regulatory cooperation to reduce the thickness of the border.

NAFTA's Aterritorial Architecture: Well Planned but Poorly Executed

Harmonization of regulatory standards or other means of removing border thickening by divergent standards, such as the granting of equivalence, cannot be achieved directly in the negotiation of trade agreements. Standards are complex, and altering them requires types of expertise that are different from that possessed by trade negotiators.

This is particularly so for standards in areas such as food safety. In the WTO, for example, while international harmonized standards are recognized as a goal in the Agreement on the Application of Sanitary and Phytosanitary Measures (SPS), the actual devising of such standards was left to pre-existing scientific organizations – the Codex Alimentarius Commission (food safety), the World Organization for Animal Health (animal health), and the International Plant Protection Convention (plant health).

The negotiators of the Canada-U.S. Trade Agreement (CUSTA), and subsequently NAFTA, anticipated the need for aterritorial institutional arrangements to address divergence in regulatory standards. They intended that these arrangements would reduce the trade difficulties associated with independent regulatory development prior to the signing of the CUSTA/NAFTA and prevent regulatory divergence in the future. Agreed in the NAFTA treaty in Article 906: Compatibility and Equivalence was:

1 Recognizing the crucial role of standards-related measures in achieving legitimate objectives, the Parties shall, in accordance with this Chapter, *work jointly* to enhance the level of safety and of protection of human, animal and plant life and health, the environment and consumers.

2 Without reducing the level of safety or of protection of human, animal or plant life or health, the environment or consumers, without prejudice to the rights of any Party under this Chapter, and taking into account international standardization activities, *the Parties shall, to the greatest extent practicable, make compatible their respective standards-related measures, so as to facilitate trade in a good or service between the Parties.*

3 Further to Articles 902 and 905, *a Party shall*, on request of another Party, *seek, through appropriate measures, to promote the compatibility of a specific standard or conformity assessment procedure that is maintained in its territory with the standards or conformity assessment procedures maintained in the territory of the other Party* (emphasis added).

In addition, under Article 913: Committee on Standards-Related Measures, to both assist in harmonizing existing regulations and to foster regulatory harmonization subsequent to NAFTA coming into force:

1 The Parties hereby establish a Committee on Standards-Related Measures, comprising representatives of each Party....

5 Further to paragraph 4, the Committee shall establish:
   (a)   the following subcommittees
         (i)   Land Transportation Standards Subcommittee, in accordance
                with Annex 913.5.a-1,
         (ii)   Telecommunications Standards Subcommittee, in accordance
                with Annex 913.5.a-2,
         (iii) Automotive Standards Council, in accordance with Annex
                913.5.a-3, and
         (iv)   Subcommittee on Labelling of Textile and Apparel Goods, in
                accordance with Annex 913.5.a-4.

In the case of agri-food trade, a Committee on Agricultural Trade (Article 706) was established that was, among other things, to put in place a Working Group on Agricultural Grading and Marketing Standards. In Article 722 a Committee on Sanitary and Sanitary Measures was also mandated, as well as several working groups: (1) Animal Health; (2) Dairy, Fruits, Vegetables, and Processed Foods; (3) Fish and Fisheries Product Inspection; (4) Food Additives and Contaminants; (5) Labelling, Packaging, and Standards; (6) Meat, Poultry, and Egg Inspection; (7) Technical Working Group on Pesticides; (8) Plant Health, Seeds, and Fertilizers; and (9) Veterinary Drugs and Feeds Working Group.

To promote cooperation in altering existing regulations or developing new regulations, set out under Article 909: Notification, Publication, and Provision of Information was:

1 Further to Articles 1802 (Publication) and 1803 (Notification and Provision of Information), each Party proposing to adopt or modify a technical regulation shall:
   (a)   at least 60 days prior to the adoption or modification of the
         measure, other than a law, publish a notice and notify in writing
         the other Parties of the proposed measure in such a manner as
         to enable interested persons to become acquainted with the proposed measure, except that in the case of any such measure relating to perishable goods, each Party shall, to the greatest extent
         practicable, publish the notice and provide the notification at
         least 30 days prior to the adoption or modification of the measure, but no later than when notification is provided to domestic
         producers;
   (b)   identify in the notice and notification the good or service to which
         the measure would apply, and shall provide a brief description of
         the objective of, and reasons for the measure; ...

(d)   without discrimination, allow other Parties and interested persons to make comments in writing and shall, on request, discuss the comments and take the comments and the results of the discussions into account.

2 Each Party proposing to adopt or modify a standard or any conformity assessment procedure not otherwise considered to be a technical regulation shall, where an international standard relevant to the proposed measure does not exist or such measure is not substantially the same as an international standard, and where the measure may have a significant effect on the trade of the other Parties:

(a)   at an early appropriate stage, publish a notice and provide a notification of the type required in paragraph 1(a) and (b); and

(b)   observe paragraph 1(c) and (d).

Thus, it is clear that those negotiating NAFTA recognized the potential for regulatory egocentrism (territoriality) among the member countries to be disruptive to international trade and took considerable and detailed care to ensure that institutional arrangements were put in place to eliminate or reduce the potential for trade-disrupting divergence in regulatory standards.

With foresight, institutions can be established with the best of intentions. They may not, however, operate as intended. While deficiencies in the institutions mandated by NAFTA pertaining to regulatory harmonization were not in the forefront of the debates surrounding the efficacy of the agreement, lack of harmonization has been a continuing observation. For example, the trade-inhibiting effect of the absence of a harmonized beef-grading system in North America was well understood prior to both CUSTA and NAFTA.[2] The Working Group on Agricultural Grading and Marketing Standards was the appropriate NAFTA institution that should have assisted in removing this impediment. Hayes and Kerr,[3] however, found that the working group had little to do with attempts to harmonize grading and that the issue remained politicized and subject to protectionist interests. As of 2020, there is still no harmonization of beef grading among the NAFTA partners. Further, Hayes and Kerr[4] found there was little indication that NAFTA partners informed or consulted with each other when new domestic regulations with potential trade effects were being developed.

USMCA provides for national treatment for grain grading standards in Article 3.A.4.1. Canadian licensing requirements for new varieties of wheat differ from those in the United States. To be registered in Canada, wheat varieties must have certain characteristics. Wheat bred and developed in the United States often does not exhibit those characteristics.

As a result, imported U.S. wheat varieties have not been recognized or graded. Under USMCA, wheat grown in either country from varieties registered in Canada and delivered to Canadian grain elevators will be eligible for grading by the Canadian Wheat Commission, addressing a longstanding trade irritant.

The Technical Working Group on Pesticides (TWGP), however, appears to have had some early success,[5] but it appears to be the exception rather than the rule. Given its apparent success compared to the other NAFTA mandated committees that were intended to foster regulatory convergence, it is worth further examining its activities.[6] In 1996, the TWGP was established under the NAFTA provisions on sanitary and phytosanitary regulations to "serve as a focal point for addressing pesticide issues arising in the context of liberalized trade among the NAFTA countries," with the key objectives being: (1) sharing information; (2) undertaking collaborative scientific work; (3) forging common data requirements; (4) collaborating on risk assessment or compliance methods; (5) carrying out joint reviews; and (5) developing common NAFTA or international standards.[7]

Membership of the TWGP comprises officials from all three countries that have expertise and a regulatory interest in pesticides. From Canada, representation is drawn from the Canadian Pest Management Regulatory Agency, which is part of Health Canada. For the United States, the Office of Pesticide Programs in the Environmental Protection Agency is the source for TWGP participants. The Mexican Ministry of Agri-foods, Aquaculture and Fisheries and Mexico's Ministry of Health jointly name participants.

While details on the first five-year period of the working group's operation are not available,[8] it would appear from subsequent publications that this initial period from 1997 to 2002 was predominantly occupied with establishing baselines for further collaboration. This involved garnering three-party agreement on principles and practices, existing standards and regulations, and clearly defining those pesticides that had been approved for use in each country. There are two reports: one examining the accomplishments of the TWGP from 2003 to 2008[9] and a 2008–13 strategic plan.[10] The assessment document for 2003–8 reports that a vision statement had been developed for the TWGP. It reads, "Canada, the United States and Mexico are striving to make the North American region a world model for common approaches to pesticide regulation and free trade in pesticides and food."[11]

To accomplish its work the TWGP annually holds a meeting of its executive board with other regulatory officials. Later, a second government-stakeholder meeting is held, including the executive board, regulatory officials, farmers, the pesticide industry, and groups from civil society.

The NAFTA-TWGP (2009) report[12] defines three objectives against which the assessment is conducted: (1) full North American collaboration in pesticide regulation, including re-assessment; (2) equal and timely access to new pest management tools; and (3) robust stakeholder participation.

Much of the 2003–8 efforts were directed towards re-evaluation and re-registration of existing pesticides. Prior to this work, the time for which approvals were valid varied. As a result of the TWGP efforts, this was standardized at fifteen years for all pesticides. Once the fifteen years has passed, each pesticide must be re-reviewed and re-approved. Considerable work was also done to harmonize maximum pesticide residue limits during the reporting period.

Simultaneous registration of new active pesticide ingredients was encouraged in all three jurisdictions. This resulted in the review assessment work being divided among the three nations with the results being shared and peer reviewed. This approach proved successful, as it is noted that by 2005 two new active ingredients were approved in fourteen and sixteen months, which represented record approval times.[13] The report acknowledges that shared review processes "resulted in increased levels of shared scientific knowledge and in increased understanding of each country's risk assessment and risk management processes. Consequently, governments have gained trust in their counter-part's regulatory decision-making."[14]

Industry stakeholders suggested that the joint review and work-sharing process should be considered as a model for international regulatory collaborations and that reaching agreement and successful operation of the TWGP was the most significant accomplishment of the entire five-year period.

The planning document (NAFTA-TWGP[15]) identified a set of common objectives for 2008–13. The underlying rationale of these objectives is to improve the coordination and cooperation in the pesticide review process and to further harmonize the regulatory frameworks. Unfortunately, no further review documentation is available for this more recent five-year period (or for the post-2013 period). The TWGP has laid out its priorities in its Five-Year Strategy 2016–21.[16] For the most part it just continues the previous commitments to move forward on the issues outlined above but with the notable addition of cooperation to protect pollinators.

The TWGP experience suggests that progress towards harmonization is possible through cooperation among regulatory regimes, particularly where aterritorial forces are aligned. What is needed is a strong economic case for why harmonization is desirable, given that one can expect some resistance to harmonization.

With the exception of the Technical Working Group on Pesticides, the NAFTA committees are primarily forums to "talk and talk" but without any mechanism to bring closure to an issue. On a more mundane level, the committees led to some cooperation on day-to-day technical issues.[17] Further, they appear to have taken on a role in assisting with the management of potential SPS crises, at times, which was not envisioned at the time of negotiation.[18] The joint Canadian-U.S. response to the threat of African swine fever (ASF) in 2019 is an example of such cooperation. On 22 May 2019 the chief veterinary officers of Canada and the United States issued a joint statement announcing that they had agreed on a plan to safeguard international trade between safe areas – areas of a country where the disease is not present in the event of an outbreak.[19]

Globally, preferential trade agreements have differing records on harmonization. NAFTA has not been particularly effective in removing technical barriers to trade and fostering regulatory harmonization. In contrast, the European Union (EU) has been very effective in its endeavours in these areas. NAFTA and the EU, however, have very different governance structures.

One major institutional deficiency in Canada-U.S. relations, inside and outside NAFTA, has been that there is no formal supranational body to foster a bilateral agenda. The United States, in particular, is suspicious of supranational institutions, largely because of concerns with the limits on sovereignty that they might impose. If one compares the EU with NAFTA, a striking difference is that there is no equivalent of the European Commission. The European Commission comprises commissioners appointed by member states. Once appointed, however, the individual commissioners are expected to adopt an EU perspective rather than to be an advocate for the member state government that appointed them. By and large, the commissioners have taken on that role – although there have been some notable exceptions. Commissioners "speak for Europe." No one in the NAFTA system is expected to "speak for North America" – one is either a Mexican, a Canadian, or an American. Of course, all of those that work in the commission also "speak for Europe." This means that at almost any meeting, conference, policy forum, or media event there is someone there to provide a European Union–wide perspective. This does two things: it forces people to consider this broader perspective and respond to it, and it keeps this broader perspective continually front and centre. The cumulative effect of these activities should not be underestimated.

The European Commission is also exclusively charged with devising European Union–wide policy proposals. Even if the proposals are rejected by the EU's Parliament or Council of Ministers, it means that proposals from such a perspective must be considered. In NAFTA/USMCA, there is no institution that plays this role. Instead, everything must be proposed and negotiated by advocates of the individual countries.

One lesson that industry needs to learn from the NAFTA experience is that while it takes resources to get the changes stakeholders would like to see into trade agreements, it is also important to commit resources to ensure their implementation. If proposals for harmonization are developed and approved, there is something to which the bureaucracy can be held accountable – unlike the vague commitments to negotiate harmonization that were, for example, the reason for mandating the NAFTA committees.

Moving beyond NAFTA

After more than fifteen years of little progress it became increasingly apparent that the well-intentioned NAFTA institutions were not functioning as envisioned. As a result, in 2011, the United States and Canada established the Regulatory Cooperation Council (RCC), to align regulatory approaches where appropriate. Food safety and biotechnology were highlighted as key areas for cooperation when the RCC was announced. There is some activity regarding food safety, but not for biotechnology.[20]

Action on improving food safety between the two countries under the RCC has focused on common approaches and testing, meat and poultry equivalence, and certification requirements and meat processing.[21] Under the RCC, work has been initiated on crop protection products, in particular pesticides. These activities indicate diligence in attempts to improve the joint submission process for pesticides, develop joint guidelines, and address obstacles to joint registration. It would appear that both Canada and the United States are now concentrating their efforts on the harmonization and possibly integration of pesticide regulations within the RCC process, largely replacing efforts preciously undertaken under the NAFTA-TWGP mechanism. Other aspects pertaining to food safety and related issues such as health claims on food product labels, which could have been dealt with by the NAFTA mandated Committee on Labelling, Packaging, and Standards, remain uncoordinated, work at cross-purposes, and act to thicken the border.

Turning to another example, the health food sector provides a good case study in regulatory disharmony and its trade impacts. Health foods include functional foods and natural health products (dietary supplements). While definitions vary internationally, in general, the term "functional food" describes enhanced foods and usually refers to food that is intended to be consumed as part of a normal diet and contains ingredients that have the potential to enhance human health or reduce the risk of disease beyond basic nutritional functions. Dietary supplements (also known as food supplements, nutraceuticals, or in Canada "natural health products") are often sold in the form of capsules, tablets, or powders and are products that have been isolated or purified from food and may include ingredients such as amino acids or vitamins.[22] In establishing regulatory frameworks for the health food sector, those developing regulatory policy need to find a workable balance between facilitating research and development leading to new health food product innovations, and the protection of consumers from false and misleading claims that may also weaken consumer confidence in the sector. While governments have moved to establish regulatory frameworks governing the use of health claims and the approval of new products with novel traits, slow and restrictive approval processes and weak protection of intellectual property rights will blunt incentives to innovate. Among broadly similar, well-established, developed markets such as the United States and Canada, significant differences in regulatory approaches continue. In the continuum of policy relations and policy instruments for regulatory cooperation outlined in chapter 1, the development of health food policies in both countries could best be characterized as an example of parallelism. The Canadian regulatory system, for example, has been criticized by the industry for being slow to adapt to the changing nature of the products in the sector.[23] Many of the disease-reduction claims permitted in Canada have been allowed for use by firms in the United States for much longer and the list of health claims allowed is much longer than that of Canada. Given the small size of the Canadian market, firms may be dissuaded from attempting to garner regulatory approval for a new health claim if this is a costly and overly slow procedure when compared with the U.S. market and its much larger potential payback.

The use of "qualified" health claims in the United States stands in stark contrast to Canada, which rejects the use of lower standards for disease-reduction claims as a result of the potential for misleading consumers. Another notable difference in the approach to health claims among countries is the approach to product-specific health claims. Generic claims (which can be used by any firm producing a product with an approved health claim) create a potential free-rider problem: many firms benefit but only one firm has to go through the application process to obtain approval for a new health claim. Canada allows product-specific health claims for natural health products but the United States does not. Nevertheless, the generic health claim systems in use for (functional) food in the United States and Canada have the advantage that more products can use approved health claims with the potential to facilitate improved consumer knowledge of the relationship between health and diet.[24]

Some countries make no regulatory distinction between dietary supplements and (functional) food. In many other countries, however, the two categories are distinct from a regulatory perspective, as in Canada, where natural health products (supplements) face a product-specific regulatory system but functional foods utilize generic health claims. The United States makes a distinction between (functional) food and dietary supplements, but does not impose significantly different regulations, allowing the same generic claims for both product categories.

In some jurisdictions, efforts are being made to harmonize regulations or to recognize equivalence in regulatory standards – Australia and New Zealand, and the development of a single regulatory approach across EU member states, being prime examples – but, for the most part, regulatory approaches in North America remain the sole responsibility of each country.[25] Establishing equivalence (e.g., recognizing U.S. health claims in Canada and vice versa) could be a possible route for functional foods. Many of the health claims approved in Canada are based on claims already approved in the United States, and the Canadian approval process for nutrient function claims specifically references the Institute of Medicine of the U.S. National Academies.[26] The regulatory approaches to dietary supplements in the United States and natural health products in Canada, however, are markedly different. For example, the categories are not even directly comparable in the sense that not all products classified as natural health products in Canada would be considered dietary supplements in the United States. Hence, it seems unlikely that they would recognize the other's regulations as equivalent.

Navigating this complex array of regulatory approaches can be a nightmare for firms wishing to access the market in the other country. Firms must determine which set of regulations apply to their products and the extent to which the same product faces different regulatory requirements in different markets. Disparate regulatory requirements for what constitutes an allowable health claim and whether firms can use an approved generic health claim or must seek approval for a product-specific health claim raise the costs of accessing markets within NAFTA/USMCA – the border is thick in this rapidly growing market in North America. The issues do not appear to even be on the radar of the appropriate NAFTA mandated committees.

The reason regulatory harmonization does not automatically follow when it is the intent of those who negotiate trade agreements is multifaceted. In the case of purposeful restrictions on trade such as tariffs and tariff-rate quotas (TRQs) implementation remains in the ministry that is responsible for trade – both regulation and negotiation. Thus, removal of tariffs on schedules agreed in the trade agreements seldom has associated implementation issues. The tariffs are removed and, if they are not, non-compliance can quickly be raised with the other party.

In the case of non-tariff barriers, removal agreed in a trade negotiation is unlikely to be within the jurisdiction of the trade ministry. In food safety, for example, jurisdiction lies with the Food and Drug Administration (FDA) or the United States Department of Agriculture (USDA) in the United States, or the Canadian Food Inspection Agency (CFIA) or Health Canada in Canada. These domestic-oriented institutions have their own agendas, priorities, and stakeholders. At some point, approval would have had to have been sought by the trade negotiators, but this does not mean that implementing commitments in trade agreements will be a priority. Removal of such SPS barriers often requires that something new must replace them, requiring changes to long established routines and additional resources to develop new protocols, and often to implement more costly regulatory activities. If the system is operating satisfactorily, there may be little enthusiasm to implement the changes agreed to in the trade negotiations, or expeditiously. If the changes require consultations and negotiations with the trading partner, these discussions can be used strategically to obtain delays.[27]

When moves towards harmonization of food safety regulations and processes are part of trade agreements, bureaucratic vested interests and other institutional impediments may come to the fore, because sanitary and phytosanitary measures and procedures can often have serious human, animal, or plant health ramifications. Further, it is a policy area that is of particular interest to politicians, given the potential negative consequences for public confidence of a major food safety incident that takes place on their watch. Hence, politicians have been particularly reluctant to give up sovereignty in this area of public policy.[28]

Consider the case of two countries, X and Y, whose regulatory regime and processes for food safety have evolved independently and differ considerably. Harmonization can follow three paths. Country X can harmonize with Country Y's regulatory regime; Country Y can harmonize with Country X's system; or the two countries can agree to jointly devise

a wholly new set of regulations.[29] From the perspective of the politicians and regulators in either country, it would be preferable if the other
country altered its regulatory regime and processes because this would
require no changes to their existing system and no additional resources.
All of the harmonization costs would be accrued by the other country.
Negotiations are likely to be protracted, with little incentive to bring
them to a conclusion. If pushed by the trade ministry to compromise,
vested interests in the regulatory system may have the option to appeal
to civil society, claiming that they are being forced to *lower food safety
standards, endangering the public as the result of pressure from trade partners.*
This is the type of story that opposition parties can use to considerable
advantage and those in government wish to avoid.[30] Thus, even where
periodic efforts towards regulatory coordination emerge, vulnerability
to vocal interest group resistance couched in the rhetoric of "lowering"
standards frequently reasserts territoriality in food safety regulations.[31]

The bottom line is that there is likely to be considerable resistance
to harmonization from those within regulatory bureaucracies faced
with change. Hence, harmonizing to one country's regulatory regime
is likely to encounter reluctance among those in the bureaucracy of the
country expected to change, or from both bureaucracies if there is to be
a move to a new, common standard.

There may also be additional vested interests resistant to changes
agreed to in trade arrangements such as NAFTA. One change agreed
to in CUSTA was the removal of routine border inspections for red
meat. These inspections had become a contentious issue raised by the
Canadian beef industry in the lead up to the negotiations.[32] The U.S.
government border inspectors faced the spectre of job losses or relocation, and the private firms that owned the border inspection facilities
were faced with the prospect of their businesses no longer being required, threatening their investments. Jointly they were able to delay
the removal of border inspections for a considerable period such that
territoriality persisted in cross-border regulation.[33] The major beef recall centred on the Canadian XL Foods packing plant in 2012 added
a further complication in removal of routine meat inspections. Hence,
there are may be substantial *fulfilment costs* for industry stakeholders
who wish to move harmonization commitments in trade agreements
to actual harmonization.[34] Resources are needed to overcome the inertia surrounding harmonization within the regulatory system and in
convincing policymakers to *make it happen.* For the most part, NAFTA
harmonization mechanisms for agri-food did not work as intended, in
part because those that would have directly benefitted from harmonization did not commit sufficient resources to the process – or they

may have naively believed that harmonization would simply *happen* because there was a commitment to it in NAFTA.

While little progress on cooperation pertaining to food safety and other food regulations was being made in the decades after NAFTA came into force, the broader food culture was evolving briskly. Food moved from having a primary role as a source of nourishment to a multifaceted phenomenon that embraced lifestyle, recreation, and health, as well as its traditional role in sustenance. Interest in multiple facets of food followed – where it comes from, how it moves through supply chains to retail shelves, what is entailed in its production and processing, what nutritional content it contains, what effect its incomplete use has on waste disposal, etc. At the top of this list sits the question of whether it is safe. With increasing globalization of food supply chains, and a growing range of imported foods, concerns over the safety of foods of foreign origin have arisen apace. Improved communication and the eventual rise of social media mean that food safety incidents are more widely publicized and commented upon. Some high-profile food safety incidents, along with the perceived rising terrorist threat to food supply chains in the wake of the 9/11 events,[35] eventually stirred governments into updating food safety regulatory regimes, including those that applied to imported food. In the United States this culminated in the Food Safety Modernization Act (FSMA), signed by President Obama on 4 January 2011. Ironically, despite the inherently aterritorial nature of consumer concerns over the provenance, content, and safety of food – issues that transcend national boundaries – the outcome has been the reinforcement of territorially specific regulatory processes.

The FSMA has been touted as the most important update of the U.S. food safety regime since the 1930s.[36] It has wide-ranging provisions for ensuring the safety of imported food that have the potential to thicken the border considerably, yet it was enacted with virtually no input from NAFTA partners and, at best, half-hearted efforts to keep them informed. This is in spite of the pledges in Article 909 of NAFTA.

The FSMA sets out wide-ranging new requirements requiring regulation, some with deadlines, that must be undertaken by the FDA to implement the Act. These tasks have proven challenging for the FDA to accomplish in a timely manner, leaving food companies, including foreign suppliers, with little clarity on what a FSMA-based regulatory regime will entail or the investments that may be required to comply. Nine years since the new Act became law, the full border-thickening force of the FSMA has yet to be fully felt.

The impetus for an updated U.S. regulatory regime for food safety arose from well-publicized outbreaks of food-borne illness that shook public confidence in the U.S. food supply system. Imports became a particular concern after widely publicized problems with food from China in 2007.[37] The United States imports food from over 150 countries, and there is a widely held public perception that the food safety standards of many countries from which imports are sourced are weak or that enforcement is lax. Imported food constitutes 15 per cent of the U.S. food supply, including 80 per cent of the seafood and approximately 60 per cent of the fresh produce that is consumed.[38] The FSMA focuses on preventing food safety–related problems rather than mitigating them.

The FMSA became law in 2011, but a considerable grace period for implementation was granted as the FDA and other U.S. regulatory agencies needed to develop new protocols and procedures, train staff, and inform domestic and foreign-origin food supply chain participants what compliance entailed. The FSMA does not regulate meat, poultry, and dairy products, which remain in the regulatory ambit of the USDA. Seafood, dietary supplements, and alcoholic beverages, however, are covered in the Act.

The following are the key policy changes in the FSMA with potential[39] implications for Canadian firms exporting agri-food products to the United States:[40]

1 *The foreign supplier verification program*: The FDA has been given the
  power to require import certification that attests that imported food
  was produced in compliance with U.S. laws and regulations. U.S.
  importers will be required to verify the activities of their foreign
  suppliers, ensuring their suppliers produce foods that comply
  with: (1) hazard analysis and preventative controls (HACCP); or,
  (2) production and harvesting standards. The FDA is to provide
  new regulations that define the required verification methods. Food
  processors, and in many cases farmers, will have to learn about
  and understand these verification methods if they wish to export to
  the United States. The FDA will determine requirements based on
  the known risks associated with the food or its geographic origin.
  Food without proper foreign supplier verification, and imported
  food without a verification program in place may result in import prohibitions or criminal prosecution. Food production facilities must inform the FDA, in writing, of all identified hazardous
  practices along their supply chains and their plans to implement
  preventive measures.
2 *Frequency of inspection*: The frequency of inspections by the FDA
  is to increase. Those facilities designated as "high risk" must be
  inspected every three years, while those designated as "low risk"

must be inspected within seven years. Both foreign and domestic facilities must be inspected. By 2011, the FDA was mandated to inspect no fewer than 600 foreign facilities, and inspections of foreign food facilities were then to double each year over the next five years. When fully implemented, inspection of foreign facilities must take place twice a year. Thus, food processing facilities in the exporting countries must prepare for inspections twice a year. Further, in an effort to improve food safety oversight, FDA offices are to be established in at least five foreign countries that export food to the United States. Thus far, the FDA has not has sufficient resources to meet these targets but it is making progress. The FDA will have the authority to review the current food safety practices of countries that wish to supply the U.S. market – and the foreign governments must prepare the required information and cooperate with the FDA if they wish to maintain and expand food product exports to the United States.

3 *Standards for on-farm production and harvesting*: Nationwide science-based mandatory standards for producing and harvesting fresh produce are to be established by the FDA. Farmers in foreign countries may have to alter their production methods to be able to access the U.S. market. For some specified vegetables and fruits as well as produce that are designated as being "high risk" – designated raw agricultural commodities – the FDA is to publish safety guidelines.

4 *Effective traceability*: In coordination with the fruit and vegetable industries, the FDA is to create a new method of effectively tracking and tracing fresh produce.

5 *Laboratory accreditation*: By early 2013, the FDA was mandated to develop a mechanism to accredit laboratories for food safety testing. The mechanism is to have model standards that include sampling and analytical procedures, internal quality controls, and training for individuals carrying out the collection of a sample and subsequent analysis. The goal is to increase the number of laboratories that qualify. Foreign laboratories are eligible for participation if they achieve the model standards. Laboratories were supposed to be required to be accredited to conduct any regulatory testing by mid-2013. Foreign governments must weigh the costs of having their own certified laboratories or relying on certified foreign laboratories. With perishable food products, the time that testing takes is important to prevent deterioration in the quality of the shipment.

6 *Third-party auditors*: The FSMA requires that the FDA establish a means to recognize accreditation bodies and third-party auditors.

Third parties can be a foreign government, a private firm, or a non-government organization (NGO). Third-party audit certifications will be used to ensure that an imported product complies with U.S. laws and regulations. Foreign governments must decide which form of institution will provide the most effective certification for its food supply chains.

7 *Mandatory registration*: A new twice-yearly registration procedure is to be established, and firms must attain compliance with updated requirements or risk suspension. Food facility registrations will need to be renewed every two years. The FDA can suspend a registration, meaning it would be impossible to import food into the United States from that facility.

8 *Agriculture and food products transportation*: Regulations on sanitary practices in transportation are to be developed by the FDA. Shippers, receivers, and others engaged in transportation of food will be required to implement the practices.

9 *The burden of costs and incentives*: The FDA may collect fees to offset importer re-inspection related costs and for administering the qualified importer program. Firms that require re-inspection or recall may be subject to a fee established by the FDA.

Taken together, this represents a massive undertaking for the FDA, and it has struggled to implement FSMA. It requires a large number of specially qualified personnel to develop the array of regulations and to engage in the monitoring programs. The FDA has struggled to find sufficient personnel. While US$1.4 billion was initially budgeted for implementation, it proved insufficient.[41] Once fully implemented, however, the FSMA will thicken the U.S. border for Canadian food exports. This thickening was established by the United States without any attempt to involve the NAFTA partners in the process.

A major revision to Canada's food safety system took place at approximately the same time in response to similar concerns among consumers and other groups in civil society. The result was the Safe Foods for Canadians Act.[42] Like the FSMA, it was developed unilaterally with little or no input from other countries, including Canada's NAFTA partners. There are, however, regulatory cooperation initiatives at the technical, rather than the political, level, such as the Food Safety Systems Recognition Arrangement signed by the CFIA and the FDA in 2016 to foster the granting of equivalence (but not harmonization) in food safety procedures. The agreement recognizes that the food safety systems in each country produce comparable public health outcomes and establishes a framework for regulatory cooperation. Another aterritorial initiative is the RCC agenda item on mutual recognition of zoning during disease outbreaks.[43] Such technical initiatives are often derailed by high-profile events such as the bovine spongiform encephalopathy (BSE) – mad cow disease – outbreak in both countries in 2003, mandatory country-of-origin labelling (MCOOL), or major food safety incidents. As Hale and Bartlett suggest, "Regulatory processes will remain primarily territorial to facilitate border enforcement by both countries."[44]

While principal responsibility for food regulations is primarily territorial, the agri-food sector is complex and multifaceted, and there is a spectrum of arrangements, from those that are strongly territorial and independent, to other examples where regulatory cooperation has emerged. In the continuum of policy relations and policy instruments outlined in chapter 1, the Technical Working Group on Pesticides features strong cooperation and even coordination. Meat inspection for domestic consumption reflects the establishment of parallel – but separate – inspection regimes; however, meat inspection for export exhibits an element of cooperation and aterritoriality. The USDA rights of inspection of Canadian plants licensed for export to the United States and the capacity for the USDA to withdraw export licences from Canadian plants is strongly aterritorial. Given the integration of cross-border food supply chains, major food recalls that affect domestic and export shipments call for a form of cooperative territoriality between food safety regulatory authorities in both countries.

Conclusions

Notwithstanding the attempts at cross-border cooperation, the record on food safety since the inception of NAFTA suggests that policymaking in this area remains overwhelmingly territorial, despite the aterritorial forces at play. This is because food safety done wrong is a toxic political issue that can harm a government and individual politicians in significant ways. The tendency towards territoriality appears to have increased, particularly given the greater public scrutiny of food safety incidents in recent years. Thus, while there is considerable recognition of the trade concerns that make food safety aterritorial, including attempts to create institutional venues where trade concerns can be brought into the policy process, real progress remains elusive. The one exception internationally has been the European Union, where considerable progress has been made in creating a single market that has harmonized food safety regulations. The unique institutional role of the European Commission may have been central to that success. However, this progress has not extended to the Comprehensive Economic and

Trade Agreement (CETA), where agreement on technical standards for additional EU meat imports remains elusive. NAFTA has not enjoyed similar success, despite the good intentions at the time it was negotiated.

Despite recent theoretical arguments for increased regulatory coordination[45] and optimistic projections of the benefits,[46] there appeared to be little interest in the issue in the re-negotiation of NAFTA. Further, on food safety issues, little will change with USMCA. This is because so little has been accomplished on this issue in the more than twenty years of NAFTA's lifespan. On food safety, the North American market is not integrated. What cooperation there is has more to do with the management of cross-border trade than any harmonization commitments in NAFTA. With USMCA a great deal of trade will continue to take place and technical issues will have to be managed. If the resources become available in the United States to fully implement the FSMA, then the border can be expected to thicken on food safety regulations – an outcome that would occur with or without USMCA.

NOTES

1 William A. Kerr, "What Is New in Protectionism? Consumers, Cranks and Captives," *Canadian Journal of Agricultural Economics* 58, no. 1 (2010): 5–22.

2 K. Gillis, C.D. White, S.M. Ulmer, W.A. Kerr, and A.S. Kwaczek, "The Prospects for Export of Primal Beef Cuts to California," *Canadian Journal of Agricultural Economics* 33, no. 2 (1985): 171–94; William Kerr, "Removing Nontariff Barriers to Trade under the Canada–United States Trade Agreement: The Case for Reciprocal Beef Grading," *Journal of Agricultural Taxation and Law* 14, no. 3 (1992): 273–88.

3 Dermot Hayes and William A. Kerr, "Progress toward a Single Market: The New Institutional Economics of the NAFTA Livestock Sectors," in *Harmonization/Convergence/Compatibility in Agriculture and Agrifood Policy: Canada, United States and Mexico,* ed. R.M.A. Loynes, R.D. Knutson, K. Meilke, and Daniel Sumner, 1–21 (Winnipeg: University of Manitoba, Texas A and M University, University of Guelph, University of California-Davis, 1997).

4 Hayes and Kerr, "Progress toward a Single Market."

5 David Freshwater, "Free Trade, Pesticide Regulation and NAFTA Harmonization," *Journal of International Law and Trade Policy* 4, no. 1 (2003): 32–57, https://ageconsearch.umn.edu/record/23817/files/04010032.pdf; Tina Green, Lynne Hanson, Ling Lee, Héctor Fanghanel, and Steven Zahniser, "North American Approaches to Regulatory Coordination," in *Agrifood Regulatory and Policy Integration under Stress,* 2nd annual North American Agrifood Market Integration Workshop, ed. Karen M. Huff,

Karl D. Meilke, Ronald D. Knutson, Rene F. Ochoa, and James Rude, 9–47 (Texas A&M University, University of Guelph, Instituto Interamericano de Cooperación para la Agricultura-México, 2006).

6 May T. Yeung, William A. Kerr, Blair Coomber, Matthew Lantz, and Alyse McConnell, *Declining International Cooperation on Pesticide Regulation: Frittering Away Food Security* (London: Palgrave-Macmillan, 2017).

7 Canada, Health Canada, "North American Free Trade Agreement Technical Working Group on Pesticides," 2015, http://www.hc-sc.gc.ca/cps-spc/pest /part/int/_nafta-alena/index-eng.php.

8 NAFTA's lack of a central bureaucracy means that there is no official website or other platform where information on the activities of any of the NAFTA mandated committees are detailed and made accessible. Thus, there is scant information on their activities. In his own research, the author (Kerr) has had to track down individuals who serve on the various committees – a job in itself – and then ask them to share their notes from meetings, if they have kept them.

9 North American Free Trade Agreement, Technical Working Group on Pesticides, *Accomplishments Report for the Period of 2003–2008*, 2009, http:// www.hc-sc.gc.ca/cps-spc/alt_formats/pacrb-dgapcr/pdf/pubs/pest /corp-plan/nafta-alena-2003-2008-eng.pdf.

10 North American Free Trade Agreement, Technical Working Group on Pesticides, *Five-Year Strategy: 2008–2013*, 2009, http://www.hc-sc.gc.ca /cps-spc/alt_formats/pacrb-dgapcr/pdf/pubs/pest/corp-plan/nafta -alena-strat-plan-eng.pdf.

11 North American Free Trade Agreement, Technical Working Group on Pesticides, *Accomplishments Report*, 1.

12 North American Free Trade Agreement, Technical Working Group on Pesticides, *Accomplishments Report*.

13 North American Free Trade Agreement, Technical Working Group on Pesticides, *Accomplishments Report*.

14 North American Free Trade Agreement, Technical Working Group on Pesticides, *Accomplishments Report*, 6.

15 North American Free Trade Agreement, Technical Working Group on Pesticides, *Five-Year Strategy*.

16 North American Free Trade Agreement, Technical Working Group on Pesticides, *Five Year Strategy 2016–2021*, https://www.canada.ca/en/health -canada/services/consumer-product-safety/reports-publications /pesticides-pest-management/corporate-plans-reports/north-american -free-trade-agreement-technical-working-group-pesticides-five-year-strategy -2016–2021.html.

17 Green et al., "North American Approaches to Regulatory Coordination," 9–47.

18 In the wake of NAFTA coming into force, some additional voluntary mechanisms came into existence. One is the Tri-National Accord, which includes agricultural officials from Canadian provincial governments and U.S. and Mexican state governments as well as federal representatives of the three countries. It has been operating since 1992. One of its aims is promotion of regulatory cooperation. As with the formal NAFTA mandated committees, the Tri-National Accord appears simply to have been a place to "talk and talk."

19 Canada, Canadian Food Inspection Agency, "Canada, United States Agree on Application of Zones to Allow Safe Trade in the Event of an African Swine Fever Outbreak," 22 May 2019.

20 Stuart J. Smyth, William A. Kerr, and Richard S. Gray, "Regulatory Barriers to International Scientific Innovation: Approving New Biotechnology in North America," *Canadian Foreign Policy Journal* 23, no. 2 (2017): 134–45.

21 Canada, "Canada's Economic Action Plan: Food and Agriculture," 2015, http://actionplan.gc.ca/en/page/RCC-CCR/agriculture-and-food.

22 Jill E. Hobbs, Stavroula Malla, Eric K. Sogah, and May T. Yeung, *Regulating Health Foods* (Cheltenham, UK: Edward Elgar, 2014).

23 Deepananda Herath, John Cranfield, Spencer Henson, and David Sparling, "Firm, Market and Regulatory Factors Influencing Innovation and Commercialization in Canada's Functional Food and Nutraceutical Sector," *Agribusiness* 24, no. 2 (2008): 207–30; Jennifer Farrell, Nola M. Ries, Natasha Kachan, and Heather Boon, "Foods and Natural Health Products: Gaps and Ambiguities in the Canadian Regulatory Regime," *Food Policy* 34, no. 4 (2009): 388–92.

24 Hobbs et al., *Regulating Health Foods.*

25 Looking more broadly in the context of other trade agreements, it does not appear that the Comprehensive Economic and Trade Agreement (CETA) between Canada and the European Union has a process for achieving mutual recognition of regulatory standards for health claims.

26 Hobbs et al., *Regulating Health Foods.*

27 Hayes and Kerr, "Progress toward a Single Market: The New Institutional Economics of the NAFTA Livestock Sectors," 1–21.

28 William A. Kerr, "Political Precaution, Pandemics and Protectionism," *Journal of International Law and Trade Policy* 10, no. 2 (2009): 1–14, https://ageconsearch.umn.edu/record/52231/files/kerr10-2.pdf.

29 Mutual recognition is an alternative approach and is known formally as "granting equivalence." The granting of equivalence has been seldom observed because if it could be granted it would also be relatively easy to fully harmonize. While granting equivalence can be done and may seem a simple (or low cost) solution, it requires monitoring to ensure that

divergent standards do actually remain equivalent over time –which takes an ongoing commitment of resources.

30 Laura Loppacher and William A. Kerr, "The Efficacy of World Trade Organization Rules on Sanitary Barriers: Bovine Spongiform Encephalopathy in North America," *Journal of World Trade* 39, no. 3 (2005): 427–43.

31 Rhetoric surrounding the imposition of mandatory country of origin labelling (MCOOL) by the United States (and opposition to its rollback) being one example.

32 William A. Kerr and Susan E. Cullen, "Canada-U.S. Free Trade: Implications for the Western Canadian Livestock Industry," *Western Economic Review* 4, no. 3 (1985): 24–36.

33 Kerr, "Removing Nontariff Barriers to Trade."

34 A full discussion of *fulfillment costs* can be found in Hayes and Kerr, "Progress toward a Single Market."

35 William A. Kerr, "Homeland Security and the Rules of International Trade," *Journal of International Law and Trade Policy* 5, no. 1 (2004): 1–10, https://ageconsearch.umn.edu/record/23852/files/05010001.pdf.

36 Tekuni Nakuja, Mahzabin Akhand, Jill E. Hobbs, and William A. Kerr, "Evolving US Food Safety Regulations and International Competitors: Implementation Dynamics," *International Journal on Food System Dynamics* 6, no. 4 (2015): 259–68.

37 Huanan Liu, William A. Kerr, and Jill E. Hobbs, "Product Safety, Collateral Damage and Trade Policy Responses: Restoring Confidence in China's Exports," *Journal of World Trade* 43, no. 1 (2009): 97–124.

38 D. Superville and M.C. Jalonick, "Obama to Sign Food Safety Bill Today," *Huffington Post*, 4 January 2011.

39 The term "potential" is used because the FDA has not had sufficient resources to implement most of the mandated changes. Over time, as the FDA is able to accomplish what has been mandated for the agency, Canadian exporters will be negatively impacted. See Nakuja et al., "Evolving US Food Safety Regulations," for a discussion of the FDA's challenges in acquiring sufficient resources to implement the provisions of the FSMA.

40 Nakuja et al., "Evolving US Food Safety Regulations."

41 Nakuja et al., "Evolving US Food Safety Regulations."

42 Geoffrey Hale and Cailin Bartlett, "Managing the Regulatory Tangle: Critical Infrastructure Security and Distributed Governance in Alberta's Major Traded Sectors," *Journal of Borderlands Studies* 34, no. 2 (2019): 257–79, doi.org/10.1080/08865655.2017.1367710.

43 For a discussion of zoning issues, see Laura Loppacher, William A. Kerr, and Richard R. Barichello, "Regional Management of Diseases and Pests to

Facilitate Canadian Exports," *Canadian Journal of Regional Science* 31, no. 1 (2008): 30–54.

44 Hale and Bartlett, "Managing the Regulatory Tangle," 18.

45 Yadira Tejeda Saldana and Erin Cheney, "NAFTA 2.0 Trade and Safe Food," *Ivey Briefing*, Agri-food at Ivey (London, ON: University of Western Ontario, 2017).

46 Rory McAlpine and Mike Robach, *Risk and Reward: Food Safety and NAFTA 2.0* (Ottawa and Washington: The Canadian Agri-food Policy Institute and the Canada Institute of the Wilson Center, 2017).

19 International Traffic in Arms Regulations and Quebec's Aerospace Industry

MATHILDE BOURGEON AND ÉLISABETH VALLET

The beginning of the 1990s has been seen as the end of history,[1] marking the disappearance of borders[2] and consecrating the obsolescence of border barriers.[3] This decade was also to signal the advent of a globalized world, in which certain authors primarily emphasized its economic dimension, with the mantras of free movement of goods, capital, and people.[4] In addition, while the Treaty of Maastricht initiated a quantum leap in European integration, the 1994 signing of the North American Free Trade Agreement (NAFTA) by the United States, Canada, and Mexico marked the deepest extent of institutional globalization in this region. However, North American integration has remained imperfect. Although NAFTA provided for the gradual thinning of trade borders, all three countries maintained exceptions for certain industries and activities such as fishing, uranium mining, domestic transportation, and water management.[5]

As globalization has not been uniform or continuous,[6] national debates over its limits and conditions remain highly visible,[7] based as they are on different theories and perceptions of economic security. These biases reflect the paradox observed by Denis Duez that borders can be open and closed at the same time.[8] Similarly, Beck finds the key to political globalization in the relationships between territory (space) and borders (frontiers).[9] Wendy Brown observes that the reinforcement of borders is based on the logic of globalization, for which it has become an essential element.[10] Thus, reshaping and delocalizing borders in a world of intensified mobility has not contributed to the deconstruction of borders as much as to their contemporary (re-)definition:[11] borders are consubstantial with globalization.

The events of 11 September 2001 brought the era of debordering to an abrupt halt, triggering instead a series of rebordering processes.[12] Initially, it was thought that these security responses ultimately would give way to the victory of free trade over state borders. However, the 9/11 effect[13] has not only persisted but has become entrenched to the point of redefining norms of international relations.[14] Economic actors on each side of borders have come to recognize not only the continuation but reinforcement of protective and protectionist measures,[15] not merely for domestic industries or markets, but for national security.[16]

American National Security in a Global Environment

At the peak of the Cold War, potential threats created by the spread of globalization and the emergence of new economic powers led advanced industrial nations, especially the United States, to attempt to protect their technological and scientific leadership by imposing strict export controls while simultaneously pursuing greater openness of international markets.[17] In 1976, the U.S. Congress adopted the Arms Export Control Act (AECA), which gave the president and subsequently, in 1997, the State Department,[18] the right to control exports of defence-related products, technologies, and services. These powers were intended to reinforce presidential capacity to respond rapidly to technological shifts and diplomatic developments during potential international crises. The Executive Branch thus acquired the power to control exports of defence-related products, which it could define through the United States Munitions List (USML). The AECA allowed these goods to be exported on condition that they did not contribute to the arms race, to the development of weapons of mass destruction, to supporting terrorism, to triggering or escalating violent conflict, or to prejudice the negotiation of bilateral or multilateral arms control or non-proliferation agreements.[19] Thus, AECA's objective was both to limit the arms race and prevent technologies developed to defend the United States from falling into the wrong hands.[20]

Until policy shifts initiated by the Trump administration,[21] the sharing of technology and other intellectual property with allied countries such as NATO members was seen to serve U.S. foreign policy and security interests.[22] At the same time, this policy entailed risks of transferring these technologies and knowledge to other countries, given differences in allies' security practices and commercial networks. The AECA provided a means of resolving this dilemma. It gave the Executive the power to allow security-sensitive transactions with allied countries, such as arms sales and military cooperation, while taking steps to prevent such benefits from accruing to "enemy" countries.[23] These principles reflect a classic, state-centred view of international relations.

This framework allowed for implementation of a set of regulations aimed at controlling defence-related technologies and associated information based on the USML in the International Traffic in Arms Regulations (ITAR) by the Directorate of Defense Trade Controls (DDTC) of the Department of State to administer export controls on otherwise unclassified American defence technologies under Executive Order 13637. These regulations were intended to control all exports of U.S. defence products and related technologies. They required aerospace industry firms, among others, to secure State Department authorization through the DDTC before exporting products on the USML.

The Impact of American Policy on Canada's and Quebec's Aerospace Industry

The United States' hegemonic position[24] has permitted it to define international norms consistent with its perceptions of the economic environment and the roles of particular actors[25] through international organizations of which it is a member and often sponsors. This influence clearly results in the elaboration of norms for the aerospace sector, broadly defined,[26] as U.S. pre-eminence within the global economy has enabled it to redefine the terms to permit trade flows in light of its narrowly defined national interest. U.S. international economic policies always have some impact on the Canadian economy, if only for reasons of geographic proximity. For example, U.S. adoption of "Buy American" measures had immediate effects on demand for certain Canadian products.[27] In the same way, the U.S. embargo on trade with Cuba came into conflict with Canada's Foreign Extraterritorial Measures Act (FEMA) so that complying with Canadian laws could place firms, including subsidiaries of U.S. firms, at risk of being fined by the U.S. Office of Foreign Assets Control (OFAC).[28]

   Like the automotive sector, American and Canadian aerospace industries are particularly interconnected, representing significant bilateral trade flows.[29] Canada's aerospace industry is the world's fifth-largest, reporting revenues of US$20.8 billion in 2016. Quebec accounts for 43 per cent of industry revenues.[30] The civil aerospace industry is closely linked to the defence sector because dual use technologies are important and many technologies and components are shared between the neighbouring countries. The Canadian aerospace industry is oriented towards U.S. markets, particularly for the purchase of components, but also for cooperation in bilateral and multinational projects.[31] In fact, Canada is the United States' fifth-largest aerospace export market with US$8.3 billion in 2015. The United States is the Canadian industry's largest supplier, providing more than half of imports, while being its largest customer, with 60 per cent of Canadian exports.[32]

In June 2019, Canada ratified the United Nations Arms Trade Treaty, which entered into force in Canada in September of the same year. This ratification creates a new legal framework in Canada on arms exports, with a mandatory licence for the export of items and technologies regulated by the Arms Trade Treaty. This licence applies regardless of the country of the end user. However, in order not to burden trade relations between the United States and Canada, the latter has implemented a general export licence. This provides a single permit for Canadian individuals and companies wishing to export treaty weapons, as well as other defence items and technologies, to the United States. In this sense, although Canada's ratification of the treaty limits in principle its exports of defence articles and technologies on a global scale, the prevalence of the American market in Canada is such that the federal government has had to provide for a special export regime to the United States.

This interdependence explains why laws governing U.S. technologies have a major impact on the Canadian aerospace industry and why the Canadian industry is particularly affected by ITAR's existence. ITAR's original intentions were to avoid uses of dual use technologies, developed or used for defence-related purposes, which would ultimately serve the interests of U.S. adversaries. But if ITAR was initially intended to protect U.S. territory and national security, it has become an instrument of commercial warfare.[33]

ITAR: Security and Economic Instrument

The ITAR system has always had a disproportionate effect due to the deep integration of dual use technologies within the U.S. economy as much as because of wider American economic hegemony.[34] However, amid the aftershocks of 9/11, its operations were subject to an important qualitative change. The State Department adapted its application of ITAR to the assessments of the Defense Department and other national security agencies, which took this opportunity to re-evaluate the nature and sensitivity of technologies they were called to assess.[35] From the early 2000s, ITAR's application evolved to include a far broader number of components. From this perspective, *every* product and technology used by the aerospace sector was subject to a tension between its use in global markets and the potential risks of its misuse by an adversary of the United States, particularly if it had been developed for military uses. ITAR thus had the potential to become *both* a tool for reinforcing U.S. security and an instrument of economic warfare.

These regulations also have an extraterritorial effect[36] to the extent that they affect not just the U.S. aerospace industry but also related sectors beyond its borders from the moment they use American technologies to develop their products. Foreign firms located outside the United States must conform to ITAR rules if they wish to import or export technologies subject to those rules, notably those listed on the USML.

These regulations, particularly if applied strictly, cover a very broad range of products and technologies, not only because of the nature of the articles included on the USML, but also the *intention* behind their initial creation. All products conceived, developed, configured, adapted, or modified for military application are subject to ITAR.[37] Therefore, the mere presence of a technology or component originally developed for military uses can fall within the ambit of these regulations. For example, the mere presence of a U.S. defence-related component or technology in a final product, such as an aircraft cockpit, creates a ratchet effect that makes aircraft designed by companies like Bombardier subject to U.S. State Department oversight. By limiting and controlling the export of U.S. defence-related items and technologies, this agency can reduce the likelihood that such technologies be misused or repurposed to threaten American security or territorial integrity. ITAR's objective is therefore to control the final destinations of articles and technologies listed on the USML to prevent them from being used by the defence industries of U.S. adversaries or by members of terrorist networks.

As a result, the strict application of ITAR regulations can preclude the sale of an entire aircraft to a third county if it contains a single component subject to its provisions.[38] For example, if an American firm exports a gas gauge to Canada, the State Department must confirm that the Canadian company that uses it to build an aircraft cockpit will not export the latter to a country subject to U.S. trade restrictions. Interviews conducted by the authors of this chapter indicate that the inclusion of components as prosaic as keyboard keys can "contaminate" an entire supply chain[39] and a fully assembled product.[40]

However, as a result of a generalized mistrust rooted in broader U.S. attitudes towards security and international relations and increased involvement of extra-governmental organizations in global markets,[41] ITAR does not apply to firms' U.S. operations in the same way that it does to those beyond its territory, where certain regulations are applied more strictly. Canada has been particularly affected as a result of the integration of North American production networks under NAFTA,[42] tougher enforcement of regulations, and broader trends towards border thickening affecting both companies and jurisdictions. These borders have become emblems of tensions between increased enforcement and efforts at facilitation.[43] As a result, the aerospace sector, particularly its operations around Montreal, has been directly affected by these rapid changes, as the functional border has shifted in ways that affect relations between aerospace firms and their clients, suppliers, and employees – leading the entire aerospace cluster to make significant adaptations to these new realities.

The Aterritoriality of the Border

For the United States, trade-related risks addressed by ITAR include Access to Information regulations, not just those involving final users. For this reason, Canadian businesses subject to these regulations must ensure that their employees with access to USML-listed technologies and data do not represent a risk to American security – as defined by the United States – even if located on Canadian soil.

Canada's initial exemption from this aspect of ITAR rules was withdrawn in 1999. Some Canadian firms had been accused of passing USML-listed technologies to China, Libya, Sudan, and Iran, even as the United States maintained trade embargoes on these countries.[44] The very fact that Canada's ITAR exemption was lifted clearly demonstrates the extraterritorial character of these rules covering the activities of foreign firms taking place beyond U.S. territory, even if consistent with the laws of such third countries. The transfers noted above were allowed under Canadian law, but were inconsistent with U.S. security interests and foreign policy.[45]

Canada's Controlled Goods Program (CGP) is a set of regulations similar to ITAR. Technologies and technical data declared as sensitive can be passed on only to individuals or businesses registered under its provisions. To this end, nationals of countries subject to U.S. embargoes are not allowed access to technologies or technical data defined by the USML unless such persons are U.S. citizens or permanent residents, or nationals of countries enjoying specific exemptions from these ITAR provisions.[46] As Canada benefits from such an exemption, Canadian firms are able to employ such individuals if they pass security screening. Even so, between 1999 and 2011, Canada did not have the advantage of exemptions covering dual citizens and permanent residents coming from countries subject to U.S. trade embargoes, placing numerous Quebec-based aerospace companies at risk of violating provincial and federal anti-discrimination laws. In effect, ITAR forbade the employment of nationals of countries subject to U.S. embargoes or considered "at risk" of coming into contact with USML-listed technologies and technical data, in whatever form. For more than a decade, Canadian firms had to come up with alternative solutions for their employees to avoid heavy penalties associated with violating ITAR provisions. These charges could have been brought under civil or criminal law, as illegal transfers of technology were seen as a major threat to American security, whether involving the companies or individuals responsible. Conviction could bring fines as high as US$100,000 and imprisonment of up to two years.

For example, the *Bell Helicopter* case, in which a Haitian-born student was refused an internship by the company because of risks of exposure to USML-listed technologies, was widely publicized in Quebec. As this case was not the only one of its kind, other companies sought to find alternative approaches that would avoid running afoul of sanctions or violations of U.S. or Canadian law. One result of this case was that a particular aerospace company felt obliged to divide its production into two separate geographic locations: one for civilian production without specific restrictions, and the other for military production whose components (and final products) were subject to ITAR, along with all relevant employees.[47] The company assigned all personnel sidelined by ITAR into its civilian production division to avoid discrimination in its employment practices while conforming to U.S. requirements. To minimize the worst impacts on their production processes, numerous firms found alternatives to ITAR-listed U.S. technologies by sourcing them in other countries, according to a 2006 study by Choi and Niculescu.[48]

With most of the industry located in Quebec, both federal and provincial governments were responsible for managing the extraterritorial aspects of ITAR. For more than a decade, Canadian aerospace firms had to find ways to adapt in the absence of an exemption, reconciling the need to respect both ITAR requirements and Canadian laws. The standardization brought about by the Canadian government confirmed the influence of its southern neighbour on Canada's economy and the defining of Canadian interests to accommodate American imperatives. ITAR's extraterritorial application also exported U.S. security and foreign policy policies to move out the border through its impact on Canadian laws.

The restoration of Canada's ITAR exemption in 2011 permitted the reintegration of Canadian aerospace firms in American markets by enabling them to export to final product users approved by the DDTC without facing double jeopardy in U.S. and Canadian laws. It also enabled employment of dual nationals and permanent residents from countries subject to U.S. embargoes if they followed strict, DDTC-approved screening. The latter condition required affected

Canadian firms to carry out their own security investigations along with those carried out by DDTC through the State Department on all employees subject to ITAR processes.[49] These requirements were a reminder of both the existence and persistence of internal borders within Quebec-based companies. This ITAR clause effectively limited the employability of certain people on the basis of their birthplaces and nationalities – symbolizing the mobility of American borders, as they did not apply to U.S. dual nationals and permanent residents. This divergence demonstrates the individualization, even "pixeliza-tion"[50] of the border, as rules do not apply in the same way in different countries, so that an American permanent resident born in Afghani-stan would have greater advantages under ITAR than a Canada-Haiti dual national.

But this aspect of ITAR regulations goes even further in applying "upstream" within aerospace supply chains. Canadian universities are not included within Canada's ITAR exemptions, thus affecting their capacity to carry out certain research projects in Canada. If ap-plied research – the sort that leads to development of specific prod-ucts or technologies subject to commercialization – is subject to ITAR regulations, leading to their inclusion on the USML, ITAR's applica-tion to basic research remains fuzzy. Universities interpret ITAR rules governing basic research loosely, with the result that some have had to fight U.S. authorities for the right to exclude such research projects from ITAR.[51]

Universities are also limited in their ability to include certain foreign students (from countries judged to pose risks to the United States) in either basic or applied research projects involving USML-related pro-jects and technologies that require security clearances. As they are not affected by ITAR exemptions, universities cannot address the problem of employability of foreign students by establishing security screening similar to that used by aerospace firms.

Therefore, Canadian universities face ITAR-related limits on the re-search projects, particularly those related to applied research. Basic and applied research are exempt only from restrictions in American univer-sities actually located in the United States. If ITAR restricts aerospace firms from doing research in Canada, Canadian-based aerospace firms risk becoming more and more dependent on American research, creat-ing a ratchet effect of making them increasingly subject to ITAR rules due to their use of USML-related intellectual property. This leads to the progression of externalization of U.S. borders within Canada, applying to businesses and individuals alike, and effectively taking precedence over Canadian jurisdictions.

National Security in the Service of American Commercial Hegemony

ITAR regulations are not intended solely to reinforce American security, even though their implementation was based on security considerations and their expansion after the 9/11 attacks were explained in similar terms. However, they also reflect asymmetries of border relations[52] and the chronic imbalance in trading relationships. ITAR has become a significant tool in commercial warfare for the U.S. government, permitting it to reinforce its hegemony in the defence, aeronautic, and aerospace sectors.

On numerous occasions, the State Department has prevented sales by U.S. trading partners to third countries, anticipating that such sales would undermine American or international security. The most recent example was a 2018 U.S. veto of SCALP cruise missiles sales to Egypt by the French company MBDA. The blocking of this sale was justified on the basis of ITAR-related components in missile design, and by American commitments to protect Israeli security interests. The U.S. government did not hesitate to encroach on French sovereignty in preventing this transaction as inconsistent with American interests and that of its ally, Israel.[53] More recently, the U.S. government had to give its approval for Canada to purchase twenty-five fighter aircraft from Australia, as they were originally built in the United States and developed with U.S. technologies.[54]

This tool is a two-edged sword. On one hand, Canadian aerospace firms have found alternatives to American components to avoid subjecting their technologies to ITAR regulations. However, the strict applications of ITAR standards have harmed both U.S. and Canadian firms on occasions by permitting European and Asian firms to compete more effectively outside North America. ITAR restrictions enabled the European firm Thales Alenia to sell its first satellite to China in 2012.[55] Not only were American firms prevented from selling this type of technology to China, but Canadian firms were precluded from competing for this contract without risking their newly restored ITAR exemption.

Indeed, China represents a major challenge for the United States in arms exports and defence technologies. Since 1989, the Chinese giant has been under an American arms embargo. Originally, the U.S. wanted to punish China for human rights violations, responding to domestic public opinion. In 1993, the terms of the embargo were revised and extended. First, it prohibited the export of items, technologies, and services contained in the USML. But its revision also prohibits the import of weapons under the ITAR regime from China and the export of dual-use technologies and satellites. However, as suggested by the theory of international sanctions, their objectives are likely to change. Thus, in the late 1990s, Washington's priorities changed, and the purpose of sanctions was no longer merely symbolic, but became strategic. The aim is then to curb the modernization of China's military arsenal. This objective requires a stricter application of the embargo, but also the involvement of a greater number of national states in it. Therefore, over decades, the United States has not only prevented EU member states from withdrawing from the embargo, but also put pressure on Israel to join it, de facto. This will help to reduce potential market share and financial losses for U.S. companies, despite the fact that they are subject to stricter regulations, particularly for dual-use technologies and satellites. Thus, American defence-related companies do not have access to the Chinese market as a result of the revision of the ITAR embargo.[56]

Today, 95 per cent of defence technologies and items are developed by private companies or global players, and only 5 per cent are the result of U.S. government R&D. Thus, the United States has gradually lost its technological superiority in the field, particularly in robotics and artificial intelligence largely dominated by its allies, and where China is progressing rapidly. As a result, foreign powers are taking advantage of the space left by the American decline to establish themselves on the world market. Finally, third countries are trying to limit their collaboration with the United States so that their technologies are not subject to the ITAR regime, and so as not to be handicapped by U.S. regulations. In this context, to avoid this imposed disengagement and to regain their place in the international defence technology market, the U.S. government has everything to gain by revising ITAR.[57]

Despite such developments, the U.S. government continues to see ITAR as a toolkit to control the export and diffusion of American technology. Moreover, the U.S. Congress has made several efforts to implement similar regulatory systems in other parts of the Executive Branch, notably the Export Control Reform Act (ECRA) incorporated within the 2018 National Defense Authorization Act (NDAA) to create a permanent system of export controls. This bipartisan project aims to strengthen U.S. national security in ways that support the international competitiveness of its industries.[58] This legislation is similar to ITAR, but it applies to export controls on dual-use technologies administered by the Department of Commerce rather than the Department of State. ECRA therefore extends and "completes" ITAR by controlling exports of dual-use technologies beyond those contained on the USML, sharing or diffusion of which might pose a risk for U.S. national security.[59] U.S. government control on exports of dual-use technologies thus appears to be unlimited. It applies not only to exports originating in the United States, but to all products and technologies having components linked in some way to American national defence, anywhere on earth. On the basis of this logic, some observers have suggested that any airline that owns a Boeing product should, in theory, demand State Department authorization before the take-off of any international flight.[60] ECRA will multiply restrictions for U.S. firms and their international customers. In light of the Trump administration's overriding emphasis on "America First" policies and presidential criticisms of allies, a mistaken analysis of restrictions implemented under these provisions under the auspices of national security and economic protectionism could weaken both U.S. defence industries and international security.[61]

Conclusion

ITAR is both the cause and consequence of the movement of borders in the twenty-first century, confirming that borders have become multidimensional and dispersed.[62] The increased enforcement of ITAR rules have led Canada and its aerospace industry to take steps to limit their impact. American political hegemony in North America and its global business and technological leadership have the capacity to reshape the national jurisdictions of other countries. The projection of American borders is such that Canadian companies often choose to submit to their jurisdiction – superimposing it on that of their own country. Rather than being exclusive, state sovereignty is called into question by external security imperatives – as if the United States could declare a state of emergency among its allies, impose security controls on foreign populations, and supervise their trade relations with other states.

   While the Canadian aerospace industry welcomed the restoration of its ITAR exemption in 2011, for Canada these measures mark the recognition of the prevalence of U.S. jurisdiction over its domestic industry. Its government was forced to accept a similar policy regime, employee security as rigorous as in the U.S., and the limitation of trade flows with third countries subject to U.S. trade sanctions. Bilateral economic integration meant that the survival of the Canadian aerospace sector depended on the acceptance of the extraterritorial application of ITAR on Canadian soil. ITAR's aterritorial application explains this evolution and projection of U.S. borders, together with ongoing tensions between simultaneous processes of bordering and "de-bordering."

NOTES

1 Francis Fukuyama, *The End of History and the Last Man* (New York: Free Press, 1992).

2 Malcolm Anderson, "Les frontières: un débat contemporain," *Cultures et conflits* 26–7 (1997).

3 Élisabeth Vallet, ed., *Borders, Fences and Walls: State of Insecurity* (New York: Routledge, 2014).

4 Christian Deblock and Michèle Rioux, *De la nationalisation du monde à la globalisation* (Québec: Presses de l'Université Laval, 2013).

5 Élisabeth Vallet and Pierre-Louis Malfatto, "Water Geopolitics in North America," in *The Geopolitics of Natural Resources*, ed. David Lewis Feldman (Aldershot: Edward Elgar Publishing, 2004).

6 Arjun Appadurai, "Disjuncture and Difference in the Global Cultural Economy," *Theory, Culture & Society* 7 (1990): 295–310.

7 Yun Zhao, "Export Controls over Space Products," *National Space Law in China* 10 (2015): 155.

8 Denis Duez and Damien Simmoneau, "Repenser la notion de frontière aujourd'hui. Du droit à la sociologie," *Droit & Société* 98 (2018): 37–52.

9 Ulrich Beck, *Pouvoir et contre-pouvoir à l'ère de la mondialisation* (Paris: Flammarion *Champs essais*, 2009).

10 Wendy Brown, *Murs: Les murs de séparation et le déclin de la souveraineté étatique* (Paris: Les Prairies Ordinaires, 2009).

11 See Mathilde Bourgeon, Thalia D'Aragon-Giguère, and Élisabeth Vallet, "Les flux migratoires à la frontière québéco-américaine," *Quebec Studies* 64 (2017): 141–56.

12 Vladimir Kolossov and James Scott, "Selected Conceptual Issues in Borders Studies," *Belgeo: Revue Belge de Géographie* (2013): 1, 6.

13 Chaire Raoul-Dandurand (collectif), *L'effet 11 septembre, 15 ans plus tard* (Quebec: Septentrion, 2016).

14 Bertrand Badie, *L'impuissance de la puissance* (Paris: Editions du CNRS, coll. Biblis. 2013); Bertrand Badie, *Nous ne sommes plus seuls au monde* (Paris: La découverte, 2016).

15 Bertrand Badie and Michel Foucher. *Vers un monde post-national* (Paris. Éditions du CNRS, 2017).

16 John R. Liebman and Kevin J. Lombardo, "Guide to Export Controls for the Non-Specialist," *Loyola of Los Angeles International and Comparative Law Review* 28 (2006): 497–599.

17 Fanny Coulomb and Jacques Fontanel, "Mondialisation, guerre économique et souveraineté nationale," in *La question politique en économie international*, ed. P. Berthaud nd G. Kebabdjian, 190–201 (Paris: La Découverte, coll. "Recherches," 2006).

18 Executive Order 11958 (1977) placed enforcement of the Arms Control Export Act under State Department authority.

19 Arms Control Export Act 22-2778, Legal Information Institute.

20 Arms Control Export Act 22-2753.

21 Keith Johnson, Dan De Luce, and Emily Tamkin, "Can the U.S.-Europe Alliance Survive Trump?," *Foreign Policy*, 18 May 2018; Ivo H. Daalder and James M. Lindsay, "The Committee to Save the World Order," *Foreign Affairs*, Nov.–Dec. 2018.

22 Arms Control Export Act 22-2753.

23 Arms Control Export Act 22-2751, Legal Information Institute, https://www.law.cornell.edu/uscode/text/22/chapter-39.

24 Charles-Philippe David and David Grondin, eds., *Hegemony or Empire? The Redefinition of US Power under George W. Bush* (Aldershot, UK: Ashgate, 2006).

25 Interview C; Alan O. Sykes, "Regulatory Protectionism and the Law of International Trade," *University of Chicago Law Review* 66, no. 1 (1999): 1–46. Interviews were conducted by the authors between October 2014 and December 2017 on a confidential basis at the request of persons interviewed, as a result of the sensitive nature of information for particular firms and institutions with which they are associated, and that continue to do business with the United States. Interviewees included academics, industry actors, and service providers within the Canadian aerospace sector.

26 Interview A.

27 Jessica LeCroy, Kathryn Friedman, Cyndee Todgham Cherniak, and Laurie Tannous, "The Canada–United States Customs Transaction: The Invisible Border?" *Canada-United States Law Journal* 36, no.1 (2012): 247.

28 LeCroy et al., "The Canada–United States Customs Transaction."

29 LeCroy et al., "The Canada–United States Customs Transaction."

30 U.S. International Trade Administration, "Canada: Civil Aviation" (Washington, DC: U.S. Department of Commerce, 28 September 2017).

31 U.S. International Trade Administration, "2016 Top Markets Report: Aircraft Parts – A Market Assessment Tool for U.S. Exports" (Washington, DC: U.S. Department of Commerce), 32.

32 U.S. International Trade Administration, "Canada: Civil Aviation."

33 Le Croy et al., "Canada–United States customs transaction," *op.cit.* 247.

34 Harrison G. Wolf, "ITAR Reforms for Dual-Use Technologies: A Case Analysis and Policy Outline," in *Aerospace Conference 2012 IEEE* (New York: IEEE, 2012), 1–12.

35 United States, Department of State, Office of Inspector General, *Memorandum Report 01-FP-M-027: Review of the U.S. Munitions List and the Commodity Jurisdiction Process,* March 2001, 2.

36 Mark Gibney, "Toward a Theory of Extraterritoriality," *Minnesota Law Review* 95 (2010–11): 81.

37 Lisa Bencivenga, "International Traffic in Arms Regulations (ITAR): Who Must Comply, What Is Controlled and Where Do We Go from Here?" (Orlando: Orlando World Marriott Center, 20–22 February 2013).

38 Michel Cabirol, "Réglementation ITAR: États-Unis, ces amis qui ne veulent pas que du bien à la France," *La Tribune*, 23 April 2018; Ordway, "U.S. International Traffic in Arms Regulations ('ITAR')."

39 Interview C.

40 Ordway, "U.S. International Traffic in Arms Regulations ('ITAR')," 3.

41 Interview A.

42 Christian Deblock and Christian Constantin. "Intégration des Amériques ou intégration à l'économie américaine?" *Notes et études du Groupe de recherche sur l'intégration continentale*, no. 2 (2000).

43 Victor Konrad, "Toward a Theory of Borders in Motion," *Journal of Borderland Studies* 30, no. 1 (2015): 6.

44 Interview B.

45 Eric Choi and Sorin Niculescu, "The Impact of US Export Controls on the Canadian Space Industry," *Space Policy* 22 (2006): 31.

46 Maroine Bendaoud, "Quand la sécurité nationale américaine fait fléchir le principe de la non-discrimination en droit canadien: le cas de l'International Traffic in Arms Regulation," *Les Cahiers de droit* 54, no. 2–3 (2013): 549–86.

47 Interview C.

48 Choi and Niculescu, "Impact of US Export Controls on the Canadian Space Industry," *op.cit.*

49 John W. Boscariol and Brenda C. Swick, "Final U.S. ITAR Rule on Dual- and Third-Country Nationals Raises New Challenges for Canadian Business," McCarthy Tetrault, 16 May 2011.

50 Anne-Laure Amilhat-Szary, *Qu'est qu'une frontière aujourd'hui* (Paris: Presses Universitaires de France, 2015), 103.

51 Interview D.

52 Victor Konrad and Heather N. Nicol, "Border Culture, the Boundary between Canada and the United States of America and the Advancement of Borderlands Theory," *Geopolitics* 16, no. 1 (2011): 77; Vladimir Kolossov, "Border Studies: Changing Perspectives and Theoretical Approaches," *Geopolitics* 10, no. 4 (2005): 611.

53 Interview B; see Defenseweb.com, "France Looking to Circumvent U.S. Components in Scalp Missile" (7 August 2018).

54 David Pugliese, "U.S. Approves Canada's Purchase of Used Australia Fighter Jets: Deal to Be Completed by End of Year," *National Post*, 21 September 2018.

55 Jim Wolf, "U.S. Squeezes French-Led Satellite Maker over China," Reuters, 9 February 2012 (accessed August 2018).

56 Yoram Evron, "The Enduring US-Led Arms Embargo on China: An Objective-Implementation Analysis," *Journal of Contemporary China* (2019). DOI : 10.1080/10670564.2019.1594099

57 Bill Greenwalt, "Competition with China Requires New Technology Transfer Rules for US Allies and Silicon Valley," *Defense News* (April 2019).

58 Farhad Jalinous et al, "Congress Finalizes CFIUS and Export Control Legislation" (Washington, DC: White & Case, 26 July 2018).

59 David Fagan et al., "Export Control Reform Act Introduced in Congress," *Global Policy Watch* (Bonn), 8 March 2018.

60 Interview C.

61 Lara Seligman, "Trump's America First' Policy Could Leave U.S. Defense Industry Behind," *Foreign Policy,* 18 July 2018, https://foreignpolicy.com/2018/07/18/trumps-america-first-policy-could-leave-u-s-defense-industry-behind/; Christopher Woody, "NATO Allies Are Talking about Breaking Away from the US, but Trump Isn't Their Only Problem, *Business Insider,* 9 August 2018.

62 Etienne Balibar, *Les frontières de la démocratie* (Paris: La Découverte, 1992).

20 Capacity for Choice? Managing International Policy Relations in a World of Shifting Borders

GREG ANDERSON AND GEOFFREY HALE

The end of the Cold War seemingly ushered in an era of declining great power conflict more amenable to the objectives of smaller, relatively trade-dependent states like Canada: expanded, rules-based market access, cooperative decision-making facilitated but not dictated by international institutions, and the legitimacy of expanded democratic governance. These processes have led to the partial redefinition of the nation state (along with associated concepts such as national sovereignty), the ongoing negotiation and sometimes relocation of its administrative boundaries (often internal as well as external), and varied processes for negotiating or projecting national or supra-national legal and administrative norms to navigate the spread of interdependence and shared vulnerabilities across nations.

However, as noted in the introduction to this volume, many of the resulting changes have been uneven in both design and implementation. This untidiness reflects patterns of differentiated integration across economic and policy sectors and subsectors shaped by the management of domestic political and institutional trade-offs to balance international cooperation with domestic political and legal accountability. Canadian policymakers have faced similar challenges in attempting to manage the risks and rewards of growing global interdependence while preserving varying degrees of policy sovereignty – or, at least, sufficient flexibility to provide "capacity for choice."[1] These tendencies have been reinforced by its North American location, and the dynamics arising from the general ambivalence of senior policymakers across party lines, especially in the United States, towards international institution-building since the 1990s. The latter reflects persistent preferences in the projection of U.S. policy goals through a mix of bilateral agreements and more informal institutions, the preservation of presidential autonomy in foreign policy and congressional autonomy over domestic policies, and an overarching aversion to supranational governance institutions and related derogations of sovereignty.[2]

As noted by authors in this volume, political, economic, and administrative elements of this balancing act have been complicated by three major realities that complicate efforts to assess, let alone quantify, interactions among territorial and aterritorial elements of governance regimes across policy fields, economic sectors, and regional contexts. First, relatively few national economic and regulatory policies in Canada can be partitioned any longer into purely "domestic" and "international" silos, although one of these categories may predominate in specific contexts, depending on the relative symmetries or asymmetries of interdependence in particular policy and/or economic areas and subsectors. Varying degrees of intermesticity have become the norm across economic and policy sectors and subsectors, reflecting the relative importance of international (as opposed to domestic) market flows and human movements, market concentration, and relative dependence on domestic (including subnational) as opposed to North American or wider international regulatory regimes and governance networks. This complexity is reinforced by differences in the relative integration (including depth of political relationships) of cross-border regions within North America, as noted in the introduction to this volume.

Second, Canada's international competitiveness has largely depended on its integration within the wider North American economy, particularly relatively barrier-free access to U.S. markets in response to the growing regionalization of the global economy during the 1980s and 1990s.[3] However, its capacity for policy choice in navigating the international projection of U.S. policy preferences and related institutional structures has depended historically either on engaging U.S. political and/or bureaucratic processes to secure recognition and accommodation of its interests, or the use of plurilateral and multinational contexts to facilitate soft law policy arrangements, which are often more adaptable to national institutions and circumstances.[4] But Canada's ability to bridge regional trading blocs or networks has generally been contingent either on the institutional structures, domestic politics, and international priorities of its major trading partners, or on shared recognition of the need to respond to international political, economic, or other shocks such as post-9/11 terrorist threats and the 2008 financial and economic crisis. The result has been limited success in reducing dependence on U.S. markets, diversifying its international trade (if not investment) relations, and navigating the growing divergence of policy goals and norms among other major trading powers. The extent of policy integration depends both on shared policy goals and relative institutional symmetry (or asymmetries) in legislative and regulatory responsibilities within national governments, and regulatory and administrative agencies. These lessons are critical for Canada's adaptation to tensions arising from growing geopolitical competition between its two largest trading partners: the United States and China.

Third, the growing fragility of domestic political support for international economic institutions (including liberalized trade and investment flows) among major democratic powers, especially the United States, has seriously undermined reciprocal policy regimes that have supported the cohesion of the international trading system and post-1990 trends towards North American economic integration. Reciprocal policy cooperation depends on relative unity of national (or, in federal systems, central and subnational) policy objectives, and on the relative priorities of national agencies whose mandates include multiple policy objectives. However, it is difficult to maintain international consensus in responding to the emergence of new domestic political forces demanding recognition or challenging a previous policy consensus that fails to accommodate their interests, priorities, or values. Such demands have been central both to the Trump administration's challenges to the international priorities of its predecessors, whether Democratic or Republican, and in those posed by some of its bitterest critics on the political left – not least, for the latter, to the cross-border integration of energy and environmental policies. Despite policy continuity in some areas, several of these policy shifts pose major challenges to Canada and its related domestic policy processes.

The tensions arising both from NAFTA's multi-stage renegotiation (including congressional amendments outside normal U.S. legal processes) and U.S. domestic economic policy debates suggest that current trends towards American unilateralism, managed trade, and other forms of protectionism within (and beyond) North America will continue beyond the 2020 presidential and congressional elections.

Previous iterations of these debates since the 1990s have contributed to the growth of functional rules-based frameworks in widely varied economic and policy sectors and subsectors, combining both aterritorial and territorial elements. The greater the extent to which debates within major economic powers such as the United States, European Union, and increasingly China focus on domestic priorities or conflicting domestic interests, thereby encouraging their governments to arrogate greater flexibility to themselves rather than commitments to reciprocity within such arrangements, the greater the need for Canadian governments to maintain their flexibility in managing the intermestic effects of such debates

Intermesticity in International Policy Relations

Contributors to this volume have explored numerous international and intermestic policies of Canadian governments in recent years across policy fields to assess the mixture of territorial or aterritorial influences that shape these policies. Table 20.1 attempts to capture their central findings on the relative primacy of territorial, aterritorial, or functionally varied forms of governance in a five-point scale across sectors and subsectors. Importantly, table 20.1 and the findings it summarizes are a snapshot in a changing international policy landscapes. Indeed, as descriptive as each contributor is about factors contributing to the evolution of "territoriality" in each policy domain, in many cases it points to a fluidity of governance that changes the current answers to the central questions about governance along the continuum of "territoriality" described here: who governs, where is governance located, and what processes are underwriting and legitimizing it?

Some, like Anderson and Jones, have emphasized the persistent role played by different federal systems in Canada and the United States, whether in accommodating diverse regional and local interests amid the interaction of integrative and fragmenting forces, or in asserting the primacy of regional interests within territorially based national and/or subnational regulatory systems. Others have pointed to the simultaneous presence of "offensive" and "defensive" policy goals and instruments within the same policy regimes. Kukucha and Anderson note the persistent commitment of Canadian governments to the pursuit of market opening and ongoing access for export-oriented sectors, especially those deeply integrated within continental and global supply chains and investment networks. At the same time, they note Canada's "defensive" efforts to preserve neo-mercantilist and protectionist regimes for selected, politically sensitive sectors, managing resulting conflicts through international dispute resolution regimes and negotiated exceptions to trade rules.

Larger economic powers including the United States, the European Union, China, and Japan continue to pursue similar strategies tailored to their perceived interests, sometimes aggressively so. However, their often competitive efforts at the aterritorial projection of power across policy sectors – clearly illustrated in Bourgeon and Vallet's analysis of U.S. International Traffic in Arms regulations (ITAR) – have prompted countervailing tendencies towards the exercise of defensive economic nationalism (or, in Europe, clashing defensive regionalisms and nationalisms), particularly since the 2008–9 financial crisis.

The default response to such tensions in Canada (and other countries) appears to have been a resort to decentralized, functional policy

Table 20.1. Relative "Territoriality" vs. "A-territoriality" of Policy Fields/Regimes Studied in this Volume

Chapter		Mainly aterritorial	Functionally varied	Mainly territorial
2	Federalism			Xe*
3	Trade	Xb		
4	Regulatory cooperation			Xd
5	Foreign investment	Xb		
6	Border management			
	* Goods trade	Xb		
	* Services trade		Xc	
	* Travel		Xc	
	* Migration			Xd
7	High-skilled temporary workers			Xe
8	Transportation			
	* Trucking			Xd
	* Rail		Xc	
	* Air	Xb		
	* Shipping			Xd
9	Cabotage			Xe
10	National security			Xd
	Critical Infrastructure			Xd
11	Environmental policies – water	Xb		
	Regional cross-border integration			
12	* Pacific NW/Prairies		Xc	
13	* U.S. Northeast/Quebec/Atlantic			Xe
14	* Buffalo-Niagara municipal			Xd
15	* North			Xd
16	Automotive	Xb		
17	Energy		Xc	
18	Food safety			Xd
19	Aerospace/ITAR	Xa		

* Growing with functional decentralization of policy fields
Xa Primarily aterritorial
Xb Primarily aterritorial with territorial elements
Xc Functionally varied
Xd Primarily territorial with aterritorial elements
Xe Primarily territorial

regimes involving formal and informal cooperation or mutual recognition in cases of broadly similar policy regimes, especially with the United States, particularly for industry or policy sectors heavily dependent on American markets. Such processes may be "bundled" by national central agencies to encourage policy coordination, as in the Canada-U.S. Regulatory Cooperation Council initiative since 2011. However, in the absence of significant *political* benefits to national leaders, not just economic benefits to other stakeholders, such measures have shown a repeated tendency to drift. In other settings, particularly when interdependence with Canada is peripheral to U.S. domestic policy debates (as it often is), Canadian governments may parallel major elements of U.S. policy initiatives while retaining sufficient discretion to accommodate particular regional or subsectoral domestic interests. Hale notes that such arrangements are not uncommon in regulatory cooperation processes.

U.S. policymakers' discretion in international and trans-governmental relations is often constrained by domestic policy, procedural considerations, and congressional engagement with details of sectoral and subsectoral policies (and related interest group networks) that often limit executive discretion in dealing with other countries, including Canada. As a result, Canadian parallelism (and often cooperation) in trucking, railway, pipelines, and other energy subsectors has been reinforced by substantial interdependence of markets and industry actors. However, the persistence of federalism, with its assortment of central and diffused subnational regulatory structures in both countries, has reinforced territorial processes for balancing domestic interests while accommodating and managing cross-border (and wider international) interdependence. This national focus is persistently reinforced by intensifying interest group (and sometimes ideological) competition over policy priorities, the related pursuit of regulatory advantage, and the allocation of scarce fiscal resources. In some sectors such as electricity production and distribution, as noted by Tomblin and Hale, it has also reflected domestic competition among provincial (and state) governments that has placed limits on their capacity (or willingness) to coordinate domestic or cross-border regional policies.

Political pressures to maintain domestic (including provincial/ state) capacity for choice are also highly visible in other policy fields addressed by contributors to this volume. Lilly's analysis of highly skilled temporary worker programs notes the impetus for such market movements rooted in the global organization and reach of larger technology firms, the mismatch between relative surpluses and shortages of specialized workers in major industrial countries, and variations in national policies, especially Canada and the United States, aimed at managing these forces. However, she also notes the increasing prevalence in both countries of domestic forces that have contributed to substantial policy divergence. Examples include domestic backlash in Canada against the real and perceived use of temporary workers to engage in abusive labour market practices at the expense of both domestic and imported workers, differences in domestic and cross-border labour markets between the two countries, and broader pressures for immigration restriction in the United States.

International networks and agreements may influence domestic immigration policies in particular areas of interdependence. Examples include formal and informal agreements governing international air travel, passport standards, offshore security screening, international agreements governing treatment of refugees, or limiting "asylum-shopping," such as the U.S.-Canada "Safe Third Country Agreement" (STCA) and related administrative arrangements. Periodic spikes in irregular migration flows highlight both the necessity and limitations of administrative cooperation with foreign governments as social media have greatly expanded opportunities for prospective migrants to share information on enforcement practices and gaps.[5] However, central elements of immigration policies remain resolutely domestic, reflecting their political sensitivity and substantial differences in domestic legal systems and administrative cultures.

Kerr and Hobbs identify similar domestic political sensitivities as a key element in preserving domestic primacy in national food safety systems, despite extensive international interdependence and extensive cross-border cooperation in monitoring and enforcement of domestic standards in international trade. Trade agreements such as NAFTA and the Canada-EU Comprehensive Economic and Trade Agreement (CETA) have been relatively ineffective in fostering expanded regulatory coordination, although some progress towards improved cooperation emerged from the work of the Canada-U.S. Regulatory Cooperation Council in 2012–16. Some agencies such as the U.S. Food and Drug Administration have pursued mutual recognition (and related information sharing) agreements with Canada and other countries as part of broader international regulatory cooperation strategies.[6] However, the diversification of Canada's agri-food relations and consumer-driven concerns for food safety in many countries, sometimes in disregard of scientific evidence, makes Canadian agri-food exporters increasingly sensitive to national market variations in consumer expectations and related domestic regulatory requirements. Such arrangements reflect both "offensive" efforts by Canadian and other governments to pursue the harmonization of regulatory standards and "defensive" efforts to protect particular domestic interests (or political sensitivities) noted in broader discussions of trade policies.

Brendan Sweeney's analysis of investment, employment, and policy trends in the automotive sector also emphasizes the interaction of evolving international systems for organizing automotive production in and beyond North America. Sweeney also notes the emergence of distinct national (and subnational) policies that have both reflected and shaped evolving differences in domestic industry organization. Canada's motor vehicle assemblers are more dependent on U.S. markets (90 per cent of total sales in 2016) than any major Canadian industry.[7] As a result, Canada typically follows developments in U.S. automotive regulatory and related environmental regimes. (Similar patterns apply in other heavily integrated sectors, such as computers, electrical and electronic equipment, and chemical products.) However, evolving patterns of foreign investment and global production networks have resulted in a fundamental restructuring of the industry, which sometimes pits Detroit-based firms against Japanese-based "transplants" with significant Canadian production, and large Canadian (and foreign-based) multinational parts suppliers against smaller competitors oriented to domestic and North American markets. These intra-industry conflicts have provided Canadian governments with greater trade policy discretion than might otherwise apply, although with differences in application between Harper and Trudeau governments.

Regulatory regimes emerging from USMCA negotiations reflected broader U.S. policy preferences for managed trade, including general, product, and labour-content based rules of origin. These rules vary significantly across individual product categories, several of which follow wording from the initial TPP agreement. However, despite their complexity – USMCA's initial chapter 4 (rules of origin) totalled 234 pages – they may provide enough flexibility for continued industry integration, if at increased costs to consumers resulting from rising compliance burdens, particularly for smaller and more specialized producers.[8]

Both energy infrastructure, as noted by Gattinger, and cabotage restrictions in transportation, discussed by Prentice and Coleman, represent examples of interest group–driven policy "bordering," which have limited transportation and energy infrastructure sectors from becoming genuinely integrated within North America. These realities are reflected in the thickening of internal bordering practices within both Canada and the United States, not just the securing of authorization from central governments.

Gattinger's discussion of energy infrastructure policies emphasizes internal, not just cross-border, dynamics of locating energy infrastructure, reflecting the interaction of wider international concerns over climate change policies with place-based environmental concerns over the potential effects of pipeline infrastructure, particularly among First Nations and Native American tribal governments. Ongoing transportation (especially pipeline) and refining bottlenecks in each country had increased price differentials between Western Canadian and broader North American markets before Alberta's government-imposed production cuts in 2018–20 and subsequent declines in market demand. These factors, added to the ongoing restructuring of North American and global energy markets and renewed geopolitical competition among major oil exporting countries, have significantly affected cross-border investment flows, with cascading economic effects. As noted by both Tomblin and Hale, these challenges are compounded by the multilevel governance of both energy and infrastructure in each country's federal system, resulting in the need for multiple federal/subnational regulatory approvals prior to the approval or substantial renewal of new infrastructure projects in each country, along with multiple opportunities for legal contestation. Just as major energy infrastructure firms have sought to manage market risks by diversifying across national borders, interest group networks have pursued parallel strategies of cooperation and sharing financial resources, leveraging regional energy interdependence to promote regulatory bottlenecks to new development.

Prentice and Coleman note the adaptation of varied transportation subsectors to the persistence of restrictions on cabotage: shipments between two domestic locations by foreign-based carriers. Just as Canada's greater interdependence with the United States often leads to market-driven emulation by the former of regulatory trends in the latter, the capacity of U.S. domestic lobbies to maintain strong restrictions against cabotage by foreign-based firms is consistently paralleled in Canada. Olson's observations on the "logic of collective action"[9] suggest that such trends are likely to continue until groups representing shippers in both countries are able to overcome historic coordination problems to mobilize public opinion in favour of more open markets in each country. Given current U.S. trends, such developments do not appear imminent.

So where does this leave the development of more cooperative or collaborative forms of policymaking addressed by authors in this volume? Deborah VanNijnatten and Carolyn Johns note the institutional legacies of cross-border environmental cooperation that persist, despite policy shifts under the Trump administration. Cross-border institutions and networks continue to shape policy governing boundary waters, particularly when sharing a focus on the management or resolution of specific environmental problems and challenges. The effectiveness of such mandates often depends on formal legislative authority, which is inherently territorial, even if requiring intergovernmental cooperation and coordination. However, public support for such measures – and their resilience when challenged – is often linked to what they call "human ecosystem interactions," both cultural and economic, rooted in the need for cooperation to manage environmental dynamics that cross natural borders. Such linkages may take different forms in different regions of both countries.

However, deliberate efforts at domestic *and* cross-border coordination are critical to reconciling multiple and competing interests for whom climate change has place-based implications vital to sectoral, environmental, and community sustainability. Such efforts have been relatively effective in the Great Lakes basin, somewhat so in engaging often competing environmental and community stakeholders of the Columbia basin across British Columbia and five U.S. Pacific Northwest states towards a possible renewal of the Columbia River Treaty, and demonstrably less so in navigating the energy transition away from fossil fuels (and, to some extent, nuclear power) in Quebec, Atlantic Canada, and the six New England states.

The broader the range of issues to be addressed – and more particularly, the broader the territorial scope of policy goals and interests to be reconciled, as with many climate change issues – the more difficult it becomes to cultivate or maintain the support of multiple governments in each country. Cross-border agencies such as the Western Climate Initiative (WCI) faced sizeable defections from U.S. states after the 2008–9 recession and 2010 election cycle. Both federal-provincial and interprovincial cooperation in Canada on climate change remains highly contested and contingent on effective coordination with energy and economic development policies.

Intermesticity and Cross-Border Subnational Regions

As noted above, Canada confronts three important realities in its international policymaking: the increasingly blurry lines between domestic and international (including cross-border) policy functions, the competitiveness of an economy firmly anchored to the United States, and a steady weakening of domestic support for, or faith in the stability of the post–Cold War global political and economic orders among major industrial and emerging nations. The competing impulses within these realities – including the concepts of intermesticity, fragmegration, and the perforated state – have encouraged analysts to pursue greater understanding of the varieties and interaction of governance processes that have evolved within the constraints of the three realities.

Several contributors to this volume have drawn upon diverse literatures highlighting the evolving varieties of governance in Canada-U.S. relations, most of which occur away from the headline-grabbing, high stakes diplomacy between Ottawa and Washington. Indeed, between spurts of high-level international diplomacy— notably Canada-U.S. Free Trade (1985–7), NAFTA (1990–3), the 2001 Smart Border Accords, the Security and Prosperity Partnership (2005–8) and the inauguration of its successors (BTB and RCC, 2011), and USMCA 2017–20) – some of the most substantive and meaningful activity in Canada-U.S. relations takes place in the context of "cross-border regionalism": a plethora of sub-federal, municipal, private, nongovernment, and epistemic governance relationships that sometimes transcend and coordinate activity across borders, but are also shaped by or even thwarted by those same borders.

In at least six of the contributions to this volume we see important tensions arising from a natural cross-border gravity towards regionalism, alternative forms of governance, and localized connections aimed at solving specific problems. Contemporary research notes that the constraining effects of national borders on cross-border economic activity are substantially different across regions, reflecting different combinations of population densities, relative proximity, composition of export commodities, relative integration of major supply chain networks, and assorted governance and regulatory factors.[10] Tensions arise because that gravity pulling stakeholders towards cross-border regionalism (as opposed to domestic market linkages) is mitigated, sometimes thwarted, by path-dependent legacy of traditional modes of territorial governance.

For instance, Patricia Dewey Lambert and Stephen Tomblin offer important insights into the contrasting trajectories of regional governance on opposite sides of the continent. Both reference important elements of a shared cross-border culture, politics, and the obvious logic of regional economic cooperation. Tomblin notes that the U.S. Northeast and Canada's maritime provinces have historically enjoyed periods of robust cross-border cooperation and engagement. More recently, however, major linkages in some sectors (particularly electricity and other energy relations) have fallen victim to a mix of political populism and jurisdictional competition rooted in traditional territorial governance structures – even as shared interests lead individual state and provincial governments to work more closely in other areas, such as trade in forest products.[11] In other words, states and provinces have reasserted their sovereign policy latitude, thwarting efforts to transcend those boundaries to address shared problems.

In the Pacific Northwest and Great Plains/Prairies regions, Lambert's focus on the Pacific Northwest Economic Region (PNWER) describes a similar set of regional cultural, political, and economic logics that call for greater coordination among a host of sub-federal, private, and nongovernment stakeholders. Although the results have been mixed, PNWER represents an important effort at aterritorial, regional governance coordination that is, at least in part, a by-product of the gravity pulling them together through functional and sector-specific networks.

In other cases, Kathryn Friedman and Debora VanNijnatten and Carolyn Johns depict strongly aterritorial impulses towards cross-border governance that regularly clash with "legacy institutions" rooted in territorially defined lines of authority. VanNijnatten and Johns point to several regional environmental governance architectures that have evolved along the Canada-U.S. border. Some, like the International Joint Commission, are anchored in formal traditional territorial governance. More recent governance models are more fluid, being anchored in networks of stakeholders, including epistemic communities, that are focused on regional challenges such as water quality in the Great Lakes watershed. But networked, aterritorial efforts function, however imperfectly, alongside other, sub-federal, yet territorial governance initiatives like the Regional Greenhouse Gas Initiative or the Western Climate Initiative.

Similarly, Friedman's examination of the potential for municipalities to reach across the Canada-U.S. border to coordinate and exchange on a host of issues ranging from transportation planning to best practices in municipal governance is routinely limited by the jurisdictional lines of traditional territorial governance structures. Once again, logics of cultural and economic gravity pulling municipalities together across borders are upset by the entrenched territoriality of provincial and state prerogatives. Much like the entrepreneurial stakeholders in networked forms of environmental governance, municipal leaders in cross-border regions like the Buffalo-Niagara-Hamilton corridor have an intuitive and practical sense of the blurring of lines between domestic and international policymaking in dealing with practical regional problems. However, the embryonic forms of aterritorial governance they initiate must still work within (or contend with) the legacy of territoriality anchored in state, provincial, or national jurisdictions.

It is important to consider the advance of modes of aterritorial, regional governance in different domains in the context of the third "reality" of contemporary Canadian policymaking: declining domestic support for the existing institutional order in several advanced industrial countries. Although none of the contributors to this volume claims a causal link between weakening support for aterritorial policy entrepreneurship and the reassertion of territorial governance, there is a clear interaction between the blurring of jurisdictional lines of authority brought about by Canada's integration into the global economy (intermesticity) and the creation of entrepreneurial space for aterritorial governance to fill underserved policy voids. Domains of cross-border regional policymaking above seem to fit this construction: aterritorial governance entrepreneurs responding to institutional limitation brought about by evolving Canadian "realities."

In a joint contribution by Barry Prentice and John Coleman, and another by Carolyn James, we can see the necessity of new governance structures rooted in the inability of traditional territorial forms of governance to respond to the needs of border communities and cross-border trade, but also the limited ability of aterritorial governance entrepreneurs to build the public or institutional support necessary to drive cohesive regional change. In examining the potential for cabotage in trucking, regional air services, or short-sea shipping to strengthen cohesiveness of cross-border regions, Prentice and Coleman point to persistent barriers to aterritorial governance initiatives to facilitate the knitting together of cross-border regions. Similarly, while the problem of climate change is global, the management of Arctic boundaries is shared and could in some policy domains be managed through cross-border governance mechanisms. James proposes a formal set of mechanisms that finesse each country's sovereign imperatives by creating an aterritorial governance body to expand coordination among different levels of government, borrowing from the experience of the International Joint Commission. Interestingly, the gravity of shared regional interests anchored in culture, economics, the environment, and geostrategic considerations looms over Arctic governance. Ironically, James's hoped-for aterritorial governance mechanism is a faint hope, reliant as it is on the traditional territorial state to set it in motion. It remains to be seen how the (often highly territorial) responses of central governments to emergency governance during the 2020 pandemic – and the limited effectiveness of transnational institutions – will affect public expectations of governments.

Bridging Differences, Managing Interdependence

Cross-border cooperation remains vital to border governance and infrastructure improvements, whether in managing sizeable if shifting flows of cross-border commuters, business travellers, and tourists, integrating security (and cross-cutting legal) considerations with the management of travel and trade flows in multiple modes of transportation, and the related task of streamlining processing and screening procedures for cross-border trade. As demonstrated by the challenge of managing the 2017–19 surge of irregular migrants[12] and the tightening of border controls during the 2020 pandemic,[13] administrative cooperation and the maintenance of mutual trust among government officials remain key factors in sustaining both physical and virtual borders while managing periodic shifts in movements of people, services, and goods. Although the persistence of such cooperation, despite the predatory unilateralism of the Trump administration in some areas, is an encouraging sign, the resilience of border processes depends heavily on reciprocity and mutual accommodation of political sensitivities and different legal systems in each country.

Ongoing processes for – and limitations on – cross-border and wider international regulatory cooperation are central to such developments. The decentralization of regulatory processes along functional lines in both Canada and the United States makes related processes heavily dependent on respective legislative mandates and procedural requirements. The greater the degree to which regulation is subject to parallel jurisdictional arrangements in each country, and the greater the degree to which relevant Canadian industry sectors or subsectors depend on U.S. markets, the greater the likelihood of policy alignment, particularly if needed for the effective integration of industry supply chains. Conversely, the greater the diffusion of responsibilities among different regulatory agencies or between central and subnational governments, or the less the extent of sectoral economic integration, the more likely that Canadian regulatory regimes will parallel broader U.S. policy goals, while allowing for significant institutional variations, differences in industry structures, or national/subnational policy priorities. Kerr and Hobbs note such parallels in food safety regulation, in which major legislative overhauls in 2009–12 occurred without much coordination while pursuing similar policy goals.

However, the greater the level of regulatory decentralization in either country, the greater the challenges facing regulatory cooperation, let alone coordination – especially in the management of major projects. Efforts to coordinate cross-border infrastructure development have been most effective in jurisdictions with simple, stable jurisdictional frameworks. However, risks of politicization and paralysis from competitive litigation increase dramatically with the number of jurisdictions and regulatory agencies sharing responsibilities for major projects. Such challenges are visible in projects ranging from the Gordie Howe International Bridge between Windsor and Detroit, in gestation since 2004 with construction beginning in 2018, and highly publicized wrangles over improvements to the Peace Bridge between Fort Erie and Buffalo,[14] to major energy pipelines in both countries and interstate transmission lines linking Canadian electricity producers with U.S. utilities and consumers.

The effectiveness of cross-border regional cooperation depends on several major factors, not least effective leadership across neighbouring provinces and states. Such leadership frequently depends on the priorities and capacity of the executive(s) of the largest province(s) and states(s) within a region, and their capacity to cultivate shared objectives and mutual accommodation among the leaders of larger and smaller jurisdictions on each side of the border. Lambert and Tomblin also emphasize the importance of institutions with politically astute administrative leadership and continuity to facilitate such cooperation, given the entrenched prerogatives of sovereign governments. Only the Pacific Northwest among cross-border regions has been able to sustain substantive cooperation in recent years, if to widely varying degrees across functional policy areas.

The incorporation of regulatory coherence chapters of the TPP (renamed CP-TPP after 2015), Canada's Comprehensive Economic and Trade Agreement (CETA) with the European Union, the USMCA, and several other agreements promote what Lin and Liu describe as "regulatory input" processes as opposed to national regulatory outputs potentially reviewable under Technical Barriers to Trade (TBT) provisions of the World Trade Organization (WTO).[15] They note that provisions for "notice and comment, public consultation, cost-benefit analyses, inter-agency coordination, and regulatory impact analysis" in participants' domestic regulatory processes reflect broader proposals advanced by groups such as the WTO's Committee on Technical Trade Barriers, the OECD, and Asia-Pacific Economic Cooperation to promote soft-law measures characterized by parallelism in national regulatory systems rather than "hard law" levels of coordination and enforceability.[16] Some measures do attempt to limit the relative autonomy of domestic regulatory processes. Unlike CETA and CPTPP, the USMCA agreement provides access to dispute resolution for "a sustained and recurring course of action or inaction ... inconsistent with its GRP provisions."[17] However, the range of measures explicitly exempted from these provisions in each country is broad enough, particularly on the activities of each country's public sector, to impose significant limits on their potential application.[18]

These measures are intended to facilitate ongoing economic integration, with its potential benefits of economic specialization, while respecting domestic autonomy within mutually agreed limits on the outcomes of regulatory processes. Measures introduced in 2019 to insulate national security exemptions for Canadian federal government procurement from trade tribunal review[19] illustrate the ongoing challenge of sustaining international regulatory cooperation in a world of growing international tensions and asymmetrical security threats. However, while Canada remains heavily dependent on U.S. export markets within cross-border and wider international supply chains, whether for trade in finished goods or intermediate products, its capacity for choice remains somewhat dependent on its capacity to diversify its trade and investment relations beyond North America.

Canada in North America and the World: Cooperation, Competition, *and* Choice?

The historic decision of the Mulroney government and Reagan administration to pursue comprehensive free trade negotiations in 1986–8 alongside multilateral negotiations under the GATT reflected complementary economic interests and political calculations. The Macdonald Royal Commission appointed by Pierre Trudeau had concluded that Canada's economic competitiveness would depend on the benefits of specialization and economies of scale associated with Canada's integration within the much larger North American market.[20] This rationale was reinforced in CUFTA's subsequent expansion to include Mexico in 1990–4, although Canada's initial participation in NAFTA talks was largely defensive and aimed at preserving CUFTA-related gains in terms (and relative security) of U.S. market access. In the first case, U.S. officials recognized the benefit of using bilateral negotiations with Canada to establish precedents and policy markers for inclusion in broader agendas for liberalizing international trade and investment rules in ways consistent with their national interests.[21] This strategy resulted initially in the 1994 Uruguay Round agreement, which created the WTO and the General Agreement on Trade in Services (GATS), and was extended subsequently through successive bilateral and plurilateral negotiations agreements described as "competitive liberalization" by Schott and others.[22]

Although Canada has been a belated participant in this game, its governments have been relatively successful in paralleling or anticipating U.S. trade agreements to maintain comparable or improved market access.[23] However, the proliferation of regional trade agreements – often described as the "spaghetti" or "noodle bowl"[24] – has increasingly complicated integration of trade relations across countries and regions, given significant differences in rules-of-origin and other technical measures across agreements. Appendix 20.1 illustrates the extent and complexity of cross-cutting trade agreements of participants in the Trans-Pacific Partnership and the United States.

Canada is typically a policy-taker in such settings, responding to the initiatives of other countries – particularly major trading powers jockeying for regulatory advantage in the international marketplace – while trying to preserve some degree of policy discretion.[25] Key examples include sectoral provisions for intellectual property rights, cross-cutting automotive and other rules of origin (ROO), and geographical indicators and trade mark regimes governing agri-food trade. The capacity to negotiate "cross-cumulation" rules is a critical factor in limiting conflicts among countries with overlapping bilateral or regional trade agreements (RTAs).[26] However, Canada's efforts to diversify trade and investment relations have often been heavily constrained by competing domestic interests within Canada and integrated North American sectors.

The Harper government (2006–15) may have attempted to balance regional interests by simultaneously pursuing transatlantic trade agreements with the EU and the Trans-Pacific Partnership (TPP) with Japan, Australia, and other Asia-Pacific nations (with or without U.S. involvement). However, implementing these cooperative efforts remain an evolving work-in-progress, as noted by Anderson in his chapter on investment policies and by Kerr and Hobbs in their discussion of food safety regimes. In any event, substantive trade diversification remains heavily dependent on the capacity and willingness of major Canadian firms to cultivate markets (and navigate different legal and regulatory systems and business cultures) outside North America.[27]

The Trudeau government's efforts to balance cross-cutting interests since 2015 have exposed the inherent frailties of Canada's efforts to project its own domestic political trade-offs as an aspiring entrepreneur of evolving international norms. Examples include efforts to manage competing objectives among domestic energy and transnational environmental interests on several fronts, discussed by Gattinger, export-dependent and sheltered domestic interests during USMCA negotiations, recognition of labour rights, further trade liberalization, and the diffusion of contemporary North American "progressive" cultural and ideological norms.

Similar challenges have arisen in maintaining access to Chinese markets. Until late 2018, Canadian policymakers struggled to balance tensions between major Canadian firms' ambitions to secure access comparable to that already obtained by Korea, Australia, and ASEAN states through bilateral agreements, and strategic and security concerns related to the potential for expanded foreign state ownership of major Canadian firms and effective limits to reciprocity in dealing with China's non-market economy.[28] Since then, the arrest and continuing extradition proceedings against a senior executive of Chinese MNC Huawei over the latter's alleged attempts to evade U.S. sanctions against Iran (and related U.S. financial sector regulations) have embroiled Canada in the widening geopolitical conflict between the United States and China.[29] As noted in chapter 10, these issues overlap with perceived security risks of Huawei's commercial involvement in the installation of advanced ("5G") wireless telecom systems across the industrial world. Chinese responses, including the arbitrary imprisonment of Canadian business travellers and a series of customs administrative actions against Canadian exporters,[30] have highlighted both physical and commercial risks to the pursuit of expanded trade relations in the absence of an independent Chinese legal system capable of providing effective legal recourse against the arbitrary exercise of state power.

These developments have rendered politically moot debates over proposed restrictions imposed by the requirements of USMCA's section 32.10 for mutual disclosure of comprehensive trade negotiations with non-market countries (read China), and the potential for U.S. withdrawal from USMCA if Canada were to negotiate such an agreement without U.S. approval.[31] They also challenge its efforts to diversify trade relations in Asia, and expand risks of a geopolitical Catch-22 in regulating Chinese influence in the development of Canadian telecommunications systems noted in chapter 10 – as with several other major industrial countries.

Canadian firms face similarly cross-cutting pressures in cultivating export markets. The distinctions can be seen clearly in traded agri-food sectors, with their wide variety of products, market differentiation, and national regulatory requirements. For example, both consumer tastes and national agri-food food safety (SPS) regulations can vary widely between North American, European, and selected Asian markets. These differences often result in the need for special handling and inspection requirements for numerous products, including wide restrictions on the use of genetically modified organisms (GMOs) and other product inputs (including pharmaceuticals) outside North America. Domestic food safety concerns arising both from domestic incidents such as "mad cow" and avian flu scares and those involving imported foods, more recently, processed pork products infected with African swine fever, have led to the introduction and growing diffusion of mandatory domestic food and input traceability systems, particularly for livestock and poultry in Canada, following other OECD and some developing countries.[32] Support of domestic producer and industry groups is critical to implementation of such policies, along with the growing use of ISO 22000 international food safety standards by food processors. These issues illustrate the growing influence of international market, regulatory, and consumer expectations beyond North America in shaping the competitive environment for growing numbers of Canadian firms.

A 2018 Export Development Corporation report suggests that "knowledge gaps" relating to the growing complexity of global markets and the institutional requirements to managing national regulatory and customs systems effectively have become among the most important barriers to exporting for companies entering new markets, especially smaller and medium-sized enterprises.[33] Such challenges tend to privilege market incumbents and large firms capable of cultivating strong networks of foreign affiliates and joint ventures to manage formal and informal bordering from the "inside," reinforcing the gravity effect of North American markets in the absence of ongoing, large-scale transactions to offset transportation, regulatory, and other transaction costs related to expanding in foreign markets. A 2015 study estimates pre-CETA border effects – a measurement of the relative level of trade between countries after controlling for the relative populations of trading partners and the geographic distance between them – on Canada's exports to the EU at five times the average border effect for trade between Canadian provinces and U.S. states in 2001–10.[34] However, as noted in table 20.2, variations across provinces were much higher.

Indeed, sales by foreign affiliates of Canadian-based firms grew at twice the rate of direct goods and services exports between 2011 and 2015. Service sector firms, particularly in the financial services sector, accounted for more than half of the total sales of foreign affiliate firms in 2015.[35] Similarly, foreign-based affiliates in Canada account for 47.9 per cent of Canadian exports in that year – although their propensity to export varies widely across industrial subsectors.[36] These data point to the complementarity of trade and investment flows, particularly for markets outside North America in which access to knowledge of local market preferences, regulatory conditions, and stable economic networks is critical to business success.

However, it may not capture the wide range of services embedded within or derived from traditional goods exports – described as the "servicification of manufacturing" – including but not limited to the role of services in manufacturing production,[37] follow-up service and maintenance contracts, revenues from intellectual property, and other benefits from incorporation into international supply chains.[38] Combined with

Table 20.2. Estimates of Aggregate Border Effects for Canadian Provinces in Trade with United States, European Union (2001–2010)

	U.S.	E.U.		U.S.	E.U.
Canada	9.0	45.7	Quebec	7.7	15.1
Alberta	3.0	140.0	Manitoba	10.6	30.1
Saskatchewan	3.0	84.9	Newfoundland and Labrador	23.8	95.0
Ontario	3.5	29.7	Nova Scotia	27.0	17.7
New Brunswick	3.8	107.4	PEI	47.5	130.9
British Columbia	4.3	37.7			

Source: Farrukh Suvankulov, "Revisiting National Border Effects in Foreign Trade in Goods of Canadian Provinces" (Ottawa: Bank of Canada, 2015), 38–9.

the findings of sector-specific chapters, these data strongly suggest the continued importance of clarifying sectoral regulatory, product- and market-specific conditions – including cross-border transfers of professional and technical staff – that facilitate or hinder trade diversification, border management, and facilitation across economic and industrial sectors.

Although these factors are important variables in structuring international and cross-border market flows and human movements, other studies demonstrate the continued importance of domestic policy frameworks as key variables in enabling international competitiveness and positioning Canada as a competitive destination for foreign investment and the attraction of skilled workers and professionals.[39] Stable, sustainable monetary, fiscal, and financial sector policies can provide anchors for investment, job creation, and the maintenance of valued public services, even if social and economic disruptions arising from the 2020 pandemic have increased the challenge of pursuing such policies. Effective education and training systems are necessary to provide citizens and businesses with the knowledge, skills, and aptitudes necessary to prosper in an evolving economy.[40] Canada's far-flung geography and integration in global markets requires continued improvements in the capacity and efficiency of transportation and other infrastructure needed to service domestic and export markets. An efficient, adaptable legal and regulatory environment helps to secure policy objectives valued by citizens while providing businesses and other market participants with reasonable predictability. As noted by Gattinger, the continuing difficulties of *domestic* energy, environmental, and infrastructure regulatory systems in integrating or reconciling these objectives remains an important weakness in Canadian domestic policies that limits Canada's leverage in wider international policies. Political shocks of recent years that have challenged the stability of the international economic system – and disrupted transportation networks in Canada – reinforce the importance of sound domestic institutions and policies – and the capacity of navigating regional differences in Canada as a practical condition for making the most of Canada's advantages in North America and the wider world.

Political Shocks and Countervailing Trends

Canada's international economic and political relationships have always been vulnerable to periodic political shocks involving its major trading partners – often followed by extended periods of adaptation to new realities, including new forms of international cooperation and managing differences. Negotiation of CUFTA in 1986–8 followed fifteen years of political and economic instability among industrial countries, resulting from the partial break-up of the post-war Bretton Woods system, successive global energy shocks, macroeconomic instability, and the emergence of new patterns of regional and global economic competition.

Political tremors arising from the global financial crisis of 2008–9 have similarly disrupted advanced industrial nations, often undermining public support for institutions and patterns of international trade, investment, and migration. Although this effect has been strongest in the United States, particularly since the election of the Trump administration in 2016, it has also been notable across Europe, parts of Latin America, and the Asia-Pacific region.[41] Canadian governments have been relatively successful in preserving what Mendelssohn, Wolfe, and Parkin have labelled the "permissive consensus" on globalization and trade policy[42] in Canadian public opinion – but as much as a reaction against Donald Trump's chauvinist posturing as an affirmation of recent policy directions.[43] However, sustaining this domestic consensus and navigating the increasingly uncertain world beyond Canada's borders will require a multi-pronged strategy of mutually reinforcing domestic and international policies – especially following the seismic shocks to national and global economies in 2020, which triggered unprecedented government interventions to prevent socio-economic breakdown.

Domestically, Canadian governments will have to reinforce public trust in its domestic institutions and policies in ways that both sustain the growth of living standards and procedural fairness in an increasingly divided society. Key aspects of such policies must include mutual accommodation among Canada's regions and competing societal interests, and substantial efforts to encourage both predictability and accountability of regulatory processes while limiting opportunities for litigation-induced paralysis within a relatively decentralized federation. These challenges are particularly daunting in the balancing and integration of energy and environmental policy objectives within Canada's decentralized constitutional regime, particularly given the very different energy endowments of Canada's provinces and regions, related differences in environmental priorities, and the cross-cutting interests and aspirations of Indigenous peoples, which must also be integrated within these processes, as noted by Gattinger. It remains to be seen how economic disruptions from the 2020 COVID-19 global pandemic – and the substantial costs of responding to them – will affect Canada's capacity to pursue and balance these objectives.

Internationally, Canadian governments will have to pursue a three-way balancing act reminiscent of earlier eras of Canadian foreign and international economic policies, while seeking to balance competing regional, economic, and societal interests in a coherently articulated but adaptable vision of the national interest. The first pillar of this strategy is dictated by geography. Canada's North American location and substantial if incrementally declining dependence on U.S. markets dictates continued engagement with the American political system on multiple fronts while cultivating relationships across what is likely to remain a deeply divided country for the foreseeable future. As always, Canadian public servants managing intermestic files will have to take the initiative in dealing with their American counterparts, while continuing to work with Mexico on issues of mutual interest, as during negotiations in 2018–19 for the removal of U.S. steel and aluminium tariffs. The continued pursuit of regulatory cooperation and reciprocal recognition of regulatory standards should remain a continued objective of Canadian policy, while recognizing that progress is likely to be incremental and conditional on compatibility with U.S. international and/ or domestic interests. U.S.-Canadian security cooperation will remain a critical element on multiple fronts, ranging from defence and cyber-security, through border management and a mix of domestic and cooperative measures to encourage legal rather than irregular migration.

The second pillar of Canada's international policies must remain the encouragement, functioning, and adaptation to new circumstances of a rules-based international legal order in cooperation with other industrial and emerging nations – particularly those committed to democratic principles, market principles, and the rule of law. Developments of recent years have suggested that such developments are likely to be incremental, segmented by policy areas, and characterized by coalitions of the willing rather than comprehensive multilateral visions. Such approaches may complement U.S. policies, as with the 2017 OECD

Multilateral Instrument, which built on previous U.S. efforts to limit undue tax avoidance by its corporate residents,[44] or parallel them as with Canada's Corruption of Foreign Public Officials Act, which reflects U.S. measures and persistent efforts by Transparency International to address both supply and demand elements of corrupt practices. Canadian officials are working with counterparts from other countries to reform the WTO dispute resolution system in response to U.S. concerns and the latter's veto on appointments which deprived the WTO Appellate Body of a functioning quorum in 2019.[45] An ad hoc EU-Canada alternative appellate process expanded in early 2020 to include at least fifteen other countries, including Australia, Brazil, China, and Mexico, pending negotiation of broader reforms.[46] However, the glacial pace of trade negotiations, whether multilateral or plurilateral (e.g., CPTPP or CETA), combined with the mix of offensive and defensive objectives pursued by Canada and other countries noted by Kukucha, inherently constrain the pace and probable extent of progress.

To compensate for these constraints, Canada will have to invest in building both policy and market development capacity in non-traditional markets, particularly in the Asia-Pacific region and Latin America, where it is seeking to deepen trade regions with the Pacific Alliance. Canada's active diplomatic and economic profile in these markets has often been episodic, taking a backseat to expansion of market access and opportunities in more traditional North American and European markets. However, Canada's ratification of CPTPP and the geopolitical chill of its relationship with an increasingly assertive, expansionist China increases the importance of increasing its persistent commitment to political and market relations in these regions. Such activities will take time and are likely to provide only partial compensation for growing barriers to Chinese markets and other forms of coercive diplomacy.[47] However, if the United States persists in pursuing bilateral rather than regional approaches to trade relations, Canada is likely to enjoy greater leverage in engaging China if doing so in concert with other major economies in the region, particularly Japan, Australia, and South Korea, rather than pursuing the chimera of a reciprocal bilateral agreement with Beijing.

Recent political conflicts, both within and beyond North America, recall similar periods of disruption during the 1970s – not least widespread discussions of the dysfunctionality of *domestic* political institutions in managing social and economic conflicts.[48] Trust in international institutions (economic or political), which, by definition, lack any basis for democratic oversight or consensus building, apart from those provided by national governments, is a direct by-product of public trust in domestic political institutions – particularly "in the face of uncertainty about or vulnerability to the action" (or inactions) "of these institutions."[49]

Key elements in maintaining public trust for political institutions include functional opportunities to hold governments, businesses, and other organized interests accountable for their actions – both in general and regarding specific institutions and functions relevant to citizens' daily lives and subjective evaluations of trends in relative economic well-being.[50] These conditions spill over into market flows related to trade and investment and to human movements, whether related to employment, travel, or migration, across Canada's borders. Public support for Canadian trade policies are closely linked to expectations of reciprocity with external trading partners: shared and comparable terms of access, opportunity, and benefit. Trade relations with major partners in which market access or benefits are seen to accrue primarily to the other party, or that are excessively subject to the discretion of the stronger partner, are likely to face much stronger public resistance. They also apply to the relative capacity of governments to enforce environmental and labour standards on foreign firms (particularly state-controlled enterprises) on terms applicable to domestic firms, along with securing comparable treatment for Canadian firms operating in other countries.

The same principles of clarity, predictability, and enforceability of rules also apply to implementations of policies governing business travel, migration, and mobility of skilled labour. Canada has generally avoided the migration-related tensions experienced by other advanced industrial countries in recent years, despite occasional incidents. However, a consistently applied rules-based system remains a central factor in securing and sustaining public acceptance of mobility rules, along with effective domestic policies that facilitate effective integration of immigrants as full members of society.

Such processes are more difficult to sustain in political environments characterized by zero-sum competition – whether among domestic interests, inward-looking nationalism, or homogenizing trends towards supra-nationalism. Ultimately, the success of Canada's international policies will hinge on the capacity of its domestic policies to provide economic and social outcomes valued by diverse groups of citizens at a price they are willing to pay, while working with like-minded nations to sustain a rules-based international system, whatever its imperfections. Canada's political and economic leaders have sustained and expanded Canada's capacity for choice by pursuing these principles in the thirty years since concluding its historic trade agreement with the United States. With sound judgment and leadership that helps Canadians in all regions adapt to the changing, unpredictable world around them, Canada's federal and provincial government can maintain this capacity for choice in the years to come.

Appendix 20.1. Framing the Spaghetti/Noodle Bowl Regional Trade Agreements Involving CP-TPP Members and United States

	Australia	Brunei	Canada	Chile	Japan	Malaysia	Mexico	New Zealand	Peru	Singapore	Vietnam	USA
Australia	2019*											
Brunei	ASEAN	NYR**										
Canada			2019*									
Chile	2009		1997	NYR**								
Japan	2015	2008		2007	2019*							
Malaysia	2013	ASEAN		2012	2006	NYR**						
Mexico			1994/9	1999	2005		2019*					
New Zealand	1983	ASEAN				2010		2019*				
Peru	2020		2009	2009	2012		2012		NYR**			
Singapore	2003	ASEAN		2009	2002	ASEAN		2008	2009	2019*		
Vietnam	ASEAN	ASEAN		2014	2009	ASEAN			2019*	2019*	2019*	
USA	2005		1989/94	2004	2019		1994		2009	2004		NYR
Argentina				1991/2019			1987					NYR
ASEAN	2010	1993			2008	1993		2010		1993	1993	
Bahrain				2004		(GCC)				(GCC)		2006
Bolivia				1993			1995/2010					
Brazil							2003					

CAFTA-DR								2006–9
China	2015	ASEAN	2011	2006	ASEAN	2008	2009	2009
Colombia				2009	1995		2009	2012
Costa Rica			2002		1995		2013	CAFTA-DR
Cuba					2001			
Ecuador				2010	1987			
EFTA‡‡			2009	2004	2001		2003	2011
EU			2017	2003	2000		2019	2019†
Eurasia††				2018			2019†	2016
GD8‡								
Gulf C.C.§					2013		2013	
Honduras			2014		2001	2017		CAFTA/DR
Hong Kong	2018	ASEAN		2014	ASEAN	2011	ASEAN	ASEAN
India		ASEAN		2007	2011		2005	ASEAN
Indonesia	2018†	ASEAN		2019	ASEAN	ASEAN	ASEAN	ASEAN
Islamic Conference					2014†			
Israel		1997			2000			1989/96
Jordan		2012						2010
Mercosur§§			1996		2006			
Morocco				2016				2006
Mongolia								
Nicaragua					1996			CAFTA-DR
Oman				GCC			GCC	2009

(Continued)

Appendix 20.1. (*Continued*)

	Australia	Brunei	Canada	Chile	Japan	Malaysia	Mexico	New Zealand	Peru	Singapore	Vietnam	USA
Pacific Alliance‖				2016			2016		2016			
Pacific Islands	2017†							1981/2017†				
Pakistan						2008						
Panama			2013	2008			1986/2015		2012	2006		2012
Paraguay							1994/2006					
Philippines					2008	ASEAN				ASEAN	ASEAN	
S. Korea	2014	ASEAN	2015	2004		ASEAN		2015	2011	2006	ASEAN	2012
Sri Lanka										2018		
Switzerland			EFTA		2007					EFTA	2015	
Taiwan								2013		2014		
Thailand	2005			2015	2007	ASEAN			2009	ASEAN	ASEAN	
Turkey				2011		2015				2017		
Ukraine			2017									
Uruguay							2004					
Venezuela				1993			1995		2013			

* In force
** Not yet ratified
† Pending ratification
†† "Eurasia": Russia, Armenia, Belarus, Kazakhstan, Kirgyz Republic

‡ Group of Eight Developing Countries (G-8): Bangladesh, Egypt, Indonesia, Iran, Nigeria, Pakistan, Turkey

‡‡ EFTA: Iceland, Lichtenstein, Norway, Switzerland

§ Gulf Cooperation Council (GCC): Bahrain, Kuwait, Oman, Qatar, Saudi Arabia, United Arab Emirates;

§§ Mercosur: Argentina, Brazil, Paraguay, Uruguay

‖ Pacific Alliance: Chile, Colombia, Mexico, Peru

Sources: Asia Regional Integration Center, "Free Trade Agreements" (Manila: Asian Development Bank, 2018), https://aric.adb.org/fta; Australia, Department of Foreign Affairs and Trade, "Australia's Free Trade Agreements" (Canberra: March 2020). https://www.dfat.gov.au/trade /agreements/Pages/trade-agreements.aspx.

Canada, Global Affairs Canada, "Trade and Investment Agreements" (Ottawa: 18 October 2018), http://www.international.gc.ca/trade-commerce /trade-agreements-accords-commerciaux/agr-acc/index.aspx?lang=eng.

Organization of American States, "Foreign Trade Information System: Trade Agreements in Force" (Washington, DC: 2020), http://www.sice .oas.org/agreements_e.asp.

United States. Office of the United States Trade Representative, "Free Trade Agreements," (Washington, DC: 2020); https://ustr.gov/trade -agreements/free-trade-agreements. Updated March 24, 2020.

NOTES

1 George Hoberg, ed., *Capacity for Choice: Canada in a New North America* (Toronto: University of Toronto Press, 2002).

2 Brian Bow and Greg Anderson, "Building without Architecture: Regional Governance in Post-NAFTA North America," in *Regional Governance in Post-NAFTA North America: Building without Architecture*, 1–18 (New York: Routledge, 2015); Isabel Studer, "Obstacles to Integration: NAFTA's Institutional Weakness," in *Requiem or Revival: The Promise of North American Integration*, ed. Isabel Studer and Carol Wise, 63–75 (Washington, DC: Brookings Institution, 2007); Robert A. Pastor and Rafael Fernandez de Castro, eds., *The U.S. Congress and North America* (Washington, DC: Brookings Institution, 1998).

3 Alan G. Rugman, *The End of Globalization* (New York: Amacom, 2000); Robert Pastor, *Toward a North American Community: Lessons from the Old World for the New* (Washington, DC: Institute of International Economics, 2001); Geoffrey Hale, *Uneasy Partnership: The Politics of Business and Government in Canada* (Toronto: University of Toronto Press, 2006), 252.

4 Edelgard Mahant and Graham S. Mount, *Invisible and Inaudible in Washington* (Vancouver: UBC Press, 1999); Geoffrey Hale, *So Near and Yet So Far: The Public and Hidden Worlds of Canada-U.S. Relations* (Vancouver: UBC Press, 2012).

5 Wael Nasser, "Irregular Border Crossings and Asylum in Canada: Study on the Irregular Migration from Nigeria to Canada" (honours undergraduate thesis, University of Lethbridge, 2018); Michelle Zilio, "Asylum-Seeker Surge at Quebec Border Chokes Refugee System," *Globe and Mail*, 12 September 2018, A1.

6 Geoffrey Hale and Cailin Bartlett, "Managing the Regulatory Tangle: Critical Infrastructure Security and Distributed Governance in Alberta's Major Traded Sectors," *Journal of Borderland Studies* 34, no. 2 (2018): 257–79, DOI: 10.1080/08865655.2017.1367710.

7 Douglas Porter, ed., *The Day after NAFTA: Economic Impact Analysis* (Toronto: BMO Capital Markets Economics, November 2017), 8.

8 Brian Bradley, "USMCA Has Give and Take but No Big Stretch," American Shipper.com, 17 October 2018, https://www.freightwaves.com/news /usmca-has-give-and-take-but-no-big-stretch; David Israelson, "How USMCA Will Affect This Canadian Company," *Globe and Mail*, 23 October 2018, B7; Gregory Spak, Francisco de Rosenzeig, Dean A. Barclay, Scott C. Lincocome, Matt Solomon, and Brian Picone, "Overview of Chapter 4 (Rules of Origin) of the US-Mexico-Canada Trade Agreement" (Washington, DC: White & Case, 25 October 2018), https://www.whitecase.com

/publications/alert/overview-chapter-4-rules-origin-us-mexico-canada
-trade-agreement.

9 Mancur Olson, *The Logic of Collective Action: Public Goods and the Theory
of Groups*, 2nd ed. (Cambridge, MA: Harvard University Press, 1971);
Olson, *The Rise and Decline of Nations* (New Haven: Yale University Press,
1986).

10 For example, see Martin A. Andresen, "The Geography of the Canada–
United States Border Effect," *Regional Studies* 44, no. 5 (2010): 579–94;
Farrukh Suvankulov, "Revisiting National Border Effects in Foreign Trade
in Goods of Canadian Provinces," Bank of Canada Working Paper 2015–28
(Ottawa: Bank of Canada, July 2015).

11 Brent Jang, "Maine Governor Seeks Exemption for N.B. Lumber Firms,"
Globe and Mail, 20 February 2018, B3.

12 Christian Leuprecht and Geoffrey Hale, "Presentation to House
of Commons Standing Committee on Citizenship and Immigration,
re: Bill C-97, An Act to Implement Certain Provisions of the Budget
Tabled in Parliament on March 19, 2019 and Other Measures: Division
16, Part 4 Amending the Immigration and Refugee Protection Act,"
7 May 2019.

13 National Post Staff, "Border Shut … Except for About 200K per Week,"
National Post, 30 June 2020, A1.

14 Anne Jarvis, "Behind the Bridge," *Windsor Star*, 28 October 2016, A1; John
Daly, "Double Crossing," *Globe and Mail*, 23 November 2017; Munroe
Eagles, "At War over the Peace Bridge: A Case Study in the Vulnerability
of Binational Institutions," *Journal of Borderland Studies* (2018): https://doi
.org/10.1080/08865655.2018.1465354.

15 Ching-Fu Lin and Han-Wei Liu, "Regulatory Rationalization Clauses in
FTAs: A Complete Survey of the US, EU and China," *Melbourne Journal of
International Law* 19, no. 1 (2018): 150; European Commission, *Good Regu-
latory Practices (GRPs) in TTIP: An Introduction to the EU's Revised Proposal*
(Brussels: 21 April 2016), http://trade.ec.europa.eu/consultations/index
.cfm?consul_id=179.

16 Lin and Liu, "Regulatory Rationalization Clauses in FTAs," 150–1, 155–61.

17 USMCA Article 28.20; Inu Manak, "Regulatory Issues in the New NAFTA,"
International Economic Law and Policy Blog, 10 October 2018, https://
worldtradelaw.typepad.com/ielpblog/2018/10/regulatory-cooperation-in
-the-new-nafta.html.

18 USMCA, Annex 28-A, Sec. 1.

19 Clifford Sosnow, Marcia Mills, Peter N. Mantas, and Andrew House,
"'Because We Said So': Removing the CITT from the Review of Canada's
'National Security Exemption'" (Toronto: Fasken, 18 June 2019).

20 Canada, Royal Commission on the Economic Union and Development Prospects for Canada. *Report*, 3 vols. Ottawa: Supply and Services Canada, 1985.

21 Mordechai Kreinin, ed. *Building a Partnership: The Canada–United States Free Trade Agreement* (East Lansing: Michigan State University Press, 2000).

22 Jeffrey J. Schott, ed., *Free Trade Agreements: US Strategies and Priorities* (Washington, DC: Institute or International Economics, 2004).

23 Geoffrey Hale, "Pulling against Gravity? The Evolution of Canadian Trade and Investment Policies in a Multipolar World," in *Canadian Foreign Policy in a Unipolar World*, ed. Christopher Kirkey and Michael Hawes, 183–203 (Toronto: Oxford University Press, 2016).

24 Jagdish Bhagwati, "U.S. Trade Policy: The Infatuation with Free Trade Agreements," Discussion Paper #726 (New York: Department of Economics, Columbia University, 1995), https://core.ac.uk/reader/161436448; Masahiro Kawai and Ganeshan Wignaraja, "Tangled Up in Trade? The 'Noodle Bowl' of Free Trade Agreements in East Asia," VoxCEPR Policy Portal, 15 September 2009, https://voxeu.org/article/noodle-bowl-free-trade-agreements-east-asia.

25 Geoffrey Hale, "Triangulating the National Interest: Getting to 'Yes' on TPP," paper presented to Fulbright Canada – CSC SUNY Plattsburgh Colloquium, "Canada, the United States, and the Trans-Pacific Partnership," University of Hawaii, Honolulu, HI, 24 February 2017.

26 Sandy Moroz, "Navigating the Maze: Canada, Rules of Origin, and the Trans-Pacific Partnership," in *Redesigning Canadian Trade Policies for New Global Realities*, ed. Stephen Tapp, Ari Van Assche, and Robert Wolfe, 423–46 (Montreal and Kingston: McGill-Queen's University Press, 2017).

27 Export Development Canada, *Evolving with the Changing Needs of Exporters: Submission to the 2018 Legislative Review of Export Development Canada* (Ottawa: October 2018), 9, 13–14; David Parkinson, "A Year Later, Canada Has Failed to Take Advantage of Trade Deal with Europe," *Globe and Mail*, 8 November 2018, B4.

28 Public Policy Forum, "Diversification, Not Dependence: A Made-in-Canada China Strategy" (Ottawa: 11 October 2018), *https://www.ppforum.ca/publications/diversification-not-dependence-a-made-in-canada-china-strategy/*; *Globe and Mail*, "The Risks of a Free Trade Deal with China," editorial, 15 October 2018, A12; Carlo Dade, "Our Country's Third Trade Deal Offers a Different Path to China," *Globe and Mail*, 17 October 2018, B4.

29 Campbell Clark, "Canada Can Expect More Chinese Wrath as U.S. Takes Broad Aim at Huawei," *Globe and Mail*, 28 January 2019, A4.

30 Doug Saunders, "How the Huawei Crisis Has Exploded Trudeau's China Policy," *Globe and Mail*, 19 January 2019, O1.

31 "Exceptions and General Provisions," USMCA Chapter 32.

32 Sylvain Charlebois, Brian Sterling, Sanaz Haratifar, and Sandi Kyaw Naing, "Comparison of Global Food Traceability Regulations and Requirements," *Comprehensive Reviews in Food Science and Food Safety* 13 (2014); Blackwell reports airport screening, including sniffer dogs, for such products in travellers' carry-on luggage. Tom Blackwell, "Asian Swine Fever Poses 'Huge Threat' to Canadian Pork," *National Post*, 3 January 2020, A6.

33 Export Development Canada, *Evolving with the Changing Needs of Exporters: Submission to the 2018 Legislative Review of Export Development Canada.* (Ottawa: October 2018), 13–14.

34 James E. Anderson, Mykyta Vesselovsky, and Yoto V. Yotov, "Gravity with Scale Effects," EconPapers, 10 March 2016, https://econpapers.repec.org /article/eeeinecon/v_3a100_3ay_3a2016_3ai_3ac_3ap_3a174-193.htm; Suvankulov, "Revisiting National Border Effects."

35 Mykyta Vesselovsky, ed., *Canada: State of Trade – Trade and Investment Update 2018* (Ottawa: Global Affairs Canada, July 2018), 136.

36 Vesselovsky, *Canada: State of Trade*, 141–3.

37 Rainer Lanz and Andreas Maurer, "Services and Global Value Chains: Servicification of Manufacturing and Services Networks," Journal of International Commerce, Economics and Policy 6, no. 3 (2015), https://doi .org/10.1142/S1793993315500143.

38 Peter Koudai, "The Service Revolution in Global Manufacturing Industries," (New York Deloitte Research, January 2006), https://www.researchgate.net /publication/267268075_The_Service_Revolution_ in Global_Manufacturing _Industries; *Vesselovsky*, Canada: State of Trade, *151–3.*

39 Wendy Dobson and Diana Kuzmanovic, "Differentiating Canada: The Future of the Canada–U.S. Relationship," SPP Research Paper 3:7 (Calgary: School of Public Policy, University of Calgary, November 2010).

40 Ari Van Assche, "Global Value Chains and the Rise of a Supply Chain Mindset," in *Redesigning Canadian Trade Policies for New Global Realities*, ed. Stephen Tapp, Ari Van Assche, and Robert Wolfe (Montreal and Kingston: McGill-Queen's University Press, 2017), 202–4.

41 Tom W.G. van der Meer, "Political Trust and the 'Crisis of Democracy,'" *Oxford Research Encyclopaedia of Politics* (January 2017); DOI 10.1093/acrefore /9780190228637.013.77.

42 Matthew Mendelssohn, Robert Wolfe, and Andrew Parkin, "Globalization, Trade Policy, and the Permissive Consensus in Canada," *Canadian Public Policy* 28, no. 3 (2002): 351–71.

43 Ipsos Canada, "Majority (72%) of Canadians Approve of Trudeau's Handling of Trump Situation, but Acknowledge (81%) That It Has Hurt Bilateral Relations" (Toronto: 16 June 2018), https://www.ipsos.com/en -ca/news-polls/Global-News-NAFTA-June-16-2018.

44 Brian J. Arnold, "Canada's International Tax System: Historical Review, Problems, and Outlook for the Future," *Canada's International Law at 150 and Beyond: Paper # 8* (Waterloo: Centre for International Governance Innovation, February 2018).

45 Robert McDougall, "Crisis in the WTO: Restoring the WTO Dispute Resolution Function," CIGI Papers #194 (Waterloo, ON: Centre for International Governance Innovation, October 2018); Robert Wolfe and Bernard Hoekman, "The WTO Still Has Vital Work to Do, Despite the Crippling of Its Appeals Court," *Globe and Mail*, 19 December 2019, B4.

46 Naomi Powell, "Trade End Run," *Financial Post*, 25 January 2020, FP1.

47 Fergus Hanson, Emilia Currie, and Tracy Beattie, *The Chinese Communist Party's Coercive Diplomacy* (Barton, ACT: Australian Strategic Policy Institute, September 2020), https://www.aspi.org.au/report/chinese-communist -partys-coercive-diplomacy.

48 For example, see Michel Crozier, Samuel P. Huntington, and Joji Watanuki, *The Crisis of Democracy: Report on the Governability of Democracies to the Trilateral Commission* (New York: New York University Press, 1975); Fareed Zakaria, "Can America Be Fixed? The New Crisis of Democracy," *Foreign Affairs* 92, no. 1 (January-February 2013): 22–33.

49 Van der Meer, "Political Trust and the 'Crisis of Democracy,'" 1.

50 Van der Meer, "Political Trust and the 'Crisis of Democracy,'" 1.

Bibliography

Abelson, Donald. "Environmental Lobbying and Political Posturing: The Role of Environmental Groups in Ontario's Debate over NAFTA." *Canadian Public Administration* 38, no. 3 (1995): 352–81.

– "From Policy Research to Political Advocacy: The Changing Role of Think Tanks in American Politics." *Canadian Review of American Studies* 25, no. 1 (1995): 93–126.

Abrams, Abigail. "U.S. Avoids Postal 'Brexit' as Universal Postal Union Reaches a Deal." *Time*, 26 September 2019.

Agnew, John. "Sovereignty Regimes: Territoriality and State Authority in Contemporary World Politics." *Annals of the Association of American Geographers* 95, no. 2 (2005): 437–61.

Aireon, Inc. "Interactive Timeline." aireon.com, accessed 10 August 2020.

Akbari, Ather H., and Martha MacDonald. "Immigration Policy in Australia, Canada, New Zealand, and the United States: An Overview of Recent Trends." *International Migration Review* 48, no. 3 (2014): 801–22. https://doi.org/10.1111/imre.12128.

Alba, Davey. "The U.S. Government Will Be Scanning Your Face at 20 Top Airports, Documents Show." *Buzzfeed News*, 11 March 2019.

Alcantara, Christopher, and Jen Nelles. "Indigenous Peoples and the State in Settler Societies: Toward a More Robust Definition of Multilevel Governance." *Publius: The Journal of Federalism* 44 no. 1 (2013): 183–204.

Alden, Edward. *The Closing of the American Border.* New York: Norton, 2008.

Alexander, Jeffrey A., Maureen E. Comfort, Bryan J. Weiner, and Richard Bogue. "Leadership in Community Collaborative Health Partnerships." *Nonprofit Management and Leadership* 12, no. 2 (2001): 159–75.

Allan, Alistair. "Canada Needs a Deal with the U.S. to Control the Northwest Passage." *National Post*, 8 August 2019. https://nationalpost.com/opinion/canada-needs-to-cut-a-deal-with-the-u-s-to-control-the-northwest-passage.

Allan, James P., and Richard Vengroff. "Paradiplomacy: States and Provinces in the Emerging Governance Structure of North America." In *North America in Question: Regional Integration in an Era of Economic Turbulence,* edited by Jeffrey Ayres and Laura Macdonald, 277–82. Toronto: University of Toronto Press, 2012.

Amazon. "Amazon HQ2 Request for Proposals." 2017. https://images-na.ssl -images-amazon.com/images/G/01/Anything/test/images/usa/RFP_3 ._V516043504_.pdf.

Amilhat-Szary, Anne-Laure. *Qu'est, qu'une frontière aujourd'hui.* Paris: Presses Universitaires de France, 2015.

Amin, Ash, and Nigel Thrift. "Neo-Marshallian Nodes in Global Networks." *International Journal of Urban and Regional Research* 16, no. 4 (1992): 571–87.

Anastakis, Dimitry. *Autonomous State: The Struggle for a Canadian Car Industry from OPEC to Free Trade.* Toronto: University of Toronto Press, 2013.

– "Industrial Sunrise? The Chrysler Bailout, the State, and the Re-industrialization of the Canadian Automotive Sector, 1975–1986." *Urban History Review/Revue d'histoire urbaine* 35, no. 2 (2007): 37–50.

Anderson, Alun. *After the Ice: Life, Death, and Geopolitics in the New Arctic.* New York: HarperCollins, 2009.

Anderson, Greg. "Can Someone Please Settle This Dispute: Canadian Softwood Lumber and the Dispute Settlement Mechanisms of the NAFTA and WTO." *World Economy* 29, no. 5 (June 2006): 585–610.

– "Expanding the Partnership? States and Provinces in U.S.-Canada Relations." *Canadian-American Public Policy* 80 (2012): 1–44.

– "Expanding the Partnership? States and Provinces in Canada-U.S. Relations 25 Years On." In *Forgotten Partnership Redux: Canada-U.S. Relations in the 21st Century,* edited by Greg Anderson and Christopher Sands, 535–74. Amherst, NY: Cambria, 2011.

– "How Did Investor-State Dispute Settlement Get a Bad Rap? Blame It on NAFTA, of Course" *World Economy* 40, no. 12 (2017): 2954–5.

– "NAFTA on the Brain: Why Creeping Integration Has Always Worked Better." *American Review of Canadian Studies* 42, no. 4 (2012): 450–9.

– "The Reluctance of Hegemons: Comparing the Regionalization Strategies of a Crouching Cowboy and a Hidden Dragon." In *China and the Politics of Regionalization,* edited by Emilian Kavalski, 91–107. Surrey, UK: Ashgate Publishers, 2009.

– "Securitization and Sovereignty in Post-9/11 North America." *Review of International Political Economy* 19, no. 5 (December 2012): 711–41.

Anderson, Greg, and Christopher Sands. "Fragmegration, Federalism, and Canada–United States Relations." In *Borders and Bridges: Canada's Policy Relations in North America,* edited by Monica Gattinger and Geoffrey Hale, 41–58. New York: Oxford University Press, 2010.

– *Negotiating North America: The Security and Prosperity Partnership.* Washington, DC: Hudson Institute, 2007.

Anderson, James E. *Public Policymaking*. 7th ed. Florence, KY: Wadsworth/ Cengage, 2011.

Anderson, James E., Mykyta Vesselovsky, and Yoto V. Yotov. "Gravity with Scale Effects." EconPapers, 10 March 2016. https://econpapers.repec.org /article/eeeinecon/v_3a100_3ay_3a2016_3ai_3ac_3ap_3a174-193.htm.

Anderson, Malcolm. "Les frontières: un débat contemporain." *Cultures et conflits* 26–7 (1997).

Anderson, Stuart. "New Evidence USCIS Policies Increased Denials of H-1B Visas." *Forbes*, 25 July 2018. https://www.forbes.com/sites/stuartanderson /2018/07/25/new-evidence-uscis-policies-increased-denials-of-h-1b-visas /#5687d8105a9f.

Anderson, William P. "The Binational Advantage," *How It Works and (And Why It Matters)*. Windsor, ON: Cross Border Institute, 2016.

Anderson, William P., Hanna F. Maoh, and Charles M. Burke. "Passenger Car Flows across the Canada-U.S. Border: The Effect of 9/11." *Transport Policy* 35 (2014): 50–6.

Andresen, Martin A. "The Geography of the Canada–United States Border Effect." *Regional Studies* 44, no. 5 (2010): 579–94.

Andreas, Peter, and Thomas J. Biersteker, eds. *The Rebordering of North America*. New York: Routledge, 2003.

Anson-Cartwright, Sarah. "Immigration for a Competitive Canada: Why Highly Skilled International Talent Is at Risk." Canadian Chamber of Commerce, 14 January 2016. http://www.chamber.ca/media/blog/160114 -immigration-for-a-competitive-canada/.

Aoyama, Yuko, James T. Murphy, and Susan Hanson. *Key Concepts in Economic Geography*. Thousand Oaks, CA: Sage, 2011.

Appadurai, Arjun. "Disjuncture and Difference in the Global Cultural Economy." *Theory, Culture & Society* 7 (1990): 295–310.

Arctic Council. "Observers" (Tromso, Norway: 31 July 2019). https://arctic -council.org/index.php/en/about-us/arctic-council/observers.

– "Permanent Participants." https://arctic-council.org/en/about/permanent -participants/.

Arnold, Brian J. "Canada's International Tax System: Historical Review, Problems, and Outlook for the Future." *Canada's International Law at 150 and Beyond: Paper #8*. Waterloo, ON: Centre for International Governance Innovation, February 2018.

Arsenault, Julien. "Hydro-Quebec Looks to Maine as Massachusetts Ends Northern Pass Hydro Project." *City News Toronto*, 28 March 2018. http:// toronto.citynews.ca/2018/03/28/hydro-quebec-looks-to-maine-as -massachusetts-ends-northern-pass-hydro-project/.

Asheim, Bjørn T., Philip Cooke, and R.L. Martin. "The Rise of the Cluster Concept in Regional Analysis and Policy: A Critical Assessment." In *Clusters and Regional Development: Critical Reflections and Explorations*, edited by Asheim, Cooke, and Martin, 1–29. New York: Routledge, 2006.

Asian Regional Integration Center. "Free Trade Agreements." Manila: Asian Development Bank, 2018. https://aric.adb.org/fta-country.

Association of American Railroads. "Welcome to AAR's Technical Services." https://www.aar.com/standards/index.html.

Atkins, Eric. "NATS Takes Equity Stake in NavCanada's Space-Based Air Traffic System." *Globe and Mail*, 17 May 2018, B2.

– NavCanada's Ambitious Plan: A Satellite Network to Track Planes around the World." *Globe and Mail*, 25 October 2017.

– "Omnitrax, Ottawa Deal Legal Blows over Manitoba Railway. *Globe and Mail*, 15 November 2017, B1.

– "With Two New Global Deals, Canada's Container Growth Is Forecast to Surpass the U.S." *Globe and Mail*, 22 February 2018, B1.

– "Transat A.T. Inc. delays closing $ 720 million takeover by Air Canada." *The Globe and Mail*, 28 July 2020.

Atkinson, Michael M. "What Kind of Democracy Do Canadians Want?" *Canadian Journal of Political Science* 27, no. 4 (1994): 717–45.

Austin, John, and Brittany Affolter-Caine. *The Vital Center: A Federal-State Compact to Renew the Great Lakes Region*. Washington, DC: Brookings, 2006.

Australia. Department of Foreign Affairs and Trade. "Australia's Free Trade Agreements." Canberra, November 2017. http://dfat.gov.au/trade/agreements/Pages/trade-agreements.aspx.

Automotive Policy Research Institute. "Canadian Manufacturing Employment by Automaker" (internal database). Hamilton, ON, (2018).

Bache, Ian, and Matthew Flinders, eds. *Multi-Level Governance*. New York: Oxford University Press, 2004.

Badie, Bertrand. *L'impuissance de la puissance*. Paris: Editions du CNRS, coll. Biblis, 2013.

– *Nous ne sommes plus seuls au monde*. Paris: La découverte, 2016.

Badie, Bertrand, and Michel Foucher. *Vers un monde post-national*. Paris: Éditions du CNRS, 2017.

Bagnall, James E. *100 Days: The Rush to Judgment That Killed Nortel*. Ottawa: Ottawa Citizen, 2013.

Balibar, Etienne. *Les frontières de la démocratie*. Paris: La Découverte, 1992.

Bank of Canada. "Canada's Economic Expansion: A Progress Report." Remarks by Timothy Lane, deputy governor of the Bank of Canada. 8 March 2018. https://www.bankofcanada.ca/wp-content/uploads/2018/03/remarks-080318a.pdf.

Banks, Kerry. "Fuelling Controversy: What Do New U.S. Fuel Efficiency Regulations Mean for Canada's Climate Goals?" *Electric Autonomy*, 31 March 2020. https://electricautonomy.ca/2020/03/31/new-us-fuel-efficiency-regulations-impact-on-canada/.

Barami, Bahar, and Michael Dyer. "Assessment of Short-Sea Shipping; Options for Domestic Applications, Appendix G." Washington, DC: Volpe National Transportation Systems Center, US Department of Transportation, 23 December 2009.

Bardach, Eugene, and Robert A. Kagan. *Going by the Book: The Politics of Regulatory Unreasonableness*. New York: Transaction, 2002.

Barnes, Robert. "Supreme Court Rules That States May Require Online Retailers to Collect Sales Taxes." *Washington Post*, 22 June 2018.

Basalisco, Bruno, Jimmy Gardebrink, Martina Facino, and Henrik Okholm. "E-commerce Imports into Canada: Sales Tax and Customs Treatment." Copenhagen: Copenhagen Economics, March 2017.

Beale, Elizabeth, Dale Beguin, Bev Dahlby, Don Drummond, Nancy Olewiler, and Christopher Ragan. *Provincial Carbon Pricing and Competitiveness Pressures: Guidelines for Business and Policymakers*. Canada's Ecofiscal Commission, 2015. http://ecofiscal.ca/wp-content/uploads/2015/11/Ecofiscal-Commission -Carbon-Pricing-Competitiveness-Report-November-2015.pdf.

Beck, Ulrich. *Pouvoir et contre-pouvoir à l'ère de la mondialisation*. Paris: Flammarion "Champs essais," 2009.

Becker, Rachel. "California to Sue after Trump Administration Eases Vehicle Emissions Standards." *Times of San Diego*, 31 March 2020. https:// timesofsandiego.com/business/2020/03/31/california-to-sue-after-trump -aministration-eases-vehicle-emissions-standards/.

Beilock, Richard, Robert Dolyniuk, and Barry Prentice. "Encouraging Development through Better Integration of U.S. and Canadian Transportation: The Open Prairies Proposal." *Regional Economic Development* 2, no. 2 (2006): 73–86.

Beitsch, Rebecca, and Rachel Frazin. "Trump Budget Slashes EPA Funding, Environmental programs." *The Hill*, 10 February 2020. https://thehill.com /policy/energy-environment/482352-trump-budget-slashes-funding-for -epa-environmental-programs.

Bellavita, Christopher. "Changing Homeland Security: What Is Homeland Security?" *Homeland Security Affairs* 4, no. 2 (2008). https://www.hsaj.org /articles/118.

Bencivenga, Lisa. "International Traffic in Arms Regulations (ITAR): Who Must Comply, What Is Controlled and Where Do We Go from Here?" Orlando, FL: Orlando World Marriott Center, 20–2 February 2013. https:// www.visiononline.org/userassets/aiauploads/file/TH_AIA_Lisa%20 _Bencivenga_distribution.pdf.

Bendaoud, Maroine. "Quand la sécurité nationale américaine fait fléchir le principe de la non-discrimination en droit canadien: Le cas de l'International Traffic in Arms Regulations." *Les Cahiers de droit* 54, no. 2–3 (2013): 549–86.

Bennett, Mary-Jane. *Airport Policy in Canada*. Winnipeg: Frontier Institute for Public Policy, August 2012.

Bercuson, David J. "Pompeo Is Blunt, But Right, about Canada's North." *National Post*, 21 May 2019, https://nationalpost.com/opinion/david-j -bercuson-pompeo-is-blunt-but-right-about-canadas-north#comments-area.

Berkes, Fikret. "Evolution of Co-management: Role of Knowledge Generation, Bridging Organizations and Social Learning." *Journal of Environmental Management* 90, no. 5 (2009): 1692–1702.

Berkes, Fikret, and Carl Folke, eds. *Linking Social and Ecological Systems: Management Practices and Social Mechanisms for Building Resilience.* Cambridge, UK: Cambridge University Press, 1998.

Berkow, Jameson. "'A Parade of Broken Promises': How CNOOC Stumbled with Its Nexen Takeover." BNN Bloomberg, 15 September 2017. https://www.bnnbloomberg.ca/a-parade-of-broken-promises-how-cnooc-stumbled -with-its-nexen-takeover-1.857533.

Bernasconi-Osterwalder, Nathalie. "Background Paper on Vattenfall v. Germany Arbitration." International Institute for Sustainable Development, July 2009, https://www.iisd.org/library/state-play-vattenfall-v-germany-ii -leaving-german-public-dark.

Berry, Dawn Alexandrea, Nigel Bowles, and Halbert Jones, eds. *Governing the North American Arctic: Sovereignty, Security, and Institutions.* New York: Palgrave Macmillan, 2016.

Bertrand, Nancy Hoi, and Pamela Hughes. "Canadian Inter-Listed Companies: Navigating the Maze of Governance Requirements." *Ivey Business Journal* (September–October 2004). https://iveybusinessjournal. com/publication/canadian-inter-listed-companies-navigating-the-maze-of -governance-requirements/.

Bhagwati, Jagdish. "U.S. Trade Policy: The Infatuation with Free Trade Agreements." Discussion Paper #726. New York: Department of Economics, Columbia University, 1995. https://core.ac.uk/reader/161436448.

Bickis, Ian. "As US Lowers Auto Emission Standards, Canada Is at a Fuel Efficiency Crossroads" *Financial Post*, 9 May 2019.

Bidler, Richard B. "The Canadian Arctic Waters Pollution Prevention Act: New Stresses on the Law of the Sea." *Michigan Law Review* 69, no. 1 (1970): 1–54.

Binational.net. "About the Great Lakes Water Quality Agreement." https://binational.net/glwqa-aqegl/.

Birkland, Thomas A. *An Introduction to the Policy Process: Theories, Concepts and Models of Public Policy Making,* 3rd ed. Armonk, NY: M.E. Sharpe, 2011.

Blackwell, Tom. "Asian Swine Fever Poses 'Huge Threat' to Canadian Pork." *National Post*, 3 January 2020, A6.

Blanchard, Emily. "Leveraging Global Supply Chains in Canadian Trade Policy." In *Redesigning Canadian Trade Policies for New Global Realities,* edited by Stephen Tapp, Ari Van Assche and Robert Wolfe, 209–28. Montreal: Institute for Research on Public Policy, 2017.

Blank, Stephen. "Infrastructure, Attitude and Weather: Today's Threats to Supply Chain Security." Calgary: Canadian Global Affairs Institute and School of Public Policy, University of Calgary, June 2016. https://papers.ssrn.com/sol3/papers.cfm?abstract_id=3079192.

Block, Walter, and George Lermer, eds. *Breaking the Shackles: The Economics of Deregulation – A Comparison of U.S. and Canadian Experience.* Vancouver: Fraser Institute, 1991.

Blondiaux, Loïc, and Yves Sintomer. "L'impératif Délibératif." *Politix* 15, no. 57 (2002): 17–35.

Bloom, Michael, and Michael Grant. *"Hollowing Out" – Myth and Reality: Corporate Takeovers in an Age of Transformation.* 3 vols. Ottawa: Conference Board of Canada, 2008.

Bloomfield, L.M., and Gerald F. Fitzgerald. *Boundary Waters Problems of Canada and the United States: The International Joint Commission (1912–1958).* Toronto: Carswell, 1958.

Bollyky, Douglas and Chad P. Bown. "The tragedy of vaccine nationalism." *Foreign Affairs,* 27 July, 2020.

Boone, Rebecca. "'Megaload' Settlement Bans New Big Truck Loads on Idaho Road." *Missoulian,* 27 January 2017.

Boscariol, John W., and Brenda C. Swick. "Final U.S. ITAR Rule on Dual – and Third-Country Nationals Raises New Challenges for Canadian Business." *McCarthy Tetrault,* 16 May 2011.

Botha, Johanu. *Boots on the Ground.* Toronto: University of Toronto Press, forthcoming.

Botts, Lee, and Paul Muldoon. *Evolution of the Great Lakes Water Quality Agreement.* East Lansing, MI: Michigan State University Press, 2005.

Boucher, Christian. *Toward North American or Regional Cross-Border Communities.* Government of Canada Policy Research Initiative Working Paper Series 002. Ottawa, June 2005.

Bound, John, Gaurav Khanna, and Nicolas Morales. "Understanding the Economic Impact of the H-1B Program on the U.S." NBER Working Paper no. 23153. February 2017. http://www.nber.org/papers/w23153.

Bourgeon, Mathilde, Thalia D'Aragon-Giguère, and Élisabeth Vallet. "Les flux migratoires à la frontière québéco-américaine." *Quebec Studies* 64 (2017): 141–56.

Bow, Brian, and Greg Anderson. "Building without Architecture: Regional Governance in Post-NAFTA North America." In *Regional Governance in Post-NAFTA North America: Building without Architecture,* 1–18. New York: Routledge, 2015.

Bown, Chad P., and Greg Autry. "Is the International Trading System Broken." *Economist,* 7 May 2018.

Bown, Chad P., Cathleen Cimino-Isaacs, and Melina Kolb. "Will Trump Invoke National Security to Start a Trade War?" Washington, DC: Petersen Institute

for International Economics, 5 July 2017. https://piie.com/blogs/trade
-investment-policy-watch/will-trump-invoke-national-security-start-trade
-war.

BP. "Natural Gas Prices." *Statistical Review of World Energy*. June 2018, https://
www.bp.com/content/dam/bp/business-sites/en/global/corporate/pdfs
/energy-economics/statistical-review/bp-stats-review-2018-full-report.pdf.

– *Statistical Review of World Energy*. 2020. https://www.bp.com/content
/dam/bp/business-sites/en/global/corporate/pdfs/energy-economics
/statistical-review/bp-stats-review-2020-full-report.pdf.

Bradbury, Susan. "Irritable Border Syndrome: The Impact of Security on
Travel across the Canada-U.S. Border." *Canadian-American Public Policy*
79 (December 2012).

Bradley, Brian. "USMCA Has Give and Take but No Big Stretch." American
Shipper.com, 17 October 2018. https://www.freightwaves.com/news/usmca
-has-give-and-take-but-no-big-stretch.

Bradsher, Keith. "France Says Tax Deal with U.S. Is Closer." *New York Times*,
22 January 2020, B6.

Bramley, Matthew. "Far from Turning the Corner: Canada's Conservative
Government Has Substantially Shifted Its Position on Climate Change, but Is
Its Policy Response Too Timid, Too Complex and Likely to Be Superseded?"
Carbon Finance, 20 June 2008. http://www.pembina.org/op-ed/1661.

Bremmer, Ian. *Us vs. Them: The Failure of Globalism*. New York: Portfolio/Penguin,
2018.

Brister, Bernard J. "The Same Yet Different: The Evolution of the Post-9/11
Canada–United States Security Relationship." In Gattinger and Hale, *Borders
and Bridges*, 82–99.

British Columbia Securities Commission. "72-701 Guide for Use of the
Multijurisdictional Disclosure System by Canadian Issuers in the U.S. Market
(Previously NIN#91/22)." Victoria: 1 June 2001.

Bronskill, Jim. "Canada Border Services Agency Sharing Information on
American Border Crossings with Homeland Security." *Toronto Star*, 31 August
2017.

Brooks, Mary R. "Mapping the New North American Reality: The Road Sector."
Policy Options, June 2004. https://policyoptions.irpp.org/wp-content
/uploads/sites/2/assets/po/north-american-integration/brooks.pdf.

Brown, Timothy M. *Market Rules: Economic Union Reform and Intergovernmental
Policy-Making in Canada*. Montreal-Kingston: McGill-Queen's University
Press, 2002.

Brown, W. Mark, "How Much Thicker Is the Canada-U.S. Border? The Cost
of Crossing the Border by Truck in the Pre- and Post 9/11 Eras." Economic
Analysis Research Paper Series 11F0027M — No. 99. Ottawa: Statistics Canada,
July 2015.

– "Overcoming Distance, Overcoming Borders: Comparing North American Regional Trade." Economic Analysis Research Paper Series 11F0027 No. 008. Ottawa: Statistics Canada, April 2003.

Brown, Wendy. *Murs. Les murs de séparation et le déclin de la souveraineté étatique.* Paris: Les Prairies Ordinaires, 2009.

Brownell, Claire. "For Fentanyl Importers, Canada Post Is the Shipping Method of Choice." *Maclean's,* 7 March 2019.

Brunet-Jailly, Emmanuel, ed. *Borderlands: Comparing Border Security in North America and Europe.* Ottawa: University of Ottawa Press, 2007.

– "Cascadia in Comparative Perspectives: Canada-U.S. Relations and the Emergence of Cross-Border Regions." *Canadian Political Science Review* 2, no. 2 (2008): 104–24.

Brunet-Jailly, Emmanuel, Susan E. Clarke, and Debora L. VanNijnatten. "The Results in Perspective: An Emerging Model of Cross-Border Regional Co-operation in North America." Leader Survey on Canada-US Cross-Border Regions: An Analysis. North American Linkages Working Paper Series 012 (2006).

Bruno, Gerry. "Connecting Canada: An Aviation Policy Agenda for Global Competitiveness and Economic Prosperity." Submission to Canadian Airports Council, Canadian Transportation Act Review, Ottawa, January 2015.

Brusco, Sebastiano. "The Emilian Model: Productive Decentralisation and Social Integration." *Cambridge Journal of Economics* 6, no. 2 (1982): 167–84.

Buffalo and Fort Erie Public Bridge Authority. "Buffalo-Niagara International Bridge Officials Call for NEXUS Reform to Improve Border Travel, Security." Buffalo–Fort Erie: 23 January 2017.

Bunch, Kevin. "New Commission Co-Chairs Focus on Climate Change and Resilient Shorelines." *Great Lakes Connection Newsletter: International Joint Commission,* 14 January 2020. https://www.ijc.org/en/new-commission -co-chairs-focus-climate-change-and-resilient-shorelines.

Burgess, Michael. *Comparative Federalism: Theory and Practice.* New York: Routledge, 2006.

Burgoon, Brian. "Globalization and Welfare Compensation: Disentangling the Ties That Bind." *International Organization* 55, no. 3 (Summer 2001): 509–51.

Burke, James. *Connections: From Ptolemy's Astrolabe to the Invention of Electricity: How Inventions Are lLnked and How They Cause Change across History.* New York: Simon and Schuster, 2007.

Burleton, Derek, and Admir Kolaj. *Canada-U.S. Border Spending: A Reversal of Fortunes.* Toronto: TD Economics, 8 February 2016.

Burns, Cliff. "U.S Threatens Thales Alenia Space over 'ITAR-Free' Satellite." Export Law Blog, 13 February 2012. https://www.exportlawblog.com /archives/3837.

Burpee, Lawrence J. "Good Neighbours." In *Contemporary Neighbours*. Toronto: Ryerson, 1940.

– "The International Joint Commission." In *Canada in World Affairs*. Canadian Institute of International Affairs. Toronto: University of Oxford Press, 1941.

Büthe, Tim, and Walter Mattli. *The New Global Rulers: The Privatization of Regulation in the World Economy*. Princeton, NJ: Princeton University Press, 2011.

Cabirol, Michel. "Réglementation ITAR: États-Unis, ces amis qui ne veulent pas que du bien." *La Tribune*, 23 April 2018.

Cairns, Alan. *Constitution, Government, and Society in Canada*. Toronto: McClelland and Stewart, 1988.

– "The Embedded State: State-Society Relations in Canada." In *State and Society: Canada in Comparative Perspective*, edited by Keith G. Banting, 53–86. Toronto: University of Toronto Press, 1986.

Caldwell, Linton. "The Ecosystem as a Criterion for Public Land Policy." *Natural Resources Journal* 10, no. 2 (1970): 203–21.

Campbell, Maureen, Matthew J. Cooper, Kathryn Friedman, and William P. Anderson. "The Economy as a Driver of Change in the Great Lakes–St. Lawrence River Basin." *Journal of Great Lakes Research* 41, no. S1 (2015): S69–S83.

Canada. *Canada's Cyber-Security Strategy: For a Stronger, More Prosperous Canada*. Ottawa: Public Safety Canada, 2010. http://publications.gc.ca/collections/collection_2010/sp-ps/PS4-102-2010-eng.pdf

– Canada Border Services Agency. "Evaluation of the Trusted Traveler Programs (Air, Land, Marine)." Ottawa: November 2016.

– "Canada's Economic Action Plan: Food and Agriculture." 2015.

– Canadian Food Inspection Agency. "Canada, United States Agree on Application of Zones to Allow Safe Trade in the Event of an African Swine Fever Outbreak." 22 May 2019. https://www.canada.ca/en/food-inspection-agency/news/2019/05/canada-united-states-agree-on-application-of-zones-to-allow-safe-trade-in-the-event-of-an-african-swine-fever-outbreak.html

– Citizenship and Immigration Canada. "Asylum Claims by Year." Ottawa: updated 17 February 2020; https://www.canada.ca/en/immigration-refugees-citizenship/services/refugees/asylum-claims.html.

– Department of Finance. *Budget 2017: Building a Strong Middle Class*. Ottawa, 2017. https://www.budget.gc.ca/2017/docs/plan/budget-2017-en.pdf.

– Department of Finance. *Jobs, Growth and Long-Term Prosperity: Economic Action Plan 2012*. Ottawa: 2012.

– Department of Finance. "Pension Plan Investment in Canada: The 30 Percent Rule." Ottawa: June 2016.

– Department of National Defence. *Strong, Secure Engaged: Canada's Defence Policy*. Ottawa: 2017. http://dgpaapp.forces.gc.ca/en/canada-defence-policy/docs/canada-defence-policy-report.pdf.

– Department of the Environment. "Canadian Environmental Protection Act,
  1999." *Canada Gazette* 143, no. 14 (2009): 840–5. http://publications.gc.ca
  /collections/collection_2009/canadagazette/SP2-1-143-14.pdf.
– Employment and Social Development Canada. "Overhauling the Temporary
  Foreign Worker Program." 2014. https://www.canada.ca/content/dam
  /canada/employment-social-development/migration/documents/assets
  /portfolio/docs/en/foreign_workers/employers/overhauling_TFW.pdf.
– Employment and Social Development Canada. "Program Requirements for
  the Global Talent Stream." https://www.canada.ca/en/employment-social
  -development/services/foreign-workers/global-talent/requirements.html.
– Environment and Climate Change Canada. "Clean Fuel Standard: Discussion
  Paper." Ottawa: February 2017. https://www.ec.gc.ca/lcpe-cepa/default
  .asp?lang=En&n=D7C913BB-1.
– Environment and Climate Change Canada. "Clean Fuel Standard: Proposed
  Regulatory Approach." Ottawa: June 2019. https://www.canada.ca/content
  /dam/eccc/documents/pdf/climate-change/pricing-pollution/Clean-fuel
  -standard-proposed-regulatory-approach.pdf.
– Environment and Climate Change Canada. *Departmental Plan 2017 to 2018
  Report, Environment and Climate Change Canada* (2018). https://www.canada
  .ca/en/environment-climate-change/corporate/archive/archived
  -departmental-plans/2017-2018.html.
– Environment and Climate Change Canada. *Environment and Climate Change
  Canada: Clean Fuel Standard Workshop.* Ottawa: 6 March 2017.
– External Advisory Committee on Smart Regulation. *A Regulatory Strategy for
  Canada.* Ottawa: September 2004.
– Global Affairs Canada. "NAFTA – Chapter 11 – Investment: Cases Filed
  against the Government of Canada." 2020. http://www.international.gc.ca
  /trade-agreements-accords-commerciaux/topics-domaines/disp-diff/gov
  .aspx?lang=eng, updated 11 May 2020.
– Global Affairs Canada. "Trade and Investment Agreements." Ottawa:
  18 October 2018. http://www.international.gc.ca/trade-commerce/trade
  -agreements-accords-commerciaux/agr-acc/index.aspx?lang=eng.
– Health Canada. "North American Free Trade Agreement Technical Working
  Group on Pesticides." 2015. http://www.hc-sc.gc.ca/cps-spc/pest/part
  /int/_nafta-alena/index-eng.php.
– Immigration, Refugees and Citizenship Canada. *Express Entry Year End
  Report 2015.* Last modified 15 March 2017. https://www.canada.ca/en
  /immigration-refugees-citizenship/corporate/publications-manuals
  /express-entry-year-end-report-2015.html.
– Immigration, Refugees and Citizenship Canada. *Express Entry Year End
  Report 2016.* https://www.canada.ca/content/dam/ircc/migration/ircc
  /english/pdf/pub/ee-2016-eng.pdf.

- Immigration, Refugees and Citizenship Canada. "International Mobility Program: Canadian Interests – Significant Benefit General Guidelines [R205(a) – C10]." Last modified 18 September 2014. https://www.canada.ca/en/immigration-refugees-citizenship/corporate/publications-manuals/operational-bulletins-manuals/temporary-residents/foreign-workers/exemption-codes/canadian-interests-significant-benefit-general-guidelines-r205-c10.html.
- Immigration, Refugees and Citizenship Canada. "International Mobility Program: Federal-Provincial or Territorial Agreements [R204(c)] (LMIA Exemption Code T13)." Last modified 13 October 2017. https://www.canada.ca/en/immigration-refugees-citizenship/corporate/publications-manuals/operational-bulletins-manuals/temporary-residents/foreign-workers/exemption-codes/federal-provincial-territorial-agreements-r204-lmia-exemption-code-t13.html.
- Immigration, Refugees and Citizenship Canada. "Passport Program Annual Report for 2014-2015." October 2016. https://www.canada.ca/en/immigration-refugees-citizenship/corporate/publications-manuals/passport-program-annual-report-2014-2015.html
- Immigration, Refugees and Citizenship Canada. "Passport Program Annual Report for 2017-2018." February 2020. https://www.canada.ca/en/immigration-refugees-citizenship/corporate/publications-manuals/passport-program-annual-report-2017-2018.html
- Immigration, Refugees and Citizenship Canada. "Speaking Notes for the Honourable Jason Kenney, P.C., M.P. Minister of Citizenship, Immigration and Multiculturalism." 2 November 2012. https://www.canada.ca/en/immigration-refugees-citizenship/news/archives/speeches-2012/jason-kenney-minister-2012-11-02.html.
- Immigration, Refugees and Citizenship Canada. *2017 Consultations on Immigration Levels, Settlement and Integration – Final Report.* Last modified 1 November 2017. https://www.canada.ca/en/immigration-refugees-citizenship/corporate/transparency/consultations/2017-consultations-immigration-levels-settlement-integration-final-report.html.
- Immigration, Refugees and Citizenship Canada. "Who Can Apply: Canadian Experience Class (Express Entry)." Last modified 24 October 2017. https://www.canada.ca/en/immigration-refugees-citizenship/services/immigrate-canada/express-entry/become-candidate/eligibility/canadian-experience-class.html?_ga=2.254487425.880335319.1512612908-900327754.1512612908.
- Industry Canada. "Trade Data Online: Trade by Product (HS) – HS Codes." http://strategis.gc.ca/sc_mrkti/tdst/tdo/tdo.php?lang=30&headFootDir=/sc_mrkti/tdst/headfoot&productType=HS6&cacheTime=962115865#tag.
- Natural Resources Canada. *Energy Fact Book* (annual). Ottawa: 2015–17.

– Natural Resources Canada. "Natural Gas Facts." Last modified 2 July 2020. https://www.nrcan.gc.ca/energy/facts/natural-gas/20067.
– Office of the Auditor General. "Report 3: Taxation of E-Commerce." Ottawa: May 2019.
– Office of the Auditor General. "Report 5 – Temporary Foreign Worker Program – Employment and Social Development Canada." *2017 Spring Reports of the Auditor General of Canada to the Parliament of Canada.* http://www.oag-bvg.gc.ca/internet/English/parl_oag_201705_05_e_42227.html.
– Office of the Auditor General, Commissioner of the Environment and Sustainable Development. *A Legacy Worth Protecting: Charting a Sustainable Course in the Great Lakes and St. Lawrence River Basin.* Ottawa: 2001.
– Office of the Auditor General, Commissioner of the Environment and Sustainable Development. Status Report, Chapter 7. *Ecosystems and Areas of Concern in the Great Lakes Basin.* Ottawa: 2008.
– Office of the Privacy Commissioner. "Privacy and Cyber Security: Emphasizing Privacy Protection in Cyber Security Activities." Ottawa: December 2014. https://www.priv.gc.ca/media/1775/cs_201412_e.pdf
– Parliament. House of Commons. Standing Committee on National Defence. *The Canada–U.S. Relationship: Perspectives on Defence, Security, and Trade.* 42nd Parliament, 1st sess., May 2017. Committee Report 5. https://www.ourcommons.ca/Content/Committee/421/NDDN/Reports/RP8977649/nddnrp05/nddnrp05-e.pdf.
– Passport Canada. "International Comparison of Passport-Issuing Authorities." Ottawa: March 2012.
– Passport Canada. (Immigration, Refugees and Citizenship Canada). Annual Reports 2003-2004 to 2012-2013. April 2013. http://publications.gc.ca/site/fra/9.506930/publication.html.
– Policy Research Institute. *The Emergence of Cross-Border Regions between Canada and the United States: Reaping the Promise and Public Value of Cross-Border Regional Relationships – Final Report.* Ottawa: Industry Canada, November 2008. http://publications.gc.ca/collections/collection_2009/policyresearch/PH4-31-2-2008E.pdf.
– Policy Research Institute. *The Emergence of Cross-Border Regions: Interim Report.* Ottawa: Industry Canada, 2005.
– Policy Research Institute. *The Emergence of Cross-Border Regions: Roundtables Synthesis Report.* Ottawa: Industry Canada, May 2006.
– Privy Council Office and United States, Office of Information and Regulatory Affairs. "Joint Action Plan for the Canada-U.S. Regulatory Cooperation Council." Ottawa and Washington: December 2011. https://www.canada.ca/en/treasury-board-secretariat/corporate/transparency/acts-regulations/canada-us-regulatory-cooperation-council/joint-action-plan.html.

– Public Safety Canada. "Canada Hosts Five Country Ministerial Meeting and Quintet of Attorneys General." News release, 28 June 2017. https://www .canada.ca/en/public-safety-canada/news/2017/06/canada_hosts _fivecountryministerialmeetingandquintetofattorneysg.html.
– Public Safety Canada. "Cross-Border Crime Forum." Ottawa: 2 December 2015.
– Public Safety Canada. "Integrated Border Enforcement Teams." Ottawa: 1 December 2015.
– Public Safety Canada. "Five Country Ministerial Meeting and Quintet of Attorneys General Concludes." Ottawa: 30 August 2018.
– Public Safety Canada. "National Strategy for Critical Infrastructure." Ottawa: 15 December 2009. http://www.publicsafety.gc.ca/cnt/rsrcs/pblctns/srtg -crtcl-nfrstrctr/srtg-crtcl-nfrstrctr-eng.pdf.
– Public Services and Procurement Canada. "Enterprise Crown Corporations and Other Government Business Enterprises." Ottawa: 2 June 2016.
– Public Works and Government Services Canada. *Opening New Markets in Europe, Creating Jobs and Opportunities for Canadians.* Ottawa: 2013.
– Royal Canadian Mounted Police. "Border Security." Ottawa: 26 August 2015. http://www.rcmp-grc.gc.ca/ibet-eipf/index-eng.htm.
– Royal Commission on the Economic Union and Development Prospects for Canada. *Report*, 3 vols. Ottawa: Supply and Services Canada, 1985.
– School of Public Service. *Advancing Canadian Interests in the United States: A Practical Guide for Canadian Public Officials.* Ottawa: 2004.
– *Securing an Open Society: Canada's National Security Policy.* Ottawa: Privy Council Office, 27 April 2004.
– Statistics Canada. "Balance of International Payments, by Current and Capital Account, Annual." CANSIM table 36-10-0007-01 (formerly CANSIM table 376-0036). Ottawa: 1 March 2019.
– Statistics Canada. *Business Enterprise In-House Research and Development Characteristics, by Industry Group Based on the North American Industrial Classification System (NAICS), Country of Control and Expenditure Type.* CANSIM table 358-0518. Ottawa: 2018.
– Statistics Canada. *Capital Repairs and Expenditures, Non-Residential Tangible Assets, by Industry and Geography.* Table 34-10-0035-01. Ottawa: 2018.
– Statistics Canada. "Electrical Power, Electrical Utilities and Industry: Annual Supply and Disposition." Table 25-10-0021-01 (formerly CANSIM table 127-0008). Ottawa: 19 July 2019.
– Statistics Canada. "International Travellers Entering or Returning to Canada, by Type of Transport." CANSIM table 427-0001. Ottawa: 17 October 2017.
– Statistics Canada. "International Travelers Entering or Returning to Canada, by Type of Transport." CANSIM table 24-10-0041-01 (formerly CANSIM table 427-0001) (July 2019).

 – Statistics Canada. "Merchandise Imports and Exports: Customs Basis."
   Table 12-10-0090-01 (formerly CANSIM table 228-0081). Ottawa: December
   2017. https://www150.statcan.gc.ca/t1/tbl1/en/tv.action?pid=1210009001
 – Statistics Canada. "Number of Vehicles Travelling between Canada and the
   United States." CANSIM table 427-0002. Ottawa: 17 October 2017.
 – Statistics Canada. "Railway Industry Summary Statistics on Freight
   and Passenger Transportation." Table 23-10-0057-01 (formerly CANSIM
   table 404-0016). Ottawa: 27 July 2019.
 – Transportation Act Review. *Pathways: Connecting Canada's Transportation
   System to the World.* Vol. 1. Ottawa: Transport Canada, October 2015.
 – Transport Canada. *Border Infrastructure Investment Program 2.0: Canada-United
   States.* Ottawa: December 2014.
 – Transport Canada. *Border Infrastructure Investment Program 3.0: Canada-United
   States.* Ottawa and Washington: August 2016.
 – Transport Canada. "Canada's Oceans Protection Plan: What It Means for
   Canada's Regions." Ottawa: 7 November 2016. https://pm.gc.ca/eng/news
   /2016/11/07/canadas-oceans-protection-plan-what-it-means-canadas-regions.
 – Transport Canada. "The National Highway System Map." Ottawa: 30 March
   2010. https://www.tc.gc.ca/eng/policy/acg-acgd-menu-highways-map
   -2151.htm.
 – Transport Canada. *Transportation in Canada: 2016.* Ottawa: 2017.
 – Treasury Board Secretariat. *Canada–United States Regulatory Cooperation
   Council E-Newsletter.* Ottawa: March 2016.
 – Treasury Board Secretariat. *Guidelines on International Regulatory Obligations
   and Cooperation.* Ottawa: 17 August 2007.
 – Treasury Board Secretariat and Newfoundland and Labrador. Canadian
   Environmental Assessment Agency and Department of Environment and
   Conservation, Joint Review Panel. *Report of the Joint Review Panel: Lower
   Churchill Hydroelectric Generation Project.* Ottawa: August 2011. http://
   publications.gc.ca/collections/collection_2011/ec/En106-101-2011-eng.pdf.
 – Treasury Board Secretariat and United States, Office of Information and
   Regulatory Affairs. "Memorandum of Understanding between the Treasury
   Board of Canada Secretariat and the United States Office of Information
   and Regulatory Affairs regarding the Canada–United States Regulatory
   Cooperation Council." Ottawa and Washington: 4 June 2018. https://www
   .canada.ca/en/treasury-board-secretariat/services/regulatory-cooperation
   /memorandum-understanding-between-canada-united-states-advance
   -regulatory-cooperation-council.html.
Canada Post Corp. *2018 Annual Report.* Ottawa: April 2019.
Canadian Architectural Licencing Authorities. "Updates re. the NCARB/
   CALA Mutual Recognition Agreement and Acceptance of the Architect
   Registration Exam (ARE)." Ottawa: 21 January 2015. http://saskarchitects

.com/wp-content/uploads/2015/03/January-2015-Bulletin-NCARB-CALA
-Mutal-Recognition-Agreement-MRA.pdf.

Canadian National Railway. *Together into Our Next Century: Investor Fact Book: 2019.* Montreal: 2019.

Canadian Pacific Railway. *Investor Fact Book: 2019.* Montreal: 2019.

Canadian Press. "Canada-EU Visa Dispute Could Impact Free-Trade Deal Vote, Ambassador Says." *Toronto Star,* 12 April 2016.

– "Hydro-Quebec and U.S. Utility Sign Agreement to Get Power to Massachusetts." *Globe and Mail,* 14 June 2018. https://www.theglobeandmail.com/amp/business/industry-news/energy-and-resources/article-hydro-quebec-and-us-utility-sign-agreement-to-get-power-to/.

– "Pompeo Says Canadian Claim to Northwest Passage Is 'Illegitimate.'" *Ottawa Citizen,* 7 May 2019, https://ottawacitizen.com/news/canada/pompeo-says-canadian-claim-to-northwest-passage-is-illegitimate/wcm/162bffe0-b328-4160-ab82-0903ba315d33.

Canadian Press and Postmedia News. "Many Visitors to Canada Will Pay $7 Fee under Security Plan." *Ottawa Citizen,* 21 June 2014.

Canner, Stephen J. "The Multilateral Agreement on Investment." *Cornell International Law Journal* 31, no. 3 (1998): 657–81.

Canso Investment Council. "Canadian Financial Markets and the Removal of the Foreign Property Rule." Richmond Hill, ON: 2005. https://www.cansofunds.com/the-canadian-financial-markets-and-the-removal-of-the-foreign-property-rule/.

Carberry, Robert. "Making a Good Thing Better: Finishing What Was Started and Leveraging NAFTA to Advance Canada-U.S. Regulatory Cooperation." Washington, DC: Wilson Center, March 2018.

Carey, Jeffrey, and John Holmes. "What Does the Trans-Pacific Partnership Agreement Portend for the Canadian Automotive Industry?" *Canadian Public Policy* 43, no. S1 (2017): S30–S42.

Carment, David, and Joe Landry. "Civil Society and Canadian Foreign Policy." In *Readings in Canadian Foreign Policy,* 3rd ed., edited by Duane Bratt and Christopher J. Kukucha, 277–89. Don Mills, ON: Oxford University Press, 2015.

Carmody, Steve. "President Trump Proposes Increase in Great Lakes Funding, Critics Point to Cuts to EPA." Michigan Radio, 10 February 2020. https://www.michiganradio.org/post/president-trump-proposes-increase-great-lakes-funding-critics-point-cuts-epa.

Carpentier, Cécile, and Jean-Marc Suret. "The Canadian and American Financial Systems: Competition and Regulation." *Canadian Public Policy* 29, no. 4 (December 2003): 431–47.

Carroll, John E. "Patterns Old and New." In *The International Joint Commission Seventy Years On,* edited by Robert Spencer, John Kirton, and Kim Richard Nossal, 43–59. Toronto: University of Toronto Press, 1981.

Carroll, William K., and Murray Shaw. "Consolidating a Neoliberal Policy Bloc in Canada, 1976–1996." *Canadian Public Policy* 27, no. 2 (2001): 1–23.

Cascadia Now. "Welcome to Cascadia." https://www.cascadianow.org/.

Cavell, Janice. "'A Little More Latitude': Explorers, Politicians, and Canadian Arctic Policy during the Laurier Era." *Polar Record* 47, no. 4 (2011): 289–309.

CBC. "B.C. Mine's Temporary Foreign Workers Case in Federal Court." CBC News, 3 April 2013. http://www.cbc.ca/news/canada/british-columbia /b-c-mine-s-temporary-foreign-workers-case-in-federal-court-1.1374502.

– "Muskrat Falls Critics Fire Back after Danny Williams Bashes Inquiry Report," CBC News, 12 March 2020.

– "'Muskrat Falls Danny's Biggest Mistake,' Says Roger Grimes." CBC News, 19 January 2017. http://www.cbc.ca/news/canada/newfoundland-labrador /roger-grimes-danny-williams-muskrat-falls-cost-1.3941494.

– "Netflix Now Charging Sask. Customers PST for Streaming Services." CBC News, 19 January 2019.

– "NL Unveils Plan to Double Oil Production by 2030, Speed Up Development Process." CBC News, 19 February 2018. http://www.cbc.ca /news/canada/newfoundland-labrador/newfoundland-oil-plan-1.4541830.

– "Ottawa, NL Agree to Cut Financing Costs of Muskrat Falls, But How Isn't Clear." CBC News, 10 February 2020.

– "PM: Dion's Carbon Tax Would 'Screw Everybody.'" CBC News, 20 June 2008. http://www.cbc.ca/news/canada/pm-dion-s-carbon-tax-would-screw -everybody-1.696762.

– "Netflix Now Charging Sask. Customers PST for Streaming Service." CBC News, 19 January 2019. https://www.cbc.ca/news/canada/saskatchewan /pst-netflix-saskatchewan-1.4985600.

– "RBC Publicly Apologizes to Employees Affected by Outsourcing Arrangement." CBC News, 6 April 2013. http://business.financialpost.com /news/fp-street/rbc-apology-outsourcing.

– "An Unrepentant Ed Martin Goes on the Attack in Response to Muskrat Falls Inquiry Report." CBC News, 11 March 2020.

CBC News. "Trudeau: 'Governments Grant Permits, Communities Grant Permission' Prime Minister Justin Trudeau Reacts to Quebec Seeking an Injunction against the Energy East Pipeline." n.d. https://www.cbc.ca/player /play/2684686536.

Ceglowski, Janet. "Has Globalization Created a Borderless World?" *Business Review* (March/April 1998): 17–27. http://people.tamu.edu/~aglass/econ452 /Ceglowski_BordlessWorld.pdf.

Centers for Disease Control and Prevention. "Drug Overdose Deaths." Atlanta: 19 March 2020. https://www.cdc.gov/drugoverdose/data/statedeaths .html.

Cepla, Zuzana. "Fact Sheet: Temporary Protected Status." Washington, DC: National Immigration Forum, 5 May 2018. https://immigrationforum.org /article/fact-sheet-temporary-protected-status/.

Chacko, Chirakaikaran Joseph. *The International Joint Commission between the United States of America and the Dominion of Canada.* New York: Columbia University Press, 1932.

Chaire Raoul-Dandurand (collectif). "L'effet 11 septembre, 15 ans plus tard." Québec: Septentrion, 2016.

Charlebois, Sylvain, Brian Sterling, Sanaz Haratifar, and Sandi Kyaw Naing. "Comparison of Global Food Traceability Regulations and Requirements." *Comprehensive Reviews in Food Science and Food Safety* 13 (2014). https:// onlinelibrary.wiley.com/doi/full/10.1111/1541-4337.12101.

Charron, Andrea. "Contesting the Northwest Passage: Four Far-North Narratives." In *Border Flows: A Century of the Canadian American Water Relationship,* edited by Lynne Heasley and Daniel Macfarlane, 87–109. Calgary: University of Calgary Press, 2016.

– "NATO, Canada and the Arctic." Canadian Global Affairs Institute, September 2017. https://d3n8a8pro7vhmx.cloudfront.net/cdfai/pages/1793/attachments /original/1505784594/NATO_Canada_and_the_Arctic.pdf?1505784594.

Chartered Professional Accountants Canada. "Canadian CPA Bodies Sign Mutual Recognition Agreement with U.S., Mexican Counterparts." Ottawa: 1 November 2017. https://www.cpacanada.ca/en/connecting-and-news /news/media-centre/2017/november/mra-with-us-and-mexico.

Chase-Lubitz, Jesse. "U.S. Hopes USMCA Wine-Win Will Break Open Canadian Market." Politico, 15 April 2019. https://www.politico.com/story /2019/04/15/usmca-wine-canada-1298447.

China Institute, University of Alberta. "State-Owned Enterprises in the Chinese Economy Today: Role, Reform, and Evolution." Edmonton: University of Alberta, 2018.

Choi, Eric, and Sorin Niculescu. "The Impact of US Export Controls on the Canadian Space Industry." *Space Policy* 22 (2006): 29–34.

Chopin, Thierry, and Jean-Francois Jamet. "The Future of Europe." Fondation Robert Schuman Policy Paper: European Issues no. 393. Paris: Fondation Robert Schuman, 24 May 2016. https://www.robert-schuman.eu/en/doc /questions-d-europe/qe-393-en.pdf.

Chorzempa, Martin. "Confronting China through CFIUS Reform: Improved but Still Problematic." Petersen Institute for International Economics, 13 June 2018. https://piie.com/blogs/trade-investment-policy-watch/confronting -china-through-cfius-reform-improved-still?utm_source=update-newsletter &utm_medium=email&utm_campaign=2018-06-18.

Chouinard, Eric, and Chris D'Souza. "The Rationale for Cross-Border Listings." *Bank of Canada Review* (Winter 2003–4): 23–30.

Christensen, Norman L., Ann M. Bartuska, James H. Brown, Stephen Carpenter, Carla D'Antonio, Rober Francis, Jerry F. Franklin, et al. "The Report of the Ecological Society of America Committee on the Scientific Basis for Ecosystem Management." *Ecological Applications* 6 (1996): 665–90.

Ciuriak, Dan. "The March into Trade Wars: US Policy Aims and the Implications for Reconciliation." Toronto: C.D. Howe Institute, 2 August 2018.

Ciuriak, Dan, Lucy Ciuriak, Ali Dadkhah, and Jingliang Xiao. "The NAFTA Renegotiation: What If the U.S. Walks Away." Toronto: C.D. Howe Institute, 28 November 2017.

Ciuriak, Dan, and Jingliang Xiao. "The Effects of Steel and Aluminum Tariffs." Toronto: C.D. Howe Institute, 6 June 2018.

Clark, Campbell. "Canada Can Expect More Chinese Wrath as U.S. Takes Broad Aim at Huawei." *Globe and Mail*, 28 January 2019, A4.

Clarkson, Stephen. "Continental Governance, Post-Crisis: Where Is North America Going?" In *North America in Question: Regional Integration in an Era of Economic Turbulence*, edited by Jeffrey Ayres and Laura MacDonald, 85–110. Toronto: University of Toronto Press, 2012.

– *Does North America Exist? Governing the Continent after NAFTA and 9/11.* Toronto: University of Toronto Press, 2008.

Cleland, Michael, and Monica Gattinger. *System under Stress: Energy Decision-Making in Canada and the Need for Informed Reform.* Ottawa: University of Ottawa, Positive Energy, 2017.

Cleland, Michael and Monica Gattinger, with Rafael Aguirre and Marisa Beck. *Durable Balance: Informed Reform of Energy Decision-Making in Canada.* Ottawa: University of Ottawa, Positive Energy, 2018.

Cleland, Michael, with Laura Nourallah and Stewart Fast. *Fair Enough: Assessing Community Confidence in Energy Authorities.* Calgary and Ottawa: Canada West Foundation and University of Ottawa (Positive Energy), 2016.

Clemens, Jason, and Brian Lee Crowley, eds. *Solutions to Critical Infrastructure Problems: Essays on Protecting Canada's Infrastructure.* Ottawa: Macdonald Laurier Institute, 2012. http://www.macdonaldlaurier.ca/files/pdf/Solutions-to-Critical-infrastructure-Problems-February-2012.pdf.

Coe, Neil M., Peter Dicken, and Martin Hess. "Global Production Networks: Realizing the Potential." *Journal of Economic Geography* 8, no. 3 (2008): 271–95.

Coe, Neil M., and Martin Hess. "Global Production Networks, Labour, and Development." *Geoforum* 44, no. 1 (2013): 4–9.

Coe, Neil M., Martin Hess, Henry Wai-chung Yeung, Peter Dicken, and Jeffrey Henderson. "Globalizing Regional Development: A Global Production Networks Perspective." *Transactions of the Institute of British Geographers* 29, no. 4 (2004): 271–95.

Coe, Neil M., and Henry Wai-chung Yeung. *Global Production Networks: Theorizing Economic Development in an Interconnected World*. London: Oxford University Press, 2015.

Coen, Ross. *Breaking Ice: The Epic Voyage of the SS* Manhattan *through the Northwest Passage*. Fairbanks: University of Alaska Press, 2012.

Cole, Devan. "Obama Slams Rollback of Vehicle Emission Standards in Rare Rebuke of Trump." CNN News, 1 April 2020. https://www.cnn.com /2020/03/31/politics/barack-obama-fuel-standard-rollback-trump -administration/index.html.

Coleman, William, and Grace Skogstad. *Public Policy and Policy Communities*. Toronto: Copp Clark, 1992.

Collins, David A. "Globalized Localism: Canada's Procurement Commitments under the CETA." 23 February 2015. http://papers.ssrn.com/sol3/papers .cfm?abstract_id=2568629.

Collins, Jeffrey F., and Scott Reid. "'No More Giveaways!' Resource Nationalism in Newfoundland: A Case Study of Offshore Oil in the Peckford and Williams Administrations." *International Journal of Canadian Studies* 41 (2015): 155–81.

Competition Policy Review Panel. *Compete to Win: Final Report*. Ottawa: Industry Canada, June 2008. Congressional Research Service. "Temporary Protected Status: Overview and Current Research Issues, RS-20844. Washington, DC: Library of Congress, 1 April 2020. https://fas.org/sgp /crs/homesec/RS20844.pdf.

Conteh, Charles, Kathryn Friedman, and Barry Wright. "Strengthening Prosperity in Binational Corridors: Public Policy Lessons on Generating Innovation and Entrepreneurship." Prepared for the ICPP 3 2017 Conference, Singapore, 2017.

Conway, Kyle, and Timothy Pasch, eds. *Beyond the Border: Tensions across the Forty-Ninth Parallel in the Great Plains and Prairies*. Montreal and Kingston: McGill-Queen's University Press, 2013.

Corcoran, Terence. "Historic Coup at CP Rail." *National Post*, 4 May 2012, FP11.

Cormier, Roland, Christopher R. Kelble, M. Robin Anderson, J. Icarus Allen, Anthony Grehan, and Ólavur Gregersen. "Moving from Ecosystem-Based Policy Objectives to Operational Implementation of Ecosystem-Based Management Measures." *ICES Journal of Marine Science* 74, no. 1 (2017): 406–13.

Coulomb, Fanny, and Jacques Fontanel. "Mondialisation, guerre économique et souveraineté nationale." In *La question politique en économie international*, edited by P. Berthaud and G. Kebabdjian, 190–201. Paris: La Découverte, 2006.

Courtney, Kristen, Frédéric Forge, Sarah Jane Fraser, and Mathieu Frigon. *Bill C-18: An Act to Reorganize the Canadian Wheat Board and Make Consequential Amendments to Certain Other Acts*. Ottawa: Library of Parliament, December 2011.

Cox, James. "Canada and the Five Eyes Intelligence Community." Calgary: Canadian Foreign Affairs and Defence Institute, 18 December 2012. https://www.opencanada.org/features/canada-and-the-five-eyes-intelligence-community/

CPCS. "Evolution of Canadian Railway Economic Regulation and Industry Performance under Commercial Freedom." Report prepared for Railway Association of Canada. Ottawa: 28 November 2014.

Craik, Neil, and Debora VanNijnatten. "'Bundled' Trangovernmental Networks, Agency, Autonomy and Regulatory Cooperation in North America." *North Carolina Journal of International Law* 41, no. 1 (2016): 1–40.

Crain, Andrew Downer. "Ford, Carter and Deregulation in the 1970s." *Journal of Telecommunication and High Technology Law* 5 (2017): 413–47.

Crona, Beatrice I., and John N. Parker. "Learning in Support of Governance: Theories, Methods, and a Framework to Assess How Bridging Organizations Contribute to Adaptive Resource Governance." *Ecology and Society* 17, no. 1 (2012): 33–49.

Cross, Philip. "Liberals Expand TFW Program as Proof of Mess Arrives." *Financial Post*, 15 June 2017. http://business.financialpost.com/opinion/philip-cross-the-liberals-expand-the-temporary-foreign-workers-program-just-as-proof-arrives-of-what-a-mess-its-become.

Crowley, Meredith, and Dan Ciuriak. *Weaponizing Uncertainty*. Toronto: C.D. Howe Institute, 2018.

Crozier, Michel, Samuel P. Huntington, and Joji Watanuki. *The Crisis of Democracy: Report on the Governability of Democracies to the Trilateral Commission*. New York: New York University Press, 1975.

Cuff, R.D., and J.L. Granatstein. *Ties That Bind: Canadian-American Relations in Wartime from the Great War to the Cold War*. Toronto: Samuel Stevens Hakkert, 1997.

Cunningham, Nick. "Natural Gas Prices Could Double Next Year." Oilprice.com, 29 March 2020. https://oilprice.com/Energy/Energy-General/Natural-Gas-Prices-Could-Double-Next-Year.html.

Curry, Bill. "Canada Post Phasing Out Controversial Shipping Discount for Chinese Goods." *Globe and Mail*, 9 November 2017.

–   "Federal Government Postpones Study of Airport Privatization." *Globe and Mail*, 21 April 2018.

–   "National Airlines Council of Canada Urges Ottawa on Decision of Privatizing Airport Security." *Globe and Mail*, 1 August 2018, A1.

–   "Opposition Parties to Force Tories to Meet Kyoto Targets." *Globe and Mail*, 16 May 2006. http://www.theglobeandmail.com/news/national/opposition-parties-to-force-tories-to-meet-kyoto-targets/article18162531/.

–   "Serious Flaws in Ottawa's Defence against Cyber Attacks: Auditor General." *Globe and Mail*, 24 October 2012, A1.

Daalder, Ivo H., and James M. Lindsay. "The Committee to Save the World Order." *Foreign Affairs*, November–December 2018. https://www.foreignaffairs.com/issues/2018/97/6.

Dade, Carlo. "Our Country's Third Trade Deal Offers a Different Path to China." *Globe and Mail*, 17 October 2018, B4.

Daly, John. "Double-crossing." *Globe and Mail*, 23 November 2017.

Davenport, Coral. "Automakers Tell Trump His Pollution Rules Could Mean 'Untenable' Instability and Lower Profits." *New York Times*, 6 June 2019. https://www.nytimes.com/2019/06/06/climate/trump-auto-emissions-rollback-letter.html.

– "U.S. to Announce Rollback of Auto Pollution Rules, a Key Effort to Fight Climate Change." *New York Times*, 30 March 2020. https://www.nytimes.com/2020/03/30/climate/trump-fuel-economy.html.

David, Charles-Philippe, and David Grondin, eds. *Hegemony or Empire? The Redefinition of US Power under George W. Bush*. Aldershot, UK: Ashgate, 2006.

Dawson, Laura R. "Labour Mobility and the WTO: The Limits of GATS Mode 4." *International Migration* 51, no. 1 (2013): 1–23.

Day, J.C., James Loucky, and Donald K. Alper. "Policy Challenges in North America's Pacific Border Regions: An Overview." In *Transboundary Policy Challenges in the Pacific Border Regions of North America*, edited by James Loucky, Donald K. Alper, and J.C. Day, 1–37. Calgary: University of Calgary Press, 2008.

De Backer, Koen, and Sébastien Miroudot. "New International Evidence on Canada's Participation in Global Value Chains." In *Redesigning Canadian Trade Policies for New Global Realities*, edited by Stephen Tapp, Ari Van Assche, and Robert Wolfe Montreal: Institute for Research on Public Policy, 2017.

Deblock, Christian, and Christian Constantin. "Intégration des Amériques ou intégration à l'économie américaine?" *Notes et études du Groupe de recherche sur l'intégration continentale*, no. 2 (2000). https://www.ieim.uqam.ca/IMG/pdf/Integration.pdf.

Deblock, Christian, and Michèle Rioux. *De la nationalisation du monde à la globalisation*. Quebec: Presses de l'Université Laval, 2013.

De Gortari, Carlos Salinas. *Mexico: The Policy and Politics of Modernization*. Mexico City: Plaza Y Janes, 2002.

De Grauwe, Paul, and Magdalena Polan. "Globalisation and Social Spending." CESifo Working Paper No. 585, March 2003.

De Martin, Edoardo. "Opinion: Why Microsoft Chose Vancouver for Its Excellence Centre." *Vancouver Sun*, 17 June 2016. http://vancouversun.com/opinion/opinion-why-microsoft-chose-vancouver-for-its-excellence-centre.

Dempsey, Paul S. "Airline Bankruptcy: The Post-Deregulation Epidemic." Montreal: Institute for Air and Space Law, McGill University, March 2011.

Derthick, Martha A. *Between State and Nation: Regional Organizations of the United States*. Washington, DC: Brookings Institute, 1974.

Deslauriers, David J. "Oil Forecasting Legend Paints Dire Energy Picture." *Hard Assets*, 6 June 2005. http://www.resilience.org/stories/2005-06-06/energy-headlines-june-6-2005-part-two.

Deutsch, Karl Wolfgang. *Political Community and the North Atlantic Area: International Organization in the Light of Historical Experience*. Princeton, NJ: Princeton University Press, 1957.

Dobner, Petra. "On the Constitutionality of Global Public Policy Networks." *Symposium: Global Constitutionalism – Process and Substance – Kandersteg, Switzerland, January 17–20, 2008* (2009): 605–20.

Dobson, Wendy. *Shaping the Future of the North American Economic Space: A Framework for Action*. Commentary #162. Ottawa: C.D. Howe Institute, 2002.

Dobson, Wendy, and Diana Kuzmanovic. "Differentiating Canada: The Future of the Canada–U.S. Relationship." SPP Research Paper 3:7. Calgary: School of Public Policy, University of Calgary, November 2010.

Doelle, Meinhard. "The Bilcon NAFTA Arbitration: The Damages Ruling." Dalhousie University Blogs,1 March 2019. https://blogs.dal.ca/melaw/2019/03/01/the-bilcon-nafta-arbitration-the-damages-ruling/.

Doern, G. Bruce, John Coleman, and Barry E. Prentice, *Canadian Multimodal Transport Policy and Governance*. Montreal and Kingston: McGill-Queen's University Press, 2019.

Doern, G. Bruce, and Monica Gattinger. *Power Switch: Energy Regulatory Governance in the Twenty-First Century*. Toronto: University of Toronto Press, 2003.

Doern, G. Bruce, and Robert Johnson. "Multilevel Regulatory Governance: Concepts, Context and Key Issues." In *Rules, Rules, Rules, Rules: Multilevel Regulatory Governance*, edited by G. Bruce Doern and Robert Johnson, 3–27. Toronto: University of Toronto Press, 2006.

Donati, Eugene. "Opposed Triangles: Policy-Making and Regulation in Canada and the United States." *Policy Options* (April 2001): 44–9.

Doran, Charles F. "Additional and Dissenting Views." *Arctic Imperatives: Reinforcing U.S. Strategy on America's Fourth Coast*, chaired by Thad W. Allen and Christine Todd Whitman 51–2. Washington, DC: Council on Foreign Relations, 2017.

– *Forgotten Partnership: US-Canada Relations Today*. Baltimore, MD: Johns Hopkins University Press, 1984.

Douglas, Mary, and Aaron Wildavsky. *Risk and Culture: An Essay on the Selection of Technological and Environmental Dangers*. Berkeley: University of California Press, 1982.

Drinkwater, Rob. "Churchill Rail Deal Boosts Hopes Shuttered Port Will Be Revitalized." *Globe and Mail*, 3 September 2018, B1.

Duchacek, Ivo D. *The Territorial Dimension: Within, among, and across Nations*. Boulder, CO: Westview, 1986.

Dudley, Susan E., and Jerry Brito. *Regulation: A Primer*. Washington, DC: Mercatus Center, George Mason University and Center for Regulatory Studies, George Washington University, 2012.

Duez, Denis, and Damien Simmoneau. "Repenser la notion de frontière aujourd'hui. Du droit à la sociologie. *Droit & Société* 98 (2018): 37–52.

Dufresne, Robert. *Controversial Canadian Claims over Arctic Waters and Maritime Zones*. (PRB 07-47E). Ottawa: Parliamentary Information and Research Service, 2007.

Duleep, Harriet, and Mark Regets. "U.S. Immigration Policy at a Crossroads: Should the U.S. Continue Its Family-Friendly Policy?" *International Migration Review* 48, no. 3 (2014): 823–45. https://doi.org/10.1111/imre.12122.

Dyck, Rand. "Political Developments in the Provinces, 2005–2015." In *Provinces: Canadian Provincial Politics*, 3rd ed., edited by Christopher Dunn, 46–88. Toronto: University of Toronto Press, 2015.

Dyson, Kenneth, and Angelos Sepos. "Differentiation as Design Principle and as Tool in the Political Management of European Integration." In *Which Europe? The Politics of Differentiated Integration*, edited by Kenneth Dyson and Angelos Sepos, 3–23. Basingstoke, UK: Palgrave Macmillan, 2010.

Eagles, Munroe. "At War over the Peace Bridge: A Case Study in the Vulnerability of Binational Institutions." *Journal of Borderland Studies*, May 2018, https://www.whitecase.com/publications/alert/overview-chapter -4-rules-origin-us-mexico-canada-trade-agreement.

Ebner, David. "Aecon Takeover by Chinese Firm Could Fail over Security Concerns, Analyst Says." *Globe and Mail*, 21 March 2018, B1.

Economic Council of Canada. *Reforming Regulation: Final Report*. Ottawa: Supply and Services Canada, 1981.

Edelman. *2017 Trust Barometer: Global Results*. Toronto: 2017. https://www .slideshare.net/EdelmanInsights/2017-edelman-trust-barometer-global -results-71035413.

Edmonston, Barry. "Canada's Immigration Trends and Patterns." *Canadian Studies in Population* 43, no. 1–2 (2016): 78–166.

Eisinger, Peter. "Imperfect Federalism: The Intergovernmental Partnership for Homeland Security." *Public Administration Review* 66, no. 4 (2006): 537–45.

Elazar, Daniel. "The United States and the European Union: Models for Their Epochs." In *The Federal Vision: Legitimacy and Levels of Governance in the United States and the European Union*, edited by Kalypso Nicolaidis and Robert Howse, 31–53. London: Oxford University Press, 2001.

Elliot-Meisel, Elizabeth B. *Arctic Diplomacy: Canada and the United States in the Northwest Passage*. New York: Peter Lang, 1998.

Emmerson, Charles. *The Future History of the Arctic*. New York: Public Affairs, 2010.

Engineers Canada. "Agreements on International Mobility." Ottawa: n.d. https://engineerscanada.ca/become-an-engineer/international-mobility -of-engineers/mutual-recognition-agreements.

European Commission. *CETA: Summary of the Final Negotiating Results.*
February 2016. http://trade.ec.europa.eu/doclib/docs/2014/december
/tradoc_152982.pdf.

– "EU Finalises Proposal for Investment Protection and Court System for
TTIP." News release, 12 November 2015.

– *Good Regulatory Practices (GRPs) in TTIP: An Introduction to the EU's Revised
Proposal.* Brussels: 21 April 2016. https://trade.ec.europa.eu/doclib/docs
/2016/march/tradoc_154381.pdf

– "Investment in TTIP and Beyond: The Path for Reform." Concept Paper, May
2015. http://trade.ec.europa.eu/doclib/docs/2015/may/tradoc_153408.PDF.

– "Modernizing VAT for Cross-Border E-Commerce." Brussels: 11 December
2018.

– "Online Public Consultation on Investment Protection and Investor-to-
State Dispute Settlement (ISDS) in the Transatlantic Trade and Investment
Partnership Agreement." Brussels: 13 January 2015. http://trade.ec.europa
.eu/consultations/index.cfm?consul_id=179.

– "Recommendation for a Council Decision, Authorising the Opening of
Negotiations for a Convention Establishing a Multilateral Court for the
Settlement of Investment Disputes." 13 September 2017.

Evron, Yoram. "The Enduring US-Led Arms Embargo on China: An
Objectives-Implementation Analysis." *Journal of Contemporary China* (2019).
DOI : 10.1080/10670564.2019.1594099.

Exner-Pirot, Heather. "Canada's Arctic Council Chairmanship (2013–2015):
A Post-Mortem." *Canadian Foreign Policy Journal* 22, no. 1 (2016): 84–96.

– "Defence Diplomacy in the Arctic: The Search and Rescue Agreement as a
Confidence Builder." *Canadian Foreign Policy Journal* 18, no. 2 (2012): 195–207.

Export Development Canada. *Evolving with the Changing Needs of Exporters:
Submission to the 2018 Legislative Review of Export Development Canada.*
Ottawa: October 2018.

Fadden, Richard. "For the Security of Canadians, Huawei Should Be Banned
from Our 5G Networks." *Globe and Mail*, 21 January 2019, A13.

Fagan, David, Peter Flanagan, Corinne Goldstein, and Peter Lichtenbaum.
"Export Control Reform Act Introduced in Congress." *Global Policy Watch*
(Bonn), 8 March 2018. https://www.globalpolicywatch.com/2018/03
/export-control-reform-act-introduced-in-congress/.

Falck, Olivier, Christian Gollier, and Ludger Woessmann, eds. *Industrial Policy
for National Champions.* Cambridge, MA: MIT Press, 2011.

Farrell, Henry, and Abraham L. Newman. "Weaponized Interdependence: How
Global Economic Networks Shape State Coercion." *International Security* 44,
no. 1 (2019): 42–79.

Farrell, Jennifer, Nola M. Ries, Natasha Kachan, and Heather Boon. "Foods
and Natural Health Products: Gaps and Ambiguities in the Canadian
Regulatory Regime." *Food Policy* 34, no. 4 (2009): 388–92.

Feehan, James P. "Challenge of the Lower Churchill." In *First among Unequals: The Premier, Politics, and Policy in Newfoundland and Labrador*, edited by Alex Marland and Matthew Kerby, 231–46. Montreal and Kingston: McGill-Queen's University Press, 2014.

– "Electricity Market Integration: Newfoundland Chooses Monopoly and Protectionism." *Atlantic Institute for Market Studies* (November 2013): 1–9.

Ferrer, Ana M., Garnett Picot, and William Craig Riddell. "New Directions in Immigration Policy: Canada's Evolving Approach to the Selection of Economic Immigrants." *International Migration Review* 48, no. 3 (September 2014): 846–67. https://doi.org/10.1111/imre.12121.

Fife, Robert, and Steven Chase. "Aecon Deal Threatened Sovereignty, PM Says." *Globe and Mail*, 25 May 2018, A1.

– "Canada Arrests Huawei's Global Financial Officer in Vancouver." *Globe and Mail*, 6 December 2018, A1.

– "Chinese Takeover Would Bar Aecon from Canada-U.S. Bridge Bidding." *Globe and Mail*, 5 April 2018, A1.

– "China Is Changing the Geopolitical Climate: Canada Has to Mitigate and Adapt." *Globe and Mail*, 17 May 2019.

– "Ottawa Won't Rush Huawei Decision Despite U.S. Ban." *Globe and Mail*, 16 May 2019, A1.

– Trudeau Cabinet Blocks Chinese Takeover of Aecon over National Security Concerns." *Globe and Mail*, 23 May 2018. https://www.theglobeandmail.com/politics/article-ottawa-blocks-chinese-takeover-of-aecon-over-national-security/.

Fife, Robert, and Greg Keenan. "Ottawa Weighs Tactics to Forestall Dumping of Steel." *Globe and Mail*, 13 March 2018, A1.

Finnemore, Martha. *National Interests in International Society*. Ithaca, NY: Cornell University Press, 1996.

Fischer, Frank. *Reframing Public Policy: Discursive Politics and Deliberative Practices*. Oxford: Oxford University Press, 2003.

Fisher, Mark P. "Remarks to the House of Commons Finance Committee, Pre-Budget 2018 Consultations." Windsor, ON: 19 October 2017.

Flesher, John. "Trump Budget Again Targets Regional Water Programs." News Chief, 13 February 2018. http://www.newschief.com/news/20180213/trump-budget-again-targets-regional-water-cleanup-programs.

Fletcher, Robson. "Alberta Invites Feedback on Plan to Relax Regulations and Lower Carbon Tax on Heavy Emitters." CBC News, 19 July 2019. https://www.cbc.ca/news/canada/calgary/alberta-ucp-tier-carbon-tax-large-emitters-regulation-1.5205646.

Florida, Richard. "Megaregions: The Importance of Place." *Harvard Business Review*, March 2008. https://hbr.org/2008/03/megaregions-the-importance-of-place.

Florida, Richard, and Aria Bendix. "Mapping the Most Distressed
  Communities in the U.S." CityLab, 26 February 2016. https://www.citylab
  .com/equity/2016/02/mapping-distressed-communities-in-the-us/471150/.
Florida, Richard, Kathrine Richardson, and Kevin Stolarick. "Locating for
  Potential: An Empirical Study of Company X's Innovation Centre in
  Vancouver, British Columbia." Martin Prosperity Institute Working Paper
  Series: Ontario in the Creative Age. REF. 2009-WPONT-020. October 2009.
  http://martinprosperity.org/media/pdfs/Locating_for_Potential
  -Richardson.pdf.
Flynn, Stephen E. "The False Conundrum: Continental Integration vs.
  Homeland Security." In *The Re-Bordering of North America? Integration
  and Exclusion in a New Security Environment*, edited by Peter Andreas and
  Thomas J. Biersteker, 110–27. New York: Routledge, 2003.
Folke, Carl, Thomas Hahn, Per Olsson, and Jon Norberg, "Adaptive
  Governance of Social-Ecological Systems." *Annual Review of Environment and
  Resources* 30 (2005): 441–73.
Foot, Rosemary, S. Neil MacFarlane, and Michael Mastanduno, eds. *U.S.
  Hegemony and International Organizations: The United States and International
  Organizations.* New York: Oxford University Press, 2003.
Forrest, Maura. "Asylum-Claim Backlog Could Hit 100,000 by 2021." *National
  Post*, 29 May 2019, A4.
Foster, Peter. "Short-Sellers Beat Regulators on Sino-Forest." *National Post*,
  31 August 2018, FP13.
Foxon, Timothy J. *A Co-Evolutionary Framework for Analysing a Transition
  to a Sustainable Low Carbon Economy.* Leeds, UK: Sustainability Research
  Institute, University of Leeds, August 2010. http://www.see.leeds.ac.uk
  /fileadmin/Documents/research/sri/workingpapers/SRIPs-22_01.pdf.
Francis, David. "The Lights Are On at the Committee on Foreign Investment
  in the United States, but Nobody's Home." *Foreign Policy*, 22 June 2017.
Freeman, Alan. "Fearing Their Legal Status Will End, Asylum Seekers in U.S.
  Flee to Canada." *Washington Post*, 4 August 2017, A8.
Freeman, Gary P. "Can Liberal States Control Unwanted Migration?"
  ANNALS of the American Academy of Political and Social Science 534, no. 1
  (1994): 17–30. https://doi.org/10.1177/0002716294534001002.
Freshwater, David. "Free Trade, Pesticide Regulation and NAFTA
  Harmonization." *Journal of International Law and Trade Policy* 4, no. 1 (2003):
  32–57. https://ageconsearch.umn.edu/record/23817/files/04010032.pdf.
Friedman, Kathryn Bryk. "New York State's Foreign Policy." In *The Oxford
  Handbook of New York State Government and Politics*, edited by Gerald
  Benjamin, 502–34. Oxford: Oxford University Press, 2012.
– "Through the Looking Glass: Implication of Canada–United States
  Transgovernmental Networks for Democratic Theory, International Law,

and the Future of North American Governance." *Alberta Law Review* 46, no. 4 (2009): 1081–97.

Friedman, Kathryn Bryk, and Kathryn Foster. *Environmental Collaboration: Lessons Learned about Cross-Boundary Collaborations.* Washington, DC: IBM Center for the Business of Government, 2011.

Friedman, Kathryn Bryk, Barry Wright, Charles Conteh, Robert Shibley, Jiyoung Park, Ha Hwang, and Carol Phillips. "Creating Prosperity in Binational Buffalo-Niagara." Unpublished presentation, 2017. https://brocku.ca/brock-news/2016/11/brock-university-at-buffalo-launch-cross-border-economic-research-project/.

Friedman, Milton, and Rose Friedman. *The Tyranny of the Status Quo.* San Diego: Harcourt, Brace, Jovanovich, 1984.

Friedman, Thomas L. *The Lexus and the Olive Tree.* New York: Farrar, Straus and Giroux, 1999.

– *The World Is Flat: A Brief History of the Twenty-First Century.* New York: Farrar, Straus and Giroux, 2005.

Fry, Earl H. "Federalism and the Evolving Cross-Border Role of Provincial, State, and Municipal Governments." *International Journal* 60, no. 2 (2005): 471–82.

– *The Politics of International Investment.* New York: McGraw-Hill, 1983.

Fukuyama, Francis. *The End of History and the Last Man.* New York: Free Press, 1992.

Gallo, William. "Why Hasn't the US Signed the Law of the Sea Treaty?" Voice of America, 6 June 2016. https://www.voanews.com/a/united-states-sign-law-sea-treaty/3364342.html.

Garreau, Joel. *The Nine Nations of North America.* Boston: Houghton Mifflin, 1981.

Gattinger, Monica. "Canada–United States Energy Relations: Making a MESS of Energy Policy." *American Review of Canadian Studies* 42, no. 4 (December 2012): 460–73.

– "Canada–United States Energy Relations: Test-Bed for North American Policy-Making?" *Canadian-American Public Policy* 77 (September 2011): 8–12.

– "The Harper Government's Approach to Energy: Shooting Itself in the Foot." *The Harper Era in Canadian Foreign Policy: Parliament, Politics, and Canada's Global Posture,* edited by Adam Chapnick and Christopher J. Kukucha, 151–66. Vancouver: UBC Press, 2016.

– "Public Confidence in Energy and Mining Development: Context, Opportunities/Challenges and Issues for Discussion." Paper presented National Workshop on Public Confidence in Energy and Mining Development, 8–9 June 2016, in preparation for the Energy and Mines Ministers Conference, Winnipeg, 2016.

Gattinger, Monica, and Rafael Aguirre. "The Shale Revolution and Canada-US Energy Relations: Game Changer or Déjà-Vu All Over Again?" In

International Political Economy, edited by Greg Anderson and Christopher J. Kukucha, 409–35. Toronto: Oxford University Press, 2015.

–, eds. *Borders and Bridges: Canada's Policy Relations in North America*. Oxford: Oxford University Press, 2010.

Gattinger, Monica, and Geoffrey Hale. "Borders and Bridges: Canada's Policy Relations in North America." In Gattinger and Hale, *Borders and Bridges*, 1–18.

Geisler, Kenneth. "Fissures in the Valley: Searching for a Remedy for U.S. Tech Workers Indirectly Displaced by H-1B Visa Outsourcing Firms." *Washington University Law Review* 95, no. 2 (2017): 465–506.

Gere, Edwin A. *Rivers and Regionalism in New England*. Amherst, MA: University of Massachusetts Press, 1968.

Germain, Randall, and Abdulghany Mohamed. "Global Economic Crisis and Regionalism in North America: Regionness in Question?" In *North America in Question: Regional Integration in an Era of Economic Turbulence*, edited by Jeffrey Ayres and Laura Macdonald, 33–52. Toronto: University of Toronto Press, 2012.

Gibbins, Roger. *Regionalism: Territorial Politics in Canada and the United States*. Toronto: Butterworths, 1982.

Gibney, Mark. "Toward a Theory of Extraterritoriality." *Minnesota Law Review* 95 (2010–11): 81–91.

Giddens, Anthony. *Consequences of Modernity*. Cambridge: Polity, 1990.

Giddens, Anthony, and Christopher Pierson. *Making Sense of Modernity: Conversations with Anthony Giddens*. Redwood, CA: Stanford University Press, 1998.

Gilbert, Liette, L. Anders Sandberg, and Gerde R. Wekerle. "Building Bioregional Citizenship: The Case of the Oak Ridges Moraine, Ontario, Canada." *Local Environment: The International Journal of Justice and Sustainability* 15, no. 5 (2009): 387–401. https://www.tandfonline.com/doi/abs/10.1080/13549830902903674?src=recsys&journalCode=cloe20.

Gill, Vijay. *Driven Away: Why More Canadians Are Choosing Cross-Border Airports*. Ottawa: Conference Board of Canada, October 2012.

Gillen, David, and William G. Morrison. "Aviation Security: Costing, Pricing, Finance and Performance." *Journal of Air Transport Management* 48 (September 2015): 1–12.

Gillis, K., C.D. White, S.M. Ulmer, W.A. Kerr, and A.S. Kwaczek. "The Prospects for Export of Primal Beef Cuts to California." *Canadian Journal of Agricultural Economics* 33, no. 2 (1985): 171–94.

Glavin, Terry. "Blocking China's Takeover of Aecon Was Trudeau's Only Defensible Option." *Maclean's*, 24 May 2018. https://www.macleans.ca/opinion/blocking-chinas-takeover-of-aecon-was-trudeaus-only-defensible-option/.

Globe and Mail. "Harper Approves Nexen, Progress Foreign Takeovers," 7 December 2012.

- "Ottawa's 'National Security Review' a Warning to Foreign Investors," 1 July 2015.
- "The Risks of a Free Trade Deal with China." Editorial, 15 October 2018, A12.

Globerman, Steven. "Canada's Foreign Investment Review Agency and the Direct Investment Process in Canada." *Canadian Public Administration* 27, no. 3 (1984): 313–28.

Globerman, Steven, and Paul Storer. "An Assessment of Future Bilateral Trade Flows and Their Implications for U.S. Border Infrastructure Investment." Research Report #21. Bellingham, WA: Border Policy Research Institute, Western Washington University, December 2014.

- "The Impacts of 9-11 on Canada-U.S. Trade." Research Report #1. Bellingham, WA: Border Policy Research Institute, Western Washington University, July 2006.

Glover, George. "Canada's Foreign Investment Review Act." *Business Lawyer* 29, no. 3 (1974): 805–22.

Gluszek, Alicja. "The Security and Prosperity Partnership and the Pitfalls of North American Regionalism." *NortéAmerica* 9, no. 1 (January 2014), 7–54. DOI: https://doi.org/10.1016/S1870-3550(14)70112-0.

Goldberg, Jeffrey. "A Senior White House Official Defines the Trump Doctrine: 'We're America, Bitch.'" *Atlantic*, 11 June 2018. https://www.theatlantic .com/politics/archive/2018/06/a-senior-white-house-official-defines-the -trump-doctrine-were-america-bitch/562511/.

Goldfarb, Danielle. "Beyond Labels: Comparing Proposals for Closer Canada-U.S. Economic Relations." *The Border Papers*: Backgrounder #76. Toronto: C.D. Howe Institute, 2003.

- "The Road to a Canada-U.S. Customs Union." *The Border Papers*, Commentary #184. Toronto: C.D. Howe Institute, 2003.

Goldsmith, Stephen, and William D. Eggers. *Governing by Network: The New Shape of the Public Sector*. Washington, DC: Brookings Institution, 2004.

Gonsales, Richard. "Trump Administration Ends Temporary Protected Status for Hondurans." National Public Radio, 4 May 2018. https:// www.npr.org/sections/thetwo-way/2018/05/04/608654408/ trump-administration-ends-temporary-protected-status-for-hondurans.

Goodchild, Anne, Steven Globerman, and Susan Albrecht. *Service Time Variability at the Blaine, Washington International Border Crossing and the Impact on Regional Supply Chains*. Research Report #3. Bellingham, WA: Border Policy Research Institute, June 2008.

Goodman, Matthew, and David Parker. "The China Challenge and CFIUS Reform." *Global Economics Forum* 6, no. 3 (2017). Washington, DC: Center for Strategic and International Studies.

Governing the Future of States and Localities. "State Marijuana Laws in 2019 Map." https://www.governing.com/gov-data/safety-justice/state -marijuana-laws-map-medical-recreational.html.

Government of Alberta. "Oil Prices." Edmonton: Alberta Energy, 2020. http://economicdashboard.alberta.ca/OilPrice.

Government of Canada. "Canada's Arctic and Northern Policy Framework." 2019. https://www.international.gc.ca/world-monde/international _relations-relations_internationales/arctic-arctique/index.aspx?lang=eng.

Graham, Andrew. *Canada's Critical Infrastructure: When Is Safe Enough Safe Enough?* Ottawa: Macdonald Laurier Institute, December 2011. http://www .macdonaldlaurier.ca/files/pdf/Canadas-Critical-Infrastructure-When-is -safe-enough-safe-enough-December-2011.pdf.

Graham, Edward M. *Fighting the Wrong Enemy: Antiglobal Activists and Multinational Enterprises.* Washington, DC: Institute for International Economics, 2000.

Graham, William C. "The Arctic, North America, and the World: A Political Perspective." In *Governing the North American Arctic: Sovereignty, Security and Institutions*, edited by Dawn Alexandrea Berry, Nigel Bowles, and Halbert Jones, 13–25. New York: Palgrave Macmillan, 2016.

Grant, Shelagh D. *Polar Imperative: A History of Arctic Sovereignty in North America.* Vancouver: Douglas & McIntyre, 2010.

Grant, Tavia. "Wave of Asylum Seekers Floods Toronto's Shelters." *Globe and Mail,* 21 June 2018, A1.

Grayson, George W. *Mexico: Narco-Violence and a Failed State?* New York: Routledge, 2010.

Green, Tina, Lynne Hanson, Ling Lee, Héctor Fanghanel, and Steven Zahniser. "North American Approaches to Regulatory Coordination." In *Agrifood Regulatory and Policy Integration under Stress*, 2nd annual North American Agrifood Market Integration Workshop, edited by Karen M. Huff, Karl D. Meilke, Ronald D. Knutson, Rene F. Ochoa, and James Rude, 9–47. Texas A&M University, University of Guelph, Instituto Interamericano de Cooperación para la Agricultura-México, 2006.

Green and Spiegel LLP, "Then and Now: Canada's Collection of Biometric Data (Toronto: 10 July 2018). https://www.gands.com/knowledge-centre /blog-post/insights/2018/07/10/then-and-now-canada-s-collection-of -biometric-data

Greitens, Thomas J., J. Cherie Strachan, and Craig S. Welton. "The Importance of Multilevel Governance Participation in the Great Lakes Areas of Concern." In *Making Multilevel Public Management Work: Success and Failure from Europe and North America*, edited by Denita Cepiku, David K. Jesuit, and Ian Roberge, 160–81. Boca Raton, FL: CRC, 2012.

Griffiths, Franklyn, Rob Huebert, and P. Whitney Lackenbauer, eds. *Canada and the Changing Arctic: Sovereignty, Security and Stewardship.* Waterloo, ON: Wilfrid Laurier University Press, 2012.

Gross, Dominique. "Temporary Foreign Workers in Canada: Are They Really Filling Labour Shortages?" C.D. Howe Institute Commentary No. 407

(April 2014). https://www.cdhowe.org/sites/default/files/attachments
/research_papers/mixed//commentary_407.pdf.

Grossman, Gene M., and Esteban Rossi-Hansberg. "Trading Tasks: A Simple
Theory of Offshoring." *American Economic Review* 98, no. 5 (2008): 1978–97.

Grover, Warren. "The Investment Canada Act." *Canadian Business Law Journal*
10 (1985): 475–82.

Grubel, Herbert. "The Flight of the Foreign Investor." *Financial Post*, 26 June
2018, FP9.

Grumbine, R. Edward. "What Is Ecosystem Management?" *Conservation Biology*
8, no. 1 (1994): 27–38.

Hache, Connie. "Financing Public Goods and Services through Taxation or
User Fees: A Matter of Public Choice?" PhD diss., University of Ottawa, 2015.

Hacker, R. Scott, Börje Johansson, and Charlie Karlsson. "Emerging Market
Economies in an Integrating Europe: An Introduction." In *Emerging Market
Economies and European Economic Integration*, edited by Hacker, Johansson,
and Karlsson, 1–27. Northampton, MA: Edward Elgar Publishing, 2004.

Haddow, George D., and Jane A. Bullock. *Introduction to Emergency
Management*. 2nd ed. Burlington, MA: Elsevier, 2006.

Hainmueller, Jens, Michael J. Hiscox, and Yotam Margalit. "Do Concerns
about Labor Market Competition Shape Attitudes toward Immigration?
New Evidence." *Journal of International Economics* 97, no. 1 (2015): 193–207.

Hale, Geoffrey. "Borders Near and Far: The Economic, Geographic and
Regulatory Contexts for Trade and Border-Related Issues in Landlocked
Alberta." *Journal of Borderland Studies* 34, no. 2 (2019): 157–80. https://doi
.org/10.1080/08865655.2017.1315609.

– "CNOOC-Nexen, State-Controlled Enterprises and Canadian Foreign
Investment Policies: Adapting to Divergent Modernization." *Canadian Journal
of Political Science* 47, no. 2 (June 2014): 349–73.

– "Cross-Border Energy Infrastructure and the Politics of Intermesticity." In
Canada among Nations 2018–2019, edited by David Carment, Inger Weibust,
and Christopher Sands, 163–92. New York: Palgrave Macmillan, 2019.

– "Emergency Management in Alberta: A Study in Multi-Level Governance."
In *Multilevel Governance and Canadian Emergency Management Policy*, edited by
Daniel Henstra, 134–89. Montreal and Kingston: McGill-Queen's University
Press, 2013.

– "'Exceptional,' Immovable, Adaptable: Congress and the Limitations of
North American Governance." In *Regional Governance in a Post-NAFTA
North America: Building without Architecture*, edited by Brian Bow and Greg
Anderson, 160–77. New York: Routledge, 2015.

– "Market Flows, Human Flows and the Canada-U.S. Border." Paper presented
to Association for Canadian Studies in the United States, Las Vegas, NV, 15
October 2015.

- "People, Politics, and Passports: Contesting Security, Trade, and Travel on the US-Canadian Border." *Geopolitics* 16, no. 1 (2011): 27–69.
- "Pulling against Gravity? The Evolution of Canadian Trade and Investment Policies in a Multipolar World." In *Canadian Foreign Policy in a Unipolar World*, edited by Christopher Kirkey and Michael Hawes, 183–203. Toronto: Oxford University Press, 2016.
- "Regulatory Cooperation in North America: Diplomacy Navigating Asymmetries." *American Review of Canadian Studies* 49, no. 1 (March 2019): 123–49.
- *So Near Yet So Far: The Public and Hidden Worlds of Canada-US Relations.* Vancouver: UBC Press, 2012.
- "State-Owned and Influenced Enterprises and the Evolution of Canada's Foreign Direct Investment Regime." In *The Changing Paradigm of State-Controlled Entities Regulation*, edited by Julien Chaisse and Jędrzej Górski. Cham, Switzerland: Springer, forthcoming.
- "Transnationalism, Transgovernmentalism, and Canada-U.S. Relations in the 21st Century." *American Review of Canadian Studies* 43, no. 4 (December 2013): 494–511.
- "Triangulating the National Interest: Getting to 'Yes' on TPP." Paper presented to Fulbright Canada – CSC SUNY Plattsburgh Colloquium, "Canada, the United States, and the Trans-Pacific Partnership," University of Hawaii, Honolulu, HI, 24 February 2017.
- *Uneasy Partnership: The Politics of Business and Government in Canada.* Toronto: University of Toronto Press, 2006.
- *Uneasy Partnership: The Politics of Business and Government in Canada.* 2nd ed. Toronto: University of Toronto Press, 2018.
Hale, Geoffrey, and Cailin Bartlett. "Managing the Regulatory Tangle: Critical Infrastructure Security and Distributed Governance in Alberta's Major Traded Sectors." *Journal of Borderlands Studies* 34, no. 2 (2019): 257–79. doi .org/10.1080/08865655.2017.1367710.
Hale, Geoffrey, and Yale Belanger. "From Social Licence to Social Partnerships: Promoting Shared Interests for Resource and Infrastructure Development." Commentary #440. Toronto: C.D. Howe Institute, December 2015.
Hale, Geoffrey, and Stephen Blank. "North American Economic Integration and Comparative Responses to Globalization: Overview." In Gattinger and Hale, *Borders and Bridges*, 21–40.
Hale, Geoffrey, and Monica Gattinger. "Variable Geometry and Traffic Circles: Navigating Canada's Policy Relations in North America." In Gattinger and Hale, *Borders and Bridges*, 362–82.
Hale, Geoffrey, and Christopher Kukucha. "Investment, Trade and Growth: Multilevel Regulatory Regimes in Canada." In Doern and Johnson, *Rules, Rules, Rules, Rules*, 180–208.

Hale, Geoffrey, and Christina Marcotte. "Borders, Trade and Travel Facilitation."
In Gattinger and Hale, *Borders and Bridges*, 100–19.

Hall, Peter A., ed. *The Political Power of Economic Ideas: Keynesianism among
Nations*. Princeton, NJ: Princeton University Press, 1989.

Halling, Michael, Marco Pagano, Otto Randl, and Josef Zechner. "Where Is the
Market? Evidence from Cross-Listings in the United States." *Review of Financial
Studies* 21, no. 2 (2008): 725–61.

Halpern, Paul, Caroline Cakebread, Christopher C. Nicholls, and Poonam
Puri. *Back from the Brink: Lessons from the Asset-Based Commercial Paper Crisis*.
Toronto: Rotman–University of Toronto Press, 2016.

Hanson, Fergus, Emilia Currie, and Tracy Beattie. *The Chinese Communist
Party's Coercive Diplomacy*. Barton, ACT: Australian Strategic Policy Institute,
September 2020. https://www.aspi.org.au/report/chinese-communist
-partys-coercive-diplomacy.

Harding, Robert. "'Fresh Start' with New IJC Commissioners, Canada-U.S.
Water Panel at Full Strength." Salmon beyond Borders, 21 May 2019. https://
www.salmonbeyondborders.org/news-articles/fresh-start-with-new-ijc
-commissioners-canada-us-water-panel-at-full-strength.

Harper, Stephen. "Address by the Prime Minister at the Canada-UK Chamber of
Commerce." London, 14 July 2006. http://www.dominionpaper.ca/articles
/1491.

Hart, Michael. *Fifty Years of Canadian Tradecraft: Canada at the GATT 1947–1997*.
Ottawa: Centre for Trade Policy and Law, 1998.

– "Risks and Rewards: The Case for accelerating Canada-U.S. Regulatory
Reform." In Doern and Johnson, *Rules, Rules, Rules, Rules*, 27–51.

– *Trade: Why Bother*. Ottawa: Centre for Trade Policy and Law, 1992.

– *A Trading Nation: Canadian Trade Policy from Colonialism to Globalization*.
Vancouver, UBC Press, 2002.

Hart, Michael, Bill Dymond, and Colin Robertson. *Decision at Midnight: Inside
the Canada-U.S. Free Trade Negotiations*. Vancouver: UBC Press, 1994.

Harwell, Drew. "Hack Pulls the Veil Off Secret Eyes at Border." *Washington
Post*, 24 June 2019, A1.

Hayes, Dermot, and William A. Kerr. "Progress toward a Single Market:
The New Institutional Economics of the NAFTA Livestock Sectors." In
*Harmonization/Convergence/Compatibility in Agriculture and Agrifood Policy:
Canada, United States and Mexico*, edited by R.M.A. Loynes, R.D. Knutson,
K. Meilke, and D. Sumner, 1–21. Winnipeg: University of Manitoba, Texas A
and M University, University of Guelph, University of California-Davis, 1997.

Haynes, Frederick. "The Reciprocity Treaty with Canada of 1854." *Publications
of the American Economic Association* 7, no. 6 (1892): 7–70.

Heasley, Lynne, and Daniel MacFarlane, eds. *Border Flows: A Century of the
Canadian-American Water Relationship*. Network in Canadian History &
Environment, 2016. http://www.greatlakeslaw.org/files/Border_Flows.pdf.

Heeney, A.D.P. "Along the Common Frontier: The International Joint Commission." *Behind the Headlines* 26, no. 5 (1967): 1–17.
– *The Things That Are Caesar's: Memoirs of a Canadian Public Servant.* Toronto: University of Toronto Press, 1972.
Heeney, A.D.P., and Livingston T. Merchant. *Canada and the United States: Principles for Partnership.* Ottawa: Queen's Printer, 1965.
Helper, Susan, Timothy Krueger, and Howard Wial. "Why Does Manufacturing Matter? Which Manufacturing Matters?" *Metropolitan Policy Program Paper* (2011): 1–53.
Henderson, Jeffrey, Peter Dicken, Martin Hess, Neil M. Coe, and Henry Wai-chung Yeung. "Global Production Networks and the Analysis of Economic Development." *Review of International Political Economy* 9, no. 3 (2002): 436–64.
Henkin, Louis. "Arctic Anti-Pollution: Does Canada Make or Break International Law?" *American Journal of International Law* 65, no. 1 (1971): 131–6.
Henstra, Daniel, ed. *Multilevel Governance and Canadian Emergency Management Policy.* Montreal and Kingston: McGill-Queen's University Press, 2013.
Herath, Deepananda, John Cranfield, Spencer Henson, and David Sparling. "Firm, Market and Regulatory Factors Influencing Innovation and Commercialization in Canada's Functional Food and Nutraceutical Sector." *Agribusiness* 24, no. 2 (2008): 207–30.
Herman, Arthur. "Crisis in the Mail: Fixing a Broken International Package System." Washington, DC: Hudson Institute, April 2017.
Herman, Lawrence. "The New Multilateralism: The Shift to Private Global Regulation." Commentary #360. Toronto: C.D. Howe Institute, August 2012.
Hesse, Martin. "WTO Faces Existential Threat in Times of Trump." *Der Spiegel,* 30 June 2018.
Hetter, Katia. "Most Americans Will Need a New ID to Fly, Starting in October." CNN News, 22 February 2020. https://www.cnn.com/travel/article/real-id-us-travel-requirements-electronic-filing/index.html.
Heyman, David, and James Jay Carafano. *Homeland Security 3.0: Building a National Enterprise to Keep America Safe, Free and Prosperous.* Washington, DC: Center for Strategic and International Studies and Heritage Foundation, 18 September 2008. http://www.heritage.org/research/reports/2008/09/homeland-security-30-building-a-national-enterprise-to-keep-america-safe-free-and-prosperous.
Heynen, Jeff, and John Higginbotham. *Advancing Canadian Interests in the United States: A Practical Guide for Canadian Public Officials.* Ottawa: Canada School of Public Service, 2004. http://publications.gc.ca/Collection/SC103-3-2004E.pdf.
Hira, Anil. "Understanding Industrial Policy." In *International Political Economy,* edited by Greg Anderson and Christopher J. Kukucha, 343-359. Don Mills, ON: Oxford University Press, 2016.

Hira, Ron. "Candidates' Plans to Change Controversial H-1B Guestworker
  Program Highlight Need for an Overhaul." *Conversation*, 1 March 2016.
  https://theconversation.com/candidates-plans-to-change-controversial-h
  -1b-guestworker-program-highlight-need-for-an-overhaul-55482.
Hira, Ronil. "Immigration Reforms Needed to Protect Skilled American
  Workers." Testimony in a hearing before the U.S. Senate Judiciary
  Committee, 17 March 2015. https://www.judiciary.senate.gov/imo/media
  /doc/Hira%20Testimony.pdf.
Hobbs, Jill E., Stavroula Malla, Eric K. Sogah, and May T. Yeung. *Regulating
  Health Foods*. Cheltenham, UK: Edward Elgar, 2014.
Hoberg, George, ed. *Capacity for Choice: Canada in the New North America*.
  Toronto: University of Toronto Press, 2002.
– "Pipeline Resistance as Political Strategy: 'Blockadia' and the Future
  of Climate Politics." Paper presented at the Annual Conference of the
  Canadian Political Science Association, 2–4 June 2015, University of Ottawa.
Hodgson, J.R.F., and Mary R. Brooks. "Towards a North American Cabotage
  Regime: A Canadian Perspective." *Canadian Journal of Transportation* 1, no. 1
  (2007): 19-35.
Hoekman, Bernard. "International Regulatory Cooperation in a Supply
  Chain World." In *Redesigning Canadian Trade Policies for a Supply Chain
  World*, edited by Stephen Tapp, Ari Van Assche, and Robert Wolfe, 365–94.
  Montreal: Institute for Research in Public Policy and McGill-Queen's
  University Press, 2017.
Hoekman, Bernard, and Michel Kostecki. *The Political Economy of the World
  Trade Organization*. 3rd ed. New York: Oxford University Press, 2011.
Holmes, John. "Whatever Happened to Canada's Automotive Trade Surplus?
  A Preliminary Note." Automotive Policy Research Centre Research Briefs.
  Hamilton, ON: March 2015.
Hooghe, Liesbet, and Gary Marks. "Unraveling the Central State, but How?
  Types of Multi-Level Governance." *American Political Science Review* 97, no. 2
  (2003): 233–43.
Hopper, Tristin. "Canada Wants to Stem Migrant Tide at Border." *National
  Post*, 2 May 2018, A1.
Hornsby, Stephen J., Victor A. Konrad, and James J. Herlan, eds. *The
  Northeastern Borderlands: Four Centuries of Interaction*. Fredericton, NB:
  Canadian-American Center, University of Maine and Acadiensis, 1989.
House, Brett, and Juan Manuel Herrera. "NAFTA: Uncertainty to Persist
  throughout 2018." Toronto: Scotiabank Global Economics, 14 December 2017.
  https://www.gbm.scotiabank.com/content/dam/gbm/scotiaeconomics63
  /2017-12-14_I&V.pdf.
House, Brett, Juan Manuel Herrera, Rene Lalonde, and Nikita Perevelov.
  "NAFTA: Steeling Ourselves for the Macro Costs of Tariffs." Toronto:
  Scotiabank Global Economics, 14 June 2018.

Howlett, Michael, Alex Netherton, and M. Ramesh. *The Political Economy of Canada: An Introduction.* 2nd ed. Don Mills, ON: Oxford University Press, 1999.

Howlett, Michael, and M. Ramesh. *Studying Public Policy: Policy Cycles and Policy Subsystems.* New York: Oxford University Press, 1995.

Huebert, Rob. "NATO, NORAD and the Arctic: A Renewed Concern." In *North of 60: Toward a Renewed Canadian Arctic Agenda,* edited by John Higginbotham and Jennifer Spence, 91–7. Waterloo, ON: Centre for International Governance Innovation, 2016.

– *The Newly Emerging Arctic Security Environment.* Ottawa: Canadian Defence and Foreign Affairs Institute, 2010.

Hufbauer, Gary, and Jeffrey Schott. *NAFTA Revisited: Achievements and Challenges.* Washington, DC: Institute for International Economics, 2005.

Hughlett, Mike. "State Regulators OK Certificate of Need for Controversial Enbridge Pipeline." *Star-Tribune*, Minneapolis, 29 June 2018.

Hugo, Graeme. "Demographic Change and International Labour Mobility in Asia-Pacific – Implications for Business and Regional Economic Integration: Synthesis." In *Labour Mobility in the Asia-Pacific Region: Dynamics, Issues and a New APEC Agenda,* edited by Graeme Hugo and Soogil Young, 1–62. Singapore: Institute of Southeast Asian Studies, 2008.

Ikenberry, G. John. "America's Liberal Hegemony." *Current History* 98 (1999): 23–8.

"Ilulissat Declaration." Ilulissat, Greenland, 27–9 May 2008. http://www .oceanlaw.org/downloads/arctic/Ilulissat_Declaration.pdf.

Immigration and Refugee Board Canada. "Refugee Protection Claims Made by Irregular Border Crossers." https://irb-cisr.gc.ca/en/statistics/Pages /irregular-border-crossers-countries.aspx.

Imperial, Mark T., Sonia Ospina, Erik Johnston, Rosemary O'Leary, Jennifer Thomsen, Peter Williams, and Shawn Johnson. "Understanding Leadership in a World of Shared Problems: Advancing Network Governance in Large Landscape Conservation." *Frontiers in Ecology and the Environment* 14, no. 3 (2016): 126–34.

Information Technology Association of Canada. "The Importance of Global Workers in Canada's ICT and Digital Media Industries." January 2014. http://theesa.ca/wp-content/uploads/2015/08/ITAC-white-paper.pdf.

Institute of Governance. "Defining Governance." Ottawa. n.d. https://iog.ca /what-is-governance/.

Internal Trade Secretariat. "Canadian Free Trade Agreement: Consolidated Version." Winnipeg: 2017.

International Energy Agency. *CO2 Emissions from Fuel Combustion: Highlights 2016.* 2015. https://emis.vito.be/sites/emis.vito.be/files/articles/3331/2016 /CO2EmissionsfromFuelCombustion_Highlights_2016.pdf.

International Joint Commission. *First Triennial Assessment of Progress on Great Lakes Water Quality.* 2017. http://ijc.org/files/tinymce/uploaded/GLWQA /TAP.pdf.

International Mobility and Transportation Corridor Program. "Cascade Gateway Monthly Volumes." Bellingham, WA: April 2016.

International Organizations of Securities Commissions. "About Us." Madrid: n.d. www.iosco.org/about.

Ipsos Canada. "Majority (72%) of Canadians Approve of Trudeau's Handling of Trump Situation, but Acknowledge (81%) That It Has Hurt Bilateral Relations." Toronto: 16 June 2018. https://www.ipsos.com/en-ca/news-polls/Global-News-NAFTA-June-16-2018.

Ircha, Michael. "Ports and Shipping: Opportunities and Challenges." Paper presented to "Borders in Globalization" conference, MacEachen Institute, Dalhousie University, Halifax, NS, March 2018.

Israelson, David. "How USMCA Will Affect This Canadian Company." *Globe and Mail*, 23 October 2018, B7.

Jackson, James. "The Committee on Foreign Investment in the United States (CFIUS)." Washington, DC: Congressional Research Service, 2018.

Jalinous, Farhad, Karalyn Mildork, Keith Schornig, Stacia J. Sowerby, Cristina Brayton-Lewis, Sandra Jorgensen, and Margaret Spicer. "Congress Finalizes CFIUS and Export Control Legislation." Washington, DC: White & Case, 26 July 2018. https://www.whitecase.com/publications/alert/congress-finalizes-cfius-and-export-control-reform-legislation.

James, Patrick. *Canada and Conflict*. Oxford: Oxford University Press, 2012.

Jang, Brent. "Maine Governor Seeks Exemption for N.B. Lumber Firms." *Globe and Mail*, 20 February 2018, B3.

Jaremko, Deborah. "Nexen Name to Disappear as Subsidiary Absorbed into CNOOC International." JWNEnergy.com, 16 January 2019. https://www.jwnenergy.com/article/2019/1/nexen-name-disappear-subsidiary-absorbed-cnooc-international/.

Jarvis, Anne. "Behind the Bridge." *Windsor Star*, 28 October 2016, A1.

Jefferd-Moore, Kaila. "Canadian Submarines Not Part of International Arctic Under-Ice Exercise." CBC News, 11 June 2018. http://www.cbc.ca/news/canada/north/canadian-submarines-not-part-of-international-arctic-under-ice-exercise-1.4699208.

Jenkins, Matthew. "Anti-Corruption and Transparency Measures in Trade Agreements." Berlin: Transparency International, 2017.

Jenkins-Smith, Hank C., and Paul A. Sabatier. "The Study of Public Policy Processes." In *Policy Change and Learning: An Advocacy Coalition Approach*, edited by Sabatier and Jenkins-Smith, 1–9. Boulder, CO: Westview, 1993.

Jepperson, Ronald L., and John W. Meyer. "The Public Order and the Construction of Formal Organizations." In *The New Institutionalism in Organizational Analysis*, edited by Walter W. Powell and Paul J. DiMaggio, 204–31. Chicago: University of Chicago Press, 1991.

Jockel, Joseph, *Canada in NORAD, 1957–2007: A History*. Montreal and Kingston: McGill-Queen's University Press, 2007.

Johns, Carolyn. "The Great Lakes, Water Quality and Water Policy in Canada." In *Water Policy and Governance in Canada*, edited by Steven Renzetti and Diane Dupont, 159–78. Basel: Springer International Publishing, 2017.

– "Transboundary Environmental Governance and Water Pollution in the Great Lakes Region: Recent Progress and Future Challenges." In *Transboundary Environmental Governance across the World's Longest Border*, edited by Stephen Brooks and Andrea Olive, 77–112. East Lansing, MI: Michigan State University Press, 2018.

– "Water Pollution in the Great Lakes Basin: The Global-Local Dynamic." In *Environmental Challenges and Opportunities: Local-Global Perspectives on Canadian Issues*, edited by Christopher Gore and Peter Stoett, 95–129. Toronto: Emond Montgomery, 2009.

Johns, Carolyn, and Mark Sproule-Jones. "Great Lakes Water Policy: The Cases of Water Levels and Water Pollution in Lake Erie." In *Canadian Environmental Policy and Politics: Prospects for Leadership and Innovation*, edited by D. VanNijnatten, 252–77. 4th ed. Toronto: Oxford University Press, 2016.

Johns, Carolyn, and Adam Thorn. "Subnational Diplomacy in the Great Lakes Region: Towards Explaining Variation between Water Quality and Quantity Regimes." *Canadian Foreign Policy Journal* 21, no. 3 (2015): 195–211.

Johnson, Keith, Dan De Luce, and Emily Tamkin. "Can the U.S.-Europe Alliance Survive Trump?" *Foreign Policy*, 18 May 2018. https://foreignpolicy.com/2018/05/18/can-the-u-s-europe-alliance-survive-trump/.

Johnson, Keith, and Elias Groll. "China Raises Threat of Rare-Earths Cutoff to U.S." *Foreign Policy*, 21 May 2019.

Johnston, Douglas M., ed. *Arctic Ocean Issues in the 1980's: Proceedings, Law of the Sea Institute, University of Hawaii and Dalhousie Ocean Studies Programme, Dalhousie University, Halifax, Nova Scotia Workshop, June 10–12, 1981, Mackinac Island, Michigan*. Honolulu: Law of the Sea Institute, University of Hawaii, 1982.

– *Canada and the New International Law of the Sea*. Toronto: University of Toronto Press, 1985.

Johnston, Richard, and Mike Percy. "Reciprocity, Imperial Sentiment, and Party Politics in the 1911 Election." *Canadian Journal of Political Science* 13, no. 4 (December 1980): 711–29.

Joskow, Paul. "Restructuring, Competition and Regulatory Reform in the U.S. Electricity Sector." *Journal of Economic Perspectives* 11, no. 3 (Summer 1997): 119–38.

J.P. Morgan. "E-Commerce Payment Trends." New York: 2019. https://www.jpmorgan.com/merchant-services/insights/reports/canada

Kahan, Dan M., Ellen Peters, Maggie Wittlin, Paul Slovic, Lisa Larrimore Ouellette, Donald Braman, and Gregory Mandel. "The Polarizing Impact of Science Literacy and Numeracy on Perceived Climate Change Risks." *Nature Climate Change* 2, no. 10 (2012): 732–35.

Kahler, Miles, and David A. Lake, eds. *Governance in a Global Economy*. Princeton, NJ: Princeton University Press, 2003.

Kalicki, Jan H., and David L. Goldwyn, eds. *Energy and Security: Strategies for a World in Transition*. 2nd ed. Washington, DC: Wilson Center Press with Johns Hopkins University Press, 2013.

Kang, C.S. Eliot. "US Politics and Greater Regulation of Inward Foreign Direct Investment." *International Organization* 51, no. 2 (1997): 301–33.

Kantor, Mark. "The New Draft Model U.S. BIT: Noteworthy Developments." *Journal of International Arbitration* 21, no. 4 (2004): 383–96.

Katz, Harry C., John Paul MacDuffie, and Frits K. Pil. "Crisis and Recovery in the U.S. Auto Industry: Tumultuous Times for a Collective Bargaining Pacesetter." In *Collective Bargaining under Duress: Case Studies of Major North American Industries*, edited by Howard R. Stanger, Paul F. Clark, and Ann C. Frost, 45–80. Ithaca, NY: Cornell University Press, 2013.

Kawai, Masahiro, and Ganeshan Wignaraja, "Tangled Up in trade? The 'Noodle Bowl'of Free Trade Agreements in East Asia." VoxCEPR Policy Portal, 15 September 2009. https://voxeu.org/article/noodle-bowl-free -trade-agreements-east-asia.

Kean, Thomas, and Lee Hamilton. *The 9/11 Commission Report: Final Report of the National Commission on Terrorist Attacks upon the United States*. Washington, DC: Government Printing Office, 2011.

Keohane, Robert O. *International Institutions and State Powers: Essays in International Relations Theory*. Boulder, CO: Westview, 1989.

Keohane, Robert O., and Joseph S. Nye Jr. "Introduction." In *Governance in a Globalizing World*, edited by Keohane and Nye Jr., 1–42. (Washington, DC: Brookings Institution, 2000).

– "Transgovernmental Relations and International Organizations." *World Politics* 27, no. 1 (October 1974): 39–62.

Kerr, William A. "Homeland Security and the Rules of International Trade." *Journal of International Law and Trade Policy* 5, no. 1 (2004): 1–10. https:// ageconsearch.umn.edu/record/23852/files/05010001.pdf.

– "NAFTA and Beyond: Challenges for Extending Free Trade in the Hemisphere." *Journal of International Law and Trade Policy* 3, no. 2 (2002): 224–38. https://ageconsearch.umn.edu/record/23916/files/03020224.pdf.

– "Political Precaution, Pandemics and Protectionism." *Journal of International Law and Trade Policy* 10, no. 2 (2009): 1–14. https://ageconsearch.umn.edu /record/52231/files/kerr10-2.pdf.

– "Removing Nontariff Barriers to Trade under the Canada–United States Trade Agreement: The Case for Reciprocal Beef Grading." *Journal of Agricultural Taxation and Law* 14, no. 3 (1992): 273–88.

– "Sanitary Barriers and International Trade Governance Issues for the NAFTA Beef Market." In *Keeping the Borders Open*, edited by R.M.A. Loyns,

K. Meilke, R.D. Knutson, and A. Yunez-Naude, 26–49. Proceedings of the Eighth Agricultural and Food Policy Systems Information Workshop, Guelph: University of Guelph, 2004.

– "What Is New in Protectionism? Consumers, Cranks and Captives." *Canadian Journal of Agricultural Economics* 58, no. 1 (2010): 5–22.

Kerr, William A., and Susan E. Cullen. "Canada-U.S. Free Trade: Implications for the Western Canadian Livestock Industry." *Western Economic Review* 4, no. 3 (1985): 24–36.

Kessel, Alan H. "Canadian Arctic Sovereignty: Myths and Realities." In *Governing the North American Arctic: Sovereignty, Security, and Institutions,* edited by Dawn Alexandrea Berry, Nigel Bowles and Halbert Jones, 242–6. New York: Palgrave Macmillan, 2016.

Kettl, Donald F. *System under Stress: Homeland Security and American Politics.* Washington, DC: Sage, 2013.

Keung, Nicholas, "U.S.-Canada asylum pact violates charter, court rules." *The Toronto Star.* 23 July 2020, A1.

Khan, Shehab. "CETA: Belgian State Rejects Controversial EU-Canada Trade Deal." *Independent*, 15 October 2016. https://www.independent.co.uk /news/world/europe/ceta-belgian-state-wallonia-eu-canada-deal -comprehensive-economic-and-trade-agreement-brexit-a7363386.html.

Kilroy, Richard C. Jr., Abelardo Rodriguez Sumano, and Todd S. Hataley. *North American Regional Security: A Trilateral Approach?* Boulder, CO: Lynne Rienner, 2013.

Kim, Eliot. "Withdrawal from the Universal Postal Union: A Guide for the Perplexed." *Lawfare*, 31 October 2018.

Kinney, Laureen. "Rail Safety in Canada since Lac Mégantic." Presentation to Oil by Rail Safety Symposium, Lakewood, WA. Seattle: Center for Regional Disaster Resilience, 27 April 2016.

Kirton, John J., and Jenilee Guebert. "Soft Law, Regulatory Coordination and Convergence in North America." In Gattinger and Hale, *Borders and Bridges*, 59–76.

Kirton, John J., and Michael J. Trebilcock, eds. *Hard Choices, Soft Law: Voluntary Governance in Global Trade, Environment and Social Governance.* Aldershot, UK: Ashgate, 2004.

Klein, Naomi. *This Changes Everything: Capitalism vs. the Climate.* Toronto: Knopf, 2014.

Klier, Thomas, and James M. Rubinstein. "Restructuring of the U.S. Auto Industry in the 2008–2009 Recession." *Economic Development Quarterly* 27, no. 2 (2013): 144–59.

Knott, Christine. "Contentious Mobilities and Cheap(er) Labour: Temporary Foreign Workers in a New Brunswick Seafood Processing Community." *Canadian Journal of Sociology* 41, no. 3 (2016): 375–98.

Kobach, Christopher, and Joe Martin. *From Wall Street to Bay Street: The Origins and Evolution of American and Canadian Finance.* Toronto: University of Toronto Press, 2018.

Koldyk, Daniel, Louis Quinn, and Todd Evans. "Chasing the Chain: Canada's Pursuit of Global Value Chains." In *Redesigning Canadian Trade Policies for New Global Realities,* edited by Stephen Tapp, Ari Van Assche, and Robert Wolfe. Montreal: Institute for Research on Public Policy, 2017.

Kolossov, Vladimir. "Border Studies: Changing Perspectives and Theoretical Approaches." *Geopolitics* 10, no. 4 (2005): 606–32.

– "Études des frontières – approches post-modernes." *Diogène* 210 (April–June 2005): 13–27.

Kolossov, Vladimir, and Scott James. "Selected Conceptual Issues in Borders Studies." *Belgeo – Revue Belge de Géographie,* 1:2013. https://journals.openedition.org/belgeo/10532.

Konrad, Victor. "Towards a Theory of Borders in Motion." *Journal of Borderland Studies* 30, no. 1 (2015): 1–17.

Konrad, Victor, and Heather N. Nicol. *Beyond Walls: Reinventing the Canada–United States Borderlands.* Aldershot, UK: Ashgate, 2008.

– "Border Culture, the Boundary between Canada and the United States of America and the Advancement of Borderlands Theory." *Geopolitics* 16, no. 1 (2011): 70–90.

Koring, Paul, and Gloria Galloway. "A Symbol of Peace, a War of Words." *Globe and Mail,* 29 May 2013, A8–9.

Koudai, Peter. "The Service Revolution in Global Manufacturing Industries." New York Deloitte Research, January 2006. https://www.researchgate.net/publication/267268075_The_Service_Revolution_ in Global_Manufacturing _Industries.

Kraft, Michael E., and Scott R. Furlong. *Public Policy: Politics, Analysis, and Alternatives.* 4th ed. Thousand Oaks, CA: Sage, 2012.

Krasner, Stephen. "Sharing Sovereignty: New Institutions for Collapsed and Failing States." *International Security* 29, no. 2 (2004): 85–120.

– *Sovereignty: Organized Hypocrisy.* Princeton, NJ: Princeton University Press, 1999.

Kreinin, Mordechai, ed., *Building a Partnership: The Canada–United States Free Trade Agreement.* East Lansing: Michigan State University Press, 2000.

Kroeger, Arthur. *Retiring the Crow Rate: A Study in Political Management.* Edmonton: University of Alberta Press, 2009.

Kukucha, Christopher J. "Canada's Incremental Foreign Trade Policy." In *The Harper Era in Canadian Foreign Policy: Parliament, Politics, and Canada's Global Posture,* edited by Adam Chapnick and Christopher J. Kukucha, 195–209. Vancouver: UBC Press, 2016.

- "Federalism and Liberalization: Evaluating the Impact of American and Canadian Sub-Federal Governments on the Negotiation of International Trade Agreements." *International Negotiation* 22, no. 2 (2017): 259–84.
- "Multilateralism and Canadian Foreign Trade Policy: A Long View." In *Seeking Order in Anarchy: Multilateralism as State Strategy*, edited by Robert W. Murray, 177–96. Edmonton: University of Alberta Press, 2016.
- *The Provinces and Canadian Foreign Trade Policy*. Vancouver: UBC Press, 2008.
Kulisch, Eric. "Canada, U.S. Push Plan for Getting Goods over Border Faster." *Automotive News Canada*, 24 February 2018. http://canada.autonews.com /article/20180224/CANADA01/302249991/canada-u.s.-push-plan-for -getting-goods-over-border-faster.
Kumar, Pradeep. *From Uniformity to Divergence: Industrial Relations in Canada and the United States*. Kingston, ON: Queen's IRC, 1993.
Kurniawati, Dyah E. "Intermestic Approach: A Methodological Alternative in Studying Policy Change." *PCD Journal* 5, no. 1 (2017): 147–73.
Lackenbauer, P. Whitney, ed. "Canadian Arctic Sovereignty and Security: Historical Perspectives." Calgary Papers in Military and Strategic Studies: Occasional Paper Number 4. Calgary, AB: Centre for Military and Strategic Studies, 2011.
- "From Polar Race to Polar Saga: An Integrated Strategy for Canada and the Circumpolar World." In *Canada and the Changing Arctic: Sovereignty, Security and Stewardship*, edited by Franklyn Griffiths, Rob Huebert, and P. Whitney Lackenbauer. Waterloo, ON: Wilfrid Laurier University Press, 2011.
Lagassé, Philippe. "A Common 'Bilateral' Vision: North American Defence Cooperation, 2001–2012." In *Game Changer: The Impact of 9/11 on North American Security*, edited by Jonathan Paquin and Patrick James, 193–211. Vancouver: UBC Press, 2014.
Lajeunesse, Adam. *Lock, Stock and Icebergs: A History of Canada's Arctic Maritime Sovereignty*. Vancouver: UBC Press, 2016.
Lake, David A. "The New Sovereignty in International Relations." *International Studies Review* 5, no. 3 (2003): 303–23.
Lam, Bourree. "Canada Wants Silicon Valley's Tech Employees." *Atlantic*, 9 May 2017. https://www.theatlantic.com/business/archive/2017/05 /canada-tech/525930/.
Lanoszka, Alexandra. "From Ottawa to Riga: Three Tensions in Canadian Defence Policy." *International Journal* 72, no. 4 (2017): 520–37.
Lanz, Rainer, and Andreas Maurer. "Services and Global Value Chains: Servicification of Manufacturing and Services Networks." *Journal of International Commerce, Economics and Policy* 6, no. 3 (2015). https://doi.org /10.1142/S1793993315500143.
Layton, Peter. "Making a Canada National Security Strategy." *On Track* (Summer 2015): 37–40.

Leahy, Mike. Presentation to Pacific NorthWest Economic Region conference, Saskatoon, SK, 22 July 2019.

Leblond, Patrick. "Boeing-Bombardier Dispute Is Trumpism at Its Worst." *Globe and Mail*, 28 September 2017, sec. B, p. 4.

LeCroy, Jessica, Kathryn Friedman, Cyndee Todgham Cherniak, and Laurie Tannous. "The Canada–United States Customs Transaction: The Invisible Border?" *Canada-United States Law Journal* 36, no. 1 (2011): 237–59.

Lester, Simon, and Inu Manak. "A Framework for Rethinking NAFTA for the 21st Century: Policies, Institutions, and Regionalism." CTEI Working Paper 2017–10. Geneva: Centre for Trade and Economic Integration, 2017.

Leuprecht, Christian. "Renewing the Social Contract: Sustainable Policy Approaches to 'Irregular Migration.'" Ottawa: Macdonald-Laurier Institute, 2019.

Leuprecht, Christian, and Geoffrey Hale. "Presentation to House of Commons Standing Committee on Citizenship and Immigration, re: Bill C-97, An Act to Implement Certain Provisions of the Budget Tabled in Parliament on March 19, 2019 and Other Measures: Division 16, Part 4 Amending the Immigration and Refugee Protection Act," 7 May 2019.

Liebman, John R., and Kevin J. Lombardo. "A Guide to Export Controls for the Non-Specialist." *Loyola of Los Angeles International and Comparative Law Review* 28 (2006): 497–599.

Lilly, Meredith B. "Advancing Labour Mobility in Trade Agreements." *Journal of International Trade Law and Policy* 18, no. 2 (2019): 58–73.

– "How Demographic Transition Can Help Predict Canada–US Trade Relations in 2042." *International Journal of Canadian Studies* 55 (2017): 67–76. https://doi.org/10.3138/ijcs.55.07.

– "The NAFTA Road Ahead: Why Canada Must Be in the Fast Lane." *Policy Magazine* 5, no. 5 (2017): 29–31.

Lin, Ching-Fu, and Han-Wei Liu. "Regulatory Rationalization Clauses in FTAs: A Complete Survey of the US, EU and China." *Melbourne Journal of International Law* 19, no. 1 (2018): 149–77.

Lipschutz, Ronnie. "Bioregionalism, Civil Society and Global Environmental Governance." In *Bioregionalism*, edited by Michael Vincent McGinnis, 101–20. London: Routledge, 1998.

Liu, Huanan, William A. Kerr, and Jill E. Hobbs. "Product Safety, Collateral Damage and Trade Policy Responses: Restoring Confidence in China's Exports." *Journal of World Trade* 43, no. 1 (2009): 97–124.

Lizza, Ryan. "Leading from Behind." *New Yorker*, 26 April 2011.

Lofstrom, Magnus, and Joseph Hayes. "H-1Bs: How Do They Stack Up to US Born Workers?" IZA Discussion Paper no. 6259, December 2011. http://ftp.iza.org/dp6259.pdf.

Long, Heather, and Steven Mufson. "Trump Trade Moves Shake Global Systems." *Washington Post*, 3 June 2018, A1.

Longworth, David. "The Global Financial Crisis and Financial Regulation: Canada and the World." In *Crisis and Reform: Canada and the International Financial System: Canada among Nations 2014*, edited by Rohinton Medhora and Dane Rowlands, 87–102. Waterloo, ON: Centre for International Governance Innovation, 2014.

López-Vallejo, Marcela. *Reconfiguring Climate Governance in North America: A Transregional Approach*. Surrey, UK: Ashgate Publishing, 2014.

Loppacher, Laura J., and William A. Kerr. "The Efficacy of World Trade Organization Rules on Barriers: Bovine Spongiform Encephalopathy in North America." *Journal of World Trade* 39, no. 3 (2005): 427–43.

Loppacher, Laura J., William A. Kerr, and Richard R. Barichello. "Regional Management of Diseases and Pests to Facilitate Canadian Exports." *Canadian Journal of Regional Science* 31, no. 1 (2008): 30–54.

Lynn, L.E. Jr., J. Heinrich, and C.J. Hill. *Improving Governance: A New Logic for Empirical Research*. Washington, DC: Georgetown University Press, 2001.

Macdonald, Laura C., and Jeffrey Ayres. "Civil Society and International Political Economy." In *International Political Economy*, edited by Greg Anderson and Christopher J. Kukucha, 329–42. Don Mills, ON: Oxford University Press, 2016.

MacKinnon, Mark, and Andy Hoffman. "Inside the Sino-Forest Storm." *Globe and Mail*, 18 June 2011, B1.

Mackrael, Kim. "Canada, U.S. Unveil Tougher Rail Safety Standards for Transporting Oil." *Globe and Mail*, 1 May 2018.

Mackrael, Kim, and Jacquie McNish. "U.S. Canada Issue Rail Safety Warnings." *Globe and Mail*, 24 January 2014, A3.

Majtenyi, Cathy. "Brock, University at Buffalo Launch Cross-Border Economic Research Project." Brock, 2 November 2016, https://brocku.ca/brock-news /2016/11/brock-university-at-buffalo-launch-cross-border-economic -research-project

Mahant, Edelgard, and Graham S. Mount. *Invisible and Inaudible in Washington*. Vancouver: UBC Press, 1999.

Maloney, Maureen, Tsur Somerville, and Brigitte Unger. *Combatting Money Laundering in BC Real Estate*. Victoria: Ministry of Finance, Government of British Columbia, 31 March 2019.

Manak, Inu. "Regulatory Issues in the New NAFTA." International Economic Law and Policy Blog, 10 October 2018. https://worldtradelaw.typepad .com/ielpblog/2018/10/regulatory-cooperation-in-the-new-nafta.html.

Manning, Bayless. "The Congress, the Executive and Intermestic Affairs: Three Proposals." *Foreign Affairs* 55, no. 2 (1977): 306–24.

Mansfield, Edward D., and Helen V. Milner. "The New Wave of Regionalism." *International Organization* 53, no. 3 (1999): 589–627.

Maoh, Hanna F., Shakil A. Khan, and William P. Anderson. "Truck Movements across the Canada-U.S. Border: The Effects of 9/11 and Other Factors." *Journal of Transport Geography* 53 (2016): 12–21.

Mapchart.net. "United States – Counties – Mapchart." 2018. https://mapchart .net/usa-counties.html.

March, James G., and Johan P. Olsen. "Institutional Perspectives on Political Institutions." Paper presented at the meeting of the International Political Science Association, Berlin, 1994.

– *Rediscovering Institutions*. New York: Free Press, 1989.

Marland, Alex. *Brand Command: Canadian Politics and Democracy in the Age of Message Control*. Vancouver: UBC Press, 2016.

Marshall, Hillary. "Toronto Needs to Become North America's Next Airport Mega Hub." *Huffington Post*, 2 March 2017.

Martin, Ron, and Peter Sunley. *Economic Geography*. London: Routledge, 2008.

Martinez, Grace. "Legal Immigrants Displacing American Workers: How U.S. Corporations Are Exploiting H-1B Visas to the Detriment of Americans." *UMKC Law Review* 86, no. 1 (2017): 209–36.

Matloff, Norman. "Immigration and the Tech Industry: As a Labour Shortage Remedy, for Innovation, or for Cost Savings?" *Migration Letters* 10, no. 2 (2013): 211–28.

Matti, Simon, and Annica Sanstrom. "The Rationale Determining Advocacy Coalitions: Examining Coordination Networks and Corresponding Beliefs." *Policy Studies Journal* 39, no. 3 (2011): 385–410.

May, Kathryn. "CFIA Move Sends Out Message." *Ottawa Citizen*, 12 October 2013, A3.

McAlpine, Rory, and Mike Robach. *Risk and Reward: Food Safety and NAFTA 2.0*. Ottawa and Washington: The Canadian Agri-Food Policy Institute and the Canada Institute of the Wilson Center, 2017.

McBride, Stephen. "Quiet Constitutionalism in Canada: The International Political Economy of Domestic Institutional Change." *Canadian Journal of Political Science* 36, no. 2 (2003): 251–73.

McCarthy, John. "Digital Tax Trends: International Plans to Tax the Digital Economy." www.taxamo.com, 26 April 2018.

McCarthy, Shawn. "Canadian Energy Industry Slams Liberals' Environmental Assessment Rules." *Globe and Mail*, 2 April 2018.

– "Federal Government's Proposed Trans-Mountain Deal Could Require Trump's OK." *Globe and Mail*, 17 July 2018, A4.

– "Ottawa Commits to 30-Per-Cent Cut in GHGs but No Regulations for Oil Sands." *Globe and Mail*, 15 May 2015. http://www.theglobeandmail.com

/news/national/ottawa-commits-to-30-per-cent-cut-in-emissions-but
-not-for-oil-sands/article24453757/.

McCarthy, Shawn, and Campbell Clark. "Ottawa Swoops in with Climate-Change
Offer." *Globe and Mail*, 6 November 2008. http://www.theglobeandmail.com
/news/world/ottawa-swoops-in-with-climate-change-offer/article17973912/.

McCarthy, Shawn, Brent Jang, and Justine Hunter. "Ottawa Clears Way for
Proposed LNG Terminal on B.C. Coast." *Globe and Mail*, 26 September 2018, A1.

McCrimmon, Karen, MP. "Speech on Bill C-49, Transportation Modernization
Act." 42nd Parliament, 1st Session. Ottawa: House of Commons, 31 October
2017.

McDerment, Mike. "Canada Can't Fall Behind in the Global Race for Tech
Talent." *Globe and Mail*, 26 June 2017. https://www.theglobeandmail.com
/report-on-business/rob-commentary/canada-cant-fall-behind-in-the
-global-race-for-tech-talent/article35459472/.

McDougall, Andrew. "How the USMCA Strengthens Canada in Future Trade
Deals." *Conversation*, 21 October 2018; https://theconversation.com/how
-the-new-usmca-strengthens-canada-in-future-trade-deals-104814; accessed
November 13, 2018.

McDougall, Robert. "Crisis in the WTO: Restoring the WTO Dispute Resolution
Function." CIGI Papers #194. Waterloo, ON: Centre for International
Governance Innovation, October 2018.

McFarland, Janet. "Canada's Cooling Experiment." *Globe and Mail*, 2 August
2017, A8–9.

McGregor, Janyce. "As U.S. Threatens Retaliation on Digital Taxes, Canada
Waits for OECD Talks." CBC.ca, 22 January 2020. https://www.cbc.ca/news
/politics/davos-digital-tax-wednesday-1.5436372.

– "EU Quietly Asks Canada to Reword Trade Deal's Thorny Investment
Clause." CBC, 21 January 2016. http://www.cbc.ca/news/politics/canada
-europe-trade-isds-ceta-1.3412943.

McGuire, Peter. "CMP Corridor Opponents Submit Signatures for Referendum
Vote." *Press-Herald*, Portland, ME, 3 February 2020.

McKenna, Barrie. "Look Out, Canada: Trump May Be about to Get Serious
about Protectionism." *Globe and Mail*, 12 January 2018. https://www
.theglobeandmail.com/report-on-business/rob-commentary/look-out
-canada-trump-may-be-about-to-get-serious-about-protectionism/article
37593133/.

McKercher, Asa. "Principles and Partnership: Merchant, Heeney, and the
Craft of Canada-US Relations." *American Review of Canadian Studies* 42, no. 1
(2012): 67–83.

McLeod, K.L., J. Lubchenco, S.R. Palumbi, and A.A. Rosenberg. *Scientific
Consensus Statement on Marine Ecosystem-Based Management*. Communication
Partnership for Science and the Sea, 2005.

McRae, Donald. "Rethinking the Arctic: A New Agenda for Canada and the United States." In *Canada among Nations 2009–2010: As Others See Us,* edited by Fen Osler Hampson and Paul Heinbeckers, 245–54. Montreal and Kingston: McGill-Queen's University Press, 2010.

Meek, Chanda L., and Emily Russell. "The Challenges of American Federalism in a Rapidly Changing Arctic." In *Governing the North American Arctic: Sovereignty, Security, and Institutions,* edited by Dawn Alexandrea Berry, Nigel Bowles, and Halbert Jones, 165–79. New York: Palgrave Macmillan, 2016.

Mendelssohn, Matthew, Robert Wolfe, and Andrew Parkin. "Globalization, Trade Policy, and the Permissive Consensus in Canada." *Canadian Public Policy* 28, no. 3 (2002): 351–71.

Mendes, Errol P. "The Canadian National Energy Program: An Example of Assertion of Economic Sovereignty or Creeping Expropriation in International Law." *Vanderbilt Journal of Transnational Law* 14, no. 3 (Summer 1981): 475–507.

Mercogliano, Salvatore R. "Fourth Arm of Defense: Sealift and Maritime Logistics in the Vietnam War." Washington, DC: Naval History and Heritage Command, Department of the U.S. Navy, 2017.

Meunier, Denis. *Hidden Beneficial Ownership and Control: Canada as a Pawn in the Global Game of Money Laundering,* C.D. Howe Commentary #519. Toronto: C.D. Howe Institute, September 2018.

Miller, Ronald E., and Peter D. Blair. *Input-Output Analysis: Foundations and Extensions.* Cambridge: Cambridge University Press, 2009.

Miller, Shannon. "Job Quality and Labour Market Demographics in Canada's Automotive Industry, 2001–2017." MA thesis, McMaster University, 2018.

Mitrany, David. "The Prospects of Integration: Federal versus Functional?" In *International Regionalism: Readings,* edited by Joseph S. Nye Jr., 43–74. Boston: Little, Brown, 1968.

– *A Working Peace System.* Chicago: Quadrangle Books, 1966.

– *A Working Peace System: An Argument for the Functional Development of International Organization.* London: Royal Institute for International Affairs, 1943.

Moens, Alexander, and Nachum Gabler. "Measuring the Costs of the Canada-U.S. Border." Vancouver: Fraser Institute, August 2012.

Monahan, Torin, and Neal A. Palmer, "The Emerging Politics of DHS Fusion Centers." *Security Dialogue* 40, no. 6 (2009): 617–636.

Monteiro, Joseph, and Benjamin Atkinson. "Cabotage: Are We Ready? How Is It Dealt With in Various Sectors of Transportation in Canada?" Canadian Transportation Research Forum, 44th Conference Proceedings, Winnipeg, 2017.

Moore, Mark H. *Creating Public Value: Strategic Management in Government.* Cambridge, MA: Harvard University Press, 1995.

– *Recognizing Public Value.* Cambridge, MA: Harvard University Press, 2013.

Morales, Isidro. "The Twilight of Mexico's State Oil Monopolism: Policy, Economic, and Political Trends in Mexico's Natural Gas Industry." Harvard Kennedy School Belfer Center for Science and International Affairs and Center for Energy Studies, Rice University's Baker Institute, December 2013.

Mordue, Gregory D. "Doors Closed and Opportunities Missed: Lessons from Failed Automotive Investment Attraction in Canada in the 1980s." *Canadian Public Policy* 43, no. S1 (2017): S43–S56.

– "Unanticipated Outcomes: Lessons from Canadian Automotive FDI Attraction in the 1980s." *Canadian Public Policy* 36 (2010): S1–S29.

Mordue, Gregory D., and Brendan Sweeney. "The Commoditisation of Automotive Assembly: Canada as a Cautionary Tale." *International Journal of Automotive Technology and Management* 17, no. 2 (2017): 169–89.

– "The Economic Contributions of the Japanese-Brand Automotive Industry to the Canadian Economy, 2001–2016." Report prepared for the Japan Automobile Manufacturer's Association of Canada, 2017.

Morgan, Geoffrey. "Enbridge Settles in U.S. over 'Humbling' Spills." *Financial Post*, 21 July 2016, FP1.

– "Ottawa Beefs Up Crude-by-Rail Rules." *Financial Post*, 21 February 2015, FP2.

Moroz, Sandy. "Navigating the Maze: Canada, Rules of Origin, and the Trans-Pacific Partnership." In *Redesigning Canadian Trade Policies for New Global Realities,* edited by Stephen Tapp, Ari Van Assche, and Robert Wolfe, 423–46. Montreal and Kingston: McGill-Queen's University Press, 2017.

Morrison, William R. *Showing the Flag: The Mounted Police and Canadian Sovereignty in the North, 1894–1925.* Vancouver: UBC Press, 1985.

Mouafo, Dieudonné. "Regional Dynamics in Canada–United States Relations." Paper presented at the Annual Meeting of the Canadian Political Science Association, Winnipeg, 3 June 2004.

Mouafo, Dieudonné, Jeff Heynen, and Nadia Ponce Morales. *Building Cross-Border Links: A Compendium of Canada-US Government Collaboration* Ottawa: Canada School of Public Service, 2004.

Mullin, Malone. "Dwight Ball Stepping Down as Newfoundland and Labrador Premier." CBC News, 17 February 2020. https://www.cbc.ca /news/canada/newfoundland-labrador/dwight-ball-steps-down-1.5466521

Nakuja, Tekuni, Mahzabin Akhand, Jill E. Hobbs, and William A. Kerr. "Evolving US Food Safety Regulations and International Competitors: Implementation Dynamics." *International Journal on Food System Dynamics* 6, no. 4 (2015): 259–68.

Nasser, Wael. "Irregular Border Crossings and Asylum in Canada: Study on the Irregular Migration from Nigeria to Canada." Honours undergraduate thesis, University of Lethbridge, August 2018.

National Conference of State Legislatures. "The Real ID: State Legislative Activity in Opposition to the REAL ID Act." Washington, DC: January 2014.

National Post. "As Ottawa dithers, telcos shun Huawei," Editorial. 6 June 2020, A16.

National Post Staff. "Border shut … except for about 200K per week," *National Post*, 30 June 2020, A1.

National Security and Intelligence Committee of Parliamentarians. *Annual Report: 2019*. Ottawa: House of Commons, 12 March 2020.

Nevitte, Neil. *The Decline of Deference: Canadian Value Change in Comparative Perspective 1981–1990*. Toronto: Broadview, 1996.

– "The Decline of Deference Revisited: Evidence after 25 Years." Paper presented at Mapping and Tracking Global Value Change: A Festschrift Conference for Ronald Inglehart, University of California, Irvine, 11 March 2011.

New England Commission. *The New England Regional Plan: An Economic Development Strategy*. Hanover, NH: University Press of New England for the New England Commission, 1981.

Newfoundland and Labrador. Board of Commissioners of Public Utilities. *Reference to the Board: Review of Two Generation Expansion Options for the Least-Cost Supply of Power to Island Interconnected Customers for the Period 2011–2067*. St. John's, NL: 30 March 2012.

Newfoundland and Labrador. Commission of Inquiry Respecting the Muskrat Falls Project. "Muskrat Falls: A Misguided Project." The Honourable Richard D. LeBlanc Commissioner. St. Johns: March 2020.

Newman, Dwight G. *Revisiting the Duty to Consult*. Saskatoon, SK: Purich Publishing, 2014.

Nielson, Julia. "Labor Mobility in Regional Trade Agreements." In *Moving People to Deliver Services*, edited by Aaditya Mattoo and Antonia Carzaniga, 93–112. Washington, DC: World Bank, 2003.

Nooruddin, Irfan, and Joel Simmons. "Openness, Uncertainty, and Social Spending: Implications for the Globalization-Welfare State Debate." *International Studies Quarterly* 53 (2009): 841–66.

North, Douglas C. *Institutions, Institutional Change and Economic Performance*. Cambridge: Cambridge University Press, 1990.

North American Competitiveness Council. "Meeting the Global Challenge: Private Sector Priorities for the Security and Prosperity Partnership of North America." Washington, Mexico City, Ottawa: August 2008.

North American Free Trade Agreement, Technical Working Group on Pesticides. *Accomplishments Report for the Period of 2003–2008*. 2009. http://www.hc-sc.gc.ca/cps-spc/alt_formats/pacrb-dgapcr/pdf/pubs/pest/corp-plan/nafta-alena-2003-2008-eng.pdf.

North American Free Trade Agreement, Technical Working Group on Pesticides. *Five-Year Strategy: 2008–2013*. 2009. http://www.hc-sc.gc.ca/cps-spc/alt_formats/pacrb-dgapcr/pdf/pubs/pest/corp-plan/nafta-alena-strat-plan-eng.pdf.

North American Free Trade Agreement, Technical Working Group on Pesticides. *Five-Year Strategy 2016–2021.* https://www.canada.ca/en/health-canada/services/consumer-product-safety/reports-publications/pesticides-pest-management/corporate-plans-reports/north-american-free-trade-agreement-technical-working-group-pesticides-five-year-strategy-2016-2021.html.

Nossal, Kim Richard. "The IJC in Retrospect." In *The International Joint Commission Seventy Years On*, edited by Robert Spencer, John Kirton, and Kim Richard Nossal, 124–30. Toronto: University of Toronto Press, 1981.

Nova Scotia, New Brunswick, and Prince Edward Island. *Report on Maritime Union.* Frederiction: Maritime Union Study, 1970.

Nuthall, Keith. "Canada Awaits Washington's Lead on Emissions Rules." Wards Auto, 13 September 2019. https://www.wardsauto.com/industry/canada-awaits-washington-s-lead-on-emissions-rules.

Nye, Joseph S., and Robert O. Keohane. "Transnational Relations and World Politics: An Introduction." *International Organization* 25, no. 3 (1971): 329–49.

Office of the United States Trade Representative. "Joint Statement by the NFTA Free Trade Commission: Building on a North American Partnership." Washington, DC: 31 July 2001. https://ustr.gov/about-us/policy-offices/press-office/press-releases/archives/2001/july/joint-statement-nafta-free-trade-commission-.

Officer, Lawrence, and Lawrence Smith. "The Canadian-American Reciprocity Treaty of 1855 to 1866." *Journal of Economic History* 28, no. 4 (1968): 598–623.

Olson, Mancur. *The Logic of Collective Action: Public Goods and the Theory of Groups.* 2nd ed. Cambridge, MA: Harvard University Press, 1971.

– *The Rise and Decline of Nations.* New Haven, CT: Yale University Press, 1986.

Omhae, Kenichi. *The Borderless World: Power and Strategy in an Interlinked Economy.* New York: Harper Business, 1999.

Ontario Ministry of Environment and Climate Change. Ontario's Great Lakes Strategy (2016). https://www.ontario.ca/page/ontarios-great-lakes-strategy.

Ontario Ministry of Finance. "International Fuel Tax Agreement." Toronto: November 2017.

Ontario Ministry of Transportation. "International Registration Plan." Toronto: October 2017.

Oostling, Jonathan. "How Lawmakers Convinced Trump to Reverse Great Lakes Funding Cut." *Detroit News,* 29 March 2019. https://www.detroitnews.com/story/news/local/michigan/2019/03/29/lawmakers-convinced-trump-reverse-great-lakes-funding-cut/3309275002/.

Ordway, John A. "U.S. International Traffic in Arms Regulations ('ITAR')." Washington, DC: Berliner, Corcoran and Rowe. http://www.cistec.or.jp/english/service/report/0802ITARarticleforCISTEC.pdf

Organisation for Economic Co-operation and Development (OECD).
   Addressing the Challenges of the Digital Economy: Action 1: 2015 Final Report.
   OECD Base Erosion and Profit Shifting Project. Paris: October 2015.
– *Globalisation and Regional Economies: Can OECD Regions Compete in Global*
   Industries? Paris: Organisation for *Economic* Co-operation and Development,
   2007. https://www.oecd.org/gov/globalisationandregionaleconomiescanoecd
   regionscompeteinglobalindustries.htm.
– *International Regulatory Cooperation: Addressing Global Challenges.* Paris: April
   2013.
– *International Regulatory Cooperation and Trade.* Paris: May 2017.
– "Regulation of Insurance Industry and Pension Fund Investment: OECD
   Report to G-20 Finance Ministers and Central Bank Governors." Paris:
   September 2015. https://www.oecd.org/g20/summits/antalya
   /Regulation-of-Insurance-Company-and-Pension-Fund-Investment.pdf.
Organisation Internationale des Constructeurs d'Automobiles. "Production
   Statistics." Paris: 2017. http://www.oica.net/category/production
   -statistics/2016-statistics/.
Organization of American States. "Foreign Trade Information System: Trade
   Agreements in Force." Washington, DC: 2018. http://www.sice.oas.org
   /agreements_e.asp.
Orloff, Greg. "CSA Group Safety Standards for Oil and Gas Pipelines: A Life-
   Cycle Approach." Toronto: CSA Group, January 2015.
O'Toole, Erin. "What Must Be Done." *National Post*, 26 February 2020, A9.
Owram, Kristine. "Toronto's Pearson Airport Wants to Be a Mega-Hub, but High
   Costs and Congestion Stand in the Way." *Financial Post*, 2 December 2016.
Pacific Institute for Climate Solutions. "Briefing Note 2010–18: BC's Low
   Carbon Fuel Standard." 9 September 2010. http://www.pics.uvic.ca
   /sites/default/files/uploads/publications/BC%27s%20low%20carbon%20
   fuel%20standard%20.pdf.
Pacific Northwest Economic Region. *Annual Report 2017.* 2017. http://www
   .pnwer.org/uploads/2/3/2/9/23295822/2017_pnwer_annual_report
   _-_reduced_size.pdf.
– *Officer's Handbook: PNWER's Organizational Structure and Delegate Responsibilities.*
   2017.
Palmer, John P. "Truck and Rail Shipping: The Deregulation Evolution." In
   Breaking the Shackles: Deregulating Canadian Industry, edited by Walter Block
   and George Lermer, 151–69. Vancouver: Fraser Institute, 1991.
Panizzon, Marion. "Standing Together Apart: Bilateral Migration Agreements
   and the Temporary Movement of Persons under 'Mode 4' of GATS." Centre
   on Migration, Policy and Society. Working Paper No. 77. Oxford: University
   of Oxford, 2010. https://www.wti.org/media/filer_public/6b/e7
   /6be710fa-b343-447a-998e-c90aeac5ea83/wp1077_marion_panizzon_2.pdf.

Parkinson, David. "A Year Later, Canada Has Failed to Take Advantage of Trade Deal with Europe." *Globe and Mail*, 8 November 2018, B4.

Pascu, Luana. "Nexus Kiosks at Canadian Airports Upgraded with Facial Recognition." Biometric Update.com, 29 October 2019. https://www.biometricupdate.com/201910/nexus-kiosks-at-canadian-airports-upgraded-with-facial-recognition.

Pastor, Robert A. "Beyond NAFTA: The Emergence and Future of North America." In *Politics in North America: Redefining Continental Relations*, edited by Yasmeen Abu-Laban, Radha Jhappan, and Francois Rocher, 461–76. Peterborough, ON: Broadview, 2008.

– *Congress and the Politics of U.S. Foreign Economic Policy.* Berkeley: University of California Press, 1980.

– "The Future of North America: Replacing a Bad Neighbor Policy." *Foreign Affairs,* July/August 2008. https://www.foreignaffairs.com/articles/north-america/2008-06-01/future-north-america.

– *The North American Idea: A Vision of a Continental Future.* Oxford: Oxford University Press, 2011.

– *Toward a North American Community: Lessons from the Old World for the New.* Washington, DC: Institute for International Economics, 2001.

Pastor, Robert A., and Rafael Fernandez de Castro, eds. *The U.S. Congress and North America.* Washington, DC: Brookings Institution, 1998.

Paterson, J.H. *North America: A Geography of the United States and Canada.* 8th ed. New York: Oxford University Press, 1989.

Peck, Jamie, and Nik Theodore. "Variegated Capitalism." *Progress in Human Geography* 31, no. 6 (2007): 731–72.

Peirce, Neal R. *The New England States: People, Politics, and Power in the Six New England States.* New York: Norton, 1976.

Pelsser, Alfred. *The Postal History of ICAO.* Montreal: ICAO, updated 30 November 2017.

Perreaux, Les. "Quebec Pre-Election Budget Contains Massive Boost in Spending." *Globe and Mail*, 28 March 2018.

Peters, B. Guy. *Institutional Theory in Political Science: The "New Institutionalism."* New York: Continuum, 2005.

Pharand, Donat. "The Arctic Waters and the Northwest Passage: A Final Revisit." *Ocean Development and International Law* 38, no. 1–2 (2007): 3–69.

Philpott, Daniel. "Sovereignty: An Introduction and Brief History." *Journal of International Affairs* 48, no. 2 (1995): 353–68.

– "Usurping the Sovereignty of Sovereignty?" *World Politics* 53 (2001): 297–324.

Physicians for Smoke-Free Canada. *The Plot against Plain Packaging.* Ottawa: April 2008. http://www.smoke-free.ca/pdf_1/plotagainstplainpackaging-apr1%27.pdf.

Piattoni, Simona. *The Theory of Multi-Level Governance: Conceptual, Empirical, and Normative Challenges*. New York: Oxford University Press, 2010.

Picciotto, Sol. "Fragmented States and International Rules of Law." *Social and Legal Studies* 6, no. 2 (1997): 259–79.

Pierce, Sarah, and Julia Gelatt. "Evolution of the H-1B: Latest Trends in a Program on the Brink of Reform." Migration Policy Institute. March 2018.

Pierson, Paul. *Dismantling the Welfare State? Reagan Thatcher and the Politics of Retrenchment*. New York: Cambridge University Press, 1994.

Plourde, André. "The Changing Nature of National and Continental Energy Markets." In *Canadian Energy Policy and the Struggle for Sustainable Development*, edited by G. Bruce Doern, 51–82. Toronto: University of Toronto Press, 2005.

Porter, Douglas, ed. *The Day after NAFTA: Economic Impact Analysis*. Toronto: BMO Capital Markets Economics, 20 November 2017. https://commercial. bmoharris.com/media/filer_public/c9/d3/c9d39192-7924-462f-a453 -2bd510460a29/appmediahero_imagebmo_economics_special_report__the _day_after_nafta_-_icb.pdf.

Porter, Michael E. *On Competition*. Boston, MA: Harvard Business Press, 1998.

Porter, Tony. "Canada, the FSB, and the International Institutional Response to the Current Crisis." In *Crisis and Reform: Canada and the International Financial System: Canada among Nations 2014*, edited by Rohinton Medhora and Dane Rowlands, 71–86. Waterloo, ON: Centre for International Governance Innovation, 2014.

Posner, Eric. "International Law and the Disaggregated State." *Florida State University Law Review* 32 (2004–5): 797–842.

Powell, Naomi. "Canada Fears Foreign Steel." *Financial Post*, 27 March 2018, FP1.

– "How Mexico's Tariffs on U.S. Pork Could Hit Canadian Producers." *Calgary Herald*, 21 June 2018, B4.

– "Steelmakers Seeking Safeguards from Metal Flooding Canadian Markets." *Financial Post*, 19 June 2018, FP1.

– "Trade End Run." *Financial Post*, 25 January 2020, FP1.

Powell, Walter W., and Paul J. DiMaggio, eds. *The New Institutionalism in Organizational Analysis*. Chicago: University of Chicago Press, 1991.

Precious, Tom, and Robert J. McCarthy. "Cuomo, Canadian Ambassador Tout Peace Bridge Deal." *Buffalo News*, 26 June 2013.

Prentice, Barry E. "Canadian Airport Security: The Privatization of a Public Good." *Journal of Air Transport Management* 48 (September 2015): 52–9.

– "Time to Think Big Picture on Churchill Rail Line." *Winnipeg Free Press*, 9 August 2018.

Preusse, Heinz G. *The New American Regionalism*. Cheltenham, UK: Edward Elgar, 2004.

Price, Kent A., ed. *Regional Conflict and National Policy*. Baltimore, MD: Johns Hopkins University Press, 1982.

PricewaterhouseCoopers LLP. *Rise in Canada's De Minimis Threshold: Economic Impact Assessment.* Toronto: Retail Council of Canada, December 2017.

Prokop, Darren. "In 1988 We Freed Trade. Now Let's Free Transport." *Policy Options* 20, no. 5 (1999): 37–40.

Pross, Paul. *Pressure Group Behaviour in Canadian Politics.* Toronto: McGraw-Hill Ryerson, 1975.

Public Health Agency of Canada. *National Report: Apparent Opioid-Related Deaths in Canada.* Ottawa: June 2019.

Public Policy Forum. "Diversification Not Dependence: A Made-in-Canada China Strategy." Ottawa: 11 October 2018. https://www.ppforum.ca /publications/diversification-not-dependence-a-made-in-canada-china -strategy/.

– "The Next-Level Border: Advancing Technology and Advancing Trade." Ottawa: 30 July 2018.

Puccio, Laura, and Roderick Harte. "From Arbitration to the Investment Court System: The Evolution of the CETA Rules." Brussels: European Parliamentary Research Service, June 2017.

Pugliese, David. "U.S. Approves Canada's Purchase of Used Australia Fighter Jets: Deal to Be Completed by End of Year." *National Post*, 21 September 2018.

Rabe, Barry G. "Building on Sub-Federal Climate Strategies: The Challenges of Regionalism." In *Climate Change Policy in North America: Designing Integration in a Regional System*, edited by Neil Craik, Isabel Studer, and Debora VanNijnatten, 69–107. Toronto: University of Toronto Press, 2013.

Rachwal, Tomasz, and Krzysztof Wiedermann. "Multiplier Effects in Regional Development: The Case of the Motor Vehicle Industry in Silesian Vivodeship (Poland)." *Quaestiones Geographicae* 27 (2008): 67–80.

Railway Association of Canada. "Canadian Transportation Act Review." Ottawa: 4 October 2014.

Raustiala, Kal. "The Architecture of International Cooperation: Transgovernmental Networks and the Future of International Law." *Virginia Journal of International Law* 43 (2002): 1–92, https://papers.ssrn.com/sol3 /papers.cfm?abstract_id=333381.

Reese, Shawn. *Defining Homeland Security: Analysis and Congressional Considerations.* CRS Report R42462. Washington, DC: Congressional Research Service, Library of Congress, 8 January 2013.

Regions for Sustainable Change. *Handbook: Tackling Climate Change by Shifting to a Low-Carbon Economy.* 2016. http://rscproject.rec.org/indicators/index .php?page=tackling-climate-change-by-shifting-to-a-low-carbon-economy.

Reich, Robert. "Who Is Us?" *Harvard Business Review* 59 (January/February 1990): 53–64.

Reinhart, Carmen and Vincent Reinhart. "The Pandemic Depression." *Foreign Affairs,* September-October 2020.

Relyea, Harold C. *Presidential Directives: Background and Overview.* CRS Report #98-611. Washington, DC: Congressional Research Service, Library of Congress, 28 November 2008. https://www.fas.org/sgp/crs/misc/98-611.pdf.

Richardson, Kathrine. "Attracting and Retaining Foreign Highly Skilled Staff in Times of Global Crisis: A Case Study of Vancouver, British Columbia's Biotechnology Sector." *Population, Space and Place* 22, no. 5 (2016): 428–40.

Riddell-Dixon, Elizabeth. *Breaking the Ice: Canada, Sovereignty, and the Arctic Extended Continental Shelf.* Toronto: Dundurn, 2017.

Riker, William. "Six Books in Search of a Subject or Does Federalism Exist and Does It Matter?" *Comparative Politics* 2, no. 1 (October 1969): 135–46.

Roach, Kent, and Craig Forcese. "The Roses and the Thorns of Canada's New National Security Bill." *Maclean's,* 20 June 2017.

Roberts, Patrick. "Dispersed Federalism as a New Regional Governance for Homeland Security." *Publius: The Journal of Federalism* 38, no. 3 (2009): 416–43.

Robertson, Colin. "CDA_USA 2.0: Intermesticity, Hidden Wiring and Public Diplomacy." In *Canada among Nations 2007: What Room for Manoeuvre?*, edited by Jean Daudelin and Daniel Schwanen, 268–310. Montreal and Kingston: McGill-Queen's University Press, 2008.

– "NAFTA: A Primer for the Montreal Round." Calgary: Canadian Global Affairs Institute, January 2018.

Robertson, Grant. "Going Dark: 'Without early warning you can't have early response': How Canada's world class pandemic alert system failed." *Globe and Mail,* 25 July 2020: A12–14.

Robichau, Robbie Waters, and Laurence E. Lynn Jr. "The Implementation of Public Policy: Still the Missing Link." *Policy Studies Journal* 37, no. 1 (2009): 21–36.

Robyn, Dorothy. "Alternative Governance Models for the Air Traffic Control System: A User Cooperative versus a Government Corporation." Washington, DC: Brookings Institution, 6 April 2015. https://www.brookings.edu/blog/fixgov/2015/04/06/alternative-governance-models-for-the-air-traffic-control-system-a-user-cooperative-versus-a-government-corporation/.

Rodrik, Dani. "Sense and Nonsense in the Globalization Debate." *Foreign Policy* 107 (Summer 1997): 19–37.

Rose, Joel. "Canada's Tech Firms Capitalize on Immigration Anxiety in the Age of Trump." NPR, 9 June 2017. https://www.npr.org/2017/06/09/532220824/canadas-tech-firms-capitalize-on-immigration-anxiety-in-the-age-of-trump.

Rosenau, James N. *Distant Proximities: Dynamics beyond Globalization.* Princeton, NJ: Princeton University Press, 2003.

– "The Governance of Fragmegration: Neither a World Republic nor a Global Interstate System." Paper presented at the Congress of the International Political Science Association, Quebec City, 2000.

– "The Governance of Fragmegration: Neither a World Republic nor a Global Interstate System." *Studia Diplomatica* 53, no. 5 (2000): 15–40.

Rothwell, Donald. "The Canadian-US Northwest Passage Dispute: A Reassessment." *Cornell International Law Journal* 26, no. 2 (1993). http://scholarship.law.cornell.edu/cilj/vol26/iss2/2.

Rottem, Svein Vigeland. "A Note on the Arctic Council Agreements." *Ocean Development & International Law* 46 (2015): 50–9.

Rudra, Nita, and Jennifer Tobin. "When Does Globalization Help the Poor?" *Annual Review of Political Science* 20 (2017): 287–307.

Rugman, Alan G. *The End of Globalization.* New York: Amacom, 2000.

Russell, Betsy Z. "441-Foot-Long Megaloads Bound for Montana through CdA." *Idaho Spokesman-Review*, Boise, ID, 20 December 2013.

Russell, Edward. "Air Canada Continues to Eye Closer United Ties." FlightGlobal.com, 27 September 2018.

Russo, Margherita. "Technical Change and the Industrial District: The Role of Interfirm Relations in the Growth and Transformation of Ceramic Tile Production in Italy." *Research Policy* 14, no. 6 (1985): 329–43.

Rutherford, Tod D., and John Holmes. "Manufacturing Resiliency: Economic Restructuring and Automotive Manufacturing in the Great Lakes Region." *Cambridge Journal of Regions, Economy and Society* 7, no. 3 (2014): 359–78.

Sabatier, Paul A., and Christopher M. Weible, eds. *Theories of the Policy Process,* 3rd ed. Boulder, CO: Westview, 2014.

Sageman, Marc. *Leaderless Jihad: Terror Networks in the 21st Century.* Philadelphia: University of Pennsylvania Press, 2008.

Salacuse, Jeswald. "BIT by BIT: The Growth of Bilateral Investment Treaties and Their Impact on Foreign Investment in Developing Countries." *International Lawyer* 24, no. 3 (1990): 655–75.

Salacuse, Jeswald, and Nicholas Sullivan. "Do BITs Really Work? An Evaluation of Bilateral Investment Treaties and Their Grand Bargain." *Harvard Journal of International Law* 46 no. 1 (Winter 2005): 67–130.

Saldana, Yadira Tejeda, and Erin Cheney. "NAFTA 2.0 Trade and Safe Food." *Ivey Briefing*. Agri-food at Ivey. London, ON: University of Western Ontario, 2017.

Sampson, Andrew, and Peter Cowan. "Ed Martin Fired after Quitting Nalcor, Triggering $1.4 Severance, Says Dwight Ball." CBC News, 24 May 2016. https://www.cbc.ca/news/canada/newfoundland-labrador/ed-martin -severance-1.3597700

Sando, Paul. "Water and Political Relations between the Upper Plains States and the Prairie Provinces: What Works, What Doesn't, and What's All Wet." In *Beyond the Border: Tensions across the Forty-Ninth Parallel in the Great Plains and Prairies,* edited by Kyle Conway and Timothy Pasch, 133–50. Montreal and Kingston: McGill-Queen's University Press, 2013.

Sands, Christopher. "Canada and the U.S. Closed the Border the Proper Way – Cooperatively and with Logical Exceptions." *Ottawa Citizen*, 19 March 2020.

– "The Canada Gambit: Will It Revive North America?" Washington, DC: Hudson Institute, 2 March 2011. http://www.hudson.org/research/7757 -the-canada-gambit-will-it-revive-north-america.

– "Fading Power or Rising Power: 11 September and Lessons from the Section 110 Experience." In *Readings in Canadian Foreign Policy*, edited by Duane Bratt and Christopher J. Kukucha, 249–64. Toronto: Oxford University Press, 2007.

Sanford, Mark. "REAL-ID Side Effects." *Washington Times*, 14 April 2014.

Saunders, Doug. "How the Huawei Crisis Has Exploded Trudeau's China Policy." *Globe and Mail*, 19 January 2019, O1.

Savoie, Donald A. *Governing from the Centre: The Concentration of Power in Canadian Politics*. Toronto: University of Toronto Press, 2000.

– *Breaking the Bargain: Public Servants, Ministers, and Parliament*. Toronto: University of Toronto Press, 2003.

Scavo, Carmine, Richard C. Kearney, and Richard J. Kilroy Jr. "Challenges to Federalism: Homeland Security and Disaster Response." *Publius: The Journal of Federalism* 38, no. 1 (2007): 81–110.

Schneekloth, Lynda H., and Robert G. Shibley. *Placemaking: The Art and Practice of Building Communities*. New York: Wiley, 1995.

Schott, Jeffrey J., ed. *Free Trade Agreements: US Strategies and Priorities*. Washington, DC: Institute for International Economics, 2004.

Schulze, Max-Stephan, and Nikolaus Wolf. "On the Origins of Border Effects: Insights from the Habsburg Empire." *Journal of Economic Geography* 9 (2009). 117–36.

Schwab, Klaus. "The Fourth Industrial Revolution: What It Means and How to Respond." *Foreign Affairs*, 12 December 2015. https://www.foreignaffairs .com/articles/2015-12-12/fourth-industrial-revolution.

Schwanen, Daniel. "Deeper, Broader: A Roadmap for a Treaty of North America." In *The Art of the State: Thinking North America* 2, no. 4, edited by Thomas J. Courchene and Donald J. Savoie, 3–36. Montreal: Institute for Research on Public Policy, 2004. https://irpp.org/wp-content/uploads /2014/08/schwanen_roadmap.pdf.

Scott, James W. "Bordering, Border Politics and Cross-Border Cooperation in Europe." In *Neighbourhood Policy and the Construction of the European External Borders*, edited by C. Celata and R. Coletti, 27–47. Basel, Switzerland: Springer International Publishing, 2015.

Seligman Lara. "Trump's 'America First' Policy Could Leave U.S. Defense Industry Behind." *Foreign Policy*, 18 July 2018. https://foreignpolicy.com /2018/07/18/trumps-america-first-poplicy-could-leave-u-s-defense -industry-behind/.

Shaw, Malcolm N. *International Law.* 7th ed. Cambridge: Cambridge University Press, 2014.

Shecter, Barbara. "Bid to Create National Markets Watchdog Advances with Supreme Court Ruling." *Financial Post*, 10 November 2018, FP1.

Shibley, Robert G., Lynda H. Schneekloth, and Bradshaw Hovey. "Constituting the Public Realm of a Region: Placemaking in the Bi-National Niagaras." *Journal of Architectural Education* 57, no. 1 (2003): 28–42.

Siegfried, André. *Canada: An International Power.* 2nd ed. Translated by Doris Hemming. New York: Duel, Sloan and Pearce, 1947.

Silcoff, Sean. "Liberals Vow to Tax Foreign Tech Giants on Digital Ads, Services." *Globe and Mail*, 30 September 2019, B1.

Slaughter, Anne-Marie. "Global Government Networks, Global Information Agencies, and Disaggregated Democracy." *Michigan Journal of International Law* 24 (2002–3): 1041–75.

– *A New World Order.* Princeton, NJ: Princeton University Press, 2004.

– "The Real New World Order." *Foreign Affairs* 76, no. 5 (1997): 183–97.

Smith, Joanna. "Trudeau Wants Canada to Play Key Role in Fighting Climate Change." *Star*, 2 March 2016. https://www.thestar.com/news/canada/2016/03/02/canada-will-play-leading-role-in-new-economy-trudeau-says.html.

Smith, Kevin B., and Christopher W. Larimer. *The Public Policy Theory Primer.* 2nd ed. Boulder, CO: Westview, 2013.

Smyth, Stuart J., William A. Kerr, and Richard S. Gray. "Regulatory Barriers to International Scientific Innovation: Approving New Biotechnology in North America." *Canadian Foreign Policy Journal* 23, no. 2 (2017): 134–45.

Snoddon, Tracy, and Debora VanNijnatten. "Carbon Pricing and Intergovernmental Relations in Canada." *IRPP Insight* 12 (November) 2016. https://irpp.org/research-studies/insight-no12/.

Snyder, J.D., Kelly Christopherson, Leslie Grimm, Anthony Orlando, and Aaron Galer. *Global Models of Binational Regional Collaboration: The Potential for Great Lakes Regional Innovation.* Lansing, MI: Michigan State University Press, 2014.

Snyder, Jim. "U.S. Aiming to Phase Out Older Oil Tanker Rail Cars." *Financial Post*, 24 July 2014, FP10.

Sokolsky, Joel J., and Philippe Lagassé. "Suspenders and a Belt: Perimeter and Border Security in Canada-US Relations." *Canadian Foreign Policy* 12, no. 3 (2006): 15–29.

Somin, Ilya. "Federalism and the Roberts Court." *Publius: The Journal of Federalism* 46, no. 3 (2016): 441–62.

Sosnow, Clifford, and Peter E. Kirby. "The USMCA: A First Look at Key Contentious Issues." *International Trade and Customs Law Bulletin* (Toronto: Fasken, 3 October 2018).

Sosnow, Clifford, Marcia Mills, Peter N. Mantas, and Andrew House. "'Because We Said So': Removing the CITT from the Review of Canada's 'National Security Exemption.'" Toronto: Fasken, 18 June 2019.

Spak, Gregory, Francisco de Rosenzeig, Dean A. Barclay, Scott C. Lincocome, Matt Solomon, and Brian Picone. "Overview of Chapter 4 (Rules of Origin) of the US-Mexico-Canada Trade Agreement." Washington, DC: White & Case, 25 October 2018. https://www.whitecase.com/publications/alert /overview-chapter-4-rules-origin-us-mexico-canada-trade-agreement.

Sparber, Chad. "Choosing Skilled Foreign-Born Workers: Evaluating Alternative Methods for Allocating H-1B Work Permits." *Industrial Relations* 57, no. 1 (2018): 3–34.

Sproule-Jones, Mark. *Restoration of the Great Lakes: Promises, Practices, and Performances.* Vancouver: UBC Press, 2002.

Stanbury, W.T. "Corporate Power and Political Influence." In *Mergers, Corporate Concentration and Power in Canada,* edited by R.S. Khemani and W.T. Stanbury, 417–31. Halifax: Institute for Research in Public Policy, 1988.

Stanford, Jim. "The Geography of Auto Globalization and the Politics of Auto Bailouts." *Cambridge Journal of Regions, Economy and Society* 3, no. 3 (2010): 383–405.

State of the Lakes Ecosystem Conference. Erie, PA: 2011. https://archive.epa .gov/solec/web/html/.

Steidler, Paul. "Canada's Strong Stand to Fix International Postal System." Arlington, VA: Lexington Institute, 15 February 2019.

Stephenson, Paul. "Twenty Years of Multi-Level Governance: 'Where Does It Come From? What Is It? Where Is It Going?'" *Journal of European Public Policy* 20, no. 6 (2013): 817–37.

Stone, Laura. "Unifor's Dias Emerges as Key Voice on NAFTA." *Globe and Mail,* 26 September 2017, A3.

Storper, Michael. *Keys to the City: How Economies, Institutions, Social Interactions, and Politics Shape Development.* Princeton, NJ: Princeton University Press, 2013.

Strange, Susan. *The Retreat of the State.* Cambridge: Cambridge University Press, 1996.

Stuart, Reginald C. *Dispersed Relations: Americans and Canadians in Upper North America.* Washington and Baltimore: Wilson Center Press and Johns Hopkins University Press, 2007.

Studer, Isabel. "Obstacles to Integration: NAFTA's Institutional Weakness." In *Requiem or Revival: The Promise of North American Integration,* edited by Isabel Studer and Carol Wise, 63–75. Washington, DC: Brookings Institution, 2007.

Stuhl, Andrew. *Unfreezing the Arctic: Science, Colonialism, and the Transformation of Inuit Lands.* Chicago: University of Chicago Press, 2016.

Sturgeon, Timothy J., and Richard Florida. *Globalization and Jobs in the Automotive Industry: Final Report to the Alfred P. Sloan Foundation.* Cambridge,

MA: International Motor Vehicle Program, Center for Technology, Policy, and Industrial Development, Massachusetts Institute of Technology, 2000.

Sturgeon, Timothy J., Johannes Van Biesebrock, and Gary Gereffi. "Value Chains, Networks, and Clusters: Reframing the Global Automotive Industry." *Journal of Economic Geography* 8, no. 3 (2008): 297–321.

Superville, D., and M.C. Jalonick. "Obama to Sign Food Safety Bill Today." *Huffington Post*, 4 January 2011.

Suvankulov, Farrukh. "Revisiting National Border Effects in Foreign Trade in Goods of Canadian Provinces." Bank of Canada Working Paper 2015–28. Ottawa: Bank of Canada, July 2015.

Svrluga, Susan. "'I'll Be in Canada': More Students Are Looking to Head North." *Washington Post*, 27 March 2017. https://www.washingtonpost.com/news/grade-point/wp/2017/03/27/ill-be-in-canada-more-students-are-looking-to-head-north/?noredirect=on&utm_term=.af606834265e.

Sweeney, Brendan. "A Profile of the Automotive Manufacturing Industry in Canada, 2012–2016." Automotive Policy Research Centre Research Briefs. Hamilton, ON, April 2017.

Sweeney, Brendan, and John Holmes. "Renegotiating NAFTA: What's at Stake for Canada's Automotive Industry?" Centre for International Governance Innovation Research Briefs. Waterloo, ON, 2017.

Sweeney, Brendan, and Greigory D. Mordue. "The Restructuring of Canada's Automotive Industry, 2005–2014." *Canadian Public Policy* 43, no. S1 (2017): S1–S15.

Sweetman, Arthur, and Casey Warman. "Canada's Temporary Foreign Workers Programs." *Canadian Issues* (Spring 2010): 19–24.

Swenson, Deborah L. "Why Do Developing Countries Sign BITs?" *U.C. Davis Journal of International Law & Policy* 12 (2005–6): 131–55.

Sykes, Alan O. "Regulatory Protectionism and the Law of International Trade." *University of Chicago Law Review* 66, no. 1 (1999): 1–46.

Tapp, Stephen, Robert Wolfe, and Ari VanAssche, eds. *Redesigning Trade Policies for New Global Realities*. Montreal: Institute for Research in Public Policy and McGill-Queen's University Press, 2017.

Tarr, G. Alan. "Laboratories of Democracy? Brandeis, Federalism, and Scientific Management." *Publius* 31, no. 1 (Winter 2001): 37–46.

Taylor, Bron. "Bioregionalism: An Ethics of Loyalty to Place." *Landscape Journal* 19, no. 1 and 2 (2000): 50–72.

Thomsen, Stephen, and Fernando Mistura. "Is Investment Protectionism on the Rise?" OECD Investment Division, 6 March 2017.

Thurow, Lester C. *Fortune Favours the Bold: What We Must Do to Build a New and Lasting Prosperity*. New York: Harper Collins, 2003.

Timmins, Thomas J., Wendy J. Wagner, and Neeta Sahadev. "The WTO Decision: What It Means for Ontario FIT 1.0 and 2.0 Projects." 3 June 2013.

http://www.mondaq.com/canada/x/242596/Renewables/The
+WTO+Decision+What+It+Means+For+Ontario+FIT+10+And+20+Projects.

Tomblin, Stephen G. "Effecting Change and Transformation through Regionalization: Theory versus Practice." *Canadian Public Administration* 50, no.1 (2007): 1–20.

– *Ottawa and the Outer Provinces: The Challenges of Regional Integration in Canada.* Toronto: Lorimer, 1995.

Tomblin, Stephen G., and Charles S. Colgan, eds. *Regionalism in a Global Society: Persistence and Change in Atlantic Canada and New England.* Toronto: University of Toronto Press, 2004.

Tomesco, Frederic. "Air Canada Hopes to Resuscitate Cross-Border Venture with United." *Bloomberg*, 27 June 2017.

Toope, Stephen. "Power but Unpersuasive? The Role of United States in the Evolution of Customary International Law." *In United States Hegemony and the Foundations of International Law,* edited by Michael Byers and Georg Nolte, 288–316. Cambridge: Cambridge University Press, 2003.

TopForeignStocks.com. "The Complete List of Canada Stocks Listed on U.S. Markets." 30 June 2020. http://topforeignstocks.com/foreign-adrs-list/the-complete-list-of-canada-stocks-trading-on-us-markets/.

Toronto Star. "Ottawa Toughens Rules for Oilsands Buyouts by Foreign State-Owned Firms." 7 December 2012.

Trautman, Laurie. "Cross-Border Collaboration in the Cascadia Region." Paper presented at the Association for Borderlands Studies World Conference, Vienna, 10 July 2018.

– "Sub-National Partnerships amidst Federal Barriers: Cross-Border Collaboration in the Cascadia Region of the Canada-U.S. Border." Paper presented to Association for Borderland Studies, San Antonio, TX, 6 April 2018.

Trew, Stuart. "Correcting the Democratic Deficit in the CETA Negotiations: Civil Society Engagement in the Provinces, Municipalities, and Europe." *International Journal* 68, no. 4 (2013): 568–75.

Trossman, Jeffrey, and Jeffrey Shafer. "The Big Shake-Up: Making Sense of the OECD Digital Tax Proposals." E-Brief #297. Toronto: C.D. Howe Institute, 12 November 2019. https://www.cdhowe.org/sites/default/files/attachments/research_papers/mixed/e-brief%20297.pdf.

Tumilty, Ryan. "Canada's airports poised to hike fees in wake of pandemic." *National Post*, 13 August 2020, A5.

Tungohan, Ethel. "From Encountering Confederate Flags to Finding Refuge in Spaces of Solidarity: Filipino Temporary Foreign Workers' Experiences of the Public in Alberta." *Space and Polity* 21, no. 11 (2017): 11–26.

Tuohy, Carolyn J. *Accidental Logics: The Dynamics of Change in the Health Care Arena in the United States, Britain, and Canada.* New York: Oxford University Press, 1999.

Turnbull, Lornet, and Katie Zezima. "Legal Pot in Canada Will Mean Legal Peril at the Border." *Washington Post*, 13 July 2018, A3.

Tuzyk, John, and Liam Churchill. "National Securities Regulator on the Ropes? Quebec Court of Appeals Rules Proposed Cooperative System Unconstitutional." *Blake's Business Class*, 15 May 2017. https://www.blakes.com/insights/bulletins/2017/national-securities-regulator-on-the-ropes-quebec.

– "Supreme Court Hearing Leaves Cooperative Capital Markets Regulatory System in Limbo." Toronto: Blake, Cassells and Graydon, 28 March 2018. https://www.lexology.com/library/detail.aspx?g=f5928b2f-d39c-440e-9baf-0a7931ba6807.

Tyson, Laura D'Andrea. *Who's Bashing Whom? Trade Conflict in High-Technology Industries*. Washington, DC: Peterson Institute for International Economics, 1993.

UNCTAD, Investment Policy Hub. "Most Recent IIAs." http://investmentpolicyhub.unctad.org/IIA.

United Nations High Commission for Refugees. "Figures at a Glance." (New York: 2020). https://www.unhcr.org/figures-at-a-glance.html.

United Kingdom. National Crime Agency. "NCA DG at Five Eyes Law Enforcement Group Meetings in Washington." London, 17 June 2016.

United Nations. *International Migration Report 2015: Highlights*. New York: United Nations Department of Economic and Social Affairs, Population Division, 2016. https://www.un.org/en/development/desa/population/migration/publications/migrationreport/docs/MigrationReport2015_Highlights.pdf.

– "United Nations Office of the Special Advisor on Prevention of Genocide." http://www.un.org/en/genocideprevention/.

United States. Bureau of Customs and Border Protection. "Internet Purchases." Washington, DC: 31 January 2018.

– Bureau of Transportation Statistics. 2018.

– Census Bureau. "2009–2013 American Community Survey 5-Year Estimates" (including surveys of Economic Characteristics, Demographics and Housing, and Households and Families, 2015). Reported in Wikipedia, "List of United States Counties by Per Capita Income." Accessed 11 February 2018. https://en.wikipedia.org/wiki/List_of_United_States_counties_by_per_capita_income.

– Citizenship and Immigration Services. "Approved H 1B Petitions (Number, Salary, and Degree/Diploma) by Employer Fiscal Year 2015." As of 30 June 2017. https://www.uscis.gov/sites/default/files/USCIS/Resources/Reports%20and%20Studies/Immigration%20Forms%20Data/BAHA/h-1b-2015-employers.pdf.

– Citizenship and Immigration Services. "Approved H 1B Petitions (Number, Salary, and Degree/Diploma) by Employer Fiscal Year 2016." As of 30 June

2017. https://www.uscis.gov/sites/default/files/USCIS/Resources/
Reports%20and%20Studies/Immigration%20Forms%20Data/BAHA/h-1b
-2016-employers.pdf.
- Citizenship and Immigration Services. "Number of H-1B Petition Filings,
Applications and Approvals, Country, Age, Occupation, Industry, Annual
Compensation ($), and Education, FY2007–FY2017." US Citizenship and
Immigration Services Claim 3. 15 November 2017.
- Citizenship and Immigration Services. "Temporary Protected Status." https://
www.uscis.gov/humanitarian/temporary-protected-status.
- Congress. House. Committee on Science and Technology. *Competitiveness
and Innovation on the Committee's 50th Anniversary with Bill Gates, Chairman of
Microsoft.* 110th Cong., 2nd sess. Serial no. 110–84. 21 March 2008.
- Congress. House. Subcommittee on Coast Guard and Maritime
Transportation of the Committee on Transportation and Infrastructure.
Implementing U.S. Policy in the Arctic. 109th Cong., 2nd sess. Serial no. 113–78.
23 July 2014.
- Department of Commerce. *U.S.-Canada RCC Work Plans: 2019–20.*
Washington, DC: International Trade Administration, 9 March 2020. https://
www.trade.gov/rcc-work-plans-2019-2020.
- Department of Homeland Security. *Container Security Initiative: In Summary.*
Washington, DC: May 2011.
- Department of Homeland Security. "Fiscal Years 2014–2018: Strategic Plan."
Washington, DC: 2014. https://www.dhs.gov/sites/default/files
/publications/FY14-18%20Strategic%20Plan.PDF.
- Department of Homeland Security. "19 CFR Chapter 1: Notification of
Temporary Travel Restrictions Applicable to Land Ports of Entry and Ferries
Service between the United States and Canada." Washington, DC: 20 March
2020.
- Department of Homeland Security. "REAL-ID." https://www.dhs.gov
/real-id.
- Department of Homeland Security. "Registration Requirement for
Petitioners Seeking to File H-1B Petitions on Behalf of Cap-Subject Aliens"
(Federal Register, 31 January 2019). https://www.federalregister.gov
/documents/2019/01/31/2019-00302/registration-requirement-for
-petitioners-seeking-to-file-h-1b-petitions-on-behalf-of-cap-subject.
- Department of Homeland Security. *The 2014 Quadrennial Homeland Security
Review.* Washington, DC: 18 June 2014. https://www.dhs.gov/sites/default
/files/publications/2014-qhsr-final-508.pdf.
- Department of Homeland Security and Canada. "United States and Canada
Sign Preclearance Agreement." Washington, DC: 16 March 2015. http://www
.dhs.gov/news/2015/03/16/united-states-and-canada-sign-preclearance
-agreement.

– Department of Homeland Security and Public Safety Canada. *United States–Canada Joint Border Risk and Threat Assessment*. Washington, DC, and Ottawa: July 2010.
– Department of Labor. "Fact Sheet #62A: Changes Made by the H-1B Visa Reform Act of 2004." Last modified July 2008. https://www.dol.gov/whd/regs/compliance/FactSheet62/whdfs62A.pdf.
– Department of State. "Nonimmigrant Visa Issuances by Visa Class and by Nationality, FY1997–2017 NIV Detail Table." *Nonimmigrant Visa Statistics*. https://travel.state.gov/content/travel/en/legal/visa-law0/visa-statistics/nonimmigrant-visa-statistics.html.
– *Report of the Advisory Committee on International Economic Policy Regarding the Draft Model Bilateral Investment Treaty*. 11 February 2004.
– *Report of the Subcommittee on Investment of the Advisory Committee on International Economic Policy (ACIEP) Regarding the Model Bilateral Investment Treaty*. September 2009.
– Department of State. Bureau of Consular Affairs. "Passport Statistics." Washington, DC: 2017. Accessed 24 November 2018. https://travel.state.gov/content/travel/en/passports/after/passport-statistics.html.
– Department of State. Office of Inspector General. "Memorandum Report 01-FP-M-027: Review of the U.S. Munitions List and the Commodity Jurisdiction Process." March 2001. https://fas.org/asmp/resources/govern/01-state-ig.pdf.
– Department of Transportation. Bureau of Transportation Statistics. "Border Crossing/Entry Data." 2014–17.
– Department of the Treasury, Committee on Foreign Investment in the U.S. *Annual Report to Congress, 2015*. https://www.treasury.gov/resource-center/international/foreign-investment/Pages/cfius-reports.aspx.
– Department of the Treasury, Committee on Foreign Investment in the U.S. "Reports and Tables." https://www.treasury.gov/resource-center/international/foreign-investment/Pages/cfius-reports.aspx.
– Energy Information Administration. *Annual Energy Outlook 2018 with Projections*. Washington, DC: 2018.
– Energy Information Administration. *Annual Energy Outlook 2020 with Projections to 2050*. Washington, DC: 2020. https://www.eia.gov/outlooks/aeo/pdf/AEO2020%20Full%20Report.pdf.
– Energy Information Administration. "Arctic Oil and Natural Gas Resources." 20 January 2018. http://www.eia.gov/todayinenergy/detail.cfm?id=4650.
– Energy Information Administration. "Monthly Energy Review, June 2020." Washington, DC: 2020.
– Energy Information Administration. "Monthly Energy Review, September 2020." Washington, DC: 2020. https://www.eia.gov/totalenergy/data/monthly/pdf/mer.pdf.

- Energy Information Administration. "US Shale Natural Gas Proved Reserves." Washington, DC: 12 December 2019. https://www.eia.gov/dnav /ng/hist/res_epg0_r5301_nus_bcfa.htm.
- Energy Information Administration. "Spot Prices (Crude Oil in Dollars per Barrel)." Washington, DC: 2018. https://www.eia.gov/dnav/pet/pet_pri _spt_s1_m.htm.
- Energy Information Administration. "Table 2.13. Electric Power Industry: US Electricity Imports from and Electricity Exports to Canada and Mexico, 2006–2016." Electricity. Data Tables. https://www.eia.gov/electricity /annual/html/epa_02_13.html.
- Energy Information Administration. "Table 2.14 Electric Power Industry: US Electricity Imports from and Electricity Exports to Canada and Mexico, 2008–2018." https://www.eia.gov/electricity/annual/html/epa_02_14 .html.
- Energy Information Administration. "US Crude Oil and Natural Gas Proved Reserves, Year-End 2018." Washington, DC: 2019. https://www.eia .gov/naturalgas/crudeoilreserves/pdf/usreserves.pdf.
- Environmental Protection Agency. "Fact Sheet: Overview of the Clean Power Plan: Cutting Carbon Pollution from Power Plants." 2015. https://19january2017snapshot.epa.gov/sites/production/files/2015-08 /documents/fs-cpp-overview.pdf.
- Federal Highway Administration. "Federal Share of State Highway Capital Expenditures, by State," table HF-202FS. Washington, DC: January 2016.
- Food and Drug Administration, Canada Food Inspection Agency, and Health Canada. "FDA - CFIA and Health Canada, Food Safety Systems Recognition Arrangement." Washington and Ottawa: May 2016. https:// www.fda.gov/InternationalPrograms/Agreements/Memorandaof Understanding/ucm498197.htm.
- Government Accountability Office. *Commercial Aviation: Bankruptcy and Pension Problems Are Symptoms of Underlying Structural Issues*, GAO-05-945. Washington, DC: September 2005.
- Government Accountability Office. "International Mail: Information on Changes and Alternatives to the Terminal Dues System." Washington, DC: October 2017.
- Internal Revenue Service. "Substantial Presence Test." Washington, DC: 3 August 2017.
- International Trade Administration. "Canada – Civil Aviation." Washington, DC: U.S. Department of Commerce, 28 September 2017.
- International Trade Commission. *Digital Trade in the U.S. and Global Economies, Part I*. USITC Publication 4415. Washington, DC: July 2013.
- National Commission on Terrorist Attacks on the United States. *The 9/11 Commission Report*. Washington, DC: 2004.

– Office of the United States Trade Representative. "Free Trade Agreements." Washington, DC: November 2018. https://ustr.gov/trade-agreements/free-trade-agreements.
– Office of the United States Trade Representative. *Summary of Objectives for the NAFTA Renegotiation.* Washington, DC: 17 July 2017.
Office of the United States Trade Representative. Agreement between the United States of America, United Mexican States, and Canada 12/13/19 Text. https://ustr.gov/trade-agreements/free-trade-agreements/united-states-mexico-canada-agreement/agreement-between
– Pipeline and Hazardous Materials Safety Administration. "Pipeline Advisory -Committees." Washington, DC: 2016. https://cms7.phmsa.dot.gov/standards-rulemaking/pipeline/pipeline-advisory-committees– White House. *National Security Strategy.* Washington, DC: February 2015. https://obamawhitehouse.archives.gov/sites/default/files/docs/2015_national_security_strategy_2.pdf Accessed August 24, 2020.
– White House. "Executive Order 13609, 'Promoting International Regulatory Cooperation.'" Washington, DC: Office of Management and Budget, 4 May 2012.
– White House. *National Security Strategy of the United States.* Washington, DC: December 2017. https://www.whitehouse.gov/wp-content/uploads/2017/12/NSS-Final-12-18-2017-0905.pdf.
– White House. "Presidential Executive Order on Buy American and Hire American." 18 April 2017. https://www.whitehouse.gov/presidential-actions/presidential-executive-order-buy-american-hire-american/.
– White House and Canada. Privy Council Office. *Beyond the Border: A Shared Vision for Economic Security and Economic Competitiveness.* Washington, DC and Ottawa: December 2011.
– White House and Canada. Privy Council Office. *Beyond the Border Implementation Report.* Washington and Ottawa: 19 December 2013. https://www.dhs.gov/sites/default/files/publications/btb-canada-us-final_-_dec19_0.pdf.
– White House and Canada. Privy Council Office. *2014 Beyond the Border Implementation Report to Leaders.* Washington and Ottawa: 13 May 2015. https://www.dhs.gov/sites/default/files/publications/15_0320_15-0745_BTB_Implementation_Report.pdf.
United States and Canada. United States-Canada Agreement on Government Procurement. February 16, 2010. Accessed July 15, 2018.
https://ustr.gov/issue-areas/government-procurement/us-canada-agreement-government-procurement.
– United States Environmental Protection Agency and Environment Climate Change Canada. *2016 Progress Report of the Parties: Pursuant to the*

Canada–United States Great Lakes Water Quality Agreement. 2017. https://binational.net/wp-content/uploads/2016/09/PRP-160927-EN.pdf.

– United States–Canada Transportation Border Working Group (TBWG). *Action Plan 2015–2017.* Washington and Ottawa: 2015.

United States and Canada, United States Environmental Protection Agency and Environment Climate Change Canada. *Progress Report of the Parties Pursuant to the Canada–United States Great Lakes Water Quality Agreement.* 2017. https://binational.net/wp-content/uploads/2016/09/PRP-160927-EN.pdf.

Usdansky, Margaret L., and Thomas J. Espenshade. "The H-1B Visa Debate in Historical Perspective: The Evolution of U.S. Policy toward Foreign-Born Workers." UC San Diego Working Papers, 2000. https://escholarship.org/uc/item/8qf435d5.

Valigra, Lori. "Opponents Take the First Formal Step to Bring CMP's Transmission Project to a Statewide Vote." *Bangor Daily News,* 30 August 2019.

Vallet, Élisabeth, ed. *Borders, Fences and Walls: State of Insecurity.* New York: Routledge, 2014.

–, ed. *La présidence des États-Unis, présidence impériale ou présidence en péril?* Quebec: Presses de l'Université du Québec, coll. Enjeux contemporains, 2005.

Vallet, Élisabeth, and Andréanne Bissonnette. "The Quebec/United States Border: Language, an Asset or Liability to Border Crossing?" Paper presented to the Western Social Science Association, San Antonio, TX, 5 April 2018.

Vallet, Élisabeth, and Pierre-Louis Malfatto. "Water Geopolitics in North America." In *The Geopolitics of Natural Resources,* edited by David Lewis Feldman, 327–51. Aldershot, UK: Edward Elgar Publishing, 2004.

Van Assche, Ari. "Global Value Chains and the Rise of a Supply Chain Mindset." In *Redesigning Canadian Trade Policies for New Global Realities,* edited by Stephen Tapp, Ari Van Assche, and Robert Wolfe, 183–208. Montreal and Kingston: McGill-Queen's University Press, 2017.

Van de Kletersteeg, Theo. "Federal Port Review: 2014–15." *Canadian Sailings,* 22 February 2017.

Vanderklippe, Nathan. "Transportation Woes Threaten to Delay Imperial Oil Sands Project." *Globe and Mail,* 28 June 2011.

Van der Meer, Tom W.G., "Political Trust and the 'Crisis of Democracy.'" *Oxford Research Encyclopaedia of Politics* (January 2017). DOI 10.1093/acrefore/9780190228637.013.77.

Vandevelde, Kenneth. "A Comparison of the 2004 and 1994 US Model BITs: Rebalancing Investor and Host Country Interests." *Yearbook on International Investment Law and Policy* 2009 (2008): 283–316.

Vandiver, John. "Largest NATO Drill in 16 Years Brings Carrier, US Forces to Norway." *Stars and Stripes,* 9 October 2018. https://www.stripes.com

/news/largest-nato-drill-in-16-years-brings-carrier-us-forces-to-norway -1.550983#.

Van Horne Institute, Prolog Canada Inc., and JRSB Logistics Consulting. "Overdimensional Loads: A Canadian Solution." Calgary: Van Horne Institute, July 2015.

Van Houtum, Henk, and Ton Van Naerssen. "Bordering, Ordering and Othering." *Journal of Economic and Social Geography* 93, no. 2 (2002): 125–36.

VanNijnatten, Debora L. "North American Environmental Regionalism: Multi-Level, Bottom-Heavy and Policy-Led." In *Comparative Environmental Regionalism*, edited by Lorraine Elliot and Shaun Breslin, 147–62. New York: Routledge, 2011..

– "Standards Diffusion: The Quieter Side of North American Climate Change Cooperation?" In *North American Climate Change Policy: Designing Integration in a Regional System*, edited by Neil Craik, Isabel Studer, and Debora VanNijnatten, 108–31. Toronto: University of Toronto Press, 2013.

– "Towards Cross-Border Environmental Policy Spaces in North America: Province-State Linkages on the Canada-US Border." *AmeriQuests: The Journal of the Center for the* Americas 3, no. 1 (2006).

VanNijnatten, Debora L., and Carolyn Johns. "The IJC and the Evolution of Environmental and Water Governance in the Great Lakes–St. Lawrence Basin: Accountability, Progress Reporting and Measuring Performance under the Great Lakes Water Quality Agreement." In *The First Century of the International Joint Commission*, edited by Murray Clamen and Daniel Macfarlane, 395–430. Calgary: University of Calgary Press, 2020.

VanNijnatten, Debora L., and Marcela López-Vallejo. "Canada–United States Relations and a Low-Carbon Economy for North America?" In *Transboundary Environmental Governance across the World's Longest Border*, edited by Stephen Brooks and Andrea Olive, 201–6. New York: SUNY Press, 2018.

Vesselovsky, Mikita, ed. *Canada: State of Trade – Trade and Investment Update 2018*. Ottawa: Global Affairs Canada, July 2018.

Victor, Daniel. "A Fiery Affront, Back in 1812: But You Can't Blame Canada." *New York Times*, 6 June 2018, A17.

Vieira, Paul. "Canada Wins in Bid to Rebuff Bank Tax." *National Post*, 24 April 2010, A14.

Vine, Doug. "Interconnected: Canadian and U.S. Electricity." Arlington, VA: Center for Energy and Climate Solutions, 2017. https://www.c2es.org/site /assets/uploads/2017/05/canada-interconnected.pdf.

Waldie, Paul. "Britain to Purge Huawei from 5G Network by 2027 in Abrupt Policy U-turn." *The Globe and Mail*, 15 July 2020.

Walther, Olivier. "Border Markets: An Introduction." *Journal of Urban Research* 10 (2014). https://journals.openedition.org/articulo/2532.

Warren, Adrienne. "NAFTA: Raising Canada's Duty-Free Threshold on E-Commerce." Toronto: Scotiabank Global Economics, 25 August 2017.

Watts, Julie R. "The H-1B Visa: Free Market Solutions for Business and Labor." *Population Research and Policy Review* 20 (2001): 143–56.

Weible, Christopher M. "Introducing the Scope and Focus of Policy Process Research and Theory." In *Theories of the Policy Process*, 3rd ed., edited by Paul A. Sabatier and Christopher M. Weible, 3–21. Boulder, CO: Westview, 2014.

Weiss, Jeff. "Good Regulatory Practice in the United States." Presentation to High-Level Symposium to Enhance Regulator Expertise on Technical Barriers to Trade, Mexico City. Washington, DC: Department of Commerce, 9 February 2016.

Weiss, Thomas G. *What's Wrong with the United Nations and How to Fix It.* Malden, MA: Polity, 2009.

Wellman, Barry, and Keith Hampton. "Living Networked On and Offline." *Contemporary Sociology* 28, no. 6 (1999): 648–54.

Wezeman, Siemon T. "Military Capabilities in the Arctic: A New Cold War in the High North." SIPRI Background Paper. October 2016. https://www. sipri.org/sites/default/files/Military-capabilities-in-the-Arctic.pdf.

Whittington, Les. "Stephen Harper Says Economy Trumps Climate Action." *Toronto Star*, 9 June 2014. http://www.thestar.com/news/canada/2014/06/09/stephen_harper_says_economy_trumps_climate_action.html.

Wilson, James Q. "The Politics of Regulation." In *The Politics of Regulation*, edited by James Q. Wilson, 357–94. New York: Basic Books, 1980.

Wolf, Jim. "U.S. Squeezes French-Led Satellite Maker over China." Reuters, 9 February 2012. https://www.reuters.com/article/us-usa-france-china-satellite/exclusive-u-s-squeezes-french-led-satellite-maker-over-china-idUSTRE8181F020120209.

Wolf, Harrison G. "ITAR Reforms for Dual-Use Technologies: A Case Analysis and Policy Outline." In IEEE *Aerospace Conference 2012*, 1–12 New York: IEEE, 2012.

Wolfe, Robert, and Bernard Hoekman. "The WTO Still Has Vital Work to Do, Despite the Crippling of Its Appeals Court." *Globe and Mail*, 19 December 2019, B4.

Wood, Levi, Allan White, and James Clements. "Bill C-49." Presentations to 22nd annual Fields on Wheels Conference, Transport Institute, University of Manitoba, December 2017.

Woodard, Colin. *American Nations: A History of the Eleven Rival Regional Cultures of North America.* New York: Penguin Books, 2011.

Woody, Christopher. "NATO Allies Are Talking about Breaking Away from the US, but Trump Isn't Their Only Problem." *Business Insider,* 9 August

2018. Accessed August 2018. https://www.businessinsider.com/europe
-looks-at-domestic-defense-industry-due-to-trump-us-regulations-2018-8.

World Bank. "Carbon Pricing." http://www.worldbank.org/en/programs
/pricing-carbon.

– "FDI Trends." *Public Policy for the Private Sector*, note no. 273, September 2004.

– "Trade (% of GDP)." Washington: 2019. https://data.worldbank.org
/indicator/ NE.TRD.GNFS.ZS.

World Trade Organization. "DS426: Canada: Measures Related to the Feed-in
Tariff Program." Implementation notified by respondent on 5 June 2014.
https://www.wto.org/english/tratop_e/dispu_e/cases_e/ds426_e.htm.

– *Agreement on Government Procurement.* https://www.wto.org/english/tratop
_e/gproc_e/gp_gpa_e.htm

– "Overview of Services in RTA's: Positive or Negative List?" Asia-Pacific
Economic Cooperation Workshop on Scheduling Services and Investment
Commitments in FTAs, Singapore, 28–9 October 2014. 2016. http://mddb
.apec.org/Documents/2014/CTI/WKSP5/14_cti_wksp5_010.pdf.

Wyonch, Rosalie. "Bits, Bytes and Taxes: VAT and the Digital Economy in
Canada." Commentary #487. Toronto: C.D. Howe Institute, August 2017.

Xu, Sheng-Jun. "Skilled Labor Supply and Corporate Investment: Evidence
from the H-1B Visa Program." SSRN. 1 March 2017. http://dx.doi.org
/10.2139/ssrn.2877241.

Yakabuski, Konrad. "Australia – Caught between the U.S. and China – Has
Banned Huawei. Why Can't Canada?" *Globe and Mail*, 17 May 2019.

– "For Quebeckers, Airline Passenger Rights Bill Is a Flight of Fancy." *Globe
and Mail*, 7 February 2018, B4.

Yates, Charlotte, and Wayne Lewchuk. "What Shapes Automotive Investment
Decisions in a Contemporary Global Economy?" *Canadian Public Policy* 43,
no. S1 (2017): S16–S29.

Yeates, Neil. *Report of the Independent Review of the Immigration Review Board:
A Systems Management Approach.* Ottawa: Immigration, Refugees and
Citizenship Canada, April 2018.

Yeung, May T., William A. Kerr, Blair Coomber, Matthew Lantz, and Alyse
McConnell. *Declining International Cooperation on Pesticide Regulation:
Frittering Away Food Security.* London: Palgrave-Macmillan, 2017.

York, Geoffrey, and Michelle Zilio. "Access Denied: Canada's Refusal Rate for
Visitor Visas Soars." *Globe and Mail*, 9 July 2018, A1.

Zakaria, Fareed. "Can America Be Fixed? The New Crisis of Democracy."
Foreign Affairs 92, no. 1 (January–February 2013): 22–33.

Zakocs, Ronda C., and Erika M. Edwards. "What Explains Community
Coalition Effectiveness? A Review of Literature." *American Journal of
Preventative Medicine* 30, no. 4 (2006): 351–61.

Zhang, Lei, Henry Kinnucan, and Jing Gao. "The Economic Contribution of Alabama's Automotive Industry to Its Regional Economy: Evidence from a Computable General Equilibrium Analysis." *Economic Development Quarterly* 30, no. 4 (2016): 295–315.

Zhao, Yun. "Export Controls over Space Products." *National Space Law in China* 10 (2015): 155–79.

Zilio, Michelle. "Asylum-Seeker Surge at Quebec Border Chokes Refugee System." *Globe and Mail*, 12 September 2018, A1.

– "Illegal Border Crossings from U.S. Increase by Nearly 23 Percent from June to July." *Globe and Mail*, 15 August 2018.

Zurcher, Anthony. "Opioid Addiction and Death Mail-Ordered to Your Door." BBC.com, 22 February 2018.

Contributors

Greg Anderson is professor of political science at the University of Alberta.

Mathilde Bourgeon is PhD candidate in political science at the Université du Québec à Montréal.

John Coleman is a senior fellow at the School of Public Policy and Administration at Carleton University, and acting executive director of the Transportation Policy Innovation Centre.

Kathryn Bryk Friedman is Global Fellow, Woodrow Wilson International Center for Scholars, and research professor of law and planning, SUNY Buffalo.

Monica Gattinger is director of the Institute for Science, Society and Policy, founding chair of Positive Energy, and professor of political studies at the University of Ottawa.

Geoffrey Hale is professor of political science at the University of Lethbridge.

Jill E. Hobbs is a professor in the Department of Agricultural and Resource Economics at the University of Saskatchewan.

Carolyn C. James, PhD, Pepperdine University.

Carolyn Johns is professor of politics and public administration at Ryerson University.

David Jones is an MA candidate in political science at the University of Alberta.

William A. Kerr is a professor in the Department of Agricultural and Resource Economics at the University of Saskatchewan.

Christopher J. Kukucha is professor of political science at the University of Lethbridge.

Patricia Dewey Lambert is professor at the School of Planning, Public Policy, and Management at the University of Oregon.

Meredith B. Lilly is Simon Reisman Chair in International Affairs and associate professor of international affairs at the Norman Paterson School of International Affairs at Carleton University.

Barry E. Prentice is professor of supply chain management at the Asper School of Business, University of Manitoba.

Brendan A. Sweeney is managing director at the Trillium Network for Advanced Manufacturing at Western University.

Stephen Tomblin is professor emeritus of political science at Memorial University of Newfoundland.

Élisabeth Vallet is director of the Centre for Geopolitical Studies and scientific director of the Raoul Dandurand Chair in Strategic and Diplomatic Studies at the Université du Québec à Montréal.

Debora VanNijnatten is professor of political science and North American studies at Wilfrid Laurier University.

Index